Jerzy J. Chruściel

Silicon-Based Polymers and Materials

DE GRUYTER

Author
Dr. Jerzy J. Chruściel
ŁUKASIEWICZ Research Network –
Textile Research Institute
Brzezińska 5/15
92-103 Łódź
Poland

ISBN 978-3-11-063993-3
e-ISBN (PDF) 978-3-11-064367-1
e-ISBN (EPUB) 978-3-11-064013-7

Library of Congress Control Number: 2021942000

Bibliographic information published by the Deutsche Nationalbibliothek
The Deutsche Nationalbibliothek lists this publication in the Deutsche Nationalbibliografie; detailed bibliographic data are available on the Internet at http://dnb.dnb.de.

Cover image: nicoolay / E+ / Getty Images
Typesetting: Integra Software Services Pvt. Ltd.
Printing and binding: CPI books GmbH, Leck

www.degruyter.com

This book has been written in memory of my excellent teachers
and enthusiastic, distinguished in a world, organosilicon chemists:

Professor Zygmunt Lasocki (1921–1993)
from Faculty of Chemistry, Łódź University of Technology (Łódź, Poland)
and
Professor Adrian G. Brook (1924–2013)
from Department of Chemistry, University of Toronto (Ontario, Canada)

Motto:

1. “When God shuts the door - He opens the window instead” (Fr. Jan Twardowski).

2. “Non omnis moriar” (Horacy).

Preface

Silicon based materials and polymers are made of silicon containing polymers, mainly polysiloxanes (silicones), polysilsesquioxanes, other organosilicon polymers (polysilanes, polycarbosilanes, polysilazanes, polysilylcarbodiimides) and copolymers (most often siloxane-organic copolymers).

Functional silanes, silane coupling agents (SCA), silane modified fillers (e.g., silica and silicates), silsesquioxanes, silicones, and other silicon polymers and copolymers find many practical applications as polymeric materials and are very useful ingredients and additives in materials science.

Silicon derived high-tech ceramics: silicon carbide, silicon oxycarbide, silicon nitride, etc. have also a very important practical meaning and a hudge number of practical applications.

Polycrystalline silicon is the basic material for large scale photovoltaic (PV) applications as solar cells. Technical applications of fully inorganic materials: crystalline (c-Si) and amorphous (a-Si) silicon, silicon nanowires (SiNWs) are still quickly growing, especially in the fields of microelectronics, optoelectronics, photonics. and photovoltaics, catalysts, and different electronic devices (e.g., sensors, thermoelectric devices).

The above mentioned polymers and materials are used in a variety of industries and products, including technical and medical applications.

In Chapter 1 have been described silicon monomers, reactive silanes, siloxanes, and carbofunctional silanes (CFS) and polysiloxanes (CFPS).

Chapter 2 concerns general properties of different kinds of silicones (polysiloxanes): silicone oils, silicone elastomers and rubbers, silicone resins, miscellaneous and composite silicone resins, poly(silsesquioxanes), including their newer technical and biomedical applications. Moreover, polymer networks enhanced with POSS molecules, effect of dissolution POSS hybrids in polymers, copolymers containing siloxane linkages, poly(siloxysilanes) have been presented. Antiadhesive, hydrophobic, and superhydrophobic properties of silicones, physiological properties of organosilicon compounds and silicones, biomedical and cosmetic applications of silicones and modified silica, applications of silane-modified thermoplastics in medicine have been also described.

In Chapter 3 modifications of organic polymers with reactive silanes (CFS and CFPS), silicones, silica, and silicates have been descibed. It includes: modification of polyolefins properties by their cross-linking with silanes and by hydrosilylation methods, copolymers containing acrylate units and reactive silyl groups or siloxane segments modified with CFS and CFPS, modification of properties of thermoplastic polymers (polyesters, polycarbonates, polyamide, polyimide, polyurethanes, polyureas), and other polymers and polymeric materials with silicones, CFS and CFPS, POSS additives and other fillers, modification of fillers properties by CFS, applications of silane-modified nanofillers in composites and nanocomposites of thermoplastic polymers, polyacrylate-based composites modified with silica or/and CFS,

https://doi.org/10.1515/9783110643671-202

modification of thermoplastic polymers (polyolefins, etc.) with functional silanes, composites based on epoxy resins modified with silanes, silicones, POSS, silica, and silicates.

In Chapters 4 and 5 have been described methods of syntheses of polysilanes and polycarbosilanes, their properties and practical applications.

Chapter 6 concerns polysilazanes, polycarbosilazanes, and polysilylcarbodiimides, methods of their syntheses. Their properties and applications were also discussed.

In Chapter 7 have been described other silicon-containing polymers: poly(methylhydrosiloxanes) (PMHS), methods of their syntheses and chemical modifications, polymer electrolytes based on polymethylsiloxane copolymers, liquid-crystalline (LC) siloxane materials derived from PMHS and other silicon polymers, comprehensive applications of PMHS, other copolymers containing siloxane segments, copolymers containing only siloxane backbone, copolymers of polysiloxanes with organic polymers (methacrylate- and acrylate-polysiloxane copolymers, polystyrene–polysiloxane copolymers, copolymers of polysiloxanes with polyurethanes, copolymers of polysiloxanes with polyimides, polysiloxane–polysilazane and polysiloxazane copolymers, miscellaneous polysiloxane-organic copolymers.

In Chapter 8 have been presented: different kinds of ceramics derived from silicon polymers: silicon carbide ceramics, Si–C–O ceramics, Si_3N_4 ceramics, Si–C–N and Si–C–N–O ceramics, carbon-rich Si–C–N ceramics derived from polysilylcarbodiimides, Si–B–C–O and Si–B–C–N ceramics.

Finally, Chapter 9 concerns preparation methods, properties and applications of polycrystalline silicon, silicon nanoparticles and silicon nanowires.

Contents

Chapter 1
Introduction

Beginning of chemistry of organosilicon compounds gave the work of F.S. Kiping, in the second half of the nineteenth century. Its stormy development occurred in the 1930s. Major achievements in this field were conducted by K.A. Andrianow, and silicone resins were produced in the former Soviet Union (USSR) on a small scale as early as in 1939. However, only independent development by E. Rochow in the United States (in early 1940s) called "direct synthesis method" of methylchlorosilanes, as starting materials (monomers) for the preparation of silicones, was a breakthrough technology. Almost at the same time the technology of the "direct synthesis" was developed in Germany, under the direction of R. Miller. A production of silicones, the first generation of these unique polymers, began during the Second World War in the United States (in Dow Corning Corporation and General Electric) [1]. Soon after the Second World War, the production of silicones started on a small scale in Europe and Asia (first in Germany, France, Great Britain, USSR, and Japan, and later in other countries such as Czechoslovakia, Poland, and Korea) [2].

Silicon is the major ingredient of sand, silicate minerals, and rocks, which are very stable inorganic materials. Over 90% of the Earth's crust is composed of the silicate minerals, and silicon is a second most abundant element (after oxygen) in the Earth's crust (with about 28% content by weight) [3]. A technological pathway from the sand to silicon and its compounds is pretty long. On the industrial scale, a metallurgical-grade silicon was prepared by the reaction of high-purity silica with wood, charcoal, and coal in an electric arc furnace using carbon electrodes. At temperatures over 1,900 °C, silica is reduced by carbon from the aforementioned materials according to the following chemical reaction:

$$SiO_2 + 2\,C \longrightarrow Si + 2\,CO \tag{1.1}$$

The metallurgical-grade silicon should be at least 98% pure [3, 4]. The solar-grade silicon has purity of 6N class (99.9999%) [5], while the very pure silicon of 9N class (at least 99.9999999%) is a basic electronic material for a production of computer chips. Single silicon crystals are most often prepared by the Czochralski method [6]. Polycrystalline silicon (commonly called "polysilicon") has found huge applications for photovoltaic (PV) solar cells. It is the purest synthetic material available on the market. Polysilicon is produced from metallurgical-grade silicon by a chemical purification in the Siemens process, which involves distillation of volatile silicon compounds, and their decomposition into silicon at high temperatures [7, 8].

Similarly to the metallurgical-grade silicon, a silicon carbide (carborundum, SiC) was also produced by the Acheson method, from a sand and an excess of carbon (from coke) [9], at 2,500 °C in the following reactions:

https://doi.org/10.1515/9783110643671-001

$$SiO_2 + C \longrightarrow SiO + CO \tag{1.2}$$

$$SiO + 2C \longrightarrow SiC + CO \tag{1.3}$$

The sintered silicon carbide forms very hard ceramics which has been widely used as abrasive materials, for fabrication of car brakes, car clutches, and ceramic plates. It can be also applied in electric systems, electronic devices, and as a component for the manufacture of light-emitting diodes (LEDs) [6].

A large and most important group of various inorganic–organic (hybrid) compounds and materials are polysiloxanes (commonly known as "silicones"), composed of silicon and oxygen atoms in their main chains, and organic substituents bound to silicon. Silicones play an important role among polymers with special properties because they have many unusual features. Even an addition of a very small amount of silicones causes a crucial improvement of properties of modified materials. Most importantly, silicones increase hydrophobicity and improve water resistance, thermal stability, and flame resistance of many materials, in some cases. Silicones exhibit excellent chemical, physical, and electrical properties. Most popular organosilicon polymers are poly(dimethylsiloxanes) (PDMS). Silicones are mainly applied as silicone oils, rubbers, and resins [10–22]. Similar positive effects on properties of polymers and other materials can be reached by the addition of reactive silanes, siloxanes, and silicates, which are also very often used in practice for the modification of different polymers, textiles, and inorganic materials.

Most silicones have a good heat resistance in the temperature range from –50 to 250 °C, materials of the silicone resin (e.g., laminates and molding compositions) – to 350 °C, and enamels with addition of aluminum dust – up to 500 °C. The properties of silicone products vary to a small degree with moderate increasing temperature – silicone rubbers and resins then retain their mechanical and dielectric properties, and the viscosity of silicone oil varies slightly. They show at the same time very good hardiness and good traits to –50 °C, and some species – up to –110 °C. They are resistant to oxidation even in the atmosphere of ozone and hydrogen peroxide. All types of silicones are characterized by chemical resistance to aqueous solutions of acids, bases, salts, oils, and some solvents. They also possess valuable physical properties, and in the case of oils, for example, lubricating properties and high compressibility. Very thin layers of silicones provide hydrophobicity (water repellency), and antiadhesive (release) properties of modified surfaces in relation to the various tackifying materials. The large surface activity of silicone oils gives them the ability to extinguishing foam. Silicones find also applications in the areas of medicine, pharmaceuticals, and cosmetics [10]. Most silicones exhibit physiological indifference.

An important practical meaning has also other organosilicon polymers (especially, polysilanes, polycarbosilanes, and polysilazanes) [23], and many functional silanes with different chemical structures, containing reactive groups, mostly bound to silicon atom but also quite often attached to carbon [24]. A continuously growing interest in applications of reactive silanes, siloxanes [25–26], all kinds of silicones [26–28], other organosilicon polymers, and silicon nanowires (SiNWs) [29] have been observed in many different fields of science (with focus on materials science) and chemical technology. A surface-modified silica with various functional silanes has found special attention as a modern filler for plastics and other polymeric materials and also as an initiator for grafting different organic monomers in polymerization processes [30].

1.1 Silicon monomers

Only most important silicon monomers are described in this book, mostly organosilicon monomers. Basic organosilicon monomers (silanes) contain organic substituents and reactive functional groups bound to silicon (halogens, alkoxy, acetoxy, hydrogen, and others). On the industrial scale, chlorosilanes are most often applied, which are prepared by the so-called direct synthesis (Müller–Rochow process), based on a copper-catalyzed reaction of silicon metal with alkyl halides [31–37]. Usually, this reaction is carried out in a fluidized bed reactor. At least 97% pure silicon is used in this process and the particle size is in the range of 45–250 µm. The best results in terms of selectivity and yield of the direct process occur with methyl chloride. Under typical conditions (250–300 °C and 1–5 bar), 90–98% conversions for silicon and 30–90% for methyl chloride were observed. The following reaction leads to methyl (chloro)silanes:

$$x\,MeCl + Si \rightarrow MeSiCl_3 + Me_2SiCl_2 + Me_3SiCl \qquad (1.4)$$

(+ minor products: $SiCl_4$, $HMeSiCl_2$, HMe_2SiCl, etc.)
(Me denotes CH_3 group)

Copper and silicon form copper silicide (Cu_3Si), which catalyzes the formation of Si–Cl and Si–Me bonds. Copper is oxidized and then reduced to regenerate the catalyst. It was proposed that the Si–Cl bond reacts with a copper–chloromethane "adduct," which allows for the formation of Me–Si–Cl units. A consecutive addition of a second MeCl molecule leads to the release of the most important organosilicon monomer: dichlorodimethylsilane (Me_2SiCl_2 – DDS). DDS is the major product of the reaction (with ~70–90% yield), while the yield of methyl(trichloro)silane ($MeSiCl_3$ – MTS) is in the range of 5–15%. Small amounts of other chlorosilanes are also formed: trimethyl(chloro)silane (Me_3SiCl, 2–4%), methyl(dichloro)silane ($MeHSiCl_2$, 1–4%), and dimethyl(chloro)silane (Me_2HSiCl, 0.1–1.0%) [34, 37]. Many different activators, mostly metals (e.g., Zn, Cd, Ca, Sn, Sb, and Bi), phosphorus (P) and their compounds are very useful as promoters in the direct process. In the presence of hydrogen chloride (HCl), higher yields of the Si–H-containing chlorosilanes [trichlorosilane ($HSiCl_3$), $MeHSiCl_2$ (MDS), and Me_2SiHCl (DMCS)] were achieved [8, 34, 38].

In the presence of hydrogen the yields of MDS and DMCS in the direct synthesis were increased (up to 2.2–33% for DMCS), depending on the reaction conditions [37, 39–41]. Since the boiling points of different methyl(chloro)silanes are quite similar (Me_2SiCl_2: 70 °C; $MeSiCl_3$: 66 °C; Me_3SiCl: 58 °C; $MeHSiCl_2$: 41 °C; Me_2HSiCl: 36 °C), they can be isolated by a fractional distillation through series of columns with high separating capacities. The purity of the monomers is crucial in a further technology of the production of siloxane polymers.

Inorganic chlorosilanes such as silicon tetrachloride ($SiCl_4$), trichlorosilane ($HSiCl_3$, b.p. 31.5–32 °C), and dichlorosilane (H_2SiCl_2, b.p. 8.3 °C) are very useful monomers as well. Trichlorosilane is prepared from the reaction of elemental silicon with gaseous HCl [11, 42]:

$$Si + 3\,HCl \longrightarrow HSiCl_3 + H_2 \tag{1.5}$$

or from a mixture of silicon, tetrachlorosilane, and hydrogen at 400 °C [10, 11]:

$$Si + 3\,SiCl_4 + 2\,H_2 \rightarrow 4\,HSiCl_3 \tag{1.6}$$

Dichlorosilane (H_2SiCl_2) can be prepared from trichlorosilane by reduction with hydrogen in the presence of aluminum [11, 43]:

$$6\,HSiCl_3 + 3\,H_2 + 2\,Al \longrightarrow 6\,H_2SiCl_2 + 2\,AlCl_3 \tag{1.7}$$

Dichlorosilane is highly flammable and easily disproportionates into monosilane (SiH_4) and tetrachlorosilane ($SiCl_4$). Monosilane (SiH_4: m.p. −185 °C, b.p. −112 °C) is extremely explosive in contact with air, but in inert atmosphere it is the precious monomer for fabrication of semiconductor-grade silicon.

A liquid star monomer tetrakis(dimethylsiloxy)silane ($Si[OSi(CH_3)_2H]_4$) is commercially available nowadays. Its derivative, star tetraarm polysiloxane $Si\{[OSi(CH_3)_2]_nOSi(CH_3)_2H\}_4$, was prepared by equilibration of $Si[OSi(CH_3)_2H]_4$ with octamethylcyclotetrasiloxane (D_4) and trifluoromethane sulfonic acid [44, 45].

Hydrosilanes can be selectively chlorinated with copper dichloride ($CuCl_2$), providing (hydro)chlorosilanes and chlorosilanes [46]. Phenylchlorosilanes can also be prepared by the direct process of chlorobenzene with silicon and copper alloy (1:1, wt/wt) in an autoclave or in a vapor phase [31]. This reaction occurs at 450–550 °C, and a main product is phenyltrichlorosilane ($PhSiCl_3$: Ph = C_6H_5, b.p. 201 °C, FTS) accompanied by diphenyldichlorosilane (Ph_2SiCl_2, DPDCS, b.p. 305 °C), and traces of phenyldichlorosilane ($PhSiHCl_2$) and tetrachlorosilane ($SiCl_4$) [1]. Triphenylchlorosilane (Ph_3SiCl) is a solid crystalline compound (m.p. 91–94 °C, b.p. 378 °C). Another route to phenylchlorosilanes is dehydrocondensation of hydrochlorosilanes with aromatic hydrocarbons, carried out at 200–440 °C and 150 atm (with $HSiCl_3$) or at 200 °C and 100 atm (with $MeHSiCl_2$), in the presence of Friedel–Craft's catalysts ($AlCl_3$, $FeCl_3$, BCl_3, BF_3, H_3BO_3, etc.) or Raney Ni [1, 6], according to the reaction schemes:

$$HSiCl_3 + C_6H_6 \longrightarrow C_6H_5SiCl_3 + H_2 \tag{1.8}$$

$$CH_3HSiCl_2 + C_6H_6 \rightarrow CH_3(C_6H_5)SiCl_3 + H_2 \tag{1.9}$$

Vinyl-substituted silanes are most often prepared by the addition reaction of $HSiCl_3$ to acetylene (occurring at 175 °C at 100 atm) [1]:

$$HC \equiv CH + HSiCl_3 \longrightarrow CH_2 = CHSiCl_3 \tag{1.10}$$

or by its thermal condensation reaction with vinyl chloride (at ~600 °C):

$$CH_2 = CHCl + HSiCl_3 \longrightarrow CH_2 = CHSiCl_3 + HCl \tag{1.11}$$

Syntheses of organosilicon monomers from Grignard and other organometallic reagents (organolithium or organozinc compounds) are mainly used on the laboratory scale [10, 11, 29, 42].

Except chlorosilanes, alkoxysilanes are also often applied in syntheses of polysiloxanes, especially on the laboratory scale. Moreover, esters of silicic acid such as tetramethoxysilane ($Si(OCH_3)_4$, TMOS, b.p. 121–122 °C), tetraethoxysilane ($Si(OC_2H_5)_4$, TEOS, b.p. 168 °C), its oligomers (e.g., so-called *ethyl silicate 40*) [13], and also triethoxysilane ($HSi(OC_2H_5)_3$ – TES, b.p. 134–135 °C) are often used as comonomers in silicone technology, sol–gel chemistry, materials science, and nanotechnology. The commercial reagent, *ethyl silicate 40* (b.p. >384 °C, m.p. −86 to −77 °C), is a mixture of linear and cyclic oligomers (polyethoxysiloxanes), with an average content of five silicon atoms in the molecule, which corresponds to 40–42 wt% content of silica [16]. It is prepared by a partial hydrolysis and condensation of TEOS.

Dehydrocoupling and alcoholysis reactions of tri-, di-, and tetra-functional hydrosilanes with alcohols (1:1.5 mole ratio), catalyzed by 2 mol% of metallocene catalysts Cp_2MCl_2/Red-Al (M=Ti, Zr; red Al – sodium bis(2 methoxyethoxy)aluminum hydride solution; 60 wt% in toluene) and metallocene catalysts Cp_2M' (M′=Co, Ni), gave poly(alkoxysilane)s in one-pot manner with a high yield, as main products, along with the respective (trialkoxy)silane as a minor product. The various aryl(hydro) silanes such as p-X-$C_6H_4SiH_3$ (X=H, CH_3, OCH_3, F), $PhCH_2SiH_3$, and $(PhSiH_2)_2$ were reacted with the following alcohols: MeOH, EtOH, iPrOH, PhOH, and $CF_3(CF_2)_2CH_2OH$. These reactions were carried out for 48 h at 50 °C, and 1.5:1 mole ratio of alcohols to hydrosilanes. The average molecular weight (M_w) of the poly(alkoxysilane)s ranged from 600 to 8,000 g/mol. The dehydrocoupling reactions of phenylsilane with ethanol (1:1.5 mole ratio) using Cp_2HfCl_2/Red-Al and phenylsilane with ethanol (1:3 mole ratio) using Cp_2TiCl_2/Red-Al gave only one product – (triethoxy)phenylsilane [47].

Alkoxysilanes are frequently applied in syntheses of polyborosiloxanes and other preceramic silicon containing polymers [48–52], SiOC glasses [53, 54], borosiloxane glasses (BSiCO) [55, 56], and siloxane-based nanomaterials [57]. (Tetraacetoxy)silane ($Si(OAc)_4$, Ac=CH_3CO group) is rarely used as the monomer, but similarly as methyl (triacetoxy)silane ($MeSi(OAc)_3$), it is applied as the cross-linking agent in one-component room temperature vulcanizing silicone sealants [10, 15, 16].

Monosilane (SiH_4) is an extremely flammable and explosive monomer, which is used for the production of solar cells, PV panels, and SiNWs [58–61]. Phenylsilane ($C_6H_5SiH_3$ – $PhSiH_3$) is also used for the preparation of SiNWs [61, 62]. $PhSiH_3$, diphenylsilane (Ph_2SiH_2), and methylphenylsilane ($MePhSiH_2$) undergo catalytic dehydrocondensation and they are useful in synthesis of polysilanes [63–65].

Siloxane oligomers (linear and cyclic) and polysiloxanes (silicones) are mainly prepared by the so-called multistep "hydrolytic polycondensation" of chlorosilanes of different functionality. This step-growth process consists of the following elementary hydrolysis and condensation reactions [7, 12, 21]:

$$\equiv Si - Cl + H_2O \longrightarrow \equiv Si - OH + HCl \tag{1.12}$$

$$\equiv Si - Cl + HO - Si \equiv \rightarrow \equiv Si - O - Si \equiv + HCl \tag{1.13}$$

$$\equiv Si - OH + HO - Si \equiv \longrightarrow \equiv Si - O - Si \equiv + H_2O \tag{1.14}$$

Alkoxysilane monomers are also very useful for this purpose, especially on the laboratory scale. They undergo similar hydrolysis and condensation reactions:

$$\equiv Si - OR + H_2O \longrightarrow \equiv Si - OH + ROH \tag{1.15}$$

$$\equiv Si - OR + HO - Si \equiv \longrightarrow \equiv Si - O - Si \equiv + ROH \tag{1.16}$$

(R – mainly alkyl groups).

Next, consecutive homocondensation reactions (1.14) of intermediate silanol groups take place.

Linear oligo- and polysiloxanes with different α,ω,-terminal or pendant functional groups (most often chlorosilyl≡Si-Cl, silanol≡Si-OH, alkoxysilyl-≡Si-OR, hydrosilyl≡Si-H, vinyl≡$SiCH=CH_2$, and allylsilyl groups ≡$SiCH_2CH=CH_2$) and functional cyclosiloxanes find many applications in further syntheses.

Cyclosiloxanes are monomers used in ionic ring-opening polymerization (ROP) methods, leading to polysiloxanes and their various copolymers. Most important cyclic siloxane monomers are octamethylcyclotetrasiloxane (($Me_2SiO)_4$, m.p. 17.5 °C, b. p. 174 °C) and hexamethylcyclotrisiloxane (($Me_2SiO)_3$, m.p. 64.5 °C, b.p. 134 °C) [66, 67]. Mixtures of ring dimethylsiloxane oligomers $(Me_2SiO)_n$; (D_n, D=Me_2SiO, $n \geq 3$; yield: 20–80%) are products of the hydrolytic polycondensation of Me_2SiCl_2. They are also formed during a thermal and catalytic depolymerization of poly(dimethylsiloxane-α,ω-diols), which are remaining components of DDS hydrolyzate [68]. Ionic catalysts are used for depolymerization of PDMS. Volatile cyclic dimethylsiloxane oligomers also find applications in commercial cosmetic formulations and are known as *dimethicones* or *cyclomethicones* [69]. The cyclosiloxanes containing other substituents than methyl group bound to silicon (mainly H, vinyl, CF_3,

$CH_2CH_2CF_3$, CH_2CH_2CN, Ph, etc.) can be used as comonomers, together with D_n, in ROP processes.

The mixture of cyclic methyl(hydro)siloxanes $(MeHSiO)_n$ ($D^H{}_n$, D^H=MeHSiO, $n \geq 3$) is prepared by the hydrolytic polycondensation of $MeHSiCl_2$, followed by the thermal depolymerization of its hydrolyzate, which is composed of Si–H functional cyclic and linear oligomers. Only traces of 2,4,6-trimethylcyclotrisiloxane ($D^H{}_3$) were present in the products of depolymerization [70]. The individual hydrogen-substituted cyclosiloxanes can be isolated by the fractionation distillation [67, 68, 71, 72]. Main pyrolysis products are 2,4,6,8-tetramethylcyclotetrasiloxane ($D^H{}_4$, b.p. 134 °C, m.p. −69 °C) and 2,4,6,8,10-pentamethylcyclopentasiloxane ($D^H{}_5$). Higher oligomers ($D^H{}_n$, $n = 5–7$) can be isolated during distillation under reduced pressure [68] ($D^H{}_5$: b.p. 168–169 °C, 54 °C/10 mm Hg, m.p. −108 °C).

Linear oligosiloxanes and especially disiloxanes are used in ionic polymerization processes as blocking agents, causing decrease in the molecular weights of synthesized polysiloxanes [73].

Chloroalkyl-substituted silanes are useful in syntheses of carbosilanes [11, 13, 28]. Major carbosilane cyclic monomers such as 1-methyl-1-silacyclobutane [c-Me(H)Si$(CH_2)_3$], 1,3-dimethyl-1,3-disilacyclobutane [c-$(MeHSiCH_2)_2$], 1,1-dimethyl-1-silacyclobutane [c-$Me_2Si(CH_2)_3$], and 1,1,3,3-tetramethyl-1,3-disilacyclobutane [c-$(Me_2SiCH_2)_2$] are usually prepared by multistep syntheses and are used in ROP reactions, leading to polycarbosilanes, containing Si–C bonds in a main chain [14, 74–78]. Synthetic routes to 1,3-disilacyclobutanes, which proceed through Grignard ring closure reactions on alkoxy-substituted chlorocarbosilanes, were improved by Interrante and coworkers [79].

1,3-Disilacyclobutane [c-$(H_2SiCH_2)_2$] was polymerized in the presence of H_2PtCl_6, providing preceramic polymer, which is a silicon analogue of linear polyethylene [74].

1,1,3,3-Tetramethyl-1,3-disilacyclobutane was prepared by the reaction of (bromomethyl)chlorodimethylsilane with lithium metal or organolithium reagents. A formation of a transient silaethylene intermediate which dimerized to 1,1,3,3-tetramethyl-1,3-disilacyclobutane was proposed. When butyllithium was used, the formation of 1,3-disilacyclobutane was accompanied by the cyclic permethyl trimer 1,3,5-trisilacyclohexane [80].

Volatile carbosilanes are precursors, mainly for silicon carbide films and buffer layers and silicon carbonitride, and were utilized in passivation of silicon-based photovoltaics (PVs). Some Si-H functional silanes and carbosilanes are commercially available: methylsilane ($MeSiH_3$), dimethylsilane (Me_2SiH_2), 1,3-disilapropane (disilamethylene, $H_3SiCH_2SiH_3$), 1,3-disilabutane (1-methyldisilamethylene, $MeH_2SiCH_2SiH_3$), 1,4-disilabutane [1,2-ethanediylbis(silane)] ($H_3SiCH_2CH_2SiH_3$), 1,3,5-trisilacyclohexane (cyclotrisilamethylene, c-$(CH_2SiH_2)_3$, 1,3,5-trisilapentane [bis(silylmethyl)silane] ($H_3SiCH_2SiH_2CH_2SiH_3$), 1,1,2,2-tetramethyldisilane ($HMe_2SiSiMe_2H$), and 2,2,3,3-tetramethyltetrasilane ($H_3Si(Me_2Si)_2SiH_3$).

Silsequioxanes are a relatively the new group of silicate monomers, which are organosilicon compounds with the empirical chemical formula $RSiO_{3/2}$, where R is hydrogen or an alkyl, alkene, aryl, or arylene group [81]. Silsesquioxanes can have cage-like structures: a cube, a hexagonal prism, the octagonal prism [82], the decagonal prism [83], the dodecagonal prism [84, 85], or even an opened cage-like structure [86]. Hydridosilsesquioxanes have only hydrogen substituents on the silicon atoms, thus they are purely inorganic compounds [87]. Initial synthetic methods of Si–H-substituted silsesquioxanes involved hydrolytic condensation of benzene solutions of trichlorosilane with concentrated or fuming sulfuric acid to yield the ${T^H}_{10}$–${T^H}_{16}$ oligomers (T^H = $HSiO_{3/2}$). The ${T^H}_8$ oligomer was also synthesized by the reaction of (trimethoxy)silane solution in cyclohexane with a mixture of acetic acid and hydrochloric acid. These H-silsesquioxanes can be used for the preparation of silica coatings for application in environmental protection, and as an interlayer dielectric for integrated circuits [88].

Monomers containing Si–N bond include mainly cyclosilazanes: hexamethylcyclotrisilazane $(Me_2SiNH)_3$, octamethylcyclotetrasilazane $(Me_2SiNH)_4$ [89], hexaphenylcyclotrisilazane $(Ph_2SiNH)_3$ [90], other alkyl- or aryl-substituted cyclosilazanes [91], Si–H-substituted cyclosilazanes: $(H_2SiNH)_4$ and $(HMeSiNH)_4$, (tetrakisamino)silane Si $(NH_2)_4$, and also other Si- and N-substituted cyclodisilazane derivatives [92, 93]. Cyclotrisilazanes $(R^1R^2SiNH)_3$ containing different substituents at silicon can be prepared from $(R^1R^2SiNH)_n$ by a ring contraction in the presence of hydrogen and VIII group metal catalysts, for example, $Ru_3(CO)_{12}$ [94]. As condensation monomer, bis(diethylamino)dimethylsilane was used, which was coupled with phenylenediamines [95]. Unsubstituted, N-substituted, or Si-functional, for example, dichlorosilyl- [96–102], (dichlorosilyl)perphenyl- [103], vinyl-, and allyl- [104], or bis(hydroxy) silyl-substituted silazane monomers [102, 103] were also described in the literature. Syntheses and properties of cyclosilazanes and cyclocarbosilazanes have been described in an excellent review article [100]. Different silylamines were deprotonated with equivalents of *n*-BuLi and then underwent a rearrangement giving 1,3-dilithium silylamides, which were reacted with 1 equivalent of chlorosilane and underwent an intramolecular cyclization into a family of new silyl-substituted cyclodisilazanes. The monolithiated derivative was reacted with next equivalent of chlorosilane to give branched silazanes [105].

Cyclodisilazane ring structures exhibit a very high thermal stability (~360–500 °C) [106].

The interesting group of organosilicon monomers are cyclosilazoxanes, especially N-phenyl-substituted [107–110], and also N-methyl-substituted ring silazoxanes [110], which are used in ROP [111].

1.2 Reactive silanes, siloxanes, and carbofunctional silanes

In organosilicon chemistry, most important substrates are silanes containing following functional groups: ≡Si-Cl, ≡Si-OR, ≡SiOCOR (R – usually alkyl group), ≡Si-H, and ≡Si-CH=CH_2. Functional organic silanes include first of all chlorosilanes, alkoxysilanes, and acetoxysilanes (e.g., dimethyldichlorosilane, methyltriethoxysilane, and methyltriacetoxysilane), and different functional and carbofunctional silanes (CFS) [10, 11, 24, 25]. Tetraalkoxysilanes such as tetramethoxysilane ($Si(OCH_3)_4$, TMOS) and tetraethoxysilane ($Si(OC_2H_5)_4$, TEOS) are very often used in sol–gel processes for preparations of silicas and for modification of properties of polymers and different kinds of materials [112]. They can be grafted on silica, silicates, and some inorganic fillers.

An important reactive cyclic phenylsiloxane oligomer *Z-6018* or *RSN-6018 Intermediate* (Dow Corning®) is a flaked, low melting solid (having softening point ~40 °C), very well soluble in many solvents (ketones, esters, chlorinated solvents, alcohols, aromatic hydrocarbons, and solvent mixtures). It is a silanol-functional, low-molecular-weight (M_n = 1,200, M_w = 2,400 g/mol) silicone intermediate (Figure 1.1) which reacts with a wide variety of organic resins and monomers [113, 114].

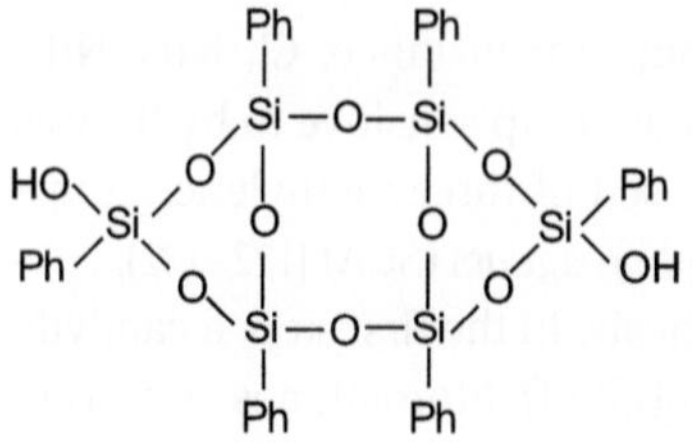

Figure 1.1: A hypothetical chemical structure of phenylsiloxane oligomer *Z-6018* Intermediate [113].

This versatile intermediate reacts with epoxides, alkyds, polyesters, phenolic resins, and other organic resinous polymers containing hydroxyl groups, which are able to react with silanol groups:

$$\equiv Si-OH + HO-C\equiv \longrightarrow \equiv Si-O-C\equiv + H_2O \quad (1.17)$$

This reaction can be carried out either in a solvent or in melt and can be catalyzed, for example, by tetrabutyl titanate. However, compatibility should be tested before use. The modification of organic resins with the *Z-6018 Intermediate* can improve weatherability and heat resistance of various organic resins. The *Z-6018 Intermediate* can be copolymerized or blended with a variety of organic resins. Copolymerization through hydroxyl groups of organic resins provides new formulations of improved compatibility with the formation of the silicone–organic bonds ≡Si–O–C≡. Sufficient excess of carbinol functionality is necessary for optimum compatibility and resin stability. Cold blending with the *Z-6018 Intermediate* limits the organic resin selection because of compatibility but offers lower processing costs in comparison with copolymerization. Solvent selection is important with cold blending the silicone with organic resins because of resin solution stability, viscosity, and paint properties.

Organic resins modified with the *Z-6018 Intermediate* show improved thermal resistance and weatherability over unmodified resins. The *Z-6018 Intermediate* is mainly applied in maintenance and architectural finishes, appliance finishes, coil coatings, and high-temperature finishes, which exhibit good chalk resistance and retain their color and gloss after years of outdoor exposure. High-temperature finishes prepared with silicone–organic copolymers and cold blends show excellent gloss and color retention and can be used on high-temperature appliances, for example, in ovens, incinerators, and space heaters. Silicone organic blends containing the *Z-6018 Intermediate* are also used in powder coating applications where improved heat resistance and weatherability are required.

1.2.1 Carbofunctional silanes (CFS)

The CFS are a large group of compounds with a general formula:

$$X_nSi(R'Y)_{4-n}$$

where R′ is an alkylene chain, Y is a functional group, for instance, Cl, NH_2, NR_2, OH, OCOR, NCO, CH_2=CH, and SH, and X is a functional group sensitive to hydrolysis (Cl, OR, OCOR). The alkylene chain R′ is often composed of three methylene groups [24, 25, 115–121]. They are usually called as *silane coupling agents* (SCA) [122–132].

The CFS are usually prepared in a three-step synthesis. In the first step, a catalytic hydrosilylation reaction of allyl chloride with hydro(alkyl)chlorosilanes is carried out; in the second step, alcoholysis of addition products takes place; and in the last step, a catalytic nucleophilic substitution of chlorine atoms occurs, for example, from $Cl(CH_2)_3SiMe_nCl_{3-n}$ chloropropyl(trialkoxy)silanes $Cl(CH_2)_3SiMe_n(OR)_{3-n}$ were prepared [133, 134].

The CFS are applied for a long time as adhesion promoters in processing of plastics and modification of properties of silica and various inorganic fillers [122, 135]. They are precursors of carbofunctional polysiloxanes. Products modified with SCA are particularly useful for bonding mineral (inorganic) additives to organic resins and the formation of *green materials*. For example, in the production of PV cells, silanes help to increase the adhesion of organic coatings applied to the layers of polysilicon. There is still a growing need in applications CFS in the tire industry – they strengthen the bond with the silica containing elastomers (e.g., polybutadiene or copolymers of styrene and maleic anhydride) and allow to increase the content of silica in rubber mixes, thus reducing rolling resistance by 30%. This leads to a reduction in fuel consumption of 3–5%. The SCA may also play a role of surfactants or dispersing and anticaking agents [135]. The so-called *Dynasylan oligomers* and *Silfin* are used for production of cables and pipes or grafting and cross-linking of polyethylene (or other polyolefins) containing mineral fillers. They can be applied for production of wind turbines, anticorrosive coatings, and other materials [136].

The following CFS are most often applied [16]:

3-Chloropropyl(trichloro)silane ($Cl(CH_2)_3SiCl_3$, d 1.36 g/cm^3, b.p. 181–182 °C)
3-Chloropropyl(trimethoxy)silane ($Cl(CH_2)_3Si(OMe)_3$, d 1.08 g/cm^3, b.p. 100 °C/40 mbar)
Octyl(triethoxy)silane ($Cl(CH_2)_3Si(OMe)_3$, d 0.87 g/cm^3, b.p. 98–99 °C/2 mbar)
Octadecyl(trichloro)silane ($CH_3(CH_2)_{17}SiCl_3$, d 0.98 g/cm^3, b.p. 223 °C/10 mbar)
Vinyl(trichloro)silane ($(CH_2{=}CH)SiCl_3$, d 1.27 g/cm^3, b.p. 88–92 °C)
Vinyl(trimethoxy)silane ($(CH_2{=}CH)Si(OMe)_3$, d 0.97 g/cm^3, b.p. 122 °C)
Vinyl(triethoxy)silane ($(CH_2{=}CH)Si(OEt)_3$, d 0.90 g/cm^3, b.p. 160 °C)
Vinyl(2-methoxyethoxy)silane ($(CH_2{=}CH)Si(OCH_2CH_2OMe)_3$, d 1.04 g/cm^3, b.p. 133 °C/5 mbar)
(*N*-2-Aminoethyl)-3-aminopropyl(trimethoxy)silane ($(CH_2{=}CH)Si(OMe)_3$, d 1.05 g/cm^3, b.p. 140 °C/15 mbar)
3-Isocyanate(trimethoxy)silane ($OCNSi(OMe)_3$, d 0.99 g/cm^3, b.p. 42–51 °C/0.5 mbar)
3-Mathacroyl(trimethoxy)silane ($CH_2{=}C(CH_3)COOSi(OCH_2CH_2OMe)_3$, d 1.04 g/cm^3, b.p. 78–81 °C/1 mbar)
3-Mercaptopropyl(trimethoxy)silane $HS(CH_2)_3Si(OCH_3)_3$, d 1.05 g/cm^3, b.p. 90 °C/10 mbar)
3-Glycidoxypropyl(trimethoxy)silane ($CH_2(O)CH(CH_2CH_2O(CH_2)_3Si(OMe)_3$, d 1.07 g/cm^3, b.p. 90 °C/10 mbar)
Bis-(triethoxypropylsilyl)tetrasulphide ($[(EtO)_3Si(CH_2)_3Si]_2S_4$, d 1.05 g/cm^3)

Allylsilanes and diallylsilanes were used as monomers for cyclopolymerization reactions [137]. The addition reactions of 3-mercaptopropyltrimethoxysilane (MPTMS) to 2,4,6-trivinyl-2,4,6-trimethyl(cyclotrisiloxane) and 2,4,6,8-tetravinyl-2,4,6,8-tetramethyl (cyclotetrasiloxane) provided new thiol-ene addition products. Their structures were confirmed by ^{29}Si-NMR and mass spectroscopy [138].

1.2.2 Synthesis of CFPS

Most often both ends of macromolecules of the carbofunctional poly(dimethylsiloxane) (CFPS) chains are terminated with functional alkylene groups such as hydroxypropyl, aminopropyl, glicydoxypropyl or methacryloxypropyl. The carbofunctional polysiloxanes may also have terminal alkene groups (e.g., allyl groups) or arylamine end groups: $-C_6H_4NH_2$ [139, 140]. Quite often for the synthesis of the carbofunctional PDMS, α,ω-dihydrosiloxanes are used. They can be synthesized, for instance, in condensation reactions of polysiloxane-α,ω-diols with chloro(hydro)dimethylsilane (Me_2SiHCl). The CFPS containing terminal carbinol groups (C–OH) were prepared by hydrosilylation reaction of different α,ω-dihydropolysiloxanes with allyl derivatives [117, 118], for example, allyloxytrimethylsilane (ATMS) [141], followed by hydrolysis reaction of alkoxysilane end groups (Scheme 1.1).

The fundamental reaction used for the preparation of the carbofunctional polysiloxanes is hydrosilylation reaction [15, 117, 118], which is usually catalyzed by platinum compounds. Speier's and Karstedt's catalysts were found to be not enough

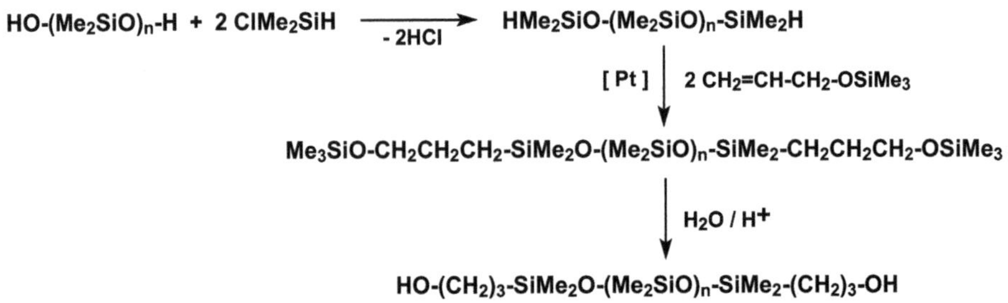

Scheme 1.1: The preparation of α,ω-di(hydroxypropyl)poly(dimethylsiloxanes). Reprinted with permission from [24]. Copyright InTech Open Sci. Publ., Croatia.

effective in the case of addition of allylamine to (tetramethyl)cyclotetrasiloxane D_4^H (D^H=MeHSiO) and gave low yield and low selectivity of products [142]. A very efficient catalyst of this reaction was platinum dioxide (PtO_2), which was resistant to "poisoning" by amine groups, giving the product with almost 100% yield and very good selectivity, determined by a ratio of isomers γ to β (93:7) [143].

A mixture of linear and cyclosiloxane oligomers containing 3-aminopropyl and [(3-aminoethylamino)propyl substituents was prepared by a hydrolytic copolycondensation of (3-aminopropyl)methyl(diethoxy)silane and [(3-aminoethylamino)propyl](methyl)(dimethoxy)silane. They were applied as a new generation of polycationic DNA supports ("*gene transfer reagents*") [144].

A very important method for the preparation of carbofunctional polysiloxanes are catalytic equilibration reactions of cyclic siloxanes, for example, octamethylcyclotetrasiloxane ($(Me_2SiO)_4$, D_4), with carbofunctional disiloxanes. These reactions were carried out in acidic or basic media (Scheme 1.2) [145, 146].

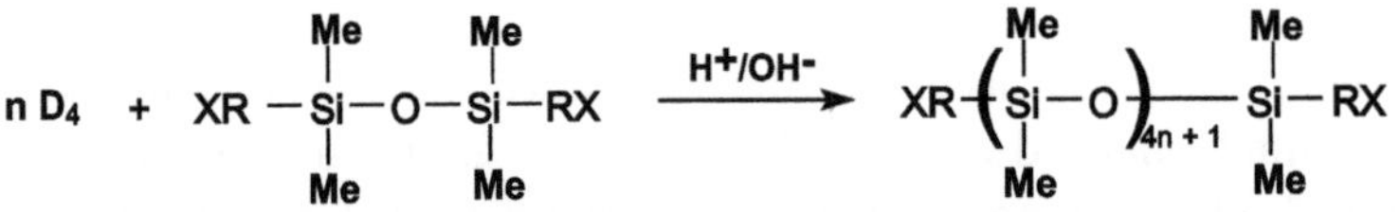

Scheme 1.2: The preparation of CFPS by a catalytic equilibration of the carbofunctional disiloxanes with D_4. Reprinted with permission from [24]. Copyright InTech Open Sci. Publ., Croatia.

For example, poly[dimethyl-*co*-(aminopropyl)(methyl)]siloxane (containing side aminopropyl groups) was obtained by the equilibration of D_4 (D=Me_2SiO) with (3-aminopropyl)(methyl)(diethoxy)silane and hexamethyldisiloxane ($Me_3SiOSiMe_3$) [147].

Poly(dimethyl-*co*-methylvinyl-*co*-diphenyl)siloxane tercopolymer containing terminal aminopropyl groups was synthesized by anionic copolymerization of D_4 with $(Ph_4SiO)_4$, $(ViMeSiO)_4$, and 1,3-bis-(3-aminopropyl)tetramethyldisiloxane (as a chain-terminating agent). This product was used for the synthesis of a segmented poly (imide-*co*-siloxane), and subsequently for the preparation of a hybrid nanocomposite, reinforced with silica [148] or titania (TiO_2) [149].

The carbofunctional polysiloxanes having alkylfunctional side groups were also prepared by the catalytic equilibration of aminopropyl functionalized cyclotetrasiloxane (D_3D^{NH2}), with decamethyltetrasiloxane (MD_2M; D = Me_2SiO, M = $Me_3SiO_{0,5}$), in the presence of tetramethylammonium trimethylsilanolate ($Me_4NOSiMe_3$, TMAS) [143, 147] (see Scheme 1.3):

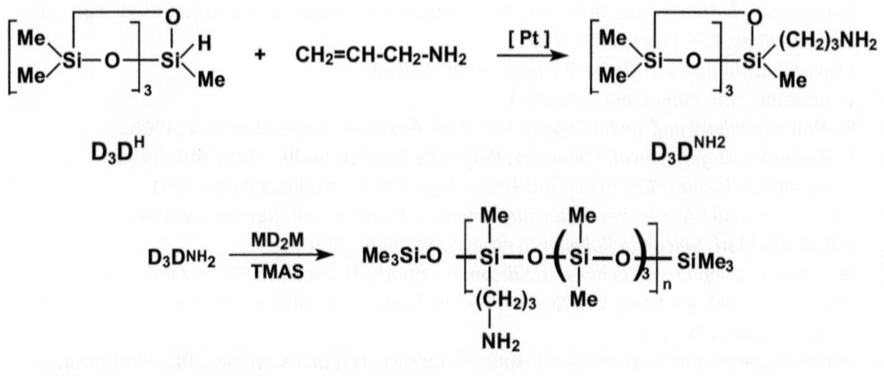

Scheme 1.3: The preparation of CFPS with pendant aminopropyl functional groups. Reprinted with permission from [24]. Copyright InTech Open Sci. Publ., Croatia.

Syntheses of carbofunctional PDMS with different photoactive groups such as benzoin [150], glicydoxyl [151], or benzylacryl groups [152] were also described in the literature.

1.3 Silicon-containing polymers

Organosilicon polymers belong to a wider group of organometallic polymers. In their chemical structures, organic substituents are directly bound to silicon atoms. Among organosilicon polymers, most important groups are polysiloxanes (*silicones*). Their inorganic chain is composed of alternating silicon and oxygen atoms. Other widely studied are preceramic organosilicon polymers (which include especially polysilanes, polycarbosilanes, polysilazanes), and also polysilthianes [1, 10–23, 68, 153–156]. Inorganic and organometallic polymers give high ceramic residue (usually at least 50 wt%) during thermal decomposition. Thus, they are excellent materials for high-temperature-resistant coatings, matrices, elastomers, and precursors for advanced (high-tech) ceramics [51, 157].

References

[1] E.G. Rochow, *Silicon and Silicones*, Springer, Berlin, Germany, 1987.

[2] P. Rościszewski, *Zastosowanie silikonów*, WNT Warszawa, 1964. [In Polish].

[3] R. Nave in *Abundances of the Elements in the Earth's Crust*, Georgia State University, 2012; http://hyperphysics.phy-astr.gsu.edu/hbase/hframe.html

[4] A.M. Mitrašinović and T.A. Utigard, *Silicon*, 2009, 1, 239.

[5] J. Wu, W. Ma, B. Yang, D. Liu and Y. Dai, *Silicon*, 2012, 4, 289.

[6] *Wikipedia, the free encyclopedia.*

[7] B. Ceccaroli, E. Ovrelid and S. Pizzini, *Solar Silicon Processes: Technologies, Challenges, and Opportunities*, CRC Press, 2016, 258.

[8] https://en.wikipedia.org/wiki/Polycrystalline_silicon

[9] G. Acheson, *U.S. Patent* 492,767, 1893.

[10] W. Noll, *Chemistry and Technology of Silicones*, Academic Press, New York, 1968.

[11] S. Pawlenko, *Organosilicon Chemistry*, Walter de Gruyter, Berlin – New York, 1986.

[12] G. Koerner, *Silicones: Chemistry and Technology*, CRC Press, Boca Raton, 1991.

[13] S.J. Clarson and J.A. Semlyen, *Siloxane Polymers*, Prentice Hall, New Jersey, 1993.

[14] H.R. Kricheldorf, *Silicon in Polymer Synthesis*, Springer, 1996.

[15] M.A. Brook, *Chapter 9: Silicones*, in *Silicon in Organic, Organometallic and Polymer Chemistry*, J. Wiley & Sons, Inc, New York – Chichester – Weinheim – Brisbane – Singapore – Toronto, 2000, 274–277.

[16] P. Rościszewski and M. Zielecka, *Silikony – właściwości i zastosowanie*, WNT, Warszawa, 2002. [In Polish].

[17] S.J. Clarson, J.J. Fitzgerald, M.J. Owen, S.D. Smith and M.E. Van Dykein *Synthesis and properties of silicones and silicone-modified materials*, ACS Symposium Series No. 838, 15 May 2003.

[18] S.J. Clarson, J.J. Fitzgerald, M.J. Owen, S.D. Smith and M.E. Van Dyke (Eds.), *Science and technology of silicones and silicone-modified materials*, ACS Symposium Series, Vol. 964, Washington, DC, American Chemical Society, 2007.

[19] F. Ganachaud, S. Boileau and B. Boury, *Silicon Based Polymers: Advances in Synthesis and Supramolecular Organization (e-Book Google)*, Springer Science & Business Media, 2008.

[20] M. Andriot, S.H. Chao, A. Colas, S. Cray, F. De Buyl, J.V. De Groot, A. Dupont, T. Easton, J.L. Garaud, E. Gerlach, F. Gubbels, M. Jungk, S. Leadley, J.P. Lecomte, B. Lenoble, R. Meeks, A. Mountney, G. Shearer, S. Stassen, C. Stevens, X. Thomas and A.T. Wolf, *Chapter 2: Silicones in industrial applications,* in *Inorganic Polymers*, D.J. Roger and G. Mario (Eds.), Nova Sci. Publishers, New York, 2007, 61–161.

[21] A.M. Muzafarov and A. Bockholt, *Silicon Polymers*, Springer Science & Business Media, 2010.

[22] R.G. Jones, W. Ando and J. Chojnowski (Eds.), *Silicon-Containing Polymers: The Science and Technology of Their Synthesis and Applications*, Springer Science & Business Media, 2013.

[23] R.D. Archer, *Inorganic and Organometallic Polymers*, Wiley-VCH, Inc, New York – Chichester – Weinheim – Brisbane – Singapore – Toronto, 2001. ISBNs: 0-471-24187-3 (Hardback); 0-471-22445-6 (Electronic).

[24] J.J. Chruściel and E. Leśniak, Chapter 9: *Modification of thermoplastics with reactive silanes and siloxanes*, in *Thermoplastic Elastomers*, A.Z. El-Sonbati (Ed.), InTech – Open Science, Rijeka, Croatia, 2012, 155–192, ISBN 979-953-307-353-7, on-line: 28.03.2012.

[25] J.J. Chruściel and E. Leśniak, *Progress in Polymer Science*, 2015, 41, 67.

[26] Q.A. Acton (Ed.) in *Siloxanes – Advances in Research and Application: 2013 Edition: Scholarly Brief*; Scholarly Edition™.

[27] F.M. Procopio, M. Cohen and A.F. Zavist, *High temperature plastics, section D: Silicones*, in *Modern Materials: Advances in Development and Applications, Vol. 3 (e-Book Google)*, H.H. Hausner (Ed), Elsevier, 2013, 97–122.
[28] http://www.newworldencyclopedia.org/entry/Silicone
[29] M. Shao, D.D.D. Ma and S.-T. Lee, *European Journal of Inorganic Chemistry*, 2010, 4264.
[30] C.M. Hui, J. Pietrasik, M. Schmitt, C. Mahoney, J. Choi, M.R. Bockstaller and K. Matyjaszewski, *Chemistry of Materials*, 2014, 26, 745.
[31] E.G. Rochow, *Journal of the American Chemical Society*, 1945, 67, 963.
[32] M. Narisawa, *Materials*, 2010, 3, 3518.
[33] C. Eaborn, *Organosilicon Compounds*, Butterworth Sci. Publ, London, 1960.
[34] L. Rösch, P. John and R. Reitmeier, Chapter: *Organic silicon compounds*, in *Ullmann's Encyclopedia of Industrial Chemistry*, John Wiley and Sons, San Francisco (CA, USA), 2003.
[35] E.G. Rochow, *The chemistry of silicon*, in Pergamon International Library of Science, Technology, Engineering and Social Studies (e-Book Google), Elsevier, 2013, 1435–1441.
[36] J. Chruściel and Z. Lasocki, Pat. PL 150 116, 1992. [In Polish].
[37] J.J. Chruściel, *Canadian Journal of Chemistry*, 2000, 78, 1405.
[38] *U.S. Pat.* 4 297 500 A (1981).
[39] N.G. Morozov, S.A. Golubtsov, L.P. Morozova-Sukreeva and I.V. Trofimova-Volkova, *Fr. Patent* 1 523 912 (1968); *Chemical Abstracts*, 1970, 72, 90613.
[40] M.G.R.T. De Cooker, J.H.N. De Bruyn and P.J. Van Der Berg, *Journal of Organometallic Chemistry*, 1975, 99, 371.
[41] D. Seyferth, *Organometallics*, 2001, 20, 4978.
[42] F.C. Whitmore, E. Pietrusza and L.H. Sommer, *Journal of the American Chemical Society*, 1947, 69, 2108.
[43] D.T. Hurd, *Journal of the American Chemical Society*, 1945, 67, 1545.
[44] G. Cai and W.P. Weber, *Polymer*, 2004, 45, 2941.
[45] M.A. Grunlan, N.S. Lee, F. Mansfeld, E. Kus, J.A. Finlay, J.A. Callow, M.E. Callow and W.P. Weber, *Journal of Polymer Science. Part A, Polymer Chemistry*, 2006, 44, 2551.
[46] A. Kunai, T. Kawakami, E. Toyoda and M. Ishikawa, *Organometallics*, 1992, 11, 2708.
[47] B.-H. Kim, M.-S. Cho, M.-A. Kim and H.-G. Woo, *Journal of Organometallic Chemistry*, 2003, 685, 93.
[48] V.V. Korshak, V.A. Zamiatina and N.I. Bekasova, *Boroorganitscheskie Polimery*, Izd. "Nauka", Moscow, 1975. [In Russian].
[49] A. Kasgöz, T. Misono and Y. Abe, *Journal of Non-Crystalline Solids*, 1999, 243, 168.
[50] R. Peña-Alonso, F. Rubio, J. Rubio and J.L. Oteo, *Journal of Analytical and Applied Pyrolysis*, 2004, 71, 827.
[51] P. Colombo, G. Mera, R. Riedel and G.D. Soraru, *Journal of the American Chemical Society*, 2010, 93, 1805.
[52] G. Ambadas, S. Packirisamy and K.N. Ninan, *Journal of Materials Science Letters*, 2002, 21, 1003.
[53] H.-J. Kleebe, C. Turquat and G.D. Soraru, *Journal of the American Ceramic Society*, 2001, 84, 1073.
[54] R. Peña-Alonso, G. Mariotto, C. Gervais, F. Babonneau and G.D. Soraru, *Chemistry of Materials*, 2007, 19, 5694.
[55] C. Gervais, F. Babonneau, N. Dallabonna and G. Sorarù, *Journal of the American Ceramic Society*, 2001, 84, 2160.
[56] M.A. Schiavon, N.A. Armelin and I.V.P. Yoshida, *Materials Chemistry and Physics*, 2008, 112(3), 1047.

[57] K. Kuroda, A. Shimojima, K. Kawahara, R. Wakabayashi, Y. Tamura, Y. Asakura and M. Kitahara, *Chemistry of Materials*, 2014, 26, 211.
[58] J.D. Carter, Y. Qu, R. Porter, L. Hoang, D.J. Masiel and T. Guo, *Chemical Communications*, 2005, 2274.
[59] V. Schmidt, J.V. Wittemann, S. Senz and U. Gösele, *Advanced Materials*, 2009, 21, 2681.
[60] H. Hamidinezhad, Y. Wahab, Z. Othaman and A.K. Ismail, *Applied Surface Science*, 2011, 257, 9188.
[61] C.K. Chan, R.N. Patel, M.J. O'Connell, B.A. Korgel and Y. Cui, *ACS Nano*, 2010, 4, 1443.
[62] H. Geaney, T. Kennedy, C. Dickinson, E. Mullane, A. Singh, F. Laffir and K.M. Ryan, *Chemistry of Materials*, 2012, 24, 2204.
[63] V.K. Dioumaev and J.F. Harrod, *Journal of Organometallic Chemistry*, 1996, 521, 133.
[64] F.-G. Fontaine, T. Kadkhodazadeh and D. Zargarian, *Chemical Communications*, 1998, 1253.
[65] N. Peulecke, D. Thomas., W. Baumann, C. Fischer and U. Rosenthal, *Tetrahedron Letters*, 1997, 38, 6655.
[66] J.A. Semlyen, B.R. Wood and P. Hodge, *Polymers for Advanced Technology*, 1994, 5, 473.
[67] S.J. Clarson, *Chapter: Cyclic polysiloxanes*, in *Cyclic Polymers*, J.A. Semlyen (Ed.), Springer, 2000, 161–183.
[68] W. Patnode and D. Wilcock, *Journal of the American Chemical Society*, 1946, 68, 358.
[69] ABCR-Gelest 2000 Catalogue; http://www.abcr.de
[70] N.N. Sokolov, *Metody Syntezy Poliorganosiloksanów*, PWN, Warszawa, 1961. [In Polish].
[71] R.D.C. Richards, J. Hollinghurst and J.A. Semlyen, *Polymer*, 1993, 34, 4965.
[72] T. Graczyk and Z. Lasocki, *Bulletin de L'Academie Polonise des Sciences, Série des Sciences Chimiques*, (a) 1978, 36, 12, 917; (b) 1979, 37, 3, 181; (c) 1979, 37, 3, 185.
[73] H. Spörk, R. Strasser, R. Riedle, W. Jaques and J. Waas, *US Pat.* 4 032 557 (1997).
[74] M. Gallei, J. Li, J. Elbert, M. Mazurowski, A. Schönberger, C. Schmidt, B. Stühn and M. Rehahn, *Polymers*, 2013, 5, 284.
[75] K. Matsumoto and T. Endo, *Macromolecular Symposia*, 2015, 349, 21.
[76] H.J. Wu and L.V. Interrante, *Chemistry of Materials*, 1989, 1, 564.
[77] L.V. Interrante, I. Rushkin and Q. Shen, *Applied Organometallic Chemistry*, 1998, 12, 695.
[78] R.J.P. Corriu, *Angewandte Chemie, International Edition*, 2000, 39, 1376.
[79] Q.H. Shen and L.V. Interrante, *Journal of Polymer Science. Part A, Polymer Chemistry*, 1997, 35, 3193.
[80] J. Chmielecka and W. Stańczyk, *Synlett*, 1990, 344.
[81] L. Guizhi, W. Lichang and C.U. Pittman Jr., *Journal of Inorganic and Organometallic Polymers*, 2001, 11(3), 123.
[82] D.B. Cordes, P.D. Lickiss and F. Rataboul, *Chemical Reviews*, 2010, 110(4), 2081.
[83] T. Jaroentomeechai, P. Yingsukkamol and C. Phurat, *Inorganic Chemistry*, 2012, 51(2), 12266.
[84] S. Chimjarn, *Inorganic Chemistry*, 2013, 52(22), 13108.
[85] R. Kunthom and P. Chancharone, *Dalton Transaction*, 2015, 44, 916.
[86] R. Sodkhomkhum, *European Journal of Inorganic Chemistry*, 2013, 19, 3292.
[87] C.L. Frye and T. Collins, *Journal of the American Chemical Society*, 1970, 92(19), 5586.
[88] G. Chandra, *Materials Research Society Symposia Proceedings*, 1991, 203, 97.
[89] D. Brewer and C.P. Haber, *Journal of the American Chemical Society*, 1948, 70, 3888.
[90] Annual Summary Report (Feb. 4, 1962 – Apr. 3, 1963), A Study of Polymers Containing Silicon-Nitrogen Bonds, NASA, Huntsville, Alabama.
[91] S.J. Grososz and J.A. Hall, *US Pat.* 2 885 370 (1959).
[92] E. Duguet, M. Schappacher and A. Soum, *Journal of Organometallic Chemistry*, 1993, 521, 9.
[93] M. Bouquey, C. Brochon, S. Bruzaud, A.F. Mingotaud, M. Schappacher and A. Soum, *Journal of Organometallic Chemistry*, 1996, 521, 21.

[94] B.C. Arkles, *Pat. EP* 0208831B1, 1991.
[95] J. Kulpiński and Z. Lasocki, *Acta Polymerica*, 1991, 42, 86.
[96] W. Fink, *Angewandte Chemie, International Edition*, 1966, 5, 760.
[97] W. Fink, *Helvetica Chimica Acta*, 1968, 51, 1011.
[98] L.W. Breed, R.L. Elliott and J.C. Wiley Jr, *Journal of Organometallic Chemistry*, 1970, 24, 315.
[99] E. Duguet, M. Schappacher and A. Soum, *Macromolecules*, 1992, 25, 4835.
[100] M. Kavala, A. Hawkins and P. Szolcsányi, *Journal of Organometallic Chemistry*, 2013, 732, 58.
[101] M. Bouquey and A. Soum, *Macromolecular Chemistry and Physics*, 2001, 202, 1232.
[102] Y. Zheng, Y. Tan, L. Dai and Z. Zhang, *Journal of Organometallic Chemistry*, 2011, 696, 3245.
[103] N. Zhou, Z.-J. Zhang, C.-H. Xu and Z.-M. Xie, *Chinese Journal of Polymer Chemistry*, 2000, 18, 551.
[104] C. Cazalis, A.-F. Mingotaud and A. Soum, *Macromolecular Chemistry and Physics*, 1997, 198, 3441.
[105] Y. Xiao and D.Y. Son, *Organometallics*, 2004, 23, 4438.
[106] L. Janiashvili, G. Andronikashvili, A. Varadashvili and M. Gagolishvili, *European Reviews of Chemical Research*, 2014, 1, 22.
[107] Z. Lasocki and M. Witekowa, *Synthesis and Reactions in Inorganic and Metallorganic Chemistry*, 1974, 4, 231.
[108] Z. Lasocki and M. Witekowa, *Macromolecular Science, Chemistry*, 1977, A 11, 457.
[109] Z. Lasocki, B. Dejak, J. Kulpiński, E. Leśniak, S. Piechucki and M. Witekowa, *ACS Symposium Series*, 1988, 360, 166.
[110] C.R. Krüger and E.G. Rochow, *Inorganic Chemistry*, 1963, 2, 1295.
[111] C. Brochon, A.F. Mingotaud, M. Schappacher and A. Soum, *Macromolecules*, 2007, 40, 3547.
[112] J. Chruściel and L. Ślusarski, *Materials Science – Poland*, 2003, 21, 461.
[113] Product Information: *Silicone Intermediates – Dow Corning® Z-6018 Intermediate*.
[114] S. Özgümüş, T.B. Ĺyim, I. Acar and E. Küçükoğlu, *Polymers for Advanced Technology*, 2007, 18, 213.
[115] B. Marciniec, J. Guliński, J. Mirecki and Z. Foltynowicz, *Polimery* (Warsaw), 1990, 35, 213. [In Polish].
[116] B. Marciniec and J. Guliński, *Polimery* (Warsaw), 1992, 37, 73. [In Polish].
[117] J. Bartz, Chapter 5: *Silany jako promotory adhezji*, in *Hydrosililowanie*, B. Marciniec (Ed.), PWN, Warszawa, 1981, 99–117. [In Polish].
[118] B. Marciniec, J. Guliński, W. Urbaniak and Z.W. Kornetka, *Comprehensive Handbook of Hydro-silylation*, B. Marciniec (Ed.), Pergamon Press, Oxford, 1992.
[119] B. Marciniec, W. Urbaniak and H. Maciejewski, *Polimery* (Warsaw), 1993, 38, 53. [In Polish].
[120] J. Chruściel, E. Leśniak and M. Fejdyś-Kaczmarek, *Polimery* (Warsaw), 2008, 53, 709. [In Polish].
[121] J. Chruściel, E. Leśniak and M. Fejdyś-Kaczmarek, *Polimery* (Warsaw), 2008, 53, 817. [In Polish].
[122] B. Arkles, *Chemtech*, 1977, 766.
[123] D. Leyden and W. Collins, *Silylated Surfaces*, Gordon & Breach, 1980.
[124] E. Plueddemann, *Silane Coupling Agents*, Plenum, 1982.
[125] D.E. Leyden, *Silanes, Surfaces and Interfaces*, Gordon & Breach, 1985.
[126] J. Steinmetz and H. Mottola, *Chemically Modified Surfaces*, Elsevier, 1992.
[127] (a) K.L. Mittal in *Silanes and Other Coupling Agents*, VSP, Utrecht, 1992; (b) K.L. Mittal in *Silanes and Other Coupling Agents, vol. 4*, CRC Press (2007).
[128] J. Blitz and C. Little, *Fundamental & Applied Aspects of Chemically Modified Surfaces*, Royal Society of Chemistry, 1999.

[129] F.D. Blum, Chapter: *Silane coupling agents*, in *Encyclopedia of Polymer Science and Technology*, 2003.
[130] B. Arkles, Y. Pan, G. Larson and M. Singh, *Chemistry – A European Journal*, 2014, 20, 9442.
[131] B. Arkles et al., www.gelest.com/wp-content/uploads/Goods-PDF-brochures-couplingagents.pdf
[132] T. Materne, F. De Buyl and G.L. Witucki, *Organosilane Technology in Coating Applications: Review and Perspectives*, Dow Corning Corporation, P.O. Box 994, Midland, MI 48640 USA.
[133] J. Guliński, B. Marciniec, H. Maciejewski, K. Pancer, I. Dąbek and R. Fiedorow, *Przemysl Chemiczny*, 2003, 82, 605. [In Polish].
[134] R.J. Hofmann, M. Vlatković and F. Wiesbrock, *Polymers*, 2017, 9, 534.
[135] M. Xanthos, *Functional Fillers for Plastics*, J. Wiley & Sons, 2010.
[136] Czynniki sprzęgające (coupling agents), *Plastics Review*, 2016, 5, 176, 60. www.plastics.pl [In Polish].
[137] D. Takeuchi, *Materials Research Society Bulletin*, 2013, 38, 252.
[138] K. Rózga-Wijas and J. Chojnowski, *Journal of Inorganic and Organometallic Polymers*, 2012, 22, 588.
[139] D.T. Liles in *The fascinating world of silicones and their impact on coatings, part 2, coatingstech*, May 2012, pp. 1–46; Presented as the Plenary Lecture at the 39th Waterborne Symposium, sponsored by the School of Polymers and High Performance Materials, The University of Southern Mississippi, February 13–17, 2012, in New Orleans, LA.
[140] Y. Kawakami, S.-P. Yu and T. Abe, *Polymer Bulletin*, 1992, 28, 525.
[141] G. Greber and S. Jäger, *Die Makromolekulare Chemie*, 1962, 57, 150.
[142] Y.Z. Wu and S.Y. Feng, *Journal of Applied Polymer Science*, 2001, 80, 975.
[143] C.J. Zhou, R.F. Guan and S.Y. Feng, *European Polymer Journal*, 2004, 40, 165.
[144] A. Kichler, N. Sabourault, R. Décor, C. Leborgne, M. Schmutz, A. Valleix, O. Danos, A. Wagner and C. Mioskowski, *Journal of Controlled Release*, 2003, 93, 403.
[145] I. Yilgor and J.E. McGrath, *Chapter: Advances in organosiloxane copolymers, Advances in Polymer Science*, 1988, 89, 1–86.
[146] V. Harabagiu, M. Pinteala, C. Cotzur and B.C. Simionescu, *The Polymeric Materials Encyclopedia; Synthesis, Properties and Applications, Vol. 4*, J.C. Salamone (Ed.), CRC Press, Boca Raton, FL, 1996, 2661.
[147] C. Yang, A. Gu, H. Song, Z. Xu, Z. Fang and L. Tong, *Journal of Applied Polymer Science*, 2007, 105, 2020.
[148] W.C. Liaw and K.P. Chen, *Journal of Applied Polymer Science*, 2007, 105, 809.
[149] W.C. Liaw and K.P. Chen, *European Polymer Journal*, 2007, 43, 1470.
[150] Y. Yagci, A. Onen, V. Harabagiu, M. Pinteala, C. Cotzur and B.C. Simionescu, *Turkish Journal of Chemistry*, 1994, 18, 101.
[151] V. Harabagiu, M. Pinteala, C. Cotzur, M.N. Holerca and M. Ropot, *Journal of Macromolecular Science – Pure & Applied Chemistry*, 1995, A32, 1641.
[152] C. Iojoiu, M.J.M. Abadie, V. Harabagiu, M. Pinteala and B.C. Simionescu, *European Polymer Journal*, 2000, 36, 2115.
[153] B. Hardman and A. Torkelson, Chapter: *Silicones*, in *Encyklopedia of Polymer Science and Engineering*, Vol. 15, 2nd ed, H. Mark, N.M. Bikales, C.G. Overberger and G. Menges (Eds.), Wiley, New York, 1989, 204–308.
[154] P.V. Wright, *Chapter: Cyclic siloxanes*, in *Ring Opening Polymerization, Vol. 2*, K.J. Ivin and T. Saegusa (Eds.), Elsevier, 1984, 1055–1133.
[155] J.M. Zeigler and F.W.G. Fearon (Eds.), *Silicon-based Polymer Science. A Comprehensive Resource, ACS Advances in Chemistry*, 1990.

[156] M. Mazurek, Chapter 23: *Polimery krzemoorganiczne*, in *Chemia Polimerów, Vol. 2*, Z. Florjańczyk and S. Penczek (Eds.), Oficyna Wydawnicza Politechniki Warszawskiej, Warsaw (Poland), 1997, 322–347. [In Polish].
[157] K. Matyjaszewski, Chapter 7: *Polimery nieorganiczne i organometaliczne*, in *Chemia Polimerów*, Vol. 3, Z. Florjańczyk and S. Penczek (Eds.), Oficyna Wydawnicza Politechniki Warszawskiej, Warsaw (Poland), 1998, 120–134. [In Polish].

Chapter 2
Silicones (polysiloxanes)

The name "silicones" is commonly used to denote polysiloxanes, which are a group of intermediate substances between organic and inorganic products and are considered as inorganic polymers. They constitute the most important group of polymers that have inorganic backbone structure consisting of repeating, alternating silicon and oxygen atoms, which contain carbon atoms from different organic groups (substituents) bound to the large number of silicon atoms. These macromolecular organosilicon compounds are basically the only inorganic–organic polymers (*hybrid*) that have gained industrial importance, although poly(phosphazenes) and carborane macromolecules also play an increasingly important role.

The most important group of organosilicon polymers are silicones, which are produced on a quite large industrial scale. Estimations of a global market of silicones reached 90 million US dollars (USD) in the year 1965, and 10 billion USD in 2000 year. World production of silicones is still growing. Annually, several thousand scientific papers from industrial laboratories and academic centers are published in this field. The number of practical applications of silicone products exceeds 150,000, and it is still expanding [1].

Dynamic development of the technology of silicones stems from the universality of silicon chemistry and the unique features offered by these materials in the most demanding applications. These data include all grades of organosilicon materials (silanes, silicates, and other silicon polymers), except metallic silicon and silicon semiconductor devices. In the last several years, there has been a lot of research work in the field of organic modification of ceramic materials and silicates. Organic silicates (e.g., (tetraethoxy)silane ($Si(OC_2H_5)_4$, TEOS) are inorganic substances, because they do not have organic groups directly linked to silicon atoms, but are linked only through oxygen atoms [2].

The development of chemistry and technology of polysiloxanes and other silicon polymers occurred gradually, as an expansion of the number of applications for this class of chemicals, so now they are versatile materials that play a key role in the development of new *high-tech* technologies and modern engineering materials. The expansion of silicones in the development of new materials technology is obvious. Currently, silicones cover numerous areas of practical application: from oils for a heating bath to materials for an artificial heart [3].

Basic components of silicones are polyorganosiloxanes, which are macromolecular compounds with siloxane skeleton, formed from silicon atoms connected by oxygen atoms. In polysiloxanes, different organic groups are bound to silicon through silicon–carbon bonds. Most of silicone products are based on polymethylsiloxanes. Other substituents bound to silicon atoms include mainly: alkyl, phenyl, vinyl, silanol, alkoxy, acyloxy groups, and hydrogen. Polysiloxanes containing phenyl or other aromatic

https://doi.org/10.1515/9783110643671-002

substituents in their structure exhibit much higher thermal stability than methyl-substituted silicones. There are also numerous known block and graft silicone–organic copolymers (e.g., silicone-glycol, -acryl, -polyurethane) and carbofunctional polysiloxanes [4–15]. In order to impart special properties, silicones may contain some additives, for example, fillers, solvents, water, or/and emulsifiers.

Table 2.1: Comparison of structure elements of organosilicon and organic polymers.

Elements of structure	Organosilicon polymers	Organic polymers
Size of atoms (Å)	Si 1.17	C 0.77
Interatomic distances (Å)	Si–O 1.63	C–C 1.43
	Si–C 1.88	
Bond angles (°)	Si–O–Si 140 ± 5	C–O–C 112
Bond energy (kJ/mol)	Si–O 445	C–O 358
	Si–C 306	C–C 346

Reprinted (adapted) with permission from [15]. Copyright 2002, by Państwowe Wydawnictwo Naukowe (PWN), Warsaw, Poland.

The chemical structure of polysiloxanes is substantially different from that in classical organic polymers. An atom volume of silicon is bigger than in carbon, and interatomic distances of Si–O and Si–C bond are bigger than in C–C bond. Bond angles of Si–O–Si linkages are much higher than in C–O–C structures, and energy of Si–O bond is significantly higher than in C–O bond. However, Si–C bond is weaker than C–C bond (see Table 2.1) [15].

An incorporation of organic groups into a strong inorganic skeleton loosens its structure, increasing the freedom of rotation about the axis of Si–O–Si linkages in siloxane chains, which affects properties of different kinds of silicones. Depending on the kind and number of organic groups attached to siloxane, different kinds of silicones with indirect properties between inorganic silicates and organic polymers can be obtained.

For shorter description of chemical structures of silicones, the following abbreviations are often used:
- **M** – $Me_3SiO_{1/2}$,
- **D** – $Me_2SiO_{2/2}$,
- **T** – $MeSiO_{3/2}$,
- **Q** – $SiO_{4/2}$.

In these siloxane units, oxygen atoms are shared with neighboring mers.

Polysiloxanes with linear structure form oils, rarely cross-linked polysiloxanes serve as elastomers and rubbers, while densely cured silicones are used as solid resins.

Linear polysiloxanes with molecular weight (MW) 10^4–2×10^5 g/mol are used as silicone oils, while with MW from (2–3) $\times 10^5$ to 10^6 g/mol they are used for elastomeric substances [15]. Long-chain poly(dialkylsiloxanes) and poly(alkylarylsiloxanes) – from several to few dozen thousand of silicon atoms in the chain show high polydispersity of MWs, for example, liquid PDMS have MWs from hundreds to 100,000 g/mol, and elastomers have MWs from hundred thousands up to few million g/mol. Linear polysiloxanes with very bulky organic substituents having up to 50 carbon atoms have the consistency of hot melt waxes [11]. Branched polysiloxanes are precursors of silicone resins.

At present, only some silicone products, mainly silicone resins and some oils, are prepared by the hydrolytic polycondensation reactions. Almost all linear polysiloxanes are synthesized by polymerization methods [16]. Many polysiloxanes (other than PDMS) with aliphatic or aromatic substituents are synthesized either by hydrolytic polycondensation or by cationic or anionic ROP. They have very interesting properties; for instance, the high MW (M_w = ~10^5–10^6 g/mol) di-*n*-alkyl-substituted polysiloxanes: poly(di-*n*-pentylsiloxane), poly(di-*n*-hexylsiloxane), and poly(di-*n*-decyl) siloxane have liquid crystalline (LC) properties and form disordered mesophases [17]. The living anionic ROP of strained cyclic siloxanes, for example, $(Me_2SiO)_3$ is used for preparation of block copolymers with organic monomers [18].

Silicones are also often prepared as dispersions, the most common being *liquid-in-liquid* (emulsion), *solid-in-liquid* (suspension), *air-in-liquid* (foam), and *solid-in-air* (powder) [19–22].

2.1 General properties of silicones

Constantly growing demand for different silicone and silicone-organic polymers results both from unique properties of silicones in comparison with organic polymers, and from the fact that even small amounts of silicones are sufficient to achieve required effects. Most important features of silicones include:
1. good thermal and oxidative stability in a wide temperature range,
2. low reactivity and good chemical resistance (with exception of strong bases and acids),
3. resistance to weathering, radiation, and UV radiation,
4. stability of physical properties in the wide temperature range,
5. very good dielectric properties,
6. liquid properties even for high MWs,
7. low free surface energy and low surface tension,
8. large free volume of macromolecules,
9. little apparent activation energy of viscous flow,
10. permeability of gases,
11. high compressibility and damping properties,
12. shear resistance,

13. low glass temperature,
14. low flammability,
15. low toxicity and little harm to the environment, and
16. insolubility in water.

The high thermal stability of silicones results from the presence of strong Si–C and Si–O bonds. Moreover, a partially anionic character of Si–O bond plays the role of a factor increasing the thermostability of silicones. The Si–C bond under anaerobic atmosphere is stable up to 500 °C. However, the resistance of this bond against oxidation is substantially lower, whereby thermal resistance of silicones in the presence of oxygen is worse and is only ~220–240 °C for PDMS. Thermostability of poly(dimethylsiloxane-α,ω-diols) is lower than in the case of trimethylsiloxy-terminated polysiloxanes MD_nM ($M = Me_3SiO_{0.5}$, $D = Me_2SiO$), and in the presence of alkalies does not exceed 175 °C [23]. The thermal and oxidative stability of silicones increases for aryl-substituted polysiloxanes. Most often, PDMS oils are used, but in some applications, when the higher thermal stability and oxidative resistance are required, dimethyldiphenylsiloxane copolymers or poly(methylphenylsiloxanes) are used. PDMS oils undergo crystallization at −40 °C and may be applied up to 200–250 °C. The addition of a few percentage of Ph_2SiO units effectively eliminates the tendency to crystallize. Thus, poly(dimethyl-*co*-diphenylsiloxanes) can be used up to ~300 °C.

The crucial property of silicones, depending on their chemical composition, is decreased flammability when compared to organic polymers. The main product of combustion of silicones is silica, which forms a protective, insulating, and ablating layer, counterforcing further burning.

Properties of silicones depend on their linearity, branching and cross-linking degrees. The siloxane chain has a special structure – it is flexible in the case of PDMS. Properties of silicones result from the character of Si–O bond and unsubstituted oxygen atoms, which enable easy rotation both around Si–O and Si–C bonds and determines values of contact angles: 143° for Si–O–Si linkages and 110° for O–Si–O moieties. Energy barriers are very low in both cases, while in polyethylene, the energy barrier of the rotation around C–C bond equals 14 kJ/mol. As a result of free rotations of segments of polysiloxanes, distances between chains in PDMS are larger than in hydrocarbons, so intermolecular forces are smaller, and, as a consequence, only small dependence of physical properties on temperature as also extremely low glass temperature ($T_g = -123$ °C) are observed. The properties of silicones significantly depend on substituents attached to silicon atoms. The more bulky alkyl or aryl substituents decrease freedom of rotation and cause increase of T_g; however, unsymmetrically substituted poly(methyl, ethyl siloxane) $-(MeEtSiO)_n-$ has lower T_g than PDMS. Polysiloxane chains in liquids and in elastomers, similar to polymer chains of other polymers, exist in an amorphic phase in the form of nests. Moreover, ease of rotation and small interactions between the substituents in the siloxane chain enable orientation of polysiloxane, depending on properties of the substrate, which leads to obtaining hydrophobic and/or antiadhesive surfaces. A very good permeability of vapors and gases, which is very characteristic of silicone polymers, primarily results from large intermolecular distances and longer chemical bonds, as compared to organic polymers [11].

Poly(dialkylsiloxanes) easily undergo crystallization, and poly(diethylsiloxanes) and poly(dipropylsiloxanes) also form liquid crystalline (LC) phases. The presence of even small content of aryl substituents in siloxane copolymers makes formation of crystalline phases impossible.

Moreover, silicones characterize very good dielectric properties, which to a small degree depend on temperature. All the above-mentioned features of silicones affect many practical applications.

Depending on their chemical structure, polysiloxanes are applied in the preparation of the following groups of liquid and solid usable products [11]:
- oils and emulsions,
- rubbers (elastomers),
- resins and varnishes,
- lubricants and pastes,
- materials for special applications.

2.2 Silicone oils

Polysiloxanes used as silicone oils usually have linear or slightly branched structures and contain a few to several thousands of alternating silicone and oxygen atoms and organic substituents attached to silicon. Siloxane chains in these oils are most often terminated with unreactive trimethylsiloxy ($Me_3SiO_{0.5}$) or reactive silanol (Si–OH) groups. Some kinds of silicone oils have branched structures with trimethylsiloxy side groups or even cyclic structures.

Most often poly(dimethylsiloxanes) (PDMS) are used as silicone oils. Other kinds of silicone oils include:
- poly[methyl(hydro)siloxanes] $-[Me(H)SiO]_n-$ (PMHS),
- poly[methyl(phenyl)siloxanes] $-[Me(Ph)SiO]_n-$,
- poly[methyl(chlorophenyl)siloxanes] $-[Me(ClC_6H_4)SiO]_n-$,
- poly[methyl(trifluoropropyl)siloxanes] $-[Me(C_3H_4F_3)SiO]_n-$,
- poly(diethylsiloxanes) $-(Et_2SiO]_n-$,
- poly[ethyl(hydro)siloxanes] $-[Et(H)SiO]_n-$,

where Ph = C_6H_5, and Et = C_2H_5.

Dynamic viscosities of silicon oils vary from few to 2 million mPa · s and only slightly depend on temperature. For chosen physical properties of some linear and cyclic oligo(permethylsiloxanes) see Table 2.2. Low MW poly(methylphenylsiloxanes) with viscosities 20–200 mPa · s and boiling points 20–200 °C/0.5 Tr are applied in diffusion pumps. Their evaporation heat is ~ 105 kJ/kg and vapor pressure: 10^{-7}–10^{-10} Tr.

Table 2.2: Physical properties of oligomers MD_nM i D_n [15, 24].

Siloxane	M.p. (°C)	B.p. (°C)	Density (g/cm³)	Refractive index $n^{20}{}_D$	Viscosity (η) at 25 °C (mPa · s)	Ignition temp. (°C)
MM	−67	100	0.7636	1.3774	0.65	−9
MDM	−80	153	0.8200	1.3840	1.04	37
MD_2M	−76	194	0.8536	1.3895	1.53	70
MD_3M	−80	229	0.8755	1.3925	2.06	94
MD_4M	−59	245	0.8910	1.3948	2.63	118
MD_5M	−78	270	0.9012	1.3965	3.24	133
MD_6M	−63	290	0.9099	1.3970	3.88	144
MD_7M		307.5	0.9180	1.3980	4.58	159
D_3	64.5	134	1.12 *			
D_4	17.5	175	0.9561	1.3968	2.30	69
D_5	−44	245	0.9593	1.3982	3.87	
D_6	−3	210	0.9672	1.4015	6.62	
D_7	−32	154	0.9730	1.4040	9.57	
D_8	31.5	290	1.1770 **	1.4040	13.23	

M.p., melting point; b.p., boiling point; * crystals; ** at 2.7 kPa (20 Tr).
Reprinted (adapted) with permission from [15]. Copyright 1997, by Oficyna Wydawnicza Politechniki Warszawskiej, Poland.

Some producers also offer low MW siloxanes: hexamethyldisiloxane ($Me_3SiOSiMe_3$, HMDS; b.p. 100 °C, m.p. −65 °C, viscosity 0.65 mPa · s) and cyclo(dimethylsiloxane) oligomers $(Me_2SiO)_n$ (D_4 and D_5, $n = 4$ and 5, b.p. 175 and 210 °C, m.p. 17.5 °C and −44 °C, respectively). The viscosities of methylsilicone oils depend on a number of Si atoms in the siloxane chain and on their number average molecular weight (Table 2.3).

Silicone oils are soluble in aliphatic and aromatic hydrocarbons, chlorinated hydrocarbons, ethers, esters, and some high MW alcohols. The low MW silicone oils with viscosity 5–20 cSt are soluble in acetone, methanol, and ethanol [5].

Silicone oils are mainly applied as
- hydraulic liquids,
- liquid shock absorbers,
- antifoaming agents,
- dielectric liquids,
- components of greases,
- components of cosmetics,
- hydrophobic agents.

Table 2.3: Dependence of viscosity of methylsilicone oils at 25 °C on a number of Si atoms in the siloxane chain and molecular weight.

Number of Si atoms in a siloxane chain	Number average molecular weight (calculated) M_n (g/mol)	Dynamic viscosity (η) (mPa · s)
2	162	0.65
50	3,700	~60
110	8,140	~140
230	17,020	~440
400	29,600	~1,440
800	59,200	~10,000
1,200	88,800	~50,000
1,400	103,600	~100,000

PDMS oils and polysiloxanes containing pendant aminoalkyl or quaternary ammonium groups are used in the form of emulsions or nanoemulsions as excellent softeners of textile materials [25, 26]. Polysiloxanes with fluoroalkyl substituents (e.g., trifluoropropyl), called "fluorosilicones," have a low surface energy and superhydrophobic properties, and thus, they find a range of important technological applications [27].

Trimethylsiloxy- or silanol-terminated low MW branched poly(methylsiloxanes) and poly(ethylsiloxanes) are useful as hydraulic liquids. A strong dependence of a free volume and viscosity on pressure results in very good hydrodynamic properties and high compressibility of silicone oils, leading to their applications in liquid shock absorbers, but they cannot be used as greases on friction surfaces under high stress. On the other hand, blends of silicone oils with fillers (e.g., colloidal silica) are good vacuum greases.

Density and a refractive index of silicone oils depend on their chemical structure. The refractive index is a good identifying factor of these oils (see data in Table 2.4).

A very good heat resistance is one of most important features of silicone oils – it is much better than for mineral and vegetable oils, and organic liquids. It results from a very good heat strength of the siloxane chain and high energy of the Si–C bonds. Additional factors affecting the excellent heat resistance are the exceptional oxidative and UV radiation resistance. The heat resistance of silicone oils depends on their chemical structure and, especially, on the kind of substituents bound to silicon atoms. It improves with replacement of alkyl substituents by aromatic groups and in anaerobic atmosphere but gets worse with increasing chain length of alkyl substituents.

The glass transition temperature (T_g) of PDMS oils was determined to be between −130 and −123 °C [28–31], while T_g of poly[methyl(hydro)siloxanes] (PMHS) is

Table 2.4: Basic physical properties of different kinds of silicone oils.

Variety of silicone oil	Viscosity (η) (mPa · s)	Density (g/cm^3)	Refractive index	Freezing point (°C)
Methylsilicone	100	0.97	1.4045	−50
Methylsilicone	1,000	0.97	1.4053	−50
Methylsilicone branched	60	0.97	1.4040	−90
Methyl(hydro)silicone	30	1.00	1.400	−55
Ethylsilicone	200	0.98	1.436	−60
Methyl(phenyl)silicone (10% Ph)	100	1.01	1.425	−80
Methyl(phenyl)silicone (~25% Ph)	100	1.07	1.500	−40
Methyl(phenyl)silicone (~45% Ph)	500	1.10	1.533	−25
Methyl(chlorophenyl)silicone	75	1.04	1.436	−65
Methyl(trifluoropropyl)silicone	300	1.25	1.381	−50
Methyl(alkylamino)silicone	25	0.97	1.411	−40

−142 °C, and T_gs of linear dimethylsiloxane copolymers containing MeHSiO units ranged from −133 to 135 °C [31].

Methylsilicone oils with medium and high viscosities can be used at temperatures from −50 to 200 °C. Oils with low viscosities and low MW branched oils do not freeze at lower temperatures: even at −100 °C. Ethylsilicone oils can be used at temperatures from −65 to 150 °C, and methyl(phenyl)- and methyl(chlorophenyl)silicone oils show much higher thermal resistance, up to 230–250 °C, and the different resistance to low temperatures: −20 to −70 °C. A specific heat of methylsilicone oils equals 1.4–1.5 kJ/kg K, and that of methyl(phenyl) silicone oils is 1.56–1.68 kJ/kg K. A thermal conductivity of methylsilicone oils with viscosities exceeding 100 mPa · s equals 0.16–0.17 W/(m K), and for methyl(phenyl)silicone oils is lower, at about 0.124–0.147 W/(m K). When heating silicone oils above 250 °C in the presence of oxygen, gradual oxidation of organic substituents takes place, leading to increased viscosity, and finally to gelation, while heating above 1,000 °C gives pure silica $(SiO_2)_n$, just as in burning of silicone oils. While heating above 300 °C in the absence of oxygen, and especially with traces of alkali, catalytic depolymerization processes occur, accompanied by decrease in viscosity and formation of appropriate volatile oligo(dialkyl)siloxanes. The ignition temperatures of silicone oils with medium and high viscosities exceed 300 °C, and self-ignition temperature is 350–400 °C. Low MW silicone oils may be flammable or difficult to ignite.

The freezing temperature of silicone oils also depends on their chemical structures. The lower freezing points are characteristic of oils with irregular structures. The lowest freezing temperatures occur for low MW branched siloxanes. Incorporation of a certain amount of phenyl groups into their structure results in decrease of freezing temperature from −50 °C to approximately −70 °C, and even −90 °C (at the same viscosity).

The chemical resistance of silicone oils is good, especially on action of oxygen, oxidizing reagents, diluted acids and bases, and solutions of salts, sulfur dioxide, phenols, and other chemicals. However, they undergo decomposition under action of water steam, concentrated alkalis and acids, chlorine, some amines, iron, aluminum, and boron chlorides.

The compressibility of silicone oils is much higher than mineral oils and organic hydraulic liquids and approximately 10 times higher than compressibility of water. Coefficients of the adiabatic compressibility of methylsilicone oils with the viscosities 50–1,000 mPa · s equal ~100×10^{-11} m^2/N, for methyl(phenyl)silicone oils ~60×10^{-11} m^2/N, while for ethylene glycol it is ~30×10^{-11} m^2/N. This feature, together with the resistance on low temperature and small change of the viscosity with temperature, characteristic of silicone oils, is very precious for application in hydraulic devices. The silicon oil with the viscosity 1 mPa · s remains liquid under pressure 4,000 MPa, compressing 35% with volume decrease, while *n*-hexadecane solidifies under pressure 42 MPa.

The surface tension (and the surface energy) of silicone oils is much lower when compared to most of mineral oils and organic liquids and depends on molecular composition, viscosity and MW of silicone oil. The surface tension of methylsilicone oils changes from 18.7 mN/m at viscosity 2 mPa · s to 20.1 mN/m at viscosity 10 mPa · s, and 21.5 mN/m at higher viscosities. The surface tension of methyl (hydro)silicone oil equals 20 mN/m, and of methyl(trifluorophenyl)silicone oils is 20–22 mN/m, depending on the content of fluoroalkyl substituents. In the case of methyl(phenyl)silicone oils, the surface tension is higher: 23–28 mN/m, which depends on the content of phenyl groups. Low values of the surface tension of silicone oils stimulate such properties as hydrophobicity, wettability of organic surfaces, and antiadhesive properties.

The radiation resistance of silicone oils is also dependent on their chemical structure. Silicone oils substituted with hydrogen, methyl, hydroxyl, and trifluoropropyl groups are less resistant on highly energetic radiation than oils substituted with aromatic moieties (and especially with phenyl groups). The radiation of silicone oils results in increase of viscosity and gradual decomposition, leading to gels.

The coefficient of a volume expansion is dependent on their chemical structure as well and has higher value for methylsilicone oils [$(9.5–10) \times 10^{-4}$ K^{-1}] than for methyl(phenyl)silicone and methyl(chlorophenyl)silicone oils [$(7.8–8.7) \times 10^{-4}$ K^{-1}], and also for methyl(trifluorophenyl)silicone oils (7.8×10^{-4} K^{-1}) – in the temperature range 25–175 °C.

2.3 Silicone elastomers and rubbers

About half of the world's production of silicones is in the form of silicone elastomers (SE) and silicone rubbers (SR). Before vulcanization, SRs have linear structures (Figure 2.1.) and contain up to several thousand silicon atoms in the siloxane chain [4, 8, 10, 11, 23, 24, 27, 32–43]. The MWs of the SR, cross-linking at ambient temperature without heating (RTV, *room-temperature vulcanizing*), are most frequently, 10^4–10^5 g/mol, while the MWs of the rubbers vulcanized at elevated temperatures (HTV, *heat temperature vulcanizing*) are much higher, generally from 3×10^5 to 8×10^5 g/mol, and sometimes even 10^6–2×10^6 g/mol. Most often, the main chains of SRs include polysiloxanes only with methyl groups at the silicon atoms.

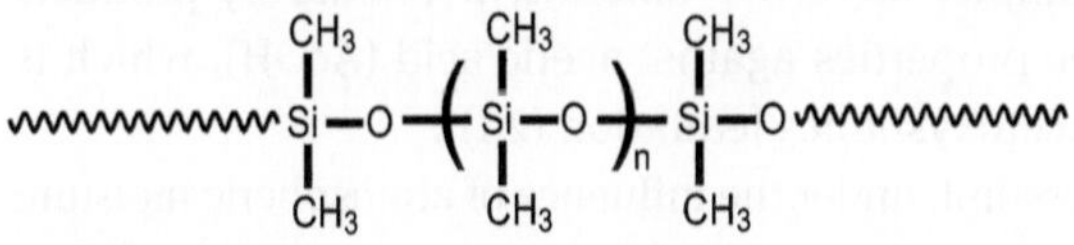

Figure 2.1: A chemical structure of a poly(dimethylsiloxanes) backbone.

The HTV SRs have the nonreactive trimethylsiloxane or reactive vinyldimethylsilyl chains ends or a very small amount of lateral vinyl groups (generally approx. 0.05–0.5 mol%), which take part in cross-linking reactions.

The RTV SRs compositions can be divided into three groups:

- one-component,
- two-component (binary),
- three-component (ternary).

The *Cold* vulcanizing SRs (RTV) are mixtures of liquid polydimethylsiloxane-α,ω-diols $HO(Me_2SiO)_nH$ (or other polysiloxanes having reactive end groups) and the organosilicon compounds containing functional groups capable of condensing with the silanol groups ≡ Si–OH (or other functional groups) after addition of the catalyst. The condensation reactions lead to the formation of three dimensional cross-linked networks. The cross-linking (vulcanization) of polysiloxanes involves the formation of a very small amount of cross-bridges between the chains, using cross-linking agents or energy (e.g., radiation), and confers the elastic properties of materials, characterized by reversible deformations under the influence of external forces.

The cross-linking processes of telechelic polysiloxanediols (PSD), most often, take place through the following condensation reactions [42]:

$$\equiv \mathrm{Si-OH} + \mathrm{AcO-Si} \equiv \longrightarrow \equiv \mathrm{Si-O-Si} \equiv + \mathrm{AcOH} \quad (2.1)$$

$$\equiv \mathrm{Si-OH} + \mathrm{RO-Si} \equiv \longrightarrow \equiv \mathrm{Si-O-Si} \equiv + \mathrm{ROH} \quad (2.2)$$

$$\equiv Si-OH + H-Si\equiv \ \rightarrow\ \equiv Si-O-Si\equiv\ + H_2 \tag{2.3}$$

$$\equiv Si-OH + R_2C=N-O-Si\equiv \ \longrightarrow\ \equiv Si-O-Si\equiv\ + R_2C=N-O-H \tag{2.4}$$

Cross-linking reactions of PSD with acetoxysilanes and alkoxysilanes [equations (2.1) and (2.2)] proceed usually in the presence of organotin catalysts of a general formula: $Bu_2Sn(RCOO)_2$. The cross-linking reactions with alkoxysilanes are catalyzed by the lithium, sodium and potassium hydroxides, amines, and ammonium salts: $(Bu_2HN^+)(^-OOCR')$, $(Me_2HN^+)(^-OOCNMe_2)$. Further acceleration of the reaction occurs in the presence of an amine catalyst which can form a complex with a tin catalyst, or can react with a carboxylic acid created during the hydrolysis of the carboxy derivative of tin catalyst [34]. In cross-linking systems based on the (oxy)iminosilanes [equation (2.4)], for example, $MeSi(ON{=}CMeEt)_3$, ketoxime by-products are formed, exhibiting no corrosive properties against acetic acid (AcOH), which is formed during cross-linking with acetoxysilanes [equation (2.1)].

One-component SR mixtures cross-link under the influence of atmospheric moisture and contain all components mixed together: polymer, filler, pigment, and cross-linker [44]. Many commercial SR compositions are prepared by vulcanization of one-part PDMS polymer with acetoxy, alkoxy, or oxime functional cross-linkers, and reinforcement fillers [45, 46]. Different physical properties of a commercial injection-molded SR and one-part SRs cross-linked with 3 various cross-linking agents (acetoxy-, alkoxy- and oxime-functionalized silanes) are presented in Table 2.5. In one-component mixtures, when methyl(triacetoxy)silane is used, trifunctional cross-links are formed, accompanied by elimination of acetic acid as the by-product of the vulcanization.

In binary SR compositions, amines or carboxylic acid salts of metals (Sn, Zn, Cd, Pb), for example, naphthenates, caprylates, or laureates are used as the catalysts. The frequently employed catalysts are dibutyltin dilaurate and stannous octoate (i.e., 2-ethylhexanoate), usually in amounts of 0.05–2 wt%, based on the weight of the rubber composition. In two-component (binary) mixtures, one part comprises functionalized PDMS, a filler and a pigment, and the other comprises the cross-linking agent and the catalyst. The two parts are mixed just before use. Depending on further use, mixtures of fluid consistency to hard sealants can be obtained. In binary compositions, the cross-linking of hydroxyl-terminated PDMS with TEOS as the cross-linking agent proceeds via condensation reactions (with an evolution of ethanol as the by-product) and leads to the formation of tetrafunctional cross-links and to gelation [43].

In ternary RTV mixtures, poly(methylhydro)siloxanes (PMHS) of linear, cyclic, and branched structures [34, 42] are used as cross-linking agents. During their vulcanization hydrogen gas is liberated which, in the case of thicker layers, causes the formation of microporous structure. Vulcanization reaction using methyl hydrogen siloxane also requires the use of catalysts (usually tin or platinum compounds).

A gelation time of binary and ternary cross-linking SRs compositions of *cold vulcanization* (RTV systems) depends on the amount of catalyst used and may range

Table 2.5: Physical properties of PDMS elastomers (SR) prepared by vulcanization of PDMS with cross-linker and reinforcement filler [45, 46].

Property	Units	Conditions	Values			
			A	B	C	D
Specific gravity	–	ASTM D 792	1.13	1.04	1.51	1.04
Viscosity	Pas	ASTM 4287, 10 s^{-1}	290	Nonflow	Nonflow	Nonflow
Extrusion rate	g/min	At 90 psi, 1/8 in orifice	100	350	110	440
Durometer (shore A)	points	ASTM D 2240	40	25	37	35
Tensile strength	MPa	ASTM D 412	9.0	2.24	1.55	1.79
Elongation at break	%	ASTM D 412	725	550	640	430
Tear strength, Die B	kN/m	ASTM D 624	37.7	4.9	6.48	5.6
Dielectric strength	kV/mm	ASTM D 149	18.5	21.7	17.4	13.5
Dielectric constant ε	–	ASTM D 150, at 100 Hz	2.98	2.8	3.69	2.77
Volume resistivity	ohm/cm	ASTM D 257	3.8×10^{34}	1.5×10^{15}	6.1×10^{14}	2.4×10^{14}
Dissipation factor	–	ASTM D 150, at 100 Hz	0.0033	0.0015	0.0022	0.0035

A. Injection-molded liquid silicone rubber, Silastic1 LSR 9280-40.
B. One-part RTV acetoxy cross-linked SR, Dow Corning1732.
C. One-part RTV alkoxy cross-linked SR, Dow Corning1737.
D. One-part RTV oxime cross-linked SR, Dow Corning1739.

from a few to several dozen hours. The *macroscopic* properties of the SRs are determined mainly by the following factors: a high energy of Si–O bond, an ionic polar nature of the main backbone, a dipole moment that shows the impact on the stability of the organic groups linked to silicon, mobility and the free movement of the polymer chain due to a wide angle of rotation around the oxygen atom of the siloxane moiety ≡Si–O–Si≡, and also a spatial arrangement of the organic groups in the chain in the form of a spiral. This explains the low level of intermolecular forces of overmolecular secondary structure, when compared to other kinds of rubbers and the much lower value of the density of cohesive energy than for other polymers [27]. Owing to the spiral chain structure, and the free rotation of the methyl groups that have the ability to even further extend the reach of the chains, the polysiloxanes have a high molar volume, which affects their compressibility, permeability to gases and vapors, and above all, their rubber-like properties, which are particularly clear after creating a three-dimensional mesh rubber. The oxygen permeability of SR is ~10 times higher than in natural rubber (NR) and low-density polyethylene (LDPE), and about 100 times greater than that of butyl rubber and nylon. Due to the low intermolecular forces, SRs do not exhibit very good physicomechanical properties, when compared to other rubbers and polymers [47–49]. In SRs with no additives, interactions between the organic substituents bonded to the silicon atoms are very weak, and therefore, their tensile strength (TS) is only 2–5 kG/cm^2 [34, 35]. Good mechanical properties of SRs can be obtained only by mixing polysiloxanes with the finely divided active fillers. The gain factor for organic rubbers is approximately 10, while for SRs is approximately 40–50 [1, 3, 4, 24]. Of great importance is the amount and type of the filler and the degree of fragmentation. The filler should be dispersed in the rubber without affecting (e.g., chemically) the decomposition of the polymer. The particle size of the fillers also affects the strength of the SRs. The best reinforcing properties were observed for fillers with a specific surface area of at least 150–200 m^2/g and a particle diameter of 10–20 μm [1, 3], and even much smaller (0.01–1 μm) [24]. On the other hand, even a very small share of the coarse filler (approximately 1 wt%) considerably worsened the mechanical properties of rubbers. The following fillers are most often used: silica, titanium white, chalk, kaolin, iron oxide red, among others. The kind of the filler also determines, to a large extent, the resistance of SRs; for example SRs filled with titanium dioxide are more resistant to weathering than those filled with the colloidal silica [1, 3, 4, 24, 34–37].

The combination of organic and inorganic components in one molecule of silicone polymer gives silicone elastomers and rubbers a number of valuable properties, such as resistance to high and low temperatures, good dielectric properties (low dependence of the dielectric properties by changes in a temperature, current frequency and humidity), good resistance to atmospheric agents, and good hydrophobic and antiadhesive properties. The SRs also have excellent the heat resistance. Short-term overheating to 300 °C, long-term work at 225 °C, and action of air almost do not affect their mechanical and electrical properties. At a temperature of −30 °C SRs retain 70% of their flexibility, they are vitrified until well below −50 °C (to −123 °C), and they burn with the secretion of the silica. In terms of the dielectric properties at room temperature, SRs behave as good electroinsulating materials, and at high temperatures, their dielectric properties are much better than in most plastics. The thermal conductivity of SRs at room temperature is about twice, and at 200 °C eight times larger, than in organic rubbers, depending on the type of SR and degree of cross-linking. Good thermal conductivity of SRs is especially valuable in electrical insulation, since it is simpler to remove heat from the windings and cables insulated with SRs. The SRs show very good antiadhesive and hydrophobic properties. They do not exhibit adhesion to the stickiest substances like epoxy resins, rubber mixes, molten polyethylene, and other materials. The SRs are practically nonwettable with water, so they can be used for coating technical textiles and paper (inter alia for the manufacture of self-adhesive labels), and for sealing construction elements.

The chemical resistance of the SRs is quite good, especially to diluted mineral acids and alkalies, salt solutions, perhydrol (30% solution of H_2O_2), SO_2, and other substances. They have the high resistance to ozone. However, they are not resistant to hydrogen fluoride, concentrated sulfuric acid, and dry hydrogen chloride. The main disadvantage of PDMS rubber, however, is also a considerable tendency to swell in organic solvents, especially aromatic hydrocarbons, petrol, halogenated derivatives, ethers, as well as esters and oils. Swelling does not damage their structure, and it can be reduced by introducing to polysiloxanes side groups linked to the silicon atoms: γ-trifluoropropyl ($-CH_2CH_2CF_3$) and β-cyanoethyl group ($-CH_2CH_2CN$).

Fluorosilicone rubbers are resistant to high temperatures (450–500 °C) and nonflammable. They are excellent material for the manufacture of military gas masks similarly to poly(phosphazene) elastomers.

Borosiloxane rubbers containing silicon atoms in tens of one boron atom have unusual rheological properties. Balls made with such rubber (so-called *jumping putty* or *sealy putty* [50, 51]) were very resilient when thrown – they bounced to a height of up to 90% of the height from which it was thrown. Left as is, they slowly take on the shape of the container in which they are kept. They are used to fill the interiors of golf balls and in training damaged muscles in hospitals.

The vulcanized SRs (without heating) are used in the following applications:

- for potting and encapsulation of electrical equipment (e.g., different transistor circuits and small motors) to protect them from external influences (moisture, climatic factors, and some chemicals) and to protect the equipment against overload gravity, shock, and vibration;
- for the manufacture of elastic molds for the processing of gypsum, wax, chemically cured polymeric materials (e.g., epoxy resins and polyurethanes (PUs)), and casting objects from the low-melting metals (e.g., lead); an advantage of forms made from SRs is their high flexibility, minimal shrinkage during vulcanization, accurate contour, lack of adhesion to the molded plastic, and high heat resistance; these features are used in reconstruction of architectural monuments and preparation of replicas in numismatics, among others;
- for impregnating and coating of a variety of fabrics, paper, wire, and cable (e.g., for automotives); impregnated and coated papers have a lot of flexibility, moisture resistance, and good dielectric properties.

The RTV SRs are also commonly used as components of an electronic equipment. They can also be used as liquid materials with high viscosity in modern railway bumpers and dampers to increase the strength and use of the ropes in the mining industry, and in various types of sealing (*sealants*), as they are resistant to the moisture and many chemicals at temperatures −50 to +225 °C [52–60]. Borosiloxanes are crucial ingredients in these compositions [54–57, 60]. Powder SEs have been used as additives for thermally and chemically cured resins (e.g., epoxides) to improve the impact strength, reduce the water content, and facilitate the separation of the molded products of the form [1]. Numerous applications of these rubbers are known in medicine. SR hoses modified inside with heparin are used for transfusion of a blood, avoiding blood clotting. With this type of the elastomer a lot of implants, and the first artificial heart were fabricated. At present, for this purpose, siloxane-urethane elastomers are used.

Most of the literature refers to the SRs containing fillers and other additives and are derived from patent literature and information materials of manufacturers. This makes it difficult to compare the mechanical properties of SEs and SRs. Furthermore, these properties to a great extent depend on the conditions of experiments. A very beneficial property of rubbers is its ability to withstand high pressure applied during processing (molding), for example, in the manufacture of injection-molded dumbbells for tensile tests, by adding a type of SEs [38]. However, in the case of elastomers cross-linked at ambient temperature and used in sealing and flexible devices (for molding chemically cured resin or gypsum), it is appropriate to prepare test samples that are as close to the procedure for the production of molds as possible. In the literature on the SRs, there is very little information about the physical properties of pure elastomers that do not contain modifying substances such as the fillers and plasticizers. This applies particularly to the kinds of the RTV SRs that are based on a poly(dimethylsiloxane-α,ω-diols) (PDMS-diol) and PMHS as the cross-linking agent.

J. Chruściel examined the influence of the chemical structure of cross-linking agents, that is, PMHS of the block and statistical structure of the siloxane chain, on some mechanical properties of the SRs obtained by cross-linking of poly(dimethylsiloxanediols). Their mechanical properties were compared with the properties of the SRs cross-linked with other conventional cross-linking agents [41].

As a base polymer for the synthesis of three-component RTV SRs were used poly(dimethylsiloxane-α,ω-diols) $HO(Me_2SiO)_nH$, containing about 150–250 silicon atoms per molecule, with different MWs ($M_n = 10^4$–2×10^4 Da; $M_w = 3.0 \times 10^4$–7.1×10^4 Da; PDI = 2.6–3.5). These polysiloxanediols were cross-linked with poly(methylhydrosiloxanes):

(1) containing single mers Me(H)SiO, of the block structure (**A**) [61, 62]:

$$\mathbf{Me_3SiO\left[(Me_2SiO)_m MeHSiO\right]_n (Me_2SiO)_m SiMe_3} \ (\mathbf{A})$$

where $m = 2, 6, 10, 14$, and $n = 5, 10, 15$;

(2) of the statistical structure (**B**) and similar functionality (the content of the Si–H groups in the macromolecule) [61]:

$$\mathbf{Me_3SiO[(Me_2SiO)_xMeHSiO]_n(M_2SiO)_ySiMe_3\ (B)}$$

where $x + y = m(n + 1)$; $m = 2, 6, 10, 14$; $n = 5, 10, 15$.

The block poly(dimethyl-*b*-methylhydro)siloxanes were prepared by heteropolycondensation of appropriate oligodimethylsiloxanediols with $MeHSiCl_2$, and next, with Me_3SiCl, in the presence of Et_3N and (4-dimethylamine)pyridine or only pyridine (for $m = 2$) [61, 62]. Random PMHS were prepared by hydrolytic polycondensation of mixtures of appropriate chlorosilane monomers. The molecular formula (**B**) corresponds to the composition of monomer mixtures used for their syntheses. The chemical structures and a microstructure of the siloxane chain were confirmed by spectroscopic methods (^{1}H- and ^{29}Si-NMR (nuclear magnetic resonance) and FTIR (Fourier-transform infrared)), gel permeation chromatography (GPC), and an elemental analysis [62].

The mixtures of the polysiloxanediols $HO(Me_2SiO)_nOH$ with different MWs and PMHS of the structures (**A**) or (**B**) were cross-linked under the influence of 1 or 2 vol% of organotin catalyst – dibutyl tin dilaurate (DBTL). The cross-linking reaction took place by condensation of silanol groups with the Si–H groups of the cross-linking agent, according to the scheme shown in Figure 2.2.

In order to obtain good quality layers of SRs for the TS and elongation at break (ε) tests, mixtures of PDMS and PHMS were evacuated in a vacuum oven under reduced pressure prior to addition of the catalyst. The removal of air (dissolved in the polysiloxane composition) ensured obtaining vulcanizates without air bubbles. In an analogous manner, polysiloxanediols mixtures containing different amounts of the colloidal precipitated silica (with a specific surface area ~150 m^2/g), relative to the weight of PDMS, were cross-linked with H-polysiloxanes. For comparison, samples of the SRs were prepared, which were cross-linked with TEOS ($Si(OEt)_4$), 1,3,5,7,9-pentamethylcyclopentasiloxane, ($D^H{}_5$, where $D^H = MeHSiO$), and PMHS of random (statistical) structures somewhat different from that of structure (**B**) and the size of the macromolecules and their functionality.

The gelation times (t_{gel}) of the mixture of the SRs cured with $MD^H{}_{35}M$ were several hours, and for the remaining H-siloxanes – from several hours to several days, depending on the type of the cross-linking agent (PMHS) and the amount of catalyst. The gelation times grew with increasing MW of PMHS and were longer for mixtures containing a filler, but were shorter with increasing MW PDMS, and the increase in the average functionality of H-siloxane (n). Knowledge of the gelation times of each composition was necessary in order to obtain layers of good quality SR for the strength tests. Vulcanizates with a smooth surface (bubble-free) were always obtained when the gelation times (t_{gel}) were not shorter than 5 h.

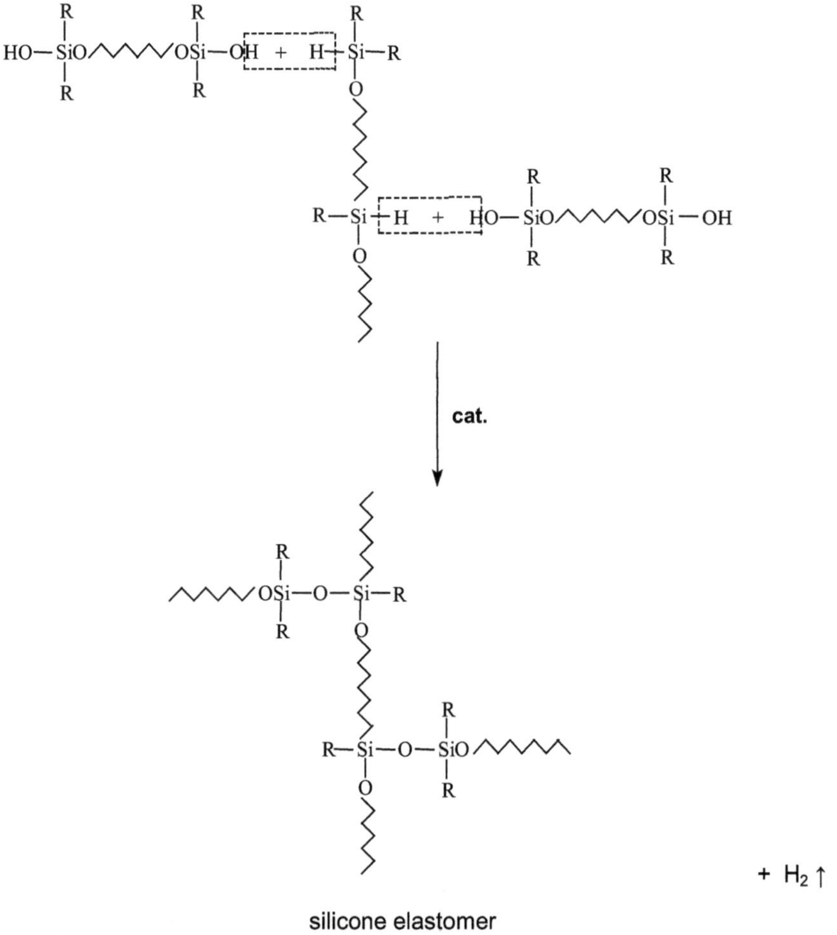

Figure 2.2: A scheme of cross-linking reactions of poly(dimethylsiloxane) elastomers via dehydrocondensation reaction.

Most important conclusions based on mechanical studies of the above-mentioned SEs and SRs were the following [41].

1. The SEs without fillers, prepared at room temperature by cross-linking polydimethylsiloxane-α,ω-diols with polymethylhydrosiloxanes (PMHS) of the block structure (**A**), random structure (B) and hydrosiloxane "$MD^{H}{}_{35}M$" (which had slightly branched structure $[M_2D^{H}{}_{34}T]_3$, where T denotes unit $MeSiO_{1.5}$, established on a basis of GPC and ^{29}Si-NMR data) showed similar, good mechanical properties (TS and relative elongation at break, ε_r), comparable to the properties of the SEs cross-linked with TEOS and $(HMeSiO)_5$. The mechanical parameters obtained were fully satisfactory for applications for manufacture of elastic molds and sealing materials and were comparable to literature data [4, 24, 35–37].

2. Applications of multifunctional PMHS, lower than stoichiometric amounts, for cross-linking polydimethylsiloxane-α,ω-diols resulted in increase of elasticity (ε) of SEs and SRs and simultaneous decrease of TS); it is presumably caused by decrease of a cross-linking density. However, at high loading of silica (30.5 wt%) in SRs mixtures and 2 wt% content of DBTL catalyst, a change in ratio of functional groups [Si–OH]/[Si–H] (1:1.27; 1:0.846, and 1:0.423) did not improve elasticity and practically had no effect on their tensile strength (TS). Thus, it seemed quite obvious that silanol groups of silica also took part in reactions with cross-linking agents. The application of half of the stoichiometric amount of low MW cross-linking agents [e.g., $Si(OEt)_4$] led to incomplete local curing in silicone vulcanizates, which caused decrease of their strength and led to sticky surfaces with good adhesion to other materials.
3. The chemical structure of the block and random PMHS similarly affected mechanical properties (TS and elongation at break ε); it improved elasticity of the SEs and SRs, especially those which were prepared from polydimethylsiloxane-α,ω-diols with relatively low MWs (e.g., for PDMS having $M_n = \sim 1{,}2\times10^4$ g/mol). The chemical structure of PMHS did not show a substantial effect on the mechanical properties of the SEs and SRs that were prepared.
4. The gelation time of the SE and SR mixtures grew:
   (a) with decrease in number of the Si–H groups in the block PMHS of the structure (**A**) and random PMHS of the structure (**B**),
   (b) with decrease of the concentration of the catalyst,
   (c) with increase of the loading of the filler.
5. With increase of the content of silica in the SRs, an enhancement of TS and decrease of ε were observed.
6. The SRs prepared from polydimethylsiloxane-α,ω-diols and PMHS of the random structures (**B**) containing 30.5 wt% of silica, cross-linked with 1 wt% of the catalyst DBTL, exhibited a very good hardness H (51–62 °Sh A) and a very good compression strength (i.e., low values of a relative stable deformation at compression, $\varepsilon_S = 5$–10%). These profitable parameters (H and ε_S) resulted from the high content of the filler.
7. The application of PMHS for cross-linking polydimethylsiloxane-α,ω-diols allowed preparation of the solid SEs and SRs (few millimeters thick) only in thin layers or microporous materials (in thicker layers or in blocks).
   The SRs with better, optimal mechanical properties can be obtained using:
   (a) silica filler with a higher surface area (>200 m^2/g),
   (b) polydimethylsiloxane-α,ω-diols with bimodal distribution of MWs [63–71],
   (c) catalysts which accelerate faster homopolycondensation reactions of PDMS-diols than their cross-linking reactions [24, 35, 72, 73].

In order to prepare the SRs with very high relative elongation at break (ε = several hundred %), it was necessary to use Sn(II) compounds as catalysts, and both reinforcing or semireinforcing fillers, together with nonreinforcing filler (e.g., $CaCO_3$), usually in moderate quantities (e.g., 10–15 wt%), and also reagents causing extension of polysiloxane chains during cross-linking processes of the SEs. Although the SRs with such compositions possess decreased strength, they are applied as sealing materials for glazing in modern buildings made of metals and glass, especially in hot climate conditions [36]. In general, the chemical structures and mechanical parameters of the SEs and SRs are programmed according to the application requirements.

Recently, ultra-high elongation SEs with elongations ~5,000%, nearly 4 times greater than any commercial elastomer, have been elaborated by B. Arkles et al. [74]. These new elastomers are based on α-vinyl-ω-hydro-PDMS (DP = 50–200, PDI < 1.3, which were obtained by anionic ROP of D_3 with $ViSiMe_2OLi$, followed by termination reaction with Me_2SiHCl), which were further polymerized by a step-growth polyhydrosilylation with Pt catalyst to extremely high MWs that were estimated to be higher than 3×10^6. Their elastomeric properties resulted from concomitant formation of intra-chain and inter-chain entanglements, without covalent cross-linking. When elastomers had elongations significantly above 1,500%, they exhibited behavior that was readily differentiated from conventional elastomers. Tear failure occurred at radically greater elongations. These SEs showed pseudo-shape memory behavior because they were able to undergo extreme multiaxial distortion and return to their original shape.

During the polymerization, the extremely flexible growing polymer chains became entangled not only with each other, but also with themselves. The segment showed multiple knots. Low levels of volatile oligomers and extractables were noticed. Moreover, these SRs showed the improved thermal stability and improved dimensional stability on aging.

A newer variation is liquid SRs (LSR) intended for injection molding or for coating, for example, textile materials. LSR parts show excellent properties of steam resistance and a low compression set and are used in different industries, particularly in high-tech fields, for instance, as automotive parts, gaskets, and hardware in consumer appliances such as microwaves. Their conductivity and fatigue resistance make them ideal for electronic interfaces on keyboards or touch pads. Their oil and the heat resistance offers long-term durability in automotive components [75–76].

The SEs and SRs find many applications in various fields. A few are described below. For example, textile materials were impregnated with an emulsion of the SEs containing polydimethylsiloxanediol having 0.02–1.7% of reactive silanol groups, with an emulsifier, and an emulsion of poly(methylhydrosiloxane) oil as the cross-linking agent. Grafting of silicones on a knitted fabric and their cross-linking during the drying and heating process was catalyzed with aqueous solutions of zinc and/or tin salts containing tertiary or quaternary aminoalcohols [77].

PEG oligomers covalently grafted onto SR surfaces, activated by plasma, and functionalized with H_2SiCl_2 gave brush-like materials with strong antifouling properties [78].

A novel finger-sensing nanocomposite (NC) with remarkable and reversible piezoresistivity was prepared by dispersing homogeneously conductive graphite nanosheets (GNs) in a SR matrix. Due to the high aspect ratio of GNs, the SR-GN NC displayed a very low percolation threshold and super-sensitive piezoresistive properties [79].

Methylvinylsilicone HTV gum filled with small amounts of Fe_2O_3 (0.5–1.5 wt%) was modified with vinyltrimethoxysilane (VTMS) and cross-linked with 2,5-bis(*tert*-butylperoxy)-2,5-dimethyl hexane at 170 °C for 10 min. and postvulcanized at 200 °C for 4 h. It showed significantly the increased tear strength and improved heat resistance [80].

The titania/SR composite fibers were obtained from $Ti(OiPr)_4$, hydroxyl-terminated PDMS (M_n 46,000), and TEOS by a combination of sol–gel and electrospinning methods, using (2-methoxy)ethanol and THF as solvents, followed by heat treatment at 250 °C for 3 h. The TS and modulus of the TiO_2/PDMS composite fibers increased gradually with increasing PDMS content (10–30 wt%). Their photocatalytic activity was strongly dependent on the TiO_2 content [81].

The TS (8.5 kg/cm^2) and elongation at break (780%) of SR composites vulcanized with dicumyl peroxide and filled with 10 wt% of nano-$CaCO_3$, surface-modified with stearic acid was significantly improved in comparison to the untreated nano- and commercial $CaCO_3$. They were a result of uniform dispersion of nanoparticles in the treated composites and their good compatibility with the SR chains [82].

A sugar-templated PDMS elastomer sponge was applied for the selective absorption of oil from water. The sugar particles were dissolved in water and the PDMS sponge were compressed repeatedly in air or liquids without collapsing. The absorbed oils and organic solvents can be removed and reused by squeezing the PDMS sponge [83].

The mixtures of bimodal telechelic divinyl-PDMS elastomers, an alkoxysilane adhesion promoter, hexamethyldisilazane (HMDS), small amount of water, that contained inorganic flame-retardant (FR) additives (fumed silica or/and alumina) were cross-linked with PMHS and Pt catalyst at 120–160 °C within 1–5 min and were used as elastic adhesives for protection and processing of wood layers [84].

The electromechanical properties of the SEs were improved by the addition of two active fillers in different ratios with complementary effects: silica, mainly as a reinforcing agent and barium titanate as dielectric permittivity enhancer. Silicone composites prepared from a high molecular mass polydimethylsiloxane-α,ω-diol (M_w = 642,000 g/mol) were further processed as films and cross-linked at high temperature [85].

Two polysiloxanes, a polydimethylsiloxane-α,ω-diol (PDMS-diol) with M_n = 370,000 g/mol, and α,ω-bis(vinyl)polydimethylsiloxane with M_n = 34,500 g/mol, were mixed together in various weight ratios (1:0.1, 1:0.2, 1:0.3, 1:0.5) and were cross-linked with TEOS/DBTL and α,ω-bis(trimethylsiloxy)poly(dimethyl-*co*-methyl-H-siloxane) in the presence of Speier's catalyst, respectively. They were cast into films and were sequentially cross-linked into IPN by different mechanisms in two steps. The homogeneous materials obtained could be used as dielectric elastomer transducers [86].

The SEs chemically modified with 3–23% of cyanopropyl groups and cross-linked into thin films showed good permittivity, which increased from 2.4 (for the silicone matrix) to 6.5 for a film containing about 23% of cyanopropyl units. They can be used as dielectric elastomer materials in electromechanical actuators [87].

Elastic SR composites were also prepared by cross-linking of polydimethylsiloxane-α,ω-diols (M_w = 139,000 g/mol) with different amounts of α,ω-bis(trimethylsiloxy)-poly(methylcyanopropyl-*co*-methylhexyl-*co*-methylhydro)siloxanes, with MWs (M_n) ranging from 4,220 to 4,830 g/mol (and containing polar CN groups) and TEOS as a cross-linking agent (CA) and silica precursor. The CA contained 0–12.5 mol% of Me(H)SiO units, 0–62 mol% of $Me[NC(CH_2)_3]SiO$ units, and 27–98.5 mol% of Me(hexyl)SiO mers. The resulting SR consisted of polar–nonpolar interconnected networks as matrices, which had 7.4 or 9.5 wt% in situ generated silica and contained up to 2.74 wt% CN groups. The dielectric and mechanical properties of the obtained SR were dependent on their compositions. A SR film with Young's modulus Y = 0.19 MPa showed the dielectric permittivity ε' = 3.6 [88].

Polysiloxanes containing CN and CF_3 groups with T_g well below room temperature (< −50 °C) were cross-linked into elastomers. A linear increase in dielectric permittivity (ε') with increasing content of polar groups was observed, with maximum values of ε' = 18 and ε' = 8.8, respectively. These materials find applications in the construction of dielectric elastomer actuators with low operation voltages [89].

Soft PDMS elastomers were prepared by cross-linking bottlebrush polysiloxanes in one step. They had lower storage moduli than typical PDMS elastomers fabricated by cross-linking linear polymers. The stiffness of soft PDMS elastomers was similar to that of hydrogels. Soft PDMS elastomers had substantially less soluble fraction and significantly lower adhesive properties. Their mechanical properties could be precisely tuned. The biocompatibility of soft PDMS elastomers could be useful in personal care products, as soft materials for biomedical research and engineering, and as materials for stretchable electronics [90].

The PDMS-based SEs are most often used in so-called soft lithography [91].

RTV SR-graphene oxide composites were prepared by casting dispersion of polydimethylsiloxane-α,ω-diol with graphene oxide (FGO), which was functionalized with γ-aminopropyl(triethoxy)silane (APTES). Next, the cross-linked composites were reduced with hydrazine hydrate to give functionalized graphene (FG)/SR composites. APTES grafted onto graphene sheets led to an increased interlayer spacing, and thermal and mechanical properties were significantly improved. Both the FGO/SR composite containing 1.0 wt% of, FGO or its reduced product showed an increase in one-step weight loss temperature of 61 and 133 °C higher (respectively) than that of the unfilled SR. The TS and the elongation at break of FG/SR composite (with 0.5 wt% FGO loading) increased by 175% and 67%, respectively, compared to those of unfilled SR [92].

Expanded graphite (EG)/PDMS composites with high thermal conductivity and high flexibility were prepared, when EG derived from natural graphite flake was infiltrated in PDMS prepolymer solution and then hot pressed in a steel mold at 80 °C for 2 h. Optical microscope and scanning electron microscopy (SEM) studies revealed the interpenetrating network structures in the EG/PDMS composites. When mass fraction of EG increased to 10.0 wt%, the thermal conductivity of EG/PDMS reached 4.70 W/m K, which was attributed to the conductive path of graphite platelets. These composites showed excellent flexibility (the compressive modulus was 0.68 MPa) because of their continuous PDMS network [93].

2.3.1 Siloxane elastomer-based healing system

A high-temperature-cured self-healing epoxy resin was obtained by incorporating micro-capsules of poly(dimethylsiloxane)diol (PDMS-diol) cross-linked with poly (diethoxysiloxane) (PDES) toward separate microcapsules containing an organotin catalyst. Healing was triggered by crack propagation through the embedded microcapsules in the epoxy matrix, which realeased the healing agents into the crack plane initiating cross-linking reactions. Healing efficiencies, based on fracture toughness recovery, ranged from 11% to 51%, depending on the MW of PDMS, quantity of healing agent delivered, and use of adhesion promoters. The PDMS elastomers prepared *via* either a tin-catalyzed polycondensation of PDMS-diol cross-linked with PDES or *via* a platinum catalyzed addition hydrosilylation reaction of Si–H bonds to C = C bonds exhibited self-healing properties, when they were incorporated into another polymer [94].

The self-healing properties of SEs, based on the tin-catalyzed polycondensation reaction of PDMS-diol and PDES were first observed by Cho et al. [95], who added the PDMS-diol and PDES healing agents as phase-separated droplets into an epoxy vinyl ester thermosetting matrix by mixing before gelation. The liquid di-*n*-butyltin dilaurate (DBTL) catalyst was co-encapsulated with chlorobenzene in separate polyurethane (PU) microcapsules made *via* interfacial polymerization. When the vinyl ester material ruptured, both liquids came into contact with one another, and the siloxane cross-linking reaction started at room temperature.

The main advantages of this chemistry were:

(1) its stability in humid or wet conditions and at elevated temperatures (up to 150 °C [96]), which allowed for a broader range of applications and use in higher curing thermosets respectively,
(2) low cost for the used components,
(3) limited amount of side reactions for the organotin catalyst, even under ambient air conditions; however, the incompatibility of the siloxane product with typical matrix materials such as epoxy and epoxy vinyl ester thermosets was a major disadvantage.

An addition of a silane adhesion promoter to the epoxy vinyl ester matrix improved wetting and bonding of the crack surface [95]. The polyhydrosilylation reaction between a vinyl-functionalized PDMS and a hydrosiloxane–PDMS copolymer toward a platinum catalyst led to gelation within 5 h at room temperature [97–99]. The non-toxic Pt(0) catalyst was used in breast implants.

The best healing efficiency was obtained with microcapsules containing PDMS with MW 49,000 g/mol, PDES, and a small amount of xylene. Silane adhesion promoters increased the compatibility of the siloxane product with the epoxy matrix, improving the healing efficiency from 34% to 52%. Self-healing of an epoxy vinyl ester thermoset was achieved with 12 wt% phase-separated PDMS-diol/PDES, 3.6 wt% DBTL PU-microcapsules, and 4 wt% of the adhesion promoter [95]. Owing to the addition of the self-healing system, the virgin fracture toughness of the material was increased by 88%. Excellent self-healing results were also observed for an elastomeric PDMS matrix cross-linked by the Pt-catalyzed hydrosilylation reaction. The PDMS chemistry showed significantly better self-healing results with matrices that were more compatible with the healed network, such as PDMS itself. The addition of adhesion promoters to the other matrices was essential to obtain improved self-healing [97, 100].

Cross-linked networks of SEs containing ethylene bridges and active silanolate end groups were prepared by ring-opening copolymerization of octamethylcyclotetrasiloxane (D_4) and bis(heptamethylcyclotetrasiloxanyl)ethane (bis-D_4), which was initiated by tetramethylammonium silanolate. The cross-link density of these living networks was controlled by the ratio of D_4 : bis-D_4. They showed self-healing behavior with restoration of the original strength of the silicone sample [101].

SEs having supramolecular structures were first prepared by controllable "salt-forming vulcanization" of poly(aminopropyl, methyl)siloxane with acids. Their structures and micrographs were analyzed by FTIR spectra, small-angle X-ray scattering (SAXS) experiments and atomic force microscopy (AFM). These SEs containing ion-association complexes that were formed during vulcanization showed self-healing and good mechanical properties, while the elastomers cross-linked with inorganic acid exhibited a higher thermal stability [102].

SEs with good mechanical and high self-healing properties were also prepared from aminopropyl-terminated PDMS and multifunctional acrylate monomer, through amino-ene Michael addition reaction of amino and acrylate groups. The addition of Me_4NOH catalyst gave the elastomeric, a dynamically cross-linked network. The TS and the elongation at break of the synthesized elastomers were ~1.1 MPa and ~206%, respectively. The fractured samples recovered 91% of their original strength after healing at 105 °C for 24 h [103].

A transparent and hydrolysis resistant self-healing SE was prepared by thiol-ene UV curing between thiol and vinyl functionalized polysiloxanes, and thermal curing of two other kinds of ingredients, that is, polysiloxanes functionalized with pendant carboxyl and amino groups. The elastomers that were obtained showed excellent healing efficiency due to reversible ionic association. The healing processes were repeatable many times over, with a recovery of 90% of virgin mechanical strengths. The reprocessed elastomers showed over 90% efficiency in repairing damages. They can be also processed via 3D printing [104].

2.3.2 Newer applications of silicone elastomers and rubbers

Aqueous dispersions of spherical colloidal silica (CS) particles with a diameter of 15 ± 5 nm were modified with three different types of monofunctional silane coupling agents to give functionalized CS (FCS) particles. The effects of the surface chemistry of FCS were studied as a function of the CS/FCS loading in the PDMS. Prepared PDMS/silica/FCS composites were investigated for their physical properties both in the cured and uncured states. The filler–filler and filler–polymer interactions changed with the type of functionalizing agent modifying the surface of the CS. The filler–filler interaction was predominant in the PDMS–CS composites, and improved filler–polymer interaction was observed in the case of the PDMS–FCS composites. The composites containing CS treated with methyl(trimethoxy)silane (MTMS) had relatively better optical and mechanical properties compared to other PDMS–FCS composites [105].

Microorganogels swollen in light mineral oil were obtained using block copolymer self-assembly at a low concentration (4–5 wt%). Their rheological properties can be tuned, enabling their use in 3D printing of silicone structures. The minimum printed feature size was controlled by the yield stress of the microorganogel medium and enabled the fabrication of numerous complex silicone structures, including branched perfusable networks and functional fluid pumps. Polystyrene-block-ethylene/propylene (SEP) diblock copolymer and a polystyrene-block-ethylene/butylene-block-polystyrene (SEBS) triblock copolymer were used as ingredients of microgels. UV-cured SE inks (vinyl-terminated PDMS) contained an addition of a low-viscosity silicone oil. [(Mercaptopropyl)methylsiloxane)]dimethylsiloxane copolymer was used as a cross-linker and 2,2-dimethoxy-2-phenylacetophenone as a photoinitiator [106].

PDMS/poly(vinylidene fluoride (PVDF), poly(phenyl methyl siloxane) (PPMS)/PVDF, poly(ethoxy methyl siloxane) (PEOMS)/PVDF, and poly(trifluropropyl methylsiloxane) (PTFMS)/PVDF composite membranes were formed on PVDF matrix reinforced with a nonwoven fiber. They were used for separation of ethanol from ethanol/water mixtures. Contact angle measurements were used for assessment of the hydrophobicity at the membrane surface. The separation properties of these membranes were strongly dependent on the SR composition. The separation factor changed in the following order: PPMS > PEOMS > PDMS > PTFMS (5 wt% ethanol at 40 °C), and the total fluxes decreased in the order: PDMS > PPMS > PEOMS > PTFMS [107].

Different PDMS nanocomoposite membranes were synthesized by incorporating various contents of trimethylsilanol-hydrophobized silica nanoparticles, which were used as the filler to improve the PDMS pervaporation performance. A uniform dispersion of silica NPs in the PDMS membranes was obtained. These membranes were used for recovery of alcohols from water [108].

Two kinds of ceramic membranes made of titania (TiO_2) were modified with methylvinyl SEs or with perfluoroalkylsilanes: $C_6F_{13}(CH_2)_2Si(OEt)_3$ (C6) and $C_{12}F_{25}(CH_2)_2Si(OEt)_3$ ($C1_2$). The poly(methylvinylsiloxanes) were cross-linked with PMHS and Pt catalysts. Surface properties of silicone membranes after modification were determined using the contact angle tests (CA). Membrane modified with C12 was more hydrophobic (CA: 148°) as compared to membrane modified with C6 (CA: 135°). Water transport was inversely proportional to silicone membrane thickness (75–195 μm). Moreover, properties of the commercial PDMS membrane were tested in pervaporation of water-ethanol, water-pentane and water-hexane systems and it was found that organic compounds were selectively transported through this membrane. Hydrophobic membranes can be used in gas sensors [109] as selective barriers.

Ceramization process of SR compositions was used for the preparation of flame-resistant composites with increased thermal stability. Properly selected fillers assured increase of mechanical properties of composites and improved their thermal resistance and barrier properties, as a result of formation of robust ceramic phase during burning of a composite [110].

SEs having polar groups (e.g., cyanopropyl groups) bound to the polysiloxane backbone showed improved dielectric permittivity and energy densities due to intrinsically increased polarizability. Many examples of polysiloxane-based LC elastomers (LCEs) were also cited by C. Racles et al. [111].

Anti-ice coatings with extremely low ice adhesion strength ($\tau_{ice} < 20$ kPa) were prepared from SR compositions filled with silica aerogels. They showed improved mechanical strength and high transparency (>90%) [112].

A new kind of SE has been developed by M.A. Brook et al., by quite an unusual method – a rapid cross-linking a variety of aminopropyl functionalized polysiloxanes (without catalysts) with aqueous solutions of the aliphatic aldehydes (formaldehyde, glyoxal, or glutaraldehyde) even underwater. Their properties were tailored by controlling the density of amino groups in the starting materials. With excess of water, a slower rate of cure was observed, but after drying, all elastomers had comparable physical properties to those cured in air. They may be 3D printed *via* two-part syringe or used both as adhesives and sealants in air or water [113].

Recently, SEs that can be reversibly and repeatedly cured and uncross-linked were elaborated by M.A. Brook et al. Thiopropyl-modified silicones were oxidized with $PhI(OAc)_2$ to elastomers that have disulfide cross-links. Their mechanical properties were tuned by varying cross-link density, while thermal stabilities in air were comparable to traditional silicone thermosets, with degradation starting only over 300 °C. Reversible uncross-linking back to the same thiopropyl-modified silicones involved reductive S–S bridge cleavage *via* a Piers–Rubinsztajn reaction with hydrosilanes (e.g., $HSiMe_2OSiMe_3$) as reducing agents and was catalyzed by $B(C_6F_5)_3$. The initially formed silicone-$(CH_2)_3$S-$SiMe_2OSiMe_3$ products were deprotected with water in isopropanol/water to completely regenerate the thiopropylsilicones. This oxidation-reduction cross-linking–uncross-linking cycle was repeated thrice, with a yield of 89% per cycle, with essentially no change in the Young's moduli of the elastomers or in ^{1}H-NMR spectra of the uncross-linked fluids after reduction. Further oxidation of disulfide groups on the elastomer surface significantly improved water wettability [114].

2.3.3 Biomedical applications of silicone elastomers and rubbers

SEs and SRs find many biomedical applications, due to their unique properties:
- good elastomeric and relatively uniform properties over a wide temperature range,
- good low-temperature resistance and stable at high temperatures,
- excellent resistance to oxidation and ultraviolet light,
- outstanding resistance to aging,
- excellent dielectric behavior over a wide range of temperatures,
- excellent resistance to biodegradation, moderate biocompatibility, and physiological indifference [111, 115, 116, 117].

RTV-vulcanizable SRs composed of PDMS-diol, poly(ethylsilicate), and aminopropyl vinyl ether (APVE) were used for immobilization of lipase, through adsorption of this enzyme onto poly(hydroxymethylsiloxane), followed by the incorporation of the formed adsorbates into RTV SRs and cross-linking with Sn(II) catalyst. The prepared lipase–SR biocomposites showed up to 54-fold enhancements of catalytic activity with respect to the native enzyme [118].

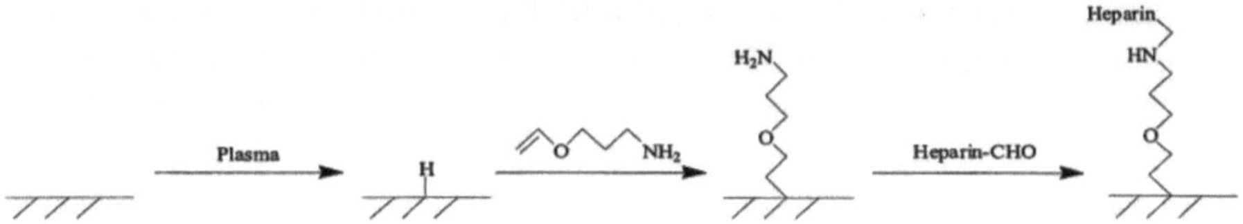

Scheme 2.1: Hydrosilylation grafting of aminopropyl vinyl ether (APVE) to Si–H groups formed by argon plasma treatment of silicone elastomer and subsequent coupling of heparin [119].

Vinyldimethylsilyl-terminated PDMS, cross-linked with PDMS-*co*-PMHS, was plasma treated and subsequently grafted with aminopropyl vinyl ether, followed by coupling with pentafluorobenzaldehyde or diazotized heparin. Thus, functionalized SEs can be promising novel biomaterials [119].

Thin films of SR, freshly activated by air plasma or 20% H_2SO_4 solution, coated with titania, by liquid phase deposition (using 0.3 M H_3BO_3 and 0.1 M $(NH_2)_2TiF_6$ in water, pH ~ 3.9) at room temperature showed decreased adhesion of gram-negative and gram-positive bacteria to a surface. This reduction was further enhanced by UV irradiation of the TiO_2 overlayer prior to introduction of the bacteria [120].

Modifications of physicochemical properties of SRs enabled their uses in different drug delivery systems and as subdermal, transdermal, or intravaginal devices [121]. PDMS-PNIPAAm IPNs membranes were used for permeation of sodium chloride and glucose. They may be further developed as ophthalmic biomaterials or corneal replacements [122].

2.4 Silicone resins

Organosilicon resins are densely branched polysiloxanes with MWs in the range 10^3–10^5 g/mol and a general formula: $R_nSiX_mO_y$, where R is organic group (usually methyl or phenyl, and X is functional group (OH, H, or OR). They are prepared from three-, tetra-, di-, and monofunctional monomers, and also from (tetraalkoxy)silanes and inorganic silicates. Water solutions of sodium or potassium silicates are usually called as *water glass*. The hydrolysis process of silanes mixtures (with average functionalities exceeding 2) leads to highly branched structures. Methyl(trichloro) silane (MTS) and phenyl(trichloro)silane (FTS), (dichloro)dimethylsilane (DDS), (dichloro)methylphenylsilane, (dichloro)diphenylsilane, methylvinyl(chloro)silanes, silicon tetrachloride, TEOS, or (tetramethoxy)silane (TMOS) have been used as basic monomers for syntheses of silicone resins.

A few examples of preparations of silicone resins, description of their properties, applications, and uses for modifications of chosen organic polymers are described in this chapter.

Many silicone resins contain silsesquioxane units $RSiO_{3/2}$ in their structure. Poly(methylsilsesquioxane) (PMSQ) and a DT-type methyl silicone resin (Me-DT) were characterized by different instrumental methods, including NMR, GC, GC-MS, and GPC. PMSQ was prepared by a hydrolytic polycondensation of methyltrichlorosilane (MTS) (in MIBK and water) and heat aging at 50 °C for 3–4 h. Although M_w of the PMSQ was ~5000 g/mol, the resin contained a significant amount of low MW siloxanes consisting of T^2 [$MeSi(OH)O_{2/2}$] and T^3 [$MeSiO_{3/2}$] units, ranging from $T^3{}_4T^2{}_3$ to $T^3{}_8T^2{}_2$, including many isomers. The isomer $T^3{}_6T^2{}_2$ was isolated, and its cage structure was determined. The other species were mainly composed of cyclotetra-, cyclopenta-, or cyclotrisiloxane rings [123].

Me-DT resin was similarly synthesized from mixtures of MTS and dimethyldichlorosilane (DDS) in MIBK and water, followed by an equilibration at 60 °C for 2 or 3 h. In Me-DT, structures, the T^2 units in the molecules from PMSQ were replaced with D^2 [$Me_2SiO_{2/2}$] units; for example $T^3{}_6D^2{}_2$ was found in products. The siloxane bond rearrangements took place during formation of the cubic structures in the equilibration (*heat aging*) step [123].

Most often, silicone resins are composed of mers D and T (resins DT) or M and Q (resins MQ), but industrial products have mixed structures (e.g. MTQ, MDT, or QDT). MQ and other silicone resins containing some reactive silanol groups (Si–OH) have good adhesive properties and tendency to increasing viscosity, and even to gelation during longer time of storage. Properties of silicone resins after curing process depend on their chemical composition and degree of cross-linking, which is characterized by the ratio **R/Si**. Tri- and tetrafunctional monomers are responsible for hardness of silicone resins, while bifunctional substrates affect their elasticity. The presence of phenyl and longer alkyl groups provides better solubility of these resins in organic solvents, while incorporation of phenyl groups into alkylsiloxane resins affects plasticization effect and improves their mechanical and dielectric properties, and heat resistance. The silicone resins prepared by the hydrolytic polycondensation of $MeSiCl_3$, $EtSiCl_3$, or $PhSiCl_3$ with excess of water were white powders. The methyl silicone resins containing 30 mol% of T units and 70% of D units are liquid materials. At a content of 50–70% of T units they are more brittle and glassy, with increasing tendency to gelation. With over 90% of T units they usually form gels. A few silicone resins synthesized from trifunctional monomers had silisesquioxane moieties, involving ladder structures [4–11, 40, 123]. They degrade easily with concentrated solutions of potassium or sodium hydroxides, forming solutions of polysiloxanolates of K or Na with similar properties as a water glass, and were useful as impregnation agents for building materials [11].

MQ silicone resins were usually prepared from the water glass or/and ethyl silicate $Si(OEt)_4$ (TEOS), as resources of tetrafunctional siloxane units (Q) [124]. TEOS is less reactive than the water glass, thus, when TEOS was applied as a starting material, the structure of the MQ resins was easier controlled and formation of a gel was not observed. However, it is important to note that TEOS is quite expensive, which limits its wide application in a manufacture of the MQ silicone resins. On the other hand, the *water glass* is readily available and inexpensive [10, 125, 126]. Moreover, hexamethyldisiloxane $Me_3SiOSiMe_3$ (HMDS) was found to be a useful monofunctional reagent (providing structural units M), when the water glass was applied as the basic starting material for the preparation of the MQ silicone resins [125–128]. Relatively a little research was devoted to increase the yield of the MQ silicone resins by controlling the formation of the gel. The fabrication of the MQ silicone resins was often described in numerous patents [129–135]. The MQ silicone resins of MW in the range of 1,000–10,000 were found to be very useful as pressure-sensitive adhesives [136–141], coatings, additives, and so on [142–149].

Vitreous solid silicone resins were prepared from phenyl(trimethoxy)silane (PTMOS) and (dimethoxy)diphenylsilane by the hydrolytic polycondensation in the sol–gel process, which was monitored by ^{29}Si-NMR. The following intermediate functional silanes were identified: $PhSi(OMe)_2(OH)$; $PhSi(OH)_3$; $Ph_2Si(OMe)(OH)$;

$Ph_2Si(OMe)(OH)$; and $Ph_2Si(OH)_2$ (where Ph = C_6H_5). The molecular structures of these silicone resins and gels were proposed, based on ^{29}Si-NMR data [150].

Classical polymer organic resins (especially alkyd and epoxy resins) are often modified with intermediate (phenyl)cyclosiloxane oligomer containing Si–OH groups, called Z-6018 (or RSN-6018; see its tricyclic structure in Chapter 1). Novel silicone-acrylic resins were prepared by the condensation reaction of the reactive siloxane intermediate Z-6018 and 2-hydroxyethyl methacrylate (HEMA), in toluene solution at 110 °C, under nitrogen atmosphere. The macromonomer so obtained was copolymerized with 2-dimethylaminoethyl methacrylate (DMAEMA) at different mole ratios (1:1, 1:3, 1:5), with benzoyl peroxide as initiator, in toluene, giving novel silicone-acrylic resins. These resins, characterized by FTIR, DSC (differential scanning calorimetry), and TGA (thermogravimetric analysis) techniques were thermally stable (their T_{10} values ranged from 446 to 500 °C in air atmosphere). All films of these copolymers were flexible and semigloss and had excellent adhesion properties [151].

Bisphenol-A-type epoxy resins were modified through a condensation reaction between the C–OH of epoxy resin and the Si–OH groups of organosilicon intermediate RSN-6018. These resins contained up to 80 wt% of solids and were cured at room temperature by a polyamide-based curing agent to yield transparent coatings. The chemical structure of the polymer was characterized by FTIR spectroscopy and NMR. The morphology of fractured surface of the cured coatings was studied by SEM. The cured coatings were stable at temperatures below 349 °C, and the film hardness increased up to 6H when the content of silicon reached 44.2%. DSC analysis tests showed that the T_g of cured coatings decreased when the ratio of RSN-6018 increased [152].

Compositions of polymethylsilsesquioxane resin (PMSQ), containing a large number of Si–OH groups were mixed with hydroxyl-terminated PDMS (PDMS-diol) in ethanol solution, and then they were first incubated in three different kinds of media (acidic, neutral, and basic). Next, they were cured into thin films that showed good thermal stability and excellent optical transparency. PDMS rubber particles formed flexible domains of about 100 nm in PMSQ matrix. The PMSQ/PDMS coatings showed good insulation properties and a high ceramic residue (85 wt%) after heating at 800 °C [153].

PMSQ was prepared in situ by dissolution of methyl(trimethoxy)silane (MTMS) in a mixture of acrylic monomers: methyl methacrylate (MMA), *n*-butyl acrylate, and acrylic acid, followed by the hydrolytic polycondensation with water at 40 °C for 12 h. The emulsified mixture of the monomers containing methylsilsesquioxane (MSQ) was next polymerized with potassium persulfate (KPS) at 80 °C for 3 h. The MSQ-polyacrylate latex particles that were obtained had a core–shell structure. The siloxane network was formed via homocondensation reactions of silanol groups Si–OH and heterocondensation reactions between Si–OH and Si–OMe groups (see Scheme 2.2 and Scheme 2.3). The MSQ particles formed the cores, while polyacrylate copolymers formed the shells. The static water contact angle measurements revealed that the incorporation of MSQ resulted in the composite latex with higher hydrophobicity [154]:

$$\mathbf{MeSi(OMe)_3 + 3\,H_2O \longrightarrow MeSi(HO)_3 + 3\,MeOH}$$

Scheme 2.2: Hydrolysis of methyl(trimethoxy)silane.

$$\mathbf{MeSi(OH)_3 + (HO)_3SiMe \longrightarrow Me(HO)_2Si\text{-}O\text{-}Si(OH)_2Me + H_2O}$$

Scheme 2.3: Homocondensation reaction of methyltrisilanol.

A silicone resin containing silphenylene units in Si–O–Si backbones was prepared by the hydrolytic polycondensation of 1,4-bis(hydroxydimethylsilyl)benzene with a mixture of chlorosilanes ($MeSiCl_3$, Me_2SiCl_2, $PhSiCl_3$, and Ph_2SiCl_2) [155]. The structure and property of this novel silicone resin were characterized by GPC, FTIR, NMR, TGA, SEM, and an electrochemical impedance spectrum (EIS). When R/Si, Ph/R, and the content of the silphenylene units were 1.3, 0.5, and 10 mol%, respectively, this silicone resin formed coatings. GPC, IR, and NMR results showed that the silphenylene units were incorporated into the silicone resin structure. The TGA of this novel silicone resin showed good heat resistance with the onset degradation temperature of ~500 °C and char residue at 900 °C of 85.6 wt%. SEM results showed that the silicone resin with silphenylene units can form full and uniform films, and its surface morphology of clear paints was protected below 350 °C. The EIS analysis revealed that clear paints of the silicone resin containing silphenylene units exhibited good resistance to corrosion [155].

A novel polysiloxane-epoxy resin (G_xD_y) with a large number of epoxy groups and flexible D segments was prepared by the hydrolytic condensation of dimethyl(diethoxy)silane $Me_2Si(OEt)_2$ and 3-glycidoxypropyl(trimethoxy)silane (GPTMS). The T_g of the epoxy-modified silicone resin was dependent on the structure of G_xD_y resin. Their chemical structures were confirmed by FTIR, ^{29}Si-NMR, and GPC. The TGA results under N_2 confirmed the improved thermal stability of the epoxy resin, which was affected by addition of the G_xD_y resin. The addition of 10 phr G_4D_6 greatly improved toughness, while preserving the transmittance of the epoxy resin (3,4-epoxycyclohexylmethyl-3,4-epoxycyclohexanecarboxylate). The G_xD_y resin was dispersed homogeneously in the epoxy resin and increased its thermal properties and toughness while simultaneously maintaining its transmittance, and it was recommended as a toughening agent for light emitting diode (LED)-packaging epoxy resins [156].

The hydrolytic condensation reaction of trifunctional organosilanes, for example, $RSiCl_3$ or $RSi(OR')_3$, leading to polyhedral oligomeric resins, including silsesquioxane structures (polyhedral oligomeric silsesquioxane, POSS), was studied by a real-time FTIR. The formation of linear siloxanes, cyclic siloxanes, and cage-like polysiloxanes was observed [157].

Products of hydrolysis and condensation of methyl(triethoxy)silane in emulsion were identified and studied by spectroscopic methods (^{29}Si-NMR, IR, and MS). It was found that after 40 days of storage at ambient conditions, the low molecular products of hydrolysis, methylsilanetriol $MeSi(OH)_3$ and its dimer $(HO)_2(Me)SiOSi(Me)(OH)_2$, showed high stability and were still present in the aqueous medium. It was concluded that condensation processes leading to the formation of polysiloxane network were slow in the emulsion and reactive silanol groups present in the system allowed for the effective hydrophobization of mineral material [158].

When the water glass and HMDS were used as the starting materials, the optimum reaction time was 30 min, and the optimal reaction temperature was 30–40 °C. A concentrated hydrochloric acid is one of the best, among various catalysts tested. On large laboratory-scale preparations (400–1,700 mL) of most MQ silicone resins, the catalyst was added first, then the water glass, and the mixture of HMDS and ethanol was added last. The structure of the MQ silicone resins was analyzed by FTIR spectroscopy, and MWs were determined by GPC. The MQ silicone resins showed narrow MW distribution, and the number average MW (M_n) of MQ silicone resin was ~ 3,000 g/mol. The silicone pressure sensitive adhesive prepared from MQ resin showed good tack and 180º peel adhesion (5.63 N/20 mm) [159].

Methylsilsesquioxane oligomers (MSSQ) with M_n = 953 g/mol and polydispersity (PDI) 1.83 were prepared by the sol–gel process from MTMS in THF. The chemical structure of the MSSQ oligomers was analyzed by FTIR. The prepared MSSQ precursors contained ~12.1% of Si–OH groups due to incomplete condensation. Hybrid and mesoporous materials were prepared from MSSQ in the medium of polystyrene-*b*-poly(2-vinyl pyridine) (PS-P2VP) block copolymers, with linear and hetero-arm star molecular architectures as the templates. The homogeneous solutions of PS-P2VP in THF were mixed together. The ratio of MSSQ to the copolymer was 60:40 (wt/wt). Then, after thermal treatment (first at 40–120 °C overnight), MSSQ underwent to some extent of the cross-linking reaction during this time. These hybrid materials were next heated under nitrogen from 120 to 400 °C at 5 °C/min in a quartz furnace. Further the cross-linking reaction took place extensively, followed by a thermal degradation. Heating was continued for 3 h at 400 °C in order to remove the copolymer species completely. As a result, the mesoporous materials were obtained, using both kinds of the templates. Two morphologies of products were observed – random spheres and cylinders were obtained. The sphere morphology was effectively preserved after the pyrolysis. It was proposed that MSSQ initially solubilized onto the surface of the P2VP copolymer micelles, then thermal curing of MSSQ occurred. This dehydration process competed with the cross-linking of MSSQ that could stabilize the structure of products [160].

Silicone resin coatings compositions were prepared from phenyl(triethoxy)silane (PhTES) and TEOS in methanol (film A) or 1-propanol (film B) as solvents on polycarbonate (PC) substrate. They contained 0–80 mol% PhTES units. The adhesion and distribution of phenyl groups were studied. Quartz crystal microbalance measurements of the alcohol evaporation rates for films A and B showed that the migration of phenyl group to the PC substrate side was strongly related to the alcohol solvent. FTIR studies for these films showed that a phase-separation between SiO_2 and $PhSiO_{3/2}$ networks took place during the alcohol evaporation [161].

The acid catalyzed hydrolytic polycondensation of $Si(OEt)_4$ (TEOS) and $Si(OMe)_4$ (TMOS) gave polyalkoxysiloxanes (PEOS and PMOS, respectively), with weight average MWs (M_w) of 1,100–12,000 or 2,700–31,000 g/mol, respectively. It was proposed that PEOS and PMOS may have cyclolinear (CL) and branched chemical structures. They were soluble in organic solvents, stable to self-condensation with high silica contents, up to 62% (in PEOS) or 72% (in PMOS). On heating at 80 °C for one to several days, PEOS and PMOS formed flexible and transparent thin hybrid coatings with TS of 1.6–5.2 or 3.6–11.8 MPa, respectively. They may find potential applications as binders as well [162].

Silicone resins containing fluorine-substituted and vinyl groups (FVi-SR), which were prepared *via* the hydrolytic condensation of (triethoxy)vinylsilane, (triethoxy)(tridecafluorooctyl)silane, TEOS, and (1,3-divinyl)tetramethyldisiloxane were further used for fabrication of a new cross-linked hybrid membrane (CHM) together with PDMS containing vinyl group and the FVi-SR as matrix materials and the poly(methylhydrosiloxane) (PMHS) as the cross-linking agent. The cross-linking hydrosilylation reactions between the vinyl group and Si–H bond and the membrane-forming process were carried out simultaneously at room temperature. The effects of the FVi-SR content in the CHMs, pressure differences, and temperature on the oxygen-enriching properties were studied. A good equilibrium between membrane-forming ability and oxygen-enriching properties was observed. Compared to cross-linked PDMS membranes without the FVi-SR, the CHMs showed better gas selectivity and similar gas permeability. Due to good oxygen solubility of the fluorine-containing component in the CHM the oxygen-enriching property was achieved [163].

Hybrid polymers with inorganic and organic structural units (so-called ORMOCER®s) can be used in preparation of new functionalized coatings on different substrates (ceramics, metals, polymers, etc.). Combined inorganic and organic network structures were formed by processing using the sol–gel method at temperatures below 150 °C. Owing to the incorporation of special organic functional groups, important applications related to their properties could be accomplished. Fabrication of ORMOCER® lacquers were achieved by nearly all conventional coating techniques (dipping, spraying, spin on, etc.). The basic application fields for ORMOCER® coatings include:
- abrasion and scratch resistance, decoration,
- barrier layers for packaging, corrosion resistant layers,
- antistatic, antireflective, antifogging, and antisoiling applications.

The combination of ORMOCER® layers with vapor deposited inorganic thin films (e.g., SiO_x) gave coatings with outstanding barrier properties, which showed synergetic effects by using vapor deposition coating techniques and wet chemical processing [164].

Monolithic crack-free hybrid glasses (**OR**ganically **MO**dified **SIL**icates or **ORMOSILs**) were prepared by the hydrolytic polycondensation of TEOS and polydimethylsiloxane-α,ω-diol (PDMS). By changing the PDMS amount and its chain length, organic–inorganic hybrid materials from brittle xerogels to soft elastomers, were obtained. Compressive stress–strain curves were correlated to materials microstructure [165]. ***Ormosils*** with different contents of Ti and Ca were prepared from PDMS and TEOS, which was first prehydrolyzed in aqueous solution containing isopropyl alcohol and the catalyst (HCl). PDMS was added to the hydrolyzed TEOS solution after reaction for 2 h and stirred for another 15 min. Tetrabuthyl orthotitanate (TBOT) and a mixture of $Ca(NO_3)_2 \cdot 4H_2O$, H_2O, and i-PrOH were added, while stirring was continued at 600 rpm for 2 h at 80 °C. The solutions were poured into plastic containers, covered, and stored at 40 °C for 1 week. The obtained crack-free gels were dried at 60 °C for 72 h and then heated 96 h at 150 °C. Monolithic discs of the ormosil materials were obtained and were cut into pieces of $10 \times 10 \times 10$ mm^3 and polished with an abrasive disc (no. 400). Small ormosil particles sieved between 200 and 300 µm were used for nitrogen adsorption measurements and other tests. The in vitro bioactivity was dependent on the polar surface characteristics of these materials. The presence of the highly polar surface with intermediate base/acid ratio and a specific roughness were necessary [166].

Thiol-functionalized gels with porous structures were prepared through a hydrolytic co-condensation of PMHS, TEOS, and 3-mercaptopropyl(trimethoxy)silane, in the absence of surfactants. The thiol-functional groups were incorporated into the porous network and their contents was controlled by adjusting the composition of starting silanes. These gels showed a large pore volume, different pore diameters, high surface areas, and good thermal stability. Selective adsorption ability toward heavy metal ions from aqueous solutions and a high capacity in adsorbing heavy metal ion of Pb^{2+} (up to 99.4%) were also observed for thiol-incorporated materials under optimized experimental conditions [167].

MTMS-based silicone aerogels with a controllable nanoporous structure were prepared in a two-step acid–base-catalyzed sol–gel process combined with supercritical drying. Their porosity and thermal conductivity were dependent on the concentration of MTMS in methanol solutions [168]. Other silicone aerogels were prepared by the sol–gel hydrolytic polycondensation of MTMS and TEOS, followed by ambient pressure drying (APD). Their properties were dependent on $NH_3 \cdot H_2O$ concentration. For example, the thermal stability reached 508 °C with the increasing $NH_3 \cdot H_2O$ concentration [169].

An allyl-terminated hyperbranched organic silicone resin was prepared by hydrosilylation of phenyltriallylsilane and 1,1,3,3-tetramethyl disiloxane with halloysite-supported platinum catalyst [170].

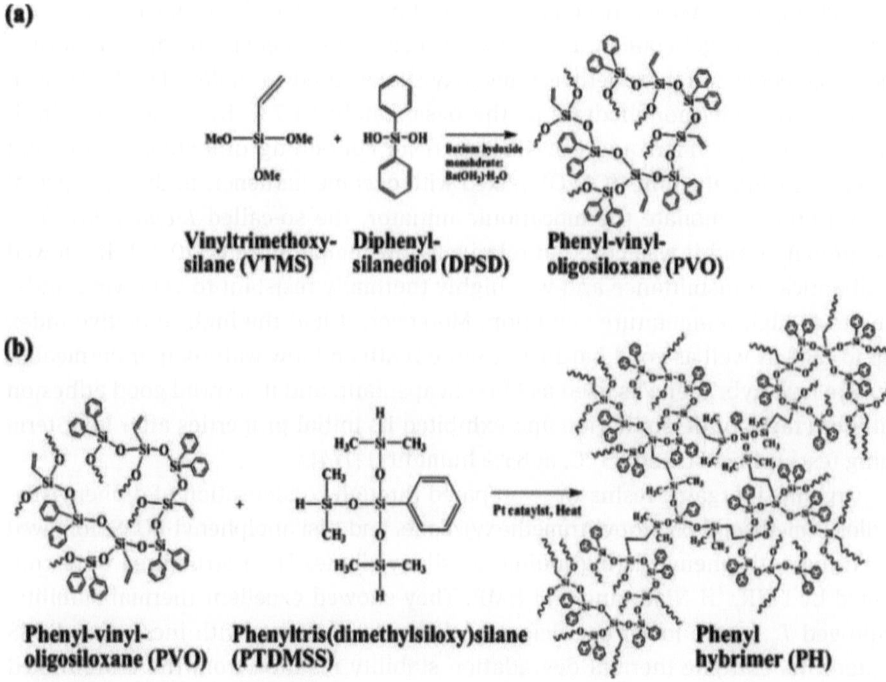

Scheme 2.4: (a) Synthesis of phenyl-vinyl-oligosiloxane (PVO) by the sol–gel condensation of ViSi$(OMe)_3$ with $Ph_2Si(OH)_2$ and (b) preparation of a phenyl hybrimer by a hydrosilylation of PVO with $PhSi(OSiMe_2H)_3$. Reprinted (adapted) with permission from [171]. Copyright 2010, American Chemical Society.

The highly branched phenyl- and vinyl-substituted siloxane oligomer resin (PVO) was prepared by a nonhydrolytic sol–gel condensation of vinyl(trimethoxy)silane $(CH_2{=}CH)Si(OMe)_3$ (VTMS) and 60 mol% excess of diphenylsilanediol $Ph_2Si(OH)_2$, in the presence of 0.1 mol% barium hydroxide monohydrate, $Ba(OH)_2 \cdot H_2O$, with respect to silane, in order to reduce the residual silanol groups. The condensation reaction (Scheme 2.4) was carried out with 20 wt% of *p*-xylene at 80 °C within 4 h. The byproduct methanol was removed by a vacuum distillation, and barium hydroxide was removed by a filtration [171].

Silicone (phenyl)siloxane oligomeric resins functionalized with Si–H groups, used in the light-emitting diode (LED) silicone encapsulant, were obtained by the nonhydrolytic sol–gel condensation of methyl(diethoxy)silane $MeHSi(OMe)_2$ and diphenylsilanediol $Ph_2Si(OH)_2$, catalyzed by acidic *Amberlite* resins. They were cured in solution by the hydrosilylation reaction with a phenyl and vinyl containing silicone resin (PVO) at 150 °C for 4 h, giving "phenyl hybrimer," which showed low shrinkage, good thermal stability (up to 440 °C), and excellent transparency (~90% at 450 nm) with a high refractive index ($n = 1.58$ at 632.8 nm) [172].

Cycloaliphatic epoxy-functionalized oligosiloxane-branched silicone resin was prepared by the sol–gel condensation reaction between diphenylsilanediol $Ph_2Si(OH)_2$ and 2-(3,4-epoxycyclohexyl)ethyl(trimethoxy)silane (used in mole ratio: 1.5:1), with barium hydroxide monohydrate as the basic catalyst [173]. By a thermally initiated, cationic polymerization of silicone resin containing diphenylsiloxane and epoxy cycloaliphatic units (CAEO), mixed with oxetane hardener, in the presence of a hexafluoroantimonate thermocationic initiator, the so-called *L-epoxy hybrimer* was prepared and it was cured at relatively low temperature (<120 °C). It showed good optical transmittance and was highly thermally resistant to yellowing under long-term high-temperature condition. Moreover, it had the high refractive index (up to 1.57) as well as good hardness (Shore D 80) and low water-vapor permeability. The new hybrimer was used as LED encapsulant, and it showed good adhesion without cracks or delamination and exhibited its initial properties after long-term aging tests (at 120 °C and 85 °C, at 85% humidity) [174].

Organic–inorganic resins were prepared through condensation of diphenylsilanediol, 3-methacryloxypropyl(trimethoxy)silane, and trisilanolphenyl-POSS, followed by curing with phenyltris(hydrodimethylsiloxy)silane. Their structures were confirmed by FTIR, ^{1}H-NMR, and ^{29}Si-NMR. They showed excellent thermal stability, improved T_g, and a lower coefficient of thermal expansion with increasing POSS content. Its extreme thermal degradation stability resulted from the cross-linked network as well as the heavily substituted aromatic ring present in the system. Incorporation of POSS substituent in methacrylate-based polysiloxane gave an excellent transparency and improved thermal discoloration resistance. This material can be used in optoelectronics [175].

A series of multifunctional polysiloxanes with different contents of epoxy groups and improved adhesion properties was prepared *via* the sol–gel condensation of methylvinyl(dimethoxy)silane, diphenylsilanediol, and 3-glycidoxypropyl(dimethoxy)methylsilane. The structures of silicone resins containing epoxy, phenyl, and vinyl groups (EPVS resin) were confirmed by ^{1}H-NMR and FTIR. The EPVS resins showed self-adhesive properties and were applied as the light emitting diode (LED) encapsulant or as an adhesion promoter for LED encapsulation materials. These materials exhibited excellent thermostability and the high refractive index of 1.55 [176].

An outstanding achievement was the elaboration of the synthetic method of thermoplastic, hot meltable silicone resins [177]. They can be melted many times without gelation and may be used in a plasticized or liquid state, mainly for fabrication of an electrical insulation. These reversibly softening silicone resins were prepared by a one-pot method: trifunctional chlorosilanes ($RSiCl_3$, where R – alkyl, alkenyl, or aryl, e.g., Me, Et, vinyl, allyl, phenyl, or naphthyl group) were partially esterified with aliphatic or aromatic monohydroxy alcohols (R'-OH, R' – alkyl or aryl, e.g., *n*-Bu, cyclohexyl, or Ph) or multihydroxy alcohols (HO-R"-OH, R" – arylene, bisphenol A moiety, or bisphenol-ether structure), using 0.1–0.9 mol of alcohol per 1 mol of $RSiCl_3$, and then were hydrolyzed in the presence of acceptors of HCl, for instance, moistened carbonates of alkali metals containing 5–30 wt% H_2O (0.6 mol per 1 mol of organochlorosilane), until pH was in the range 3–7. Next, prepared silicone resins were extracted with aromatic solvents, preferably with toluene, which was further removed by distillation under reduced pressure.

Alternatively, direct hydrolytic polycondensation of mixtures of $RSiCl_3$ and R_2SiCl_2, with mole ratio 1:1–4:1 was carried out, in the presence of moistened carbonates of alkali metals. All hydrolytic polycondensations were usually carried out at temperatures 120–180 °C.

The obtained silicone resins were hard, glassy and transparent solids, melting at 40–150 °C, depending on the type and mole ratio of monomers used. Some of their properties were similar to organic paraffin waxes and they could be called *silicone waxes*. These new resins, containing approximately 10–30 wt% of Si, with average MWs 1,500–4,000 g/mol, showed basic properties characteristic of silicone polymers, for example, electrical, hydrophobic, and antiadhesive properties. They also exhibited self-extinguishing properties and could be applied in fire resistant cable filling and electroinsulating materials. They could be also useful for fabrication of insulation in electric systems working at moderate temperatures, making repair or replacement of used elements possible in electric systems by local melting of the resin, which could be impossible in the case of chemically hardened and thermosetting resins, for example, epoxides [177].

Organic-silicone resins were prepared from multifunctional alkoxy- or aryloxysilanes of the general structure $R_nSi(OSiOR')_{4-n}$ (where R was Me or Ph, R' – alkyl with chain length C1–C4, and $n = 1–3$) or mixtures of different organoxysilanes or their mixture with organochlorosilanes of the general formula R_nSiCl_{4-n} ($n = 1, 2$), by reacting with dihydroxy ethers of bisphenol A [i.e., 1,1′-isopropylidene-bis-(*p*-phenylene-oxy)diethanol-2 or 1,1′-isopropylidene-bis-(*p*-phenylene-oxy)dipropanol-2] at 120–190 °C with vigorous stirring, followed by addition of the filler (Al or Zn powder) and cooling down to room temperature (RT). Thus, thermoplastic and reversibly softening copolymers were obtained. Alternatively, after cooling down to RT, reaction products were diluted with organic solvents (xylene, toluene, hexane, gasoline, or their mixture), then a cross-linking catalyst (preferably the solution of potassium butoxide) and possibly, also the fillers, were added, providing silicone coating materials.

The prepared silicone copolymers had melting points in the range 40–100 °C, which were plastified before processing at temperatures 40–70 °C higher than their softening temperatures. However, their heating at 200–250 °C caused irreversible curing. These new materials showed good mechanical properties (hardness, elasticity, adhesion, and cohesion of coatings), very good abrasive properties, very good dielectric properties (dielectric constant $\varepsilon = 2.7–2.9$ at 100 Hz at 25 °C), and specific resistance ($\geq 1 \times 10^{18}$ Ω), and breakdown voltage [$(15–20) \times 10^6$ V/m at 20 ± 5 °C] was even higher than in SEs. Their ignition temperatures were in the range 440 °C, and burning temperatures exceeded 500 °C. They also showed self-extinguishing properties.

Their absorbability of water was 2–3 times lower than for methylsilicone resins. They may find many perspective technical applications [178].

Similar silicone resins were synthesized from chlorosilanes R_nSiCl_{4-n} (where R – H, alkyl, or aryl radical, $n = 1–30$), by a partial alcoholysis with aliphatic or cycloaliphatic alcohols (MeOH, EtOH, i-PrOH, cyclohexanol; 0.5–1.5 mol/1 mol of Si), followed by a partial hydrolytic condensation at temperatures 50–100 °C within 0.5–3 h, using 0.1–0.9 mol H_2O per 1 mol of alcohols (0.01–0.80 mol H_2O/1 mol of Si), and transesterification at enhanced temperature of 100–200 °C, preferably with 0.19–0.30 mol (per 1 mol Si) of dihydroxy ethers of bisphenol A. Products of addition of (4,4′-dihydroxydiphenyl)propane with ethylene oxide or propylene oxide, containing at least 70 wt% of 1,1′-isopropylidene-bis(*p*-phenyleneoxy)diethanol-2 or 1,1′-isopropylidene-bis(*p*-phenyleneoxy)dipropanol-2 were formed. This synthetic procedure allowed preparation of silicone resins with elimination of large amounts of organic solvents and arduous liquid phase. The modified silicone resins that were obtained showed good electrical and adhesive properties, high elasticity, good mechanical properties, and thermal resistance. They were cured at 200–220 °C within 2–4 h without catalyst or at 160 °C within approx. 20 min in the presence of 0.02 wt% of *n*-BuOK (or *n*-BuONa) [179].

The reversibly softening thermoplastic polymers and copolymers of phenylsilsesquioxane were prepared in one-pot manner by hydrolytic polycondensation of phenyl(trichloro)silane $PhSiCl_3$ or its copolycondensation with diphenyl(dichloro)silane Ph_2SiCl_2 (carried out at room temperature or at 50 °C), using moistened or hydrated salts: (1) moistened Na_2CO_3, K_2CO_3, (2) $Na_2B_4O_7 \cdot 10H_2O$ (borax), which also served as acceptors of HCl. In syntheses (1) and (2) were obtained reversibly softening and soluble materials, and NaCl and KCl were the only by-products [180].

The hydrolytic polycondensation of 0.5 mol of $PhSiCl_3$ (0.5 mol) with 0.2 mol of diphenylsilanediol $Ph_2Si(OH)_2$ was carried out with an excess of water, using Et_3N and DMAP. The polysilsesquioxanes were characterized by elemental analysis and spectroscopic methods (^{1}H- and ^{29}Si-NMR, FTIR), SEC, DSC, and TGA. The content of Si–OH groups in these POSS resins ranged from 1.2% to 7.3%. With Na_2CO_3 or K_2CO_3, the reactions were carried out at room temperature, they proceeded relatively fast, and hydrolysis was complete at room temperature. The condensation reaction was controlled with monomer concentration, and the MWs of products were higher than those obtained in other systems. The MW of polymers obtained from $PhSiCl_3$ was higher than that of copolymers from $PhSiCl_3$ and Ph_2SiCl_2 obtained under the same conditions. The presence of a large amount of cyclic siloxanes, $[Ph_2SiO]_3$ and $[Ph_2SiO]_4$, in the copolymers increased their polydispersity and decreased their average MW. The decrease in the content of diphenylsiloxane unit in copolymers caused an increase in their softening temperature.

Polymers and copolymers with boron atoms incorporated into polymer chains were prepared in the presence of borax. Soluble poly(borosiloxane) resins with an average molar ratio of Si:B = 7–10 were obtained. In this system, considerable amounts of oligomers with cage structures, mainly T_8 and T_{12}, were formed, and the value of T_g and softening points of the materials obtained were higher than for products prepared with moistened Na_2CO_3.

In the presence of $Na_3PO_4 \cdot 12H_2O$, from toluene solution of $PhSiCl_3$ or a mixture of $PhSiCl_3$ and Ph_2SiCl_2, by heating at 180 °C under vacuum, crosslinked polymer and soluble fractions (with yields: 50–79 wt%; M_w/M_n ~ 3; M_n = 1200–3200 Da) were obtained by extraction. The soluble polyphenylsilsesquioxanes obtained by polycondensation of $PhSiCl_3$ were brittle. However, their copolymers with Ph_2SiCl_2 were more flexible. The presence of polyhedral phenylsilsesquioxanes, T_8 and T_{12}, decreased the elasticity of these resins, while cyclic siloxanes, $[Ph_2SiO]_3$ and $[Ph_2SiO]_4$, gave increased elasticity. Similarly, poly(phenylsilsesquioxane-co-diphenylsiloxane)s were prepared by the hydrolytic polycondensation in the presence of $Na_2SO_4 \cdot 12\,H_2O$.

The phenylsilsesquioxane resins obtained in the carbonate and borax systems showed reversible softening after heating and cooling, whereas the values of softening points were lower as compared to products prepared with $Na_3PO_4 \cdot 12\,H_2O$. The softening points (melting temperature, T_m) of the POSS resins prepared with moistened carbonates and borax were in the ranges 170–250 °C and 205–310 °C, respectively, and did not change after cyclical (10×) heating to 250 °C, whereas for the resins obtained using $Na_3PO_4 \cdot 12\,H_2O$, T_m values were different: 260–280 °C, 90–165 °C, and 90–170 °C (depending on their chemical compositions), and after cyclical (10×) heating to 250 °C, the T_m values were: 270–280 °C, 130–170 °C, and 150–170 °C, respectively.

The prepared phenylsilsesquioxane resins showed a weight loss of 5% (T_5) in the temperature range of 410–440 °C. The weight loss of 50% (T_{50}) was observed from 620 to 680 °C. The presence of the B-O-Si units further increased the thermal resistance of these materials by 40–100 °C. The obtained POSS resins had glass temperatures (T_g) in the range 57–88 °C. However, in contrast to the highly flexible PDMS chains with $T_g = -123$ °C, poly(diphenylsiloxane) chains with $T_g = 40$ °C were rigid [181]. Ph_2SiO units introduced flexibility to these POSS resins without decreasing their thermostability [180].

Copolymers of dimethyldichlorosilane (DDS) and phenyltrichlorosilane (PhTS) were prepared by the hydrolytic polycondensation of 1–50 mol DDS and 1 mol PhTS, in the presence of toluene or xylene dispersions of moistened carbonates or hydrocarbonates of sodium or potassium (as acceptors of HCl), containing 5–20 wt% of water. Solutions of methylphenyl(hydroxy)polysiloxanes were separated from precipitates of NaCl or KCl; aromatic solvents and low MW cyclosiloxanes were removed by distillation at 120–200 °C. Poly(methyl, phenyl, hydroxy)polysiloxanes were obtained as hazy, slightly white oils and were used as antifoaming agents [182].

Silicone resins of different structure and different silanol content were also cured with poly(methylsilazanes) (PMSZ) at ambient temperature. The cross-linking degree of the silicone resins was dependent on the structure of PMSZ and its loading in the resin, which should be high enough. Silicone resins cured with PMSZ showed improved thermostability in nitrogen, if PMSZ content reached 5 wt%

(T_5 = 370 °C), as compared to that of heat-cured compositions. Ceramic yield was also higher (75 wt%) [183].

Thin coatings made by curing ethanol solution of branched phenylsilicone resins (containing 6% of silanol group) with a mixture of PDMS and silane-modified silica were dried at 90 °C and used as flame-resistant materials for protecting steel [184].

A vinylphenyl-containing MQ silicone resin with a high average MW (M_w ~ 30,000 g/mol) was synthesized *via* the hydrosilylation reaction of vinyl-containing the MQ silicone resin and linear poly(diphenylsiloxane) terminated with two Si–H functional groups. These resins showed better thermal stability and a higher refractive index than methyl silicone resins. These inorganic/organic hybrid materials can be used as a component for heat-resistant adhesives and coatings for electronics and a high-performance liquid silicone rubber (LSR) [185].

Curing of the silicone resins is based most often on the homocondensation reactions of silanol groups or their coupling with alkoxy substituents at enhanced temperature (>200 °C) and gives densely cross-linked networks. These processes are accelerated by different types of curing agents: metal salts of fatty acids, titanates, amines, aminealcohols, amine salts with organic acids, quarternary ammonium and phosphonium salts, arylsulfonic acids, and hydrophosphates. Curing of the silicone resins by dehydrocondensation of silanol functionalities with hydrosiloxanes or hardening by the hydrosilylation reaction unsaturated C=C bonds with PMHS occur mostly in the presence of Pt compounds. Novel types of the reactive silicone resins containing methacryloxy substituents were very quickly cured at 100–150 °C in the presence of peroxide initiators or under UV radiation, within tens of seconds. The liquid resins and their solutions and mixtures with different unsaturated monomers or organic and organosilicon polymers can serve as coating materials, binders, and modifiers of polymers [186], or coatings used for protection of photovoltaic solar cells [187].

Core–shell silicone-acrylic emulsions were prepared by seeded polymerization of (3-methacryloxypropyl)trimethoxysilane (MPTMS) (at concentrations 1–2%), with addition of a small amount of octadecyl acrylate, using polymerizable maleate surfactant, which was incorporated into the copolymer. The copolymer particle size ranged from 166 to 243 nm, depending on MPTMS concentration [188]. These polymethylsilsesquioxane resins (PMSQ) were prepared in situ by the hydrolytic polycondensation of MTMS in a mixture of acrylic monomers (MMA, *n*-butyl acrylate, and acrylic acid), which were further emulsified and polymerized in emulsion at 70 °C for 3 h. Transmission electron microscopy (TEM) images indicated that the latex particles that were obtained had a core–shell structure and showed higher hydrophobicity (based on water contact angle measurements). According to the X-ray photoelectron spectroscopy (XPS) results, the cores were composed of methylsilsesquioxane (MSQ) units, and the shells were made up of polyacrylate [188].

The soluble, highly branched, phenylsilsesquioxane-silicate copolymers, functionalized with methoxy groups, were prepared by the dehydrocarbon polycondensation of phenylsilane $PhSiH_3$ with (tetramethoxy)silane (TMOS) ($PhSiH_3$: TMOS < 0.9 mole/mole) in toluene solution, catalyzed by tris(pentafluorophenyl)borane [B $(C_6F_5)_3$]. A side reaction – a disproportionation between Si–H and Si–OMe groups was observed. Molecular weights ($M_w = (3.4–81) \times 10^3$ g/mol) and polydispersities (M_w/M_n = 1.9–8.9) were dependent on molar ratio of monomers and their concentration in solution. However, at higher contents of phenylsilane ($PhSiH_3$: TMOS = 0.8–0.9) gelation occurred. The ^{29}Si NMR analysis of the copolymer product proved that extensive cyclization occurred during the polycondensation, most likely giving some ladder- and cage-like structures. This model reaction has been a new nonhydrolytic route to silicone resins with TQ structures [$T = RSiO_{3/2}$ (R - organic group), $Q = SiO_{4/2}$], which did not contain reactive and hydrophilic Si–OH groups and should have better miscibility with some organic polymers and improved stability (i.e., a *shelf life*) [190].

Polysiloxane microspheres functionalized with silanol, 3-aminopropyl, 3-(glycidyloxy)propyl, or vinyl groups were prepared in a multistep process. Four reactions of poly(methylhydrosiloxane) (PMHS), simultaneously catalyzed by a Pt(0) Karstedt's complex in an aqueous emulsion [containing *iso*-propanol (*i*-PrOH)], gave polysiloxane microspheres containing a large number of silanol groups. These reactions carried out with mechanical stirring, included: (1) hydrosilylation of vinyl group of a cross-linker with the Si–H group on the polysiloxane, (2) hydrolysis, and in some cases alcoholysis, of the Si–H bond in the PMHS and (3) dehydrocondensation of the Si–OH group formed in the hydrolysis with the Si–H group. The fourth reaction of the Si-H group with *i*-PrOH led to a small amount of *i*-PrO-Si≡ groups [191]. In most cases, preliminary hydrosilylation was carried out before emulsification, which led to the grafting of vinylsiloxane on PMHS. Alternatively, a solution of PMHS with divinyltetramethyldisiloxane (DVTMS) cross-linker and the catalyst was mechanically emulsified with water. In the emulsion process, a large number of Si–H groups from PMHS in the bulk and on the surface of microspheres were transformed into Si–OH hydrophilic groups. The density of grafting, Si–OH content, and the particle size were dependent on reaction conditions. The cross-linking of PMHS occurred as a result of hydrosilylation of vinyl groups grafted on the polymer and by a dehydrogenative condensation of the Si–H groups with the silanol groups, which were formed on PMHS. The prepared microspheres were modified by the condensation reactions with organofunctional silanes. Hydrophobic microspheres were obtained by coupling the Si–OH groups in reaction with Me_3SiCl [191].

The obtained microspheres were also further modified by reactions with other reactive organofunctional silanes [192].

The synthetic procedure, based on cross-linking of poly(methylhydrosiloxane) with 1,3-divinyl-tetramethyldisiloxane, carried out in emulsion, involving hydrolysis of Si–H groups to ≡Si–OH silanols and their next dehydrocoupling, gave soft polysiloxane microspheres and microcapsules with controlled hydrophobic–hydrophilic balance and varied porosity. The presence of $(CH_3)_3Si-$ or $\equiv SiOCH(CH_3)_2$ groups was responsible for hydrophobic properties, while siloxane moieties were hydrophilic. These particles were functionalized with amine, epoxide and vinyl functions as well as hard ceramic microspheres. The synthetic routes were modified leading to the polysiloxane core or shell hybrid particles. In the latter case, the particles had the structure of polysiloxane microcapsules containing encapsulated inorganic or organic material. Preparation of composed microspheres, which in addition to polysiloxane contain a significant fraction of organic material, was also described. Further modification of the pristine cross-linked polysiloxane microspheres containing Si–H and Si–OH functions during their reactions with alkoxy- or chlorosilanes bearing various functional groups, allowed for synthesis of functionalized microparticles. These particles were used as carriers of proteins as well as Pt and Pd catalysts. The cross-linked polysiloxane microspheres gave also microcapsules loaded with *n*-eicosane. These microcapsules could be applied as phase-change materials appropriate for maintaining temperature within the range from 22 to 42 °C. The cross-linked polysiloxane microspheres can also serve as attractive preceramic material and can be thermally modified to the Si–O–C microspheres, as potential fillers for fabrication of ceramic materials [192].

N-substituted imidazole groups were grafted on polysiloxane microspheres containing a large number of Si–OH groups by their silylation with *N*-[γ-(dimethylchlorosilyl)propyl]imidazole hydrochloride. They were used as a support for a palladium catalyst $PdCl_2(PhCN)_2$, for the preparation of a new heterogenized catalyst, Pd(II) complex with imidazole ligands, which was immobilized on polysiloxane microspheres and showed a high catalytic activity in the hydrogenation of cinnamaldehyde [193].

Poly(silsesquioxane)methylsiloxane resins (PSMSR) were prepared from (trichloro)vinylsilane and (dichloro)dimethylsilane through the hydrosilylation and hydrolytic polycondensation reactions. Poly(fluorosilsesquioxane)methylsiloxane (FPSMSR) were also synthesized – *via* a single electron transfer addition reaction of PSMSR with perfluorobutyl iodide, followed by reduction of iodide with Zn powder. The relative static contact angles of PSMSR and FPSMSR films with water were 105.1° and 119.3°, respectively. The FPSMSR showed a better water repellent property because of the migration of fluoroalkyl chains to their surface [194].

A few types of solventless silicone resins, highly viscous liquids or powders with glass temperatures (T_g) in a range 50–80 °C are also known:
- poly(methylphenylsiloxanes) cured at higher temperatures with peroxides,
- poly(methylphenylsiloxanes) and poly(vinyl, hydro)siloxanes cured by addition method,
- well-soluble, low-MW hot-melt oligomers and resins, used as binders for electronics, an adhesive sealant to holders of bulbs, and in mixtures with organic resins.

The powder silicone and silicone-epoxy resins having $T_g > 70$ °C are excellent heat resistant materials (up to 300 °C and above), which are applied as coatings in pans, stoves, grills, toasters, etc. Their scratch resistance and an abrasion resistance are better than Teflon materials [4, 11, 16].

Silicone resins are often used as excellent electrically insulating lacquers for coating glass fabrics and shirts. Their dielectric properties are good in a wide range of temperatures and depend very little on a humidity and a current frequency.

Different kinds of the silicone resins are available on a market: powder and solvent compositions, water dispersions, coating materials, binders for paints, molding compounds and laminates, glues, and hydrophobization materials. They show good adhesion to glass, ceramics and metals, except copper and its alloys. Coatings from methylsilicone resins can serve at 200 °C up to 10,000 h, while coatings from methylphenyl silicone resins – at least 10 times longer, at 230–250 °C. They have good chemical resistance to mineral and vegetable oils, fats, and food products, and good resistance to climatic agents: UV radiation, humidity, and waterproof in cold water. The coatings based on the silicone resins and zinc or aluminum powder as the filler were heat resistant up to 500–600 °C [11].

2.4.1 Miscellaneous and composite silicone resins

Oligoborosiloxane resins, prepared by the condensation of phenyl(trimethoxy)silane (PTMOS) and phenyl(triethoxy)silane with boric acid in diglyme (under stirring at 150–160 °C for 3 h), both in the presence and absence of hydrochloric acid as catalyst, gave 64–75% yield of ceramic residue (at 900 °C in inert atmosphere). These borosiloxane oligomers were soluble in common organic solvents. Upon their pyrolysis at 1,500 °C amorphous ceramic was formed which crystallized into β-SiC on heating at 1,650 °C [195].

A novel hyperbranched polysiloxane (HPSi) with a great number of epoxy groups was used as a compatibilizer of the epoxy resin (diglycidyl ether of bisphenol A, DGEBA) with a commercial methyl phenyl silicone resin (Si603) blend. It was prepared by the hydrolytic polycondensation of γ-(2,3-epoxypropoxy)propyl(trimethoxy) silane and (dimethoxy)dimethylsilane (DEDMS), which was carried out in diluted ammonia solution at 50 °C. The compatibility of diglycidyl ether of bisphenol A (DGEBA)/HPSi/Si603/DDM (4,4-diaminodiphenylmethane) was characterized by DSC and SEM, and the results showed that HPSi significantly improved the compatibility of DGEBA/Si603. The commercial epoxy resin (DGEBA) was cured with DDM. The value of a limiting oxygen index (LOI) of the DGEBA resin modified with 10% HPSi and 30% Si603 was 31 (about 1.4 times the corresponding value of the original DGEBA resin), and resulted in V-1 test of the UL-94 vertical burning. Its combustion residue at 800 °C was about 2.24 times higher than for the original DGEBA resin. These composites also showed good mechanical properties. The addition of 10 wt% of HPSi also enhanced the thermostability of the DGEBA/DDM system [196].

Silicone resins are also applied as components of waterborne silicate dispersions, which serve as paints having very good properties. They contain silicone resins, silica, silicates, other fillers ($CaCO_3$, dolomite, TiO_2), pigments, and additives. The silicate dispersions are especially useful for renovations of old coatings and cementous and gypsum surfaces [197].

Commercial compositions of silicone resins with aluminum or zinc powders find many practical applications: mainly in anticorrosive protection of metals used at high temperatures, at least up to 400 °C (e.g., engines, exhaust pipes except catalysts), in household equipment (for furnaces, stoves, heaters, and other heating elements). Due to excellent resistance against atmospheric conditions and hydrophobic properties they can be applied as coatings for rough casts, concretes, wood elements, or steel sheets coated with zinc [198].

Compositions of varnishes and paints based on alkyd resins modified with silicone resins are heat-resistant up to 600–650 °C for a long time of use, and to 750 °C for a short time of heating. They can be applied as superior coatings on engines, exhaust pipes, aircraft exhaust equipment, fire-places, heaters, space heaters and incinerators, pipes for steam, kettle drums, stoves, furnaces, among others [199–201].

Water-repellent UV-curable materials were prepared from PDMS-based macromonomers (i.e., grafted with carbinol and ureaethylene-acrylic groups) and acrylic-modified melamine (AM). UV-curable acrylic-melamine resins containing siloxane segments (SiAMs) were synthesized by condensation of (hexamethylol) melamine, 2-hydroxyethyl acrylate (HEA), and carbinol-modified PDMS. The SiAMs films with a 0.3 wt% PDMS segment that were cured by UV irradiation showed higher transparency than that of a blended sample. An examination with SEM showed that the blend sample had cohesion of the silicone segments, while such a phenomenon was not observed in the composite SiAMs samples. An aggregation of the siloxane segments was avoided by introducing the siloxane segments into the acrylic melamine. By incorporation of 0.3 wt% of PDMS segments into AM, good watershedding properties were obtained, with preservation of several of the most attractive features of AM, such as hardness, adhesion, and refractive index [202].

The acrylic-melamine-silicone hybrid resins (prepared by the condensation of hexa(methylol)melamine, 2-hydroxyethyl acrylate, and carbinol-modified PDMS) were rapidly cured at low temperature by UV irradiation through acryloyl groups. They showed high hardness, high water contact angles, high refractive indices, high transparency (as compared to a mixture of acrylic melamine and silicone resins), and good adhesion to poly(ethylene terephthalate) (PET) films [202].

Phenolic-formaldehyde (PF) resin containing boron and silicon atoms was prepared by copolymerization of phenol, formaldehyde, phenyl(triethoxy)silane (PTES), and boric acid (BA). It showed an extremely high thermal decomposition temperature and a char yield. The PTES and BA-modified PF (BSPF) resin thermoset gave the char yield of 77.0%, when the boron and silicon contents were only 1.27 and 1.7 wt%, respectively. Compared to the unmodified PF resin, the temperature at the maximum decomposing rate of the BSPF increased by 84 °C and its charring yield by 15.0% [203].

2.5 Poly(silsesquioxanes)

Silsesquioxanes (POSS) have been known for over 100 years [204]. They are very important 3D oligomers that can exist as nanoscale cage or semicage (incompletely condensed), random, ladder, and branched polymeric structures. POSS are mainly prepared by condensation methods [205]. They contain the trifunctional unit $RSiO_{1.5}$ (T):

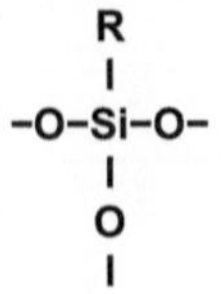

where R is, most often, organic group or H.

POSS are hybrid, spherical organic–inorganic molecules with an organic functional group in each octant and a general formula $(RSiO_{1.5})_n$ (T_n), where R can be suitable for polymerization or grafting (Fig. 2.3) [206–222]. A basic method of synthesis of oligosilsesquioxanes is the hydrolytic polycondensation of trifunctional monomers $RSiX_3$ (R = alkyl, aryl, H; X = Cl, OR', OAc), according to a general equation:

$$n\,RSiX_3 + 1.5\,n\,H_2O \longrightarrow (RSiO_{1.5})_n + 3n\ HX \qquad (2.5)$$

$$(\text{most often}, n = 6-12)$$

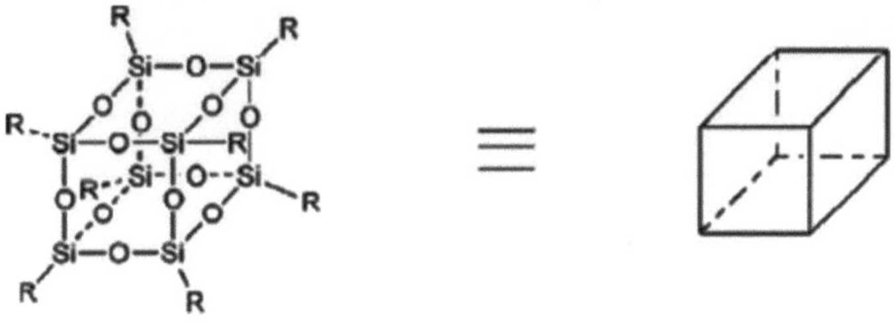

Figure 2.3: Chemical structure of cubic silsesquioxanes T_8. Reprinted (adapted) with permission from [212]. Copyright 2012, InTech Open Science, Croatia.

Oligosilsesquioxanes are also prepared through hydrosilylation of olefins with oligo (hydrosilsesquioxanes) or thermal cracking of poly(silsesquioxanes). The structure and chosen physical properties of methyl- and phenyl-substituted POSS oligomers are listed in Table 2.6 They may be used as adsorbents, hybrid heterogenized catalysts, and permeable membranes [206].

Table 2.6: Physical properties of selected oligosilsesquioxanes $(RSiO_{1.5})_n$.

R	n	Melting point (°C)	Temp. of sublimation (°C/mm Hg)	Density (g/cm^3)
H	8	250	–	1.88
CH_3	6	209–210	86–120/0.7–0.9	–
CH_3	8	415 (decomp.)	200–210/1	1.49
CH_3	10	332.6–334.0	–	–
CH_3	12	270	–	–
C_2H_5	8	282–285	160–180/1	1.31
C_3H_7	8	219–220; 175	–	1.08
$CH(CH_3)_2$	8	295.5–296.5	–	1.18
C_4H_9	8	190–195	–	1.10
C_5H_{11}	8	77–82	It does not sublime	–
C_6H_{13}	6	281–285	–	1.035
C_7H_{15}	6	292–298	–	1.012
C_8H_{17}	6	319–323	–	1.010
i-C_9H_{19}	8	–	–	0.9865
c-C_6H_{11}	8	400	–	1.174
$CH{=}CH_2$	8	–	150–220/0.1–0.001	1.374
C_6H_5	8	500 (decomp.)	250–300/*	1.3
C_6H_5	10	415–418	–	–
C_6H_5	12	385	300/*	–
C_6H_5	6	188–193	–	–

* - Under vacuum (a pressure was not given).

Most often trichloro- or trialkoxysilanes were used as monomers in syntheses of POSS hybrids. It is a three-step process usually. In the first step, the hydrolytic condensation of monomer with excess of water takes place, leading to a hydrolyzate with $M_n < 1{,}000$ g/mol, which is equilibrated with KOH at 100 °C in the second step, giving prepolymer of increased MW ($M_n < 10^4$ g/mol). Next, an equilibration of a concentrated solution (90%) of the prepolymer was continued at temperatures exceeding 100 °C and yielded polysilsesquioxanes with $M_n \sim 10^5$ g/mol.

Polymeric silsesquioxane materials exhibit outstanding thermal properties. The initial decomposition temperature (in air) corresponding to 5% mass loss (T_{d5}) for poly(methylsilsesquioxanes) (PMSQ) is much lower than for poly(phenylsilsesquioxanes) (PPSQ), which is ~500 °C (see Table 2.7).

Table 2.7: Initial decomposition temperatures of POSS.

Structure of $(RSiO_{1.5})_n$	Initial decomposition temperature, T_{d5} (°C)	Ref.
$(PhSiO_{1.5})_n$	525 (air, 20 °C/min)	43
$(PhSiO_{1.5})_n$	505 (air, 1 °C/min)	3
$(PhSiO_{1.5})_n$ with Si–OH groups	525 (air, 20 °C/min)	74
$(PhSiO_{1.5})_n$ branched	517 (air, 20 °C/min)	74
$(PhSiO_{1.5})_n$ chlorinated	458 (air, 20 °C/min)	74
$(PhSiO_{1.5})_n$ cross-linked	550 (air, 20 °C/min)	10
$(PhSiO_{1.5})_n$, $M_n = 10^3$ g/mol	460 (air)	3
$(MeSiO_{1.5})_n$	400 (air)	3
$(MeSiO_{1.5})_n$	660 (nitrogen, 50 °C/min)	3

Reprinted (adapted) with permission from [207]. Copyright 2001, *Polimery* (Warsaw, Poland).

Many poly(silsesquioxanes) with a formula $(RSiO_{1.5})_n$ (T_n) have regular ladder structures (**I**). In the case of random distribution of T units cross-linked products belong to silicone resins (structure **II**) (Fig. 2.4).

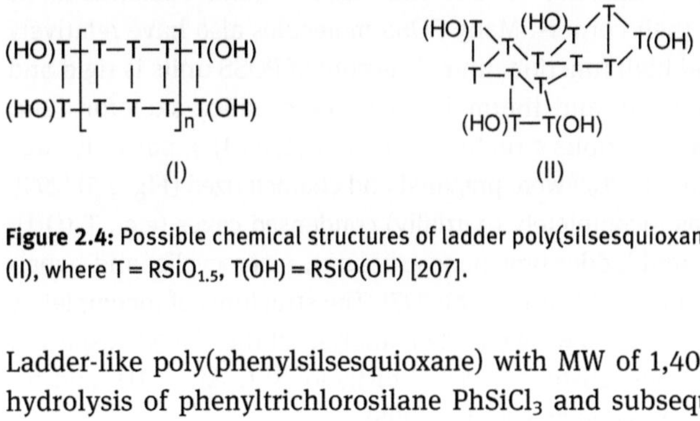

(I) (II)

Figure 2.4: Possible chemical structures of ladder poly(silsesquioxanes) (I) and silicone resins (II), where $T = RSiO_{1.5}$, $T(OH) = RSiO(OH)$ [207].

Ladder-like poly(phenylsilsesquioxane) with MW of 1,400 g/mol was prepared by hydrolysis of phenyltrichlorosilane $PhSiCl_3$ and subsequent condensation of the resulting intermediate silanols in the presence of a catalytic amount of hydrochloric acid in methyl isobutyl ketone. The progress of the condensation reactions leading to poly(phenylsilsesquioxane) was studied by ^{29}Si-NMR spectroscopy. The possible mechanism for the formation of poly(phenylsilsesquioxane) was proposed (Scheme 2.5) [213].

The size of a silsesquioxane cage is approximately 0.5–0.7 nm [217], and the size of their molecules is approximately 1–3 nm in diameter. Cubic silsesquioxanes have

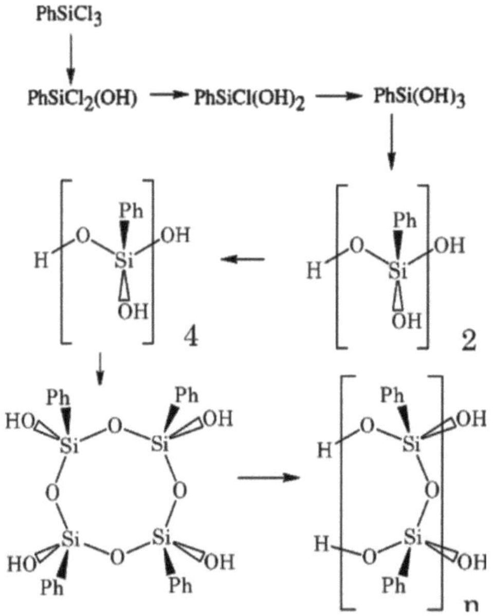

Scheme 2.5: A possible stepwise pathway for the hydrolysis of $PhSiCl_3$ and the consecutive condensation reactions of the resulting hydroxy-substituted siloxanes. Reprinted (adapted) with permission from [213]. Copyright 2004, American Chemical Society.

perfect or nearly perfect 3D symmetry usually with eight organic substituents or other functional groups in their corners. Many POSS molecules also have relatively narrow size distribution and high functionalities. The core of POSS units is rigid and is responsible for heat capacity and thermal resistance of silica skeleton [214, 223–292]. Silsesquioxane with various structures – ladder ($T_8(OH)_4$), partially cage ($T_8(OH)_2$), and cage structures (T_8–T_{16}) were prepared and characterized (Fig. 2.5) [293]. Silsesquioxanes also exist as incompletely (partially) condensed cages (e.g., $T_8(OH)_2$ or $T_7(OH)_3$, irregular cages, and ladder-type nanostructures, macrocyclic, and hyperbranched structures containing POSS cubes [293–297]. The structures of incompletely and fully condensed cyclic T_2 dimers were based on analysis of the ^{29}Si-NMR spectra. Incompletely condensed methyl-substituted POSS: disilanol $Si_8O_{11}Me_8(OH)_2$ and trisilanol $Si_7O_9Me_7(OH)_3$, were prepared by Lee et al. [298].

A huge number of papers concerning preparation of functional silsesquioxanes and their applications for fabrication of different NCs were published in a literature [267–276]. Oligosilsesquioxanes containing various aromatic substituents exhibit a high thermal stability (>500 °C), for example, octaphenylsilsesquioxane $(C_6H_5SiO_{1.5})_8$. PMSQ-based resins were prepared from incompletely condensed methyl-substituted POSS (IC-Me-POSS) and 1,3,5,7-tetramethyl-1,3,5,7-(tetrahydroxy)cyclosiloxane. Their thermal, mechanical, and electrical properties were studied as a function of the POSS content. By incorporation of IC-Me-POSS hybrids (see Scheme 2.6) at the molecular

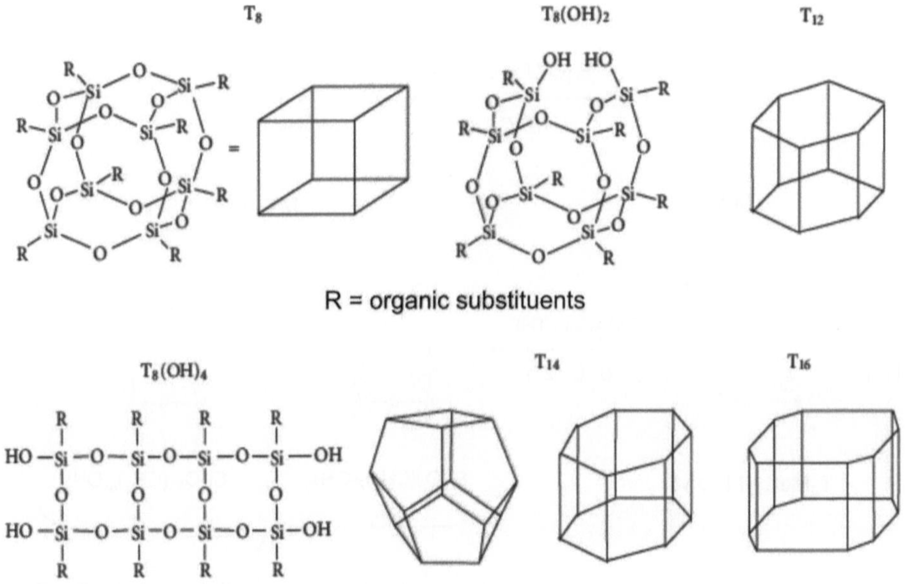

Figure 2.5: Various silsesquioxane structures: ladder [$T_8(OH)_4$], partially cage [$T_8(OH)_2$], and cage (T_8–T_{16}) structures. Reprinted (adapted) with permission from [293]. Copyright 2012, Hindawi Publishing Corporation.

level as sol–gel precursors, polymers with good mechanical properties (elastic modulus 44.0 GPa), exceptional thermo-stability (>700 °C), and an ultra-low dielectric constant ($k = 1.8$) were formed. These *spin-on-glass* materials were applied for fabrication of integration circuits, showing that the ultra low-*k*, mechanically strong were resistant during harsh wet chemical and dry etching, and under chemical mechanical planarization (CMP) processing [298]. 1,3,5,7-Tetramethyl-1,3,5,7-(tetrahydroxy)cyclosiloxane was synthesized by hydrolysis of $(MeHSiO)_4$ with Pd/C carried out in THF at −15 °C. The mixture of incompletely condensed methyl-substituted silsesquioxane disilanol and trisilanol compounds (IC-methyl-POSS) was prepared via hydrolysis of the siloxane Si–O–Si bonds [294].

Si–H functional solid silsesquioxanes (“spherosiloxanes,” T^H_n clusters, T^H=$HSiO_{1.5}$) – T^H_8, T^H_{10}, T^H_{12}, T^H_{14}, and T^H_{16} were prepared in methanol solution, in the presence of $FeCl_3$, by hydrolytic polycondensation of trichlorosilane ($HSiCl_3$) [299–301] and from (trimethoxy)silane ($HSi(OCH_3)_3$), in the presence of concentrated H_2SO_4 [302–304] or saturated anhydrous HCl in acetic acid solution [231, 267, 271, 305].

Small oligomeric hydrogen silsesquioxanes (HSQ or also called as a H-resin) with a structure $(HSiO_{3/2})_{2n}$ (T^H_{2n}, $n = 4$–16), were identified by GC-MS. However, larger HSQ molecules were not detected, presumably due to their nonpolar properties. A number of *soft* ionization techniques, such as field desorption (FD), desorption chemical ionization (DCI), and matrix-assisted laser desorption/ionization (MALDI), were applied for

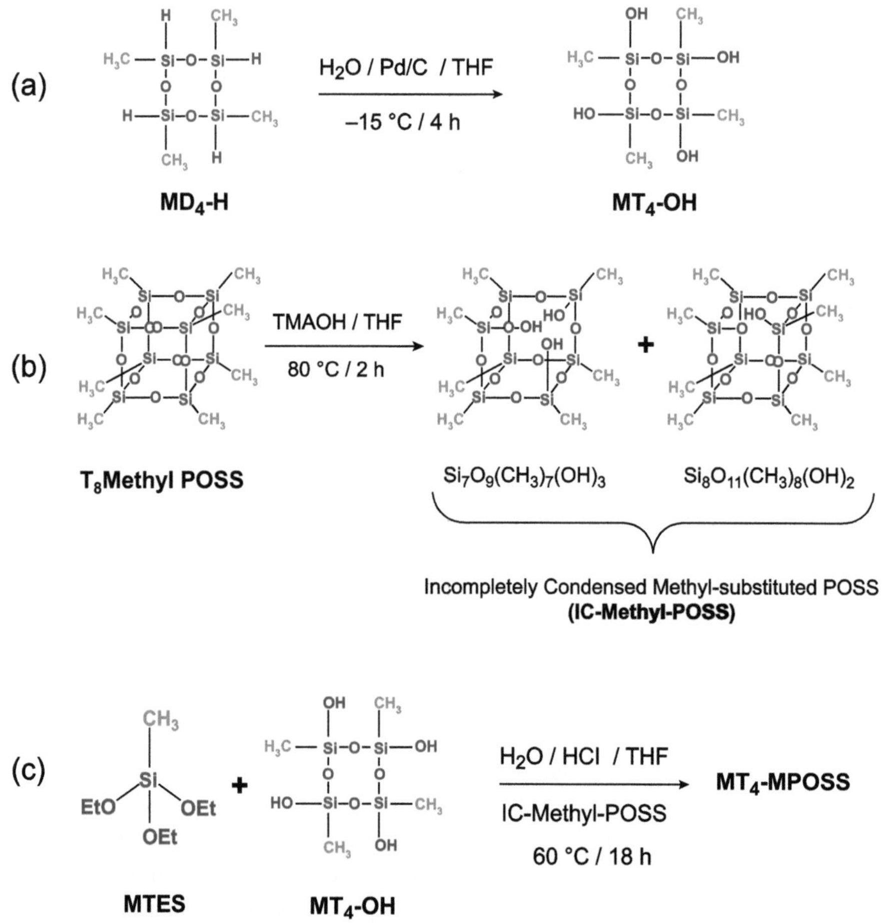

Scheme 2.6: Synthesis of (a) MT_4-OH through hydrolysis of MD_4-H, (b) incompletely condensed methyl-substituted POSS (IC-methyl-POSS), and (c) MT_4-MPOSS spin-on-glass resins. Reprinted (adapted) with permission from [298]. Copyright 2015, Royal Chemical Society.

their characterization. The $[T^H{}_{2n} - H]^{+\cdot}$ ions were observed by FD-MS, while $[T^H{}_{2n} - NH_4]^+$ and $[T^H{}_{2n} - Na]^+$ ions were found in DCI and MALDI spectra, respectively [306].

The H-silsesquioxane gels were prepared from trichloro- or trialkoxysilane at high temperature under argon, air, and ammonia atmosphere. They were analyzed by a TGA method and mass spectrometry. Under argon, these gels thermally decomposed in two possible different mechanisms: (1) the cleavage of Si–H bonds led to a loss of hydrogen, (2) a disproportionation reaction of Si–H and Si–O bonds caused the elimination of SiH_4. The latter reaction, possibly catalyzed by residual hydroxyl groups, involved the formation of SiH_2 groups as was evidenced from IR and solid-state NMR spectra. It was consistent with the thermograms obtained under air and ammonia [307].

Ladder structures of hydrogen (**H-T**)- or allyl (**Allyl-T**)-substituted ladder silsesquioxanes were prepared by a new method: the coupling of trichlorosilane ($HSiCl_3$) or allyl(trichloro)silane with 1,4-phenylenediamine (PDA), respectively, followed by hydrolysis and polycondensation reactions [308–310]. **Allyl-T** was obtained as a white solid with MW of 4,000 Da. **H-T** POSS with MW of 3000 Da was prepared as a toluene solution of 8.0×10^{-3} g/mL [311].

Hydrogen silsesquioxane (HSQ) [$(HSiO_{3/2})_8$, $T^H{}_8$] is commercially available in the form of colorless low viscous solutions, mostly in methyl isobutyl ketone (MIBK) [312–314]. The cage HSQ is a silicon-rich polymer that was first developed as a silicon-oxide-based floatable insulating layer for complementary metal-oxide semiconductor (CMOS) technologies [315]. The HSQ resins are an important class of siloxanes that are commonly used as spin-on dielectrics in the electronic industry [306]. HSQ are low density and low dielectric constant ($k \sim 2.8$) materials [315, 316], also called *spin on glass* materials [317], and find numerous applications in lithography as negative photoresists for nanolithography [318], submicron, and nanometer electronic devices [308]. HSQ is sensitive to UV and visible light, and electrons [319–321]. HSQ can also be plasma [316, 322, 323] or X-ray [315] treated or selectively etched in cryogenic systems, for example, with SF6. A low dielectric constant material, obtained from HSQ was etched in fluorocarbon (CHF_3 and CF_4) plasmas with different rates [322].

Most often, HSQ films, usually 20–100 nm thick, were etched by electron beam (EB) [314, 318, 321–328]. High-resolution negative resists with an established resolution in the range of 4.5–250 nm-wide lines were obtained by EB lithography [315, 318, 326–333]. The O_2 plasma affects increase of k [315] due to more hydrophilic surface of the resists, which leads to an increased moisture absorption. A solution of KOH [318] or a 25% tetramethylammonium hydroxide (TMAH) [333] was used as developer after EB treatment of HSQ.

The HSQ films formed from MIBK solutions were also thermally cured at 25–180 °C under 2–2.5 MPa [313]. The good performance of HSQ film for interlayer dielectric applications was observed at the high Si–H bond density [334]. It was found that the thermal processing of HSQ must be carried out carefully in order to get good film properties [335]. Many parameters (soak temperature, time, and oxygen concentration) of curing process are important during the fabrication of integrated circuits from hydrogen silsesquioxane at 350–400 °C [333, 336].

Three kinds of poly(silsesquioxane) (POSS) films were obtained from hydrogen silsesquioxane (HSQ), methyl silsesquioxane (MSQ), and poly(hydro)methylsiloxane (HMSP). The cage structures of these POSS were transformed into network structures by thermal curing. Three precursors of POSS – HSQ (18% content, in methyl isobutyl ketone as a solvent), MSQ (21% content), and HMSP (14% content, MIBK solvent) were spun coated on a wafer using a spin-on-glass coater. First, they were cured on a hot plate, and next. in a quartz furnace under nitrogen atmosphere at different temperatures of 200, 250, 300, and 350 °C, within 0–60 min. The ratio of network/cage transformation after thermal curing and the order of the refractive indices was as follows: HSQ > MSQ > HMSP. The structure and properties of the POSS film were strongly dependent on the conditions of thermal curing [312].

The HSQ resins were coated with gold nanowires by using a nanoimprinting technique. The Au nanowires were obtained by immersion plating and were studied by SEM, AFM, and TEM methods. The concentration of HF in the electroplating solution affected the nucleation and growth of the Au. Nanowires of width ~197 nm and microwires of width ~2 µm were coated onto HSQ substrates [337].

HSQ and zirconia (ZrO_2) NC was used for the preparation of a nanopatterned roll stamp. The HSQ as a *spin-on-glass* (SOG) material, was used for direct printing process with PDMS mold. Mechanical properties of SOG material were enhanced by addition of ZrO_2 nanoparticles, which were dispersed with HSQ solution. After direct printing process of composite material, annealing process was carried out to convert the polymeric HSQ structure into silica. The chemical, mechanical, and optical properties of the HSQ/ZrO_2 composite were evaluated by FTIR, a refractive index measurement, a nanoindentation test, and a pencil adhesion test. The composite material showed good adhesion properties to a glass substrate and a high hardness; thus, it can be used as a master cylindrical stamp in a roll-to-roll process [338].

Octakis(dimethylsiloxy)octasilsesquioxane [$(HMe_2SiO)SiO_{1.5}]_8$ ($Q_8M^H{}_8$) (Fig. 2.6) was prepared in reaction of Me_2HSiCl with octakis(tetramethylammonium)octasilsesquioxane in high yields (up to 85%) [240, 339–341]. At present $Q_8M^H{}_8$ is commercially available [228, 263, 342–347].

The synthesis of methacrylate-functionalized POSS (T_8-POSS-MA) was reported by J.D. Lichtenhan et al. T_8-POSS-MA can be used as new classes of monomers and additives, and are useful in modification of polymers [348].

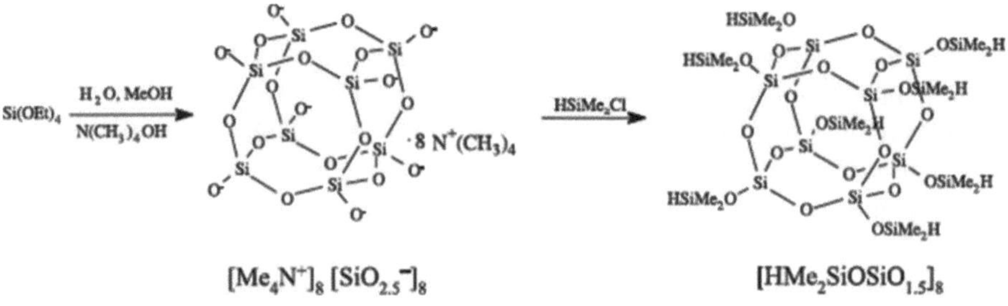

Figure 2.6: Reaction scheme of preparation of the octaanion solution, $[Me_4N^+]_8[SiO_{2.5}^-]_8$, and octa-[hydrodimethylsiloxy]octasilsesquioxane, $[HMe_2SiOSiO_{1.5}]_8$. Reprinted (adapted) with permission from [341]. Copyright Sociedade Brasileira de Química.

Novel SE NCs were prepared through the hydrosilylation reaction of vinyl-terminated PDMS with Si–H functional polyhedral octakis(hydrodimethylsiloxy)octasilsesquioxane ($Q_8M^H{}_8$), as the cross-linker and the filler. Moreover, a larger "tricubic" super-POSS cross-linker, composed of two pendant monovinylhepta(isobutyl)-POSS molecules attached to a central octasilsesquioxane core ($Q_8M^H{}_8$), was also used for the preparation of the PDMS composites. The PDMS NCs prepared with the octafunctional $Q_8M^H{}_8$ cross-linker showed enhanced mechanical properties with respect to polymer systems prepared with the tetrafunctional star (tetrakisdimethylsiloxy) silane ($Si(OSiMe_2H)_4$, TDSS) cross-linker. The PDMS elastomers obtained from the larger three-cage "super-POSS molecule" showed improved mechanical properties relative to both the TDSS and OS-POSS composites, owing to the increased content of the POSS filler in the polymer matrix [349].

Mono- ($Q_8M^H{}_7OSiMe_2Fc$), and octa-ferrocenyl-substituted octakis(dimethylsiloxy) octasilsesquioxanes [$Q_8(OSiMe_2Fc)_8$], were prepared through the hydrosilylation reaction of vinyl-ferrocene with octakis(hydrodimethylsiloxy)octasilsesquioxane ($Q_8M^H{}_8$), which was catalyzed by Karstedt's catalyst. Hydrosilylation reactions of $Q_8M^H{}_8$ with 1,1′-divinylferrocene and 1,1′-divinyl-(octamethyl)ferrocene gave appropriate poly(ferrocenyloctasilsesquioxanes). The thermal properties of these ferrocenyl-substituted polymers were analyzed by DTA, TGA, and DTG techniques. Poly(ferrocenyloctasilsesquioxane showed higher thermostability than POSS derivative prepared from 1,1′-divinyl(octamethyl)ferrocene and under N_2 atmosphere, gave char yields of 61.1 and 50.2 wt%, respectively. Electroactive polymer films obtained from 1,1′-divinyl(octamethyl)ferrocene and $Q_8M^H{}_8$ were used for preparation of electrodes [350].

Octakis (hydridodimethylsiloxy)octasilsesquioxane $(HSiMe_2O)_8Si_8O_{12}$ ($Q_8M^H{}_8$) was functionalized by the hydrosilylation reaction with allyl-terminated substrates (including polyethylene oxide component, PEO), in the presence of Pt catalyst. The MWs of the viscous products (single ion conductors), which contained between 1 and 8 labile lithium ions, ranged from ~3,000 to 5,000 g/mol [351].

A hybrid inorganic–organic, anhydrous proton-conducting polymer electrolyte containing oxyethylene units bound to silicon atoms [$(MePEG_3SiO_{3/2})_n$, where PEG=$-OCH_2CH_2$] was prepared by the hydrosilylation of appropriate allyl ether with $HSi(OEt)_3$, followed by the sol–gel hydrolytic condensation. This POSS copolymer was mainly composed of incompletely condensed T_8 silsesquioxane clusters (as was confirmed by the ^{29}Si-NMR spectra at chemical shifts between −62 and −70 ppm), but also contained a small amount of completely condensed T_6 silsesquioxane clusters, observed at −55.5 ppm, and incompletely and fully condensed T_2 dimers with chemical shifts at −48 to −50 ppm. End-group analysis showed the presence of 0.67 uncondensed Si–OH groups per silicon, indicating the presence of ladder-type and/or incompletely condensed structures [297].

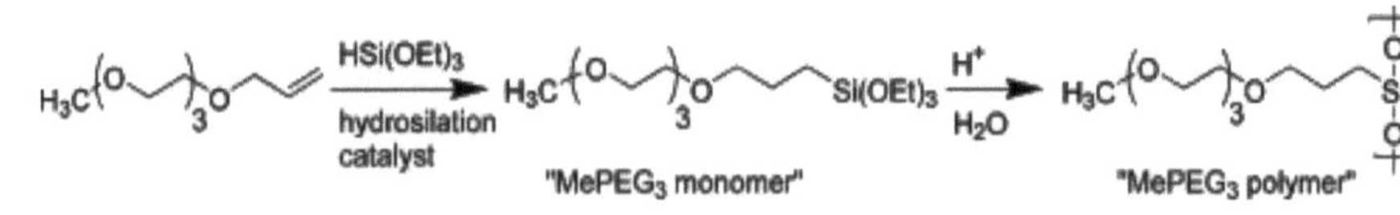

Scheme 2.7: Synthesis of $(MePEG_3SiO_{3/2})_n$. Reprinted (adapted) with permission from [297]. Copyright 2006, American Chemical Society.

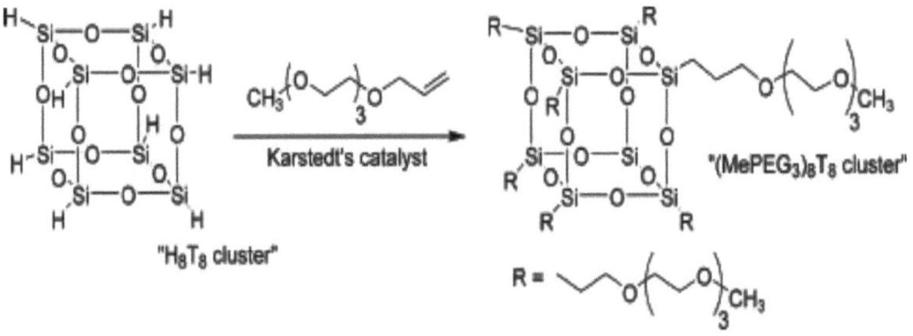

Scheme 2.8: Synthesis of $(MePEG_3SiO_{3/2})_8$ cluster. Reprinted (adapted) with permission from [297]. Copyright 2006, American Chemical Society.

The completely condensed POSS cluster $(MePEG_3SiO_{3/2})_8$ was prepared as a model compound by the hydrosilylation of (octahydro)octasilsesquioxane $H_8Si_8O_8$ ($T^H{}_8$) with allyl ether, showing ^{29}Si-NMR signals between −65.5 and −69 ppm. The ^{29}Si-NMR spectra of products of the hydrolytic condensation of $MePEG_3Si(OEt)_3$ and $(MePEG_3SiO_{3/2})_n$ contained Q-type structures at −80 to −90 and −101 ppm, respectively, which were also formed from $HSi(OEt)_3$ by oxidation or coupling, followed by the hydrolytic condensation [297].

POSS molecules substituted with PEG arms of different lengths have been used, in recent years, for preparation of NC electrolytes, as components of lithium batteries. Some POSS-PEG systems, for example, based on a blend of POSS-(benzyl)$_7$-$(BF_3Li)_3$ hybrids and POSS-PEG$_8$ formed viscous liquids that did not flow under their own weight. The PEG side chains decreased the glass transition (T_g) and melt (T_m) temperatures and –Si–O–BF_3Li ionic groups of POSS-benzyl$_7(BF_3Li)_3$ clustered on one side of the $SiO_{1.5}$ cube. At applied molar ratio of ether oxygen atoms to Li = 16:1, Li^+ cations were solvated by the POSS-PEG$_8$ and had good conductivity [352].

Solid polymer electrolytes (SPEs), which were often applied in Li batteries, were also prepared by mixing methyl cellulose (MC) and $LiClO_4$ with silsesquioxane functionalized with PEG side chains (POSS-PEG) on the $SiO_{1.5}$ cube. Above T_g POSS-PEG/$LiClO_4$ blends were highly viscous materials. Their moduli increased with MC content. Compositions with 80/20, 70/30, and 60/40 contents of POSS-PEG/MC and $LiClO_4$ (O:Li = 16:1) ranged from "tacky" with low amounts of MC to "hard" with high percentages of MC, and showed conductive properties [353].

Similar NC solid polymer electrolytes based on a blend of poly(ethylene oxide) (PEO) with POSS (POSS-PEG), were prepared by A.R. Polu et al. by grafting oligoethylene glycol ($HO(CH_2O)_nH$) ($n = 13.3$) on polyhedral silsesquioxane, followed by complexation with $LiN(SO_2CF_3)_2$. The solid polymer electrolyte with 30 wt% content of POSS-PEG showed an ionic conductivity of 5.05×10^{-5} S/cm at room temperature. The effect of POSS-PEG hybrid nanoparticles on structural, mechanical, thermal, and ionic conductivity properties was studied. The addition of POSS-PEG caused increase of the Young's modulus from 0.076 to 0.19 MPa. [354].

Conductive solid polymer electrolyte membranes (SPEMs) with star-shaped structure were prepared by free radical polymerization of (octavinyl)octasilsesquioxane (OV-POSS) with poly(ethylene glycol) (PEG) methylether methacrylate (MEM), followed by doping with lithium bis(trifluoromethane sulfonamide) (LiTFSI) in THF solution. PEG-MEM served as the polymer matrix to dissolve lithium ion. The star-shaped SPEM containing 5.1 mol% POSS exhibited the highest ionic conductivity of 1.13×10^{-4} S/cm at 25 °C, which was about two times of that of SPEM with linear structure and the same POSS content of PEG-MEM segments and OV-POSS units. Moreover, the Li/$LiFePO_4$ cell assembled with the above star-shaped SPEM revealed the highest discharge capacity at 25 °C [355]. Conductive NC solid polymer electrolyte membranes were also prepared from blends of poly(ethylene oxide) (PEO), silsesquioxane functionalized with oligo(ethylene glycol) $HO(CH_2O)_nH$ ($n = 4$) (POSS-PEG, $n = 4$), and lithium difluoro(oxalato)borate. The incorporation of POSS-PEG greatly enhanced their ionic conductivity, mechanical integrity, and compatibility [356].

Hybrid inorganic–organic polymers were prepared by the sol–gel hydrolytic polycondensation of (triethoxy)silane functionalized octahedral POSS monomers – $H_8Si_8O_8$ (HSQ) and $(HMe_2SiO)_8Si_8O_{12}$ ($Q_8M^H{}_8$)]. The properties of new siloxane networks (density and porosity, and the electronic and mechanical properties of the condensed films) were tuned, depending on a structure of monomers. The cubic cages were responsible for their stiffness, while the spacers determined the flexibility and the packing density of these polymers. The dielectric constant and the hardness of the polymeric materials were in the range $k = 2.4–3.0$ and 0.2–0.9 GPa, respectively. With the aid of a porogen, the dielectric constant was further reduced to 1.9. These materials with low-k values may be useful in semiconductor industry and optics applications [357].

A solid four-membered ring silsesquioxane $(PhSiO_{1.5})_8(MeHSiO)_2$ (so-called *double-decker-shaped-silsesquioxane*, DDSQ) (see Fig. 2.7) with 20% yield was synthesized from a reaction of $MeHSiCl_2$ with a byproduct obtained *via* a condensation of phenyl(trimethoxy)silane with NaOH [358].

Soluble multifunctional poly(hydrogensilsesquioxanes) (PHSSQ) of combined cage-like and branched network structures were prepared by condensation of (triethoxy)silane $HSi(OC_2H_5)_3$, catalyzed by HCl in the solution of THF with methylisobutyl ketone (MIBK) at 60 °C. PHSSQ were functionalized with propargyl methacrylate

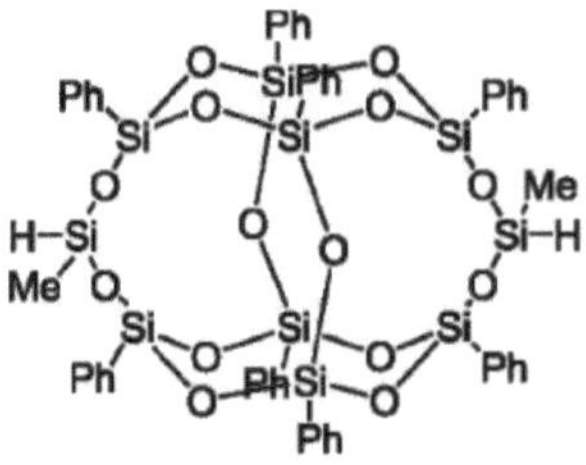

Figure 2.7: Chemical structure of DDSQ. Reprinted (adapted) with permission from [358]. Copyright 2006, American Chemical Society.

(PMA) by hydrosilylation and then formulated with acrylate monomer mixtures to yield the new acrylic/silsesquioxane hybrid photocurable materials. The mechanical and thermal properties of the parent acrylic polymers were distinctly improved by incorporating nanosized silsesquioxane molecules. With increasing silsesquioxane content, the refractive index and optical loss were reduced. The photocured new hybrid materials that were obtained may find potential applications in optoelectronic devices and patterned electronics [359].

Octahedral methylsilsesquioxanes $(CH_3SiO_{3/2})_8$ $(T^{Me}{}_8)$ were prepared in high yield from a mixture of 2,4,6,8-tetraethoxy-2,4,6,8-tetramethylcyclotetrasiloxane isomers $(D^{OEt}{}_4)$ by the hydrolytic condensation, in the presence of tetrabutylammonium fluoride (TBAF) as an ionic catalyst. The reaction was carried at room temperature, in solvents of a different polarity. Almost completely pure $T^{Me}{}_8$ was formed in THF, whereas in other solvents mixtures of $T^{Me}{}_8$ and PMSQ of different structures were obtained [360].

Polycondensation reactions of 2,4,6,8-tetraethoxy-2,4,6,8-tetramethylcyclotetrasiloxane isomers, $D_4(OEt)_4$, carried out in the presence of Bu_4NF or HCl or NH_4OH as catalysts gave xerogel materials. Depending on the reaction conditions, the structure of the resulting materials varied from octahedral cage unit (T_8) through a ladder chain to random structure. Prepared polysilsesquioxane precursors were used for preparation of silicon oxycarbide (SiC_xO_y) and silica glasses with specific nanostructure. Xerogels were heated at 800–1,000 °C. In argon, they were pyrolyzed into SiC_xO_y glass materials, while in air atmosphere, they ceramized into silica glass. Their properties were dependent on the pyrolysis conditions and on the xerogel structure [361].

The cross-linked polysiloxanes were prepared by anionic ring-opening copolymerization of octamethylcyclotetrasiloxane (D_4) with octaisobutyl-POSS, catalyzed by potassium hydroxide (KOH) or tetramethylammonium siloxanolate. Gelation time decreased with the addition of DMAc in the bulk polymerization. The cross-linked polysiloxanes showed excellent thermal stability and distinct T_g, about −119 °C [362].

Linear oligomethyl(alkoxy)siloxanes were also used for preparation of the novel type, ladder-like polymeric silsesquioxane-siloxane materials. The effect of distribution of alkoxysilyl groups along the polymer chain on the properties of silsesquioxane materials obtained by condensation of ethoxy groups was studied. Siloxanes of a regular structure with formula [$M(D_2D^{OR})_{10}D_2M$ and $M(D_2D^{OR}{}_2)_{10}D_2M$] and siloxane oligomers of random distribution of MeSi–OR (D^{OR}) units along the main siloxane chain [$MD_{(1-x)}D^{OR}{}_xM$ ($x = 0.3, 0.5, 1.0$)] were applied. Alkoxy-functionalized oligosiloxane precursors were cross-linked under hydrolytic condensation conditions catalyzed by HCl and NH_4OH or nucleophilic catalysts: TBAF and tetrabutylammonium hydroxide (TBAH). A series of thermally stable (up to 400 °C) ladder-like polysiloxane materials, densely cross-linked via ≡Si–O–Si≡ linkages, was prepared. The relationship between the structure of the siloxane chain (the amount of Si–OR units and their distribution along the polymer backbone) and the properties of preceramic materials that were obtained was mainly studied by FTIR, and NMR, XRD, DTA methods. The surface area (177–459 m^2/g) and pores distributions of the studied samples were dependent on the molecular structure of polysiloxane substrates and were determined using nitrogen adsorption methods (BET) [31].

Poly(silsesquioxane) molecules (Scheme 2.9) or mesoporous silica material (Scheme 2.10) were prepared from alkoxy-substituted polysiloxanes and cyclosiloxanes or from (octahydro)silsequioxane ($T^H{}_8$), respectively, by the hydrolytic polycondensation reactions catalyzed with TBAF or TBAH. The structure and properties of the prepared ladder-like materials were mainly dependent on the reaction conditions. Linear oligo[methyl(ethoxy)siloxanes] of the structure $Me_3SiO[Me(EtO)SiO]_nSiMe_3$ and isomers of 2,4,6,8-tetraethoxy-2,4,6,8-tetramethylcyclotetrasiloxane ($D^{OEt}{}_4$) gave similar xerogel materials [363]:

$$(MeHSiO)_4 + EtOH \longrightarrow [Me(EtO)SiO]_4 + H_2$$

$$[Me(EtO)SiO]_4 \longrightarrow \text{poly(silsesquioxane) xerogel materials}$$

Scheme 2.9: Preparation of poly(silsesquioxane) xerogel materials from functional cyclosiloxanes.

$$H_8Si_8O_8 \longrightarrow \text{mesoporous silica}$$

Scheme 2.10: Preparation of mesoporous silica from $H_8Si_8O_8$ ($T^H{}_8$).

Mesoporous organosilica materials were prepared without using any template or surfactant. The mesoporous structures were formed due to the unique structure of starting oligosiloxane or silsequioxane molecules and the specific interactions in the catalyst/solvent systems. However, after ceramization in the air, the mesoporous structure vanished. In the case of $T^H{}_8$ precursor, without any template or surfactant, the condensation process gave directly mesoporous silica material. By pyrolysis of dried poly(silsesquioxanes) at the temperature of 600 °C in argon atmosphere, SiC_xO_y glass materials were obtained as a result of ceramization [363].

Octavinyl-substituted silsesquioxanes $(ViSiO_{1.5})_8$ (T^{Vi}_8, T^{Vi}=CH_2=$CHSiO_{1.5}$) and $[(ViMe_2SiO)SiO_{1.5}]_8$ were fully and partially epoxidized with *m*-chloroperoxybenzoic acid (*m*-CPBA). These monomers easily underwent polymerization toward Lewis acids or with amines, and they can serve as coupling agents for preparations of novel inorganic–organic hybrid materials [364]. Functionalized cubic octavinylsilsesquioxanes and monovinyl(heptaisobutyl)silsesquioxanes were synthesized through highly stereoselective cross-metathesis and silylative coupling with substituted styrenes [365].

Synthesis of star POSS [253] and their octaphenol [262], octaisocyanate [286], oligoethylene oxide derivatives [287], LC POSS materials [288], alkyl, halo, nitro, amino, $[PhSiO_{1.5}]_8$, other aryl substituted, and T_{12} silsesquioxanes $(R_xPhSiO_{1.5})_{12}$ was also described in the literature [274, 279–283, 289].

Octakis(3-chloropropyl)octasilsesquioxane [(T-PrCl)$_8$] was prepared by hydrolytic condensation of 3-chloropropyl(trimethoxy)silane $Cl(CH_2)_3Si(OCH_3)_3$ in methanol solution (Scheme 2.11). In the first step, acid catalyzed hydrolysis in the presence of concentrated HCl was carried out, and in the second step, condensation reactions occurred in the presence of di-*n*-butyltin dilaurate as a catalyst. This enabled the selective formation of $(T^{PrCl})_8$ more rapidly (4 days), when compared to older methods (5 weeks), while the yield (35%) was comparable [366].

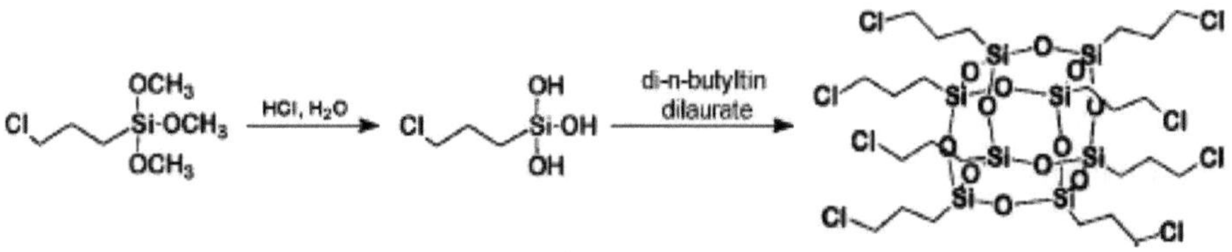

Scheme 2.11: Two-step synthesis of octakis(3-chloropropyl)silsesquioxane $(T^{PrCl})_8$. Reprinted (adapted) with permission from [366]. Copyright 2008, American Chemical Society.

Functional silsesquioxanes: octakis(3-acetoxypropyl)-, octakis[3-(phenylamino)propyl]-, octakis-[3-(methacryloyloxy)propyl]-, octakis[3-(4-methylpiperazin-1-yl)propyl]octasilsesquioxane, and octakis{3-[(2-hydroxyethyl)dimethylammonio]propyl}octasilsesquioxane chloride were prepared by nucleophilic substitution of chlorine atom in octakis (3-chloropropyl)octasilsesquioxane (T-PrCl)$_8$ with nucleophilic reagents. They were characterized by spectroscopic methods and an elemental analysis [367]. The octa(3-chloropropyl)octasilsesquioxane (ClPr-POSS) was also functionalized with the 4,5-diphenyl-2-imidazolethiol (DIT). The new compound (DIT-POSS) was characterized by several spectral methods (FTIR, solid state ^{13}C- and ^{29}Si-NMR), SEM, energy-dispersive X-ray spectroscopy (EDS), X-ray diffraction (XRD), and TGA. Its thermal stability in nitrogen atmosphere was greater than in air [368]. Aromatic nitro-, aldehyde-, and bromo-functionalized oligomeric polyhedral silsesquioxanes were easily synthesized via nucleophilic substitution reactions between octakis(3-chloropropyl)octasilsesquioxane and phenoxide derivatives. Phenoxide anions in these organic–inorganic hybrid nanobuilding blocks caused a cage-rearrangement, which led to the formation of octa-, deca-, and dodecahedral silsesquioxane structures [369].

Novel functionalized silsesquioxane nanostructures containing an encapsulated fluoride anion (from TBAF) in a POSS cage were prepared from alkoxysilanes, EWG-$(CH_2)_n$-$Si(OEt)_3$ ($n = 1 - 3$), having electron-withdrawing groups (EWG) (e.g., methacryloxymethyl, acethoxymethyl, 3-cyanopropyl). These complexes were studied by X-ray crystallography and spectroscopic methods (^{19}F- and ^{29}Si-NMR, MALDI-TOF, and ESI-MS) [370].

A great number of octafunctional POSS molecules including mixed substituents $Q_8(SiOMe_2R)_8$ (R = H, vinyl, epoxy, methacrylate, etc.) and various aryl-substituted silsesquioxanes $Q_8(C_6H_4R)_8$ [R = alkyl, alkene, alkyne, R'X (X = halogen, amine, epoxy, etc.; R' = R or mixed] was also synthesized. Janus-like symmetric, cubic monofunctional, and statistically bifunctional octasilsesquioxanes molecules, were prepared by R.M. Laine et al. They included octa(triethoxysilylethyl)- (OTSE), tetra(dimethylsiloxy)tetra(triethoxysilylethyl)- (TTSE), tetra[chloropropyl)dimethylsiloxy)tetra (triethoxysilylethyl)- (TCPTSE), and other functionalized silsesquioxane derivatives. These materials formed layer-by-layer coatings with controlled contact angles ranging from 50 to 120 °C, with a high pencil hardness of 6H. Triethoxysilyl groups easily hydrolyzed under acidic medium to silanol groups, which underwent intra- and inter-molecular condensation reactions and easily binded especially to different hydrophilic surfaces, for example, aluminum. Some new POSS derivatives contained different functional groups, for example, C=C, epoxy, NCO, oxetane, -$O(CH_2)_2OH$, and methacryloxy [371]. Many bifunctional POSS derivatives were prepared from octakis(3-azidopropyl)octasilsesquioxane through cycloaddition with alkyne, catalyzed by copper (I) salts [372].

A rod-like polysilsesquioxane (POSS-COO^-Na^+) with a hexagonally stacked parallel structure was prepared by the sol–gel hydrolytic polycondensation of 2-cyanoethyl(triethoxy)silane (CE-TEOS) in sodium hydroxide aqueous solution, which was confirmed by the TEM image [373]. A new inorganic-organic hybrid dendritic network structure containing both POSS and carborane clusters was prepared by the hydrosilylation of $(HSiMe_2O)_8Si_8O_{12}$ ($Q_8M^H{}_8$) with the vinyl-terminated carboranylene-siloxane monomer. Moreover, the POSS-carborane network was reinforced with additional ≡Si–O–Si≡ cross-links which were formed during the reaction of $Q_8M^H{}_8$ by the hydrolysis, oxidation, and condensation reactions of its Si–H bonds in the presence of the Karstedt catalyst and Et_3N/H_2O cocatalysts. The POSS-carborane films were characterized by TGA in air and in N_2 and showed superior thermal properties. Char yields at 1,000 °C were 77% in N_2 and 91% in air. The highest decomposition temperature, T_5 (which corresponds to temperature at 5% weight loss), for the POSS-carborane hybrids reached 525–575 °C. Their potential applications include: a new generation of advanced high-temperature materials, for example, photoresistive coatings for electronic and optical devices and as membranes for gas separation and neutron absorption [374].

Two novel spherosilicates composed of an octahedral Si_8O_{12} core, substituted with malonic acid $[Si_8O_{12}][OSiMe_2CH_2CH_2CH_2CH(COOH)_2]_8$ and bis-[trisvinyl(siloxycarboxy)butyl(dimethylsiloxy)] silyl ester groups $[Si_8O_{12}]\{OSiMe_2CH_2CH_2CH_2C[(COOSi(CH{=}CH_2)_3]_2\}_8$ were prepared by the hydrosilylation reaction of $[Si_8O_{12}](OSiMe_2H)_8$, followed by hydrolysis and condensation reactions. The first structure containing 16 peripheral carboxyl groups can be used in metal binding, crystal engineering, among others. The spherosilicate, having 48 vinyl groups, was cross-linked with bis(*p*-dimethylsilyl)benzene 1,4-$C_6H_4(SiMe_2H)_2$, providing soluble nanospheres with a narrow size distribution. Their silyl ester bonds can easily hydrolyze under mild conditions with formation of carboxylic COOH and silanol Si–OH functionalities that can bind metals (e.g., Pd) and other molecules for catalysis or delivery [375].

It has been well known that nanosilica particles, containing many silanol groups on a surface, are well soluble in hot methanol and even in benzene. Hydroxyl-functionalized and quaternized cationic silsesquioxane nanoparticles were also found to be water-soluble. They were prepared through the one-step condensation of the bulky (triethoxy)silane precursors [293].

Monomers functionalized with $Si(OEt)_3$ groups were prepared by the addition reaction of aminopropyl(triethoxy)silane to 2-hydroxyethyl acrylate (HEA), 2-(dimethylamino)ethyl acrylate, 1H,1H,5H-octafluoropentyl acrylate, or 2,2,2-trifluoroethyl acrylate. The hydrolytic condensation of these precursors in homogeneous systems gave, almost quantitatively, a variety of functional silsesquioxane hybrids (Fig. 2.8). The $RSiO_{1.5}/SiO_2$ and $RSiO_{1.5}/TiO_2$ hybrids were also obtained by the hydrolytic co-condensation of a (triethoxy)silane precursor with metal alkoxides. Different kinds of silsesquioxane-based hybrids were prepared – HO-functionalized cationic nanoparticles, fluorinated, and amphiphilic. Stimuli-responsive hybrids were prepared from the water-soluble silsesquioxane nanoparticles and fluorinated amphiphilic silsesquioxane hybrids [293].

Functional mesoporous silsesquioxanes with high concentrations of amine groups were prepared by co-condensation of bridged bis(trimethoxy)silanes with TEOS. Polycondensation of bis(trimethoxy)silanes containing amine groups was carried out in acidic, neutral, and basic media, resulting in high yields of solid bridged silsesquioxanes. The largest agglomerates (up to 50 μm) were obtained in basic media. Two-level fractal structures composed of aggregated 6.5–10.5 nm particles. In the presence of a surfactant, mesoporous products were prepared, which reversibly absorbed water and CO_2 at temperatures below 120 °C and were thermally stable up to 260 °C. The condensation of the bridged precursor without a linker resulted in formation of a highly functionalized mesoporous material [376].

Acetoxyphenyl-functionalized POSS compounds were synthesized through metathesis of $[vinylSiO_{1.5}]_8$ and $[vinylSiO_{1.5}]_{10/12}$ mixtures with *p*-acetoxystyrene (P1) or metathesis with *p*-bromostyrene followed by Heck's coupling with *p*-acetoxystyrene

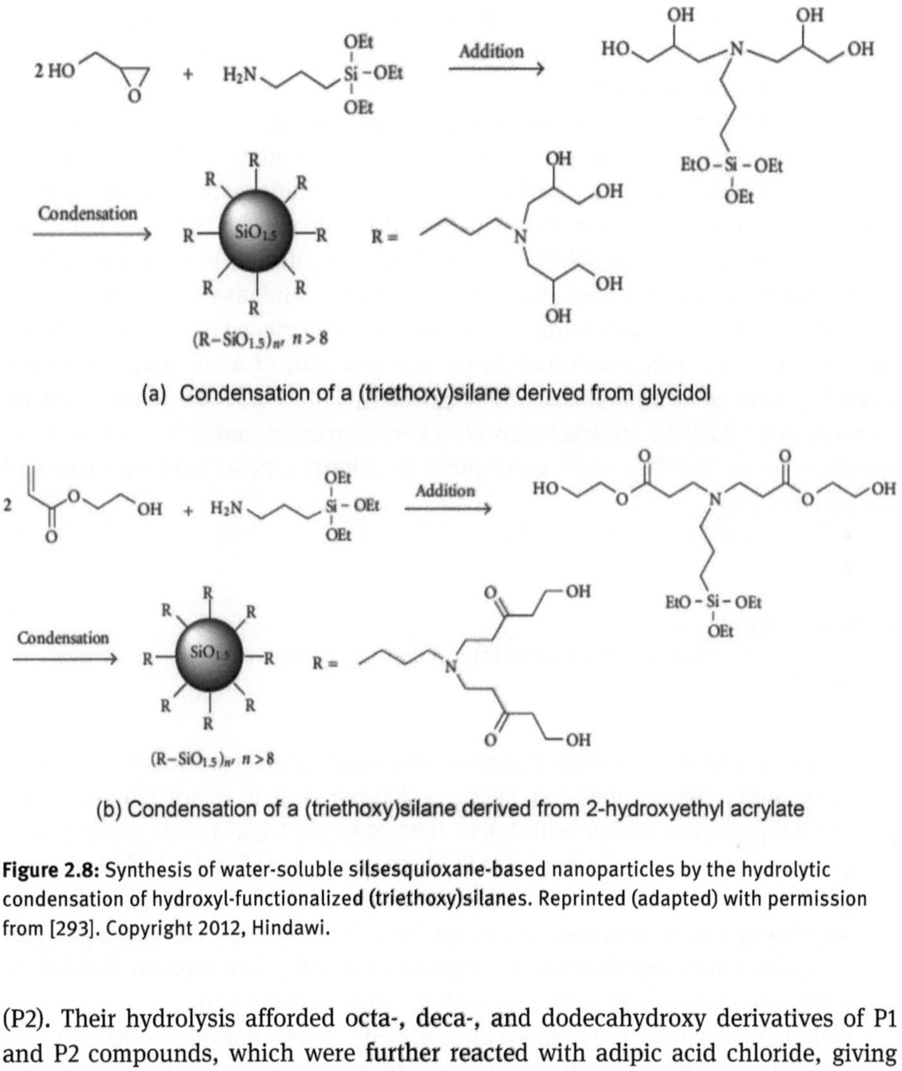

Figure 2.8: Synthesis of water-soluble silsesquioxane-based nanoparticles by the hydrolytic condensation of hydroxyl-functionalized (triethoxy)silanes. Reprinted (adapted) with permission from [293]. Copyright 2012, Hindawi.

(P2). Their hydrolysis afforded octa-, deca-, and dodecahydroxy derivatives of P1 and P2 compounds, which were further reacted with adipic acid chloride, giving highly cross-linked polyester copolymers based on silsesquioxanes. All new products and the cross-linked polymers were characterized by different spectroscopic methods. Their specific surface areas ranged from 5 to 25 m^2/g. These polyhydroxyl-substituted materials may find numerous applications as photolithographic materials, intermediates for the synthesis of porous cross-linked polymers, drug delivery, and media for molecular separations, and so on [377].

POSS functionalized with imidazolium moieties (POSS-Imi) was prepared in a two-step procedure from octavinyl-substituted silsesquioxane (Vi_8-POSS). It was characterized via ^{1}H-NMR, ^{13}C-NMR, and IR spectroscopy as well as combustion chemical analysis, MS, and TEM. The reaction of Vi_8-POSS with 3-chloro-1-propane-thiol via a thiol-ene reaction in the presence of 2,2-azobisisobutyronitrile (AIBN) gave POSS-Cl, which was further reacted with 1-methylimidazole, giving imidazolium-functionalized silsesquioxane. POSS-Imi was applied as the catalyst for the conversion of CO_2 and epoxides into cyclic carbonates [378].

A well-defined, high-molecular-weight ladder polyphenylsilsesquioxane (Ph-LPOSS) was prepared via a new three-step method – monomer self-organization in solution, lyophilization, and a surface-confined polycondensation. A ladder structure was self-assembled from the 1,3-diphenyl(tetrahydroxy)disiloxane monomer (Figure 2.9) in acetonitrile solution. Next, it was lyophilized to form a thin layer on the inner surface of a flask, followed by the polycondensation of the ordered monomeric thin layer, which was carried out in the presence of Et_3N. Thus, polycondensation side reactions (cyclization and gelation) of monomers containing silanol groups were minimized, and regularity of the ladder structures was increased. The ^{29}Si-NMR analysis showed a very narrow signal at $\delta = -78.5$ (corresponding to the $PhSiO_{3/2}$ unit), confirming the highly regular ladder structure of the polymer [379].

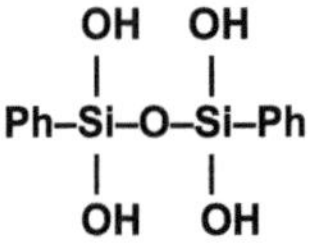

Figure 2.9: The chemical structure of 1,3-diphenyl(tetrahydroxy)disiloxane [379].

Pentacyclic and heptacyclic oligosilsesquioxanes with ladder structures were prepared on a laboratory scale via multistep condensation reactions of (1,3-dichloro-1,3-diisopropyl-1,3-diphenyl)disiloxane with 1,3,5,7-tetraisopropyl-1,3,5,7-tetrahydroxycyclotetrasiloxane, followed by chlorination of Si-Ph groups with $AlCl_3$/HCl, hydrolysis, and condensation with 1,3-diisopropyl-1,3-diphenyl)disiloxanediol. The higher MW ladder polysilsesquioxane was obtained by the hydrolytic condensation reactions from (*cis–trans–cis*-tetrabromo)(tetramethyl)cyclotetrasiloxane. The thermal stability of these POSS derivatives was increasing with the ring number [380].

Well-defined ladder poly(phenyl-*co*-methacryl silsesquioxane)s (PhMeSQ-MA) with MWs (M_w) ~13,000–40,000 g/mol, containing 9–89 mol% of methacryloxypropyl groups were prepared by the hydrolytic polycondensation of methacryloxypropyl(trimethoxy)silane and PTMOS, in the presence of K_2CO_3 at 25 °C. These copolymers showed mainly ladder-like structure and were photocured in the absence of any initiators. Their physical properties were analyzed by nanoindentation before and after photocuring. A surface modulus reached 8 GPa, and a hardness of thin films increased from 100 to 400 MPa. PhMeSQ-MA exhibited the high thermal resistance ($T_{d2\%} = 410$–454 °C) [381].

The low MW poly(phenylsilsesquioxanes) (Ph-PSQ, $M_w = 2{,}600$–10,300, $M_n =$ 1,200–3,700 g/mol; PDI = 1.2–1.6) were prepared through direct hydrolytic condensation of phenyl(triethoxy)silane $PhSi(OEt)_3$ at 30–36 °C, using sequential one pot, two-step reactions in the mixture of THF and H_2O, in the presence of an excess of potassium carbonate as the base catalyst. Two phase products (a liquid and a wax) were obtained. The liquid phases of Ph-PSQ were soluble in methylene chloride and chloroform and were further precipitated from methanol. The structure and properties of the obtained Ph-PSQ products were analyzed by means of NMR, TGA, SEC, XRD, and FTIR methods. The poly(phenyl silsesquioxanes) that was obtained had mostly closed siloxane structures and showed excellent thermal stability up to 400 °C. The ladder structure of Ph-PSQ was analyzed by XRD and was confirmed by analyses of FTIR spectra and ab initio calculations [382]. Amounts of random network structures in reaction mixtures of different POSS syntheses were dependent on reaction conditions [383].

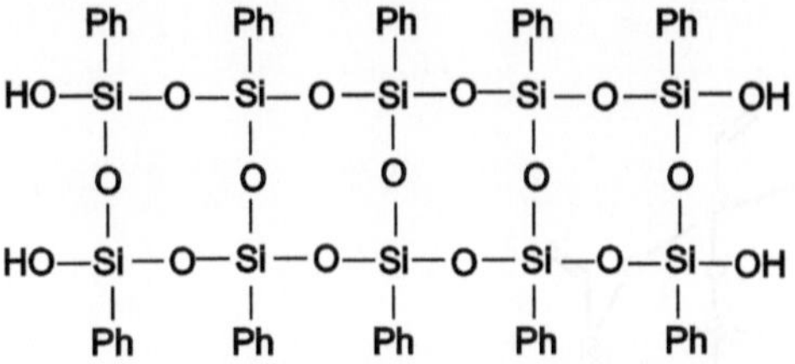

Figure 2.10: A proposed structure of ladder-like oligo[(tetrahydroxy) decaphenylsilsesquioxane] [382].

Recently, based on TGA, DTG, DSC, and FTIR studies, it was found that a thermal degradation of oligomeric ladder-like vinyl- and phenylsilsesquioxanes proceeded at relatively low temperatures with evaporation of the shortest oligomers, accompanied by partial redistribution of side groups and by partial cross-linking of organic substituents [384].

By a selective hydrosilylation reaction of 4-vinyl-1-cylcohexene with $[HMe_2SiOSiO_{1.5}]_8$, tetra- and octa(2-cyclohexenylethyl)octasilsesquioxanes were prepared. The tetrabifunctional $[(cyclohexenylethyl)Me_2SiOSiO_{1.5}]_4[HMe_2SiOSiO_{1.5}]_4$ (m.p. ~ 80 °C) was cast and then cured (by the thermal hydrosilylation) into transparent NCs. The obtained material was air stable to temperatures ≥400 °C and showed dielectric constants of 2.8–2.9 at 100 kHz to 3 MHz. The octa(cyclohexenylethyl) POSS derivative (m.p. ~ 120 °C) was copolymerized with hydrosiloxanes and provided NCs with lower thermal stability [385].

The syntheses of the 12- and 24-membered vinylsilyl- and hydrosilyl-functionalized macrocyclic ring-silsesquioxanes: hexavinylhexa(dimethylhydro)cyclohexasilsesquioxane (**A**) and (dodecavinyl)dodeca(dimethylhydro)cyclododecasilsesquioxane, (**B**) (with yields: 96 and 94.5%, respectively), as viscous liquids, with the general formula of $[vinylSi(OSiMe_2H)O]_n$ (n = 6, 12, Scheme 2.12) were first reported. These silsesquioxane ring compounds were prepared through the reactions between the alkaline metal or copper derivatives of silsesquioxanes and (chloro)dimethylsilane and were characterized by spectroscopic methods (FTIR, ^{1}H-, ^{13}C-, and ^{29}Si-NMR) [386].

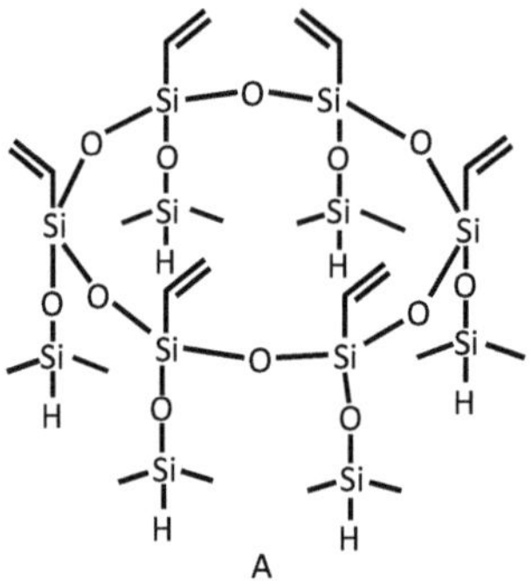

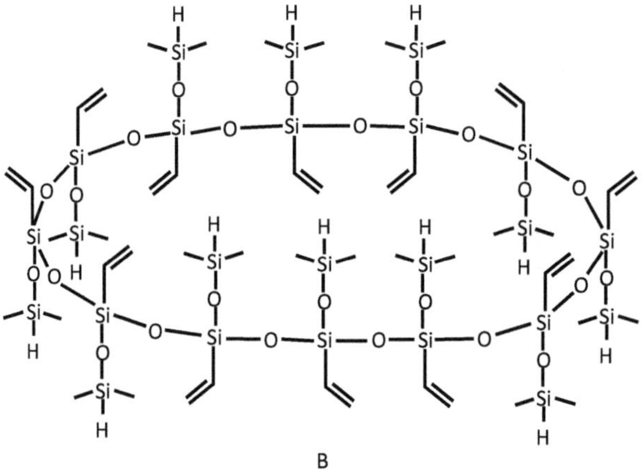

Scheme 2.12: Structures of silsesquioxanes (**A**) and (**B**). Reprinted (adapted) with permission from [386]. Copyright 2004, American Chemical Society.

The addition polyhydrosilylation of the macrocyclic (dodecavinyl)dodeca(dimethylhydro)cyclododecasilsesquioxane led to cross-linked polysilsesquioxane networks (Scheme 2.13) that exhibited high porosity with a specific surface area of 308 m^2/g (determined by the BET method) and a pore volume of 0.21 mL/g. However, the cross-linked polymer obtained from (hexavinyl)hexa(dimethylhydro)cyclohexasilsesquioxane did not show a significant porosity, presumably due to the difference in size of macrocyclic structures for the silsesquioxane macromers [386].

The POSS are molecules and not particles, because [387]:

- they have well defined MWs,
- they are well soluble in solvents and can be recrystallized from the melt or from solution,
- some POSS derivatives are liquids (particles are not liquid at 100% concentration),
- they can be characterized by a solution NMR (solids have no solution NMR spectrum),

- soluble POSS molecules were distinguished from POSS agglomerated into domains by dielectric spectroscopy (DS) [388],
- soluble POSS molecules were recognized from POSS domains by wide-angle X-ray diffraction (WAXD) [389].

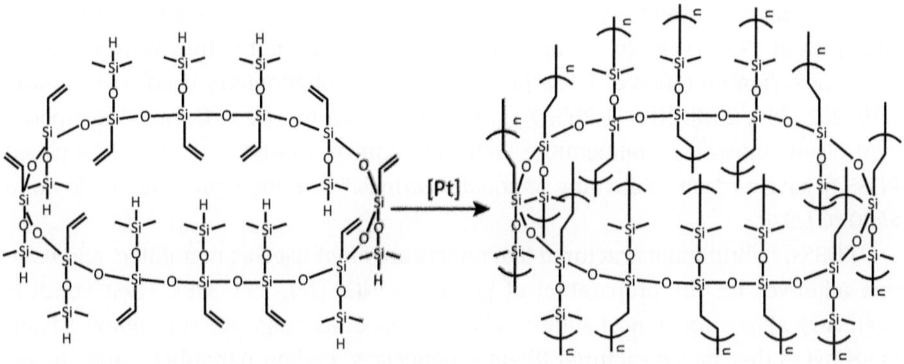

Scheme 2.13: A hydrosilylative polyaddition of ring-silsesquioxane **B**. Reprinted (adapted) with permission from [386]. Copyright 2004, American Chemical Society.

Low-cost POSS were also prepared directly from an agricultural waste product, a rice hull ash that was bifunctionalized, with about four of each type of functional group. These bifunctional POSS offer the potential to: (1) serve as superior silanizing agents, (2) build layer-by-layer structures at nanometer length scales, (3) build graded property thin films, and also (4) build porous, high strength interlayer dielectrics [390, 391].

2.5.1 Applications of POSS-modified thermoplastic polymers

Many POSS derivatives find numerous practical applications in electronic industry [392]:
- in liquid systems,
- in electroluminescent (EL) materials and light emitting devices (LEDs), EL polymers, iridium complexes,
- as nonlinear optic (NLO), optical limiting (OL) and lasers,
- in a lithography,
- in sensor systems (for gas and vapor detection, conducting composites, and electrochemical sensors),
- in fuel cells,
- in batteries,
- as lubricants.

Silsesquioxanes are useful for fabrication of different polymer NCs (e.g., epoxy NCs) [272, 277–282, 284, 285]. Mechanical and other properties of NCs containing POSS units depend on the structure of components and processing parameters [269–279, 291]. For instance, coefficients of a thermal expansion of epoxy resin composites were tailored from 25 to 250 ppm/°C [290], and the O_2 barrier properties were programmed to equal commercial systems (<1 cc · 20 lm/m^2 · day · atm), and were superior to polysiloxanes at 19,000 cc · 20 lm/m^2 · day · atm) [291]. Optimal values were obtained from epoxy/amine ratios of 1:1 rather than the 1:2 stoichiometry used traditionally [279–282, 289, 290]. $[(NO_2)_2PhSiO_{1.5}]_8$ was stable up to ≈400 °C, but decomposed explosively above this temperature [289]. In contrast, *i*-butyl and *i*-hexyl alkylated POSS behave as "self-lubricating nanoball bearings," forming liquids air stable up to >400 °C [292].

POSSs, hybrid nanostructured macromers are often used as nanofillers and polymer additives for the preparation of polymeric NCs [221, 393–398]. Their versatile chemistry allows for almost infinite chemical modifications, offering many advantages over other nanostructured fillers – nanoclays, carbon nanofibers, and carbon nanotubes. Depending on their functionality, 3D network, bead, or pendant type-POSS based POSS hybrids can be used in the preparation of the numerous polymeric NCs that have the potential to be designed for products with specific nanostructures for specific end-use applications. The POSS macromers are mainly used for the modification of viscoelastic and thermal properties of different polymers [393].

It was shown by Lee, Y.J. et al. that the POSS NCs have the ability to self-assemble in mixtures with other polymers [398]. Mixtures of POSS with polymers are often transparent. Their transparency was retained up to 5–10 wt% (or vol%), because the silsesquioxanes are soluble. However, the cage POSS molecules, due to their cubic (spherical) shape, can be considered as particles that resemble the smallest possible form of silica with the organic moieties built on a surface and surrounding the central core. A huge number of examples of polymers modifications with different POSS molecules have been described in a literature, and some of them are presented in this section.

The surfaces of siliceous fillers are very often modified with organosilanes in order to improve their performance in polymer composites and coatings. Trialkoxysilanes (e.g., *iso*-octyl(trimethoxy)silane) react with silanol groups on the surface of silica, while the *iso*-octyl groups present on the surface, avoid agglomeration of adjacent particles through sterical hindering. Similarly, the surfaces of nonreactive POSS types such as octa-*iso*butyl POSS, octa-*iso*-octyl POSS, and octaphenyl POSS substrates behave as hydrophobic layers of dispersant. Thus, the POSS additives can be easily dispersed in solvents, coating, and polymers. The siliceous fillers (silica, kaolin, mica, talc, wollastonite, and others) are modified on the surface with reactive coupling agents (SCA). Trialkoxy groups of SCA react with hydroxyl groups of the filler surface, while a second functional group can react with the polymer matrix or coating. Typical functionalities include different organic moieties – amines, alcohols, carboxylates, epoxides, urethanes, thiols, and so on. After exposure to humidity and heat, SCAs improve dispersion of fillers, enhance mechanical strength, cause increase of the filler-matrix adhesion, and even retention of the matrix adhesion. Conventional coupling agents self-condense giving an amorphous resin, but they can also form POSS cages. Many POSS functional derivatives were prepared from organosilane coupling agents (SCA), containing acrylate, methacrylate, vinyl, epoxy, thiol, amino, and other functional groups. During POSS synthesis, silane coupling agents self-condense into ordered cage structures leading to high yields of single cage types quite often. However, early syntheses took a rather long time and gave low yields of cubic structures, and product mixtures were hard to purify. The POSS compounds, composed of molecular silica surrounded by a shell of coupling agent, easily react and bond with polymers and coatings [399, 400]. The properties of POSS such as density, modulus, and refractive index are between the properties of organic polymers and silica. For instance, polyethylene has Young's moduli in the range of 1–2 GPa, whereas silica has a value closer to 70 GPa, and an intermediate calculated value for octacyclopentyl POSS was 11.8 GPa [401]. POSS with shorter alkyl groups are expected to have higher moduli, whereas POSS with long organic groups such as octa *iso*-octyl POSS and PEG POSS are liquids at room temperature, leading to a rather low modulus.

Inorganic–organic hybrid POSS molecules can react with different thermosetting polymers. Many POSS molecules dissolve in thermosets and can be diluted by common solvents. POSS molecules can improve properties of polymers, especially their thermostability and resistance to environmental degradation [399]. For instance, multifunctional POSS epoxy derivatives have numerous advantages. They are soluble in epoxy monomers and additives, and show enhanced thermomechanical properties – oxidative, hydrolytic, and corrosion resistance, as well as improved hydrophobicity and oleophobicity [394]. Thermal and mechanical properties of organic–inorganic hybrid materials are between properties of ceramics and organic materials [402, 403].

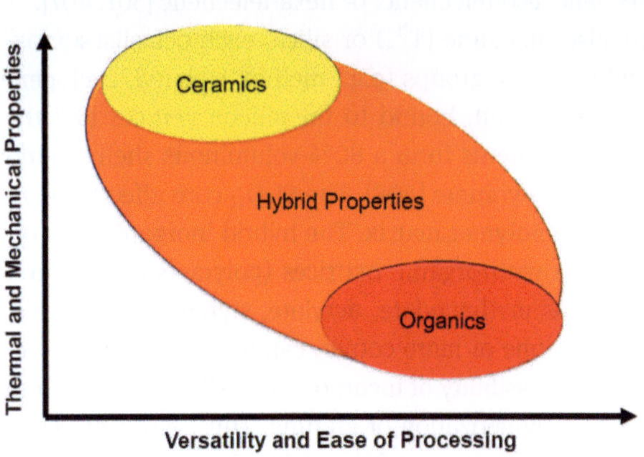

Figure 2.11: Properties of ceramics, polymers, and organic–inorganic hybrid materials [403].

Poly(silsesquioxanes) (POSS) were also used as nanofillers or modifiers to elastomers: methylvinylsilicone rubber (MVSR) and hydrogenated butadiene-acrylonitrile rubber (HNBR). Functionalized silsesquioxanes took part in the cross-linking of rubbers and increased their mechanical strength, especially at low temperature of −50 °C. They also increased the surface hydrophobicity and reduced the ageing effects, while, at the same time, affected oxygen barrier properties. The addition of POSS compounds containing acidic or basic substituents to rubber vulcanizates gave materials with self-healing properties [404, 405] and affected the barrier properties, mechanical properties at room temperature, or the relaxation rates of the MVSR vulcanizates. The content of POSS and the method of composites preparation affected the amount of ionic bonds, as indicated by measurements of equilibrium swelling in toluene. The self-healing effect was best visualized by SEM images of samples after destruction and conditioning. The most significant self-healing effect was observed in systems containing amic acid-isobutyl POSS/aminopropylisobutyl-POSS [405].

A series of self-healing, multicross-linkable ladder polysilsesquioxane hybrid materials with enhanced mechanical properties were also prepared, both through UV-curing and Diels-Alder chemistry. The incorporation of UV-curable acryl- or epoxy groups allowed for a higher degree of cross-link density, while close presence of the inorganic backbones stimulated highly efficient self-healing properties. The obtained POSS hybrid materials showed optical transparency (>95%), excellent thermostability (>400 °C), solution processability, and solid surface mechanical properties in bulk and nanoscale (pencil hardness 6H, elastic modulus >9 GPa, respectively). These properties were recovered through mild and rapid thermal treatment. These materials belong to the next generation of hybrid smart coatings with potential application in optoelectronic devices [406].

POSS hybrids are very useful in preparation of various kinds of copolymers, exhibiting valuable properties. These copolymers have different chemical structures: for example, hemi-telechelic, di-telechelic, eso-telechelic, or hexa-telechelic [301, 407].

Unlike silicones, (octahydro)silsesquioxane ($T^H{}_8$) or silica, each octasilsesquioxane molecule is connected to eight organic groups (e.g., methyl, isobutyl, cyclopentyl, cyclohexyl, phenyl, aniline, and others), bound to the silicon vertices and the surrounding Si_8O_{12} cage. Organic substituents form a 80%-voluminous shell around the Si_8O_{12} core, with respect to the POSS volume [408]. The Si_8O_{12} core affects the interactions between POSS moieties and polymer matrix. The hybrid inorganic core-organic shell architecture is compatible with organic matrices (polymers and natural biomaterials). Organic groups (such as methacrylate, acrylate, styrene, norbornene, amine, epoxy, alcohol, and phenol), in one or more corner, can react with other functional groups [409, 410], providing the possibility of incorporation POSS molecules into a polymer chain or network through polymerization or grafting. Thus, different POSS-polymer architectures can be obtained. Random grafting of POSS along the polymer chain was achieved through free radical polymerization [349, 411, 412] and step-growth polyadditions of copolymers [243]. Polymerizations and reactions with POSS also lead to POSS moiety at a single end of the polymer chain, giving a hemi-telechelic POSS polymer, at both ends to yield POSS telechelic molecules, or grafted to a single block of block copolymers or multiblock polymers [413].

2.5.1.1 Polymer networks enhanced with POSS molecules

The most important features of POSS hybrids built in polymer networks are the following [404]:
- dispersion at the molecular level,
- control of interactions,
- control of higher structure,
- property enhancement.

An incorporation of POSS units into copolymer structure affects improvement of high-temperature properties as follows:
- increase of decomposition temperature T_d,
- increase of glass temperature T_g,
- improvement of a melt flow,
- improvement of the filler dispersion,
- increase of TS,

enabling their use at higher temperatures and increasing environmental resistance as listed below:
- oxidation and corrosion resistance,
- solvent resistance,
- radiation resistance,
- reduced flammability,
- hydrophobicity and oleophobicity.

Other uses of POSS-modified polymers include [404]:
- low friction materials,
- catalysts,
- permeation control (e.g., in packaging materials).

Owing to unusual properties of POSS molecules, they have been used in fabrication of new class of polymers, copolymers, and composite materials with outstanding properties.

POSS molecules have sufficient MW to retard evaporation and migration, but have low enough MW to dissolve molecularly into polymer melts and in the final composite. The solubility of POSS in polymers and coatings depends on the polarity of cages. The typical solubility of a POSS with a polarity similar to the host polymer is ~5 wt%, which might be sufficient for many polymer additive applications, including plasticizers, antioxidants, dispersants, flow aids, and lubricants. Similar to dendritic and hyperbranched higher MW antioxidants that showed enhanced stabilization [414], POSS antioxidants were effective [415]. POSS additives are less expensive than the dendritic siloxanes and have other advantages such as stability during high-temperature processing. Degradation temperature of POSS is usually close to 500 °C [399].

The effect of POSS on properties of polymer blends and composites depends on many factors. However, the solubility and reactivity of POSS molecules have been most important factors.

1. Addition of up to ~5 wt% of a soluble POSS may improve melt flow, without decrease of modulus and strength [POSS]. Although, in some cases, the POSS additive acted as an antiplasticizer [416, 417], filling free volume and raising modulus, and the molecularly dissolved POSS systems retain full transparency.
2. Addition of ≥5–10 wt% soluble POSS (or less for an insoluble POSS) resulted in the formation of POSS domains with a loss in transparency, but the domains on the surface changed the contact angle [418] and caused significant reduction of friction [419].
3. The T_g of chemically bound POSS to a polymer was increased, as also the heat distortion and Vicat temperatures, because the rigid, bulky POSS cubes hindered chain mobility [411].

2.5.1.2 Effect of dissolution POSS hybrids in the polymers

The solubility of an unreactive POSS normally may reach ~5–10 wt% POSS. Higher loading can cause precipitation and cloudiness. At low content (<5 wt%), POSS remains molecularly dissolved; it acts as an antiplasticizer and can increase modulus, somewhat. The POSS cubes are molecularly dissolved and fill in the free volume in the polymer, thereby increasing stiffness. Above the solubility limit, the POSS forms domains of insoluble material that have low cohesive strength, the POSS acts as a plasticizer, and thus softens the polymer.

At low concentrations, the soluble POSS molecules protect transparency even if the refractive indexes of the POSS and the polymer are different, because the POSS molecules, being only 1–2 nm in diameter, are too small to scatter visible light (having wavelengths in the range ~380–750 nm). Above the solubility limit, at the higher POSS concentration, turbidity (haziness) appears. In other cases, POSS acts as the plasticizer. It was observed that effectiveness of a POSS additive was similar to conventional plasticizers, for example, dioctyl phthalate in poly(vinyl chloride) (PVC) [420] and PMMA [389]. However, the POSS hybrids could find application as a nonvolatile plasticizer, more so in specialty applications, since they are more expensive. Moreover, many POSS derivatives have been chemically incorporated into polymer matrices via copolymerization or grafting [421, 422].

The POSS molecules are fully soluble in epoxy monomers and diluents [404]. Dissolved POSS acted as a process aid in poly(phenylene oxide) (PPO). Addition of POSS not only improved the melt flow, but also mechanical properties of a composite. Moreover, the flame retardance was enhanced, and the color of the PPO was significantly reduced [423]. POSS hybrids also improved the melt flow of other high performance polymers: poly(phenylene sulfide) (PPS), polyetherimide (PEI), polyetheretherketone (PEEK), and cyclic olefin copolymer (COC) without worsening mechanical properties or degradation of MW [399, 424].

Single POSS compounds formed optical transparent films due to their high symmetry and crystallinity. It was assumed that the lower symmetries of the POSS derivatives should decrease their crystallinity and provide optical transparent film forming properties. These materials are thermoplastic hybrids and have low-k or refractive index. In order to reduce symmetries, dumbbell-shaped trifluoropropyl-substituted POSS derivatives linked by linear aliphatic chains (ethane, propane, hexane) to reduce their symmetries were prepared. These new compounds formed optical transparent films depending on their aliphatic linkages under low temperature and could be used as coatings on various thermally unstable materials [425].

Most of the POSS molecules have symmetrical structures and therefore do not exhibit surfactant properties. Although partially condensed cages, for example, POSS trisilanols are surface active due to the highly polar silanol groups Si–OH (and the less polar alkyl groups) and are effective dispersants for fillers and pigments – for many different types of fillers (calcium carbonate, silica, magnetite, titanium dioxide, titanium carbide, boron nitride, bismuth subcarbonate, and yttrium oxide) [399, 426, 427].

Due to the high acidity of the POSS trisilanols, they are effective when blended with the basic fillers, including carbonates. The silanol functional groups, the POSS trisilanols, can react with silanol groups on the surface of metal oxides, silica, and the silicate fillers. Thus, the POSS trisilanols containing three reactive groups per molecule can chelate to surfaces forming bonds stable even at elevated temperatures and may be used as dispersants of polymer compositions (during processing). Other POSS hybrids expected to show strong surface activity are the new POSS trisulfonic acids [228]. However, a series of giant macrocyclic surfactants with complex shapes and architectures based on POSS-tethered cyclic polymers were prepared by the sequential "click" chemistry [428, 429].

By addition of small amounts (5 wt%) of heptaisobutyl-substituted cubic oligo-silsesquioxane containing amino-group, the performance of poly(phenylene ether) (PPE) was significantly improved, especially anti-flammability and the melt processability of PPE, while retaining its excellent thermal stability. Excellent FR effect of amino-functionalized POSS additive was also observed in polycarbonates (PC) [243, 430].

Hydrolysis of octakis(dimethylsiloxy)octasilsesquioxane $[(HMe_2SiO)SiO_{1.5}]_8$ ($Q_8M^H{}_8$) and consecutive condensation reactions between Si–H and Si–OH groups to Si–O–Si units were carried out in the presence of Karstedt's catalyst and triethylamine (Et_3N) as the cocatalyst, in toluene at ~60 °C, and gave highly branched, densely cross-linked polysilsesquioxane network (poly-POSS). It showed excellent thermal stability when compared to $Q_8M^H{}_8$. About 5% weight loss was observed at ~500 °C (T_5), and ~10% weight loss at ~1,000 °C (T_{10}), – when poly-POSS was heated in air, while thermal decomposition of $Q_8M^H{}_8$ began at ~200 °C both in air and nitrogen [431].

Novel tricomponent amphiphilic PEG/POSS/PDMS membranes composed of hydrophilic PEG and hydrophobic PDMS segments were cocross-linked and reinforced by $Q_8M^H{}_8$ cubes with Karstedt's catalyst, Et_3N cocatalyst, and water. Diallyl-terminated PEG (A-PEG-A) and divinyl-telechelic PDMS (V-PDMS-V) were used as monomers and pentamethylpentacyclosiloxane ($D^H{}_5$) as a cross-linking agent. The cross-linking of $Q_8M^H{}_8$ achieved by the same catalyst system was studied in separate experiments. In addition to cocross-linking, $Q_8M^H{}_8$ domains also served as a covalently bound reinforcing network filler. Films were prepared by pouring components (A-PEG-A/V-PDMS-V) in molds, and cross-linking with $D^H{}_5$ and $Q_8M^H{}_8$ was carried out by heating at 60 °C for several hours. The films were characterized by extent of cross-linking, contact angle hysteresis, oxygen permeability, TGA, sol fractions, and equilibrium swelling (both in hexane and water), and mechanical properties [431].

The effect of POSS nanoparticle inclusion in cellulose acetate (CA) membranes was studied by C.H. Worthley et al. Membranes were fabricated with the POSS or CA–POSS as an additive at different weight percent loadings, or with no additive. Nanoparticle dispersion was studied by SEM. The mechanical strength of the NC membranes was determined by a dynamic mechanical analysis. The effect of nanoparticle inclusion on the filtration properties of the membranes was also studied. The NC membranes exhibited excellent flux and compaction resistance, and decreased salt rejection and mechanical strength when compared to the CA control [432].

Another crystalline NC membrane was prepared in solution by blending of poly (vinyl alcohol) (PVA) with two amphiphilic POSS bearing eight PEG arms (-CH_2CH_2 $(OCH_2CH_2)_mOCH_3$, $m = \sim 13$) or anionic (tetramethylammonium) side groups (Octa-TMA). The TEM analysis revealed that dispersion of POSS in the PVA matrix was uniform. The mechanical, dynamic mechanical, and dielectric properties of PVA were dependent on the weight fraction of POSS in NCs. The incorporation of POSS in the PVA matrix remarkably increased the Young's modulus of the matrix. Owing to the plasticizing nature of the low MW PEG group, the PEG-POSS decreased the crystallinity of PVA, while Octa-TMA-POSS enhanced the PVA crystallinity [433].

The methacrylate-substituted POSS cubes were prepared through Pt-catalyzed hydrosilylation of propargyl methacrylate with $(HSiO_{1.5})_8$ and $(HMe_2SiOSiO_{1.5})_8$. Propargyl groups reacted in preference to methacrylate groups in Pt-catalyzed hydrosilylation with $(HSiO_{1.5})_8$ and $(HMe_2SiOSiO_{1.5})_8$, while $(HSiO_{1.5})_8$ was much less reactive than $(HMe_2SiOSiO_{1.5})_8$, as it reacted at the higher temperature required (80 °C vs 40 °C). In the presence of the Karstedt's catalyst instead of dicyclopentadiene platinum(II) dichloride, Pt(dcp), both reactions gave cross-linked gels. These new methacrylate/POSS hybrids were clear liquids that can be easily manipulated. They were recommended as thermosets or photocurable monomers [268].

Monomethacrylate-functionalized, fully condensed octasilsesquioxanes (T_8-POSS) were prepared by corner-capping of incompletely condensed the POSS trisilanols, $R_7T_7(OH)_3$ with (trichlorosilylpropyl)methacrylate. Other functionalized POSS R_7Ts (acrylic, -α-olefin, and –silane) were also similarly synthesized. The T_8-POSS hybrids were polymerized into novel linear silsesquioxane-based transparent, brittle materials. They showed high thermostability, decomposing at 388 °C without melting, while depolymerization of PMMA into MMA began above 200 °C and quantitatively gave monomer at 350–450 °C [348].

Octa-arm star poly(methyl methacrylate) (PMMA), with a wide range of MWs (up to ~10^5 g/mol), grafted from cubic octakis(hydrodimethylsiloxy)octasilsesquioxane ($Q_8M^H{}_8$), were synthesized by atom transfer radical polymerization (ATRP) in the presence of CuCl. The prepared PMMA can be considered as "NC consisting of dispersed, nanosized hard particles in a thermoplastic matrix." Moreover, star MMA-2-hydroxyethyl methacrylate (HEMA) block copolymers were prepared by "living" polymerization of HEMA, initiated by star PMMA, "when MMA conversion was more than 90%." Thus, stars with a hydrophobic core and a hydrophilic shell were obtained [275].

Camelina oil (CO) was modified through epoxidation and subsequent methacrylation reactions with methacrylic acid, followed by copolymerization reaction with hydrophilic poly(ethylene glycol) dimethacrylates as macromonomers (PEG MWs: 300, 550, and 750 g/mol) under visible light. The chain length of methacrylate functionalized PEG directly affected crucial properties of oil-based networks. Their curing performance was studied by a gel fraction analysis. The hydrophilicity of these copolymers, measured as water uptake capacity, was strongly dependent on the internal arrangement of components. Next, they were blended with a mixture of 5 wt% of POSS comonomers containing one (POSS-1) or eight methacrylic groups (POSS-8) and sonicated for 30 min using an ultrasound bath. The formulations obtained were photocured, giving new hybrid NCs, reinforced with POSS hybrids. Well-defined oil-derived NCs from a hard material (with a short chain PEG) to a soft and flexible one (with long chain PEG) with controllable compressive strength were prepared Their mechanical and thermal properties were studied in compression tests, by DMA, TGA, and SEM. These NCs may be used in a variety of industrial areas [434].

Ladder-like poly(epoxysiloxanes) were cured with 1,3-bis)aminopropyl)tetramethyldisiloxane and in the absence of this curing agent [435].

Octaglycidyl and octaethylcyclohexenyl epoxide silsesquioxanes ("with a 0.53 nm body-diagonal and a spherical radius of ~1.2 nm including the dimethylsiloxy units and a core structure similar to that of crystobalite") were blended with DGEBA ($M_w = 340$, $n = 1–2$) and cured with diaminodiphenyl methane (DDM) at 140–150 °C. Tensile properties of these composites were comparable to cured epoxy resins [436].

Two kinds of newer NCs were obtained by curing octaglycidyldimethylsiloxyoctasilsesquioxane (glycidyl-$Me_2SiOSiO_{1.5})_8$ (OG) and octa(ethylcyclohexylepoxide)-dimethylsiloxy silsesquioxane (OC, 1.3 nm diameter) with DDM. Dissolution of cubes with HF followed by GPC analysis showed that only linear tethers were formed in OC/DDM. TEM studies revealed that these NCs were homogeneous at the nanometer scale. Thermomechanical properties of NCs were modified by changing the tether architecture and rigidity. The OC/DDM resin exhibited excellent elastic moduli and thermostability, but a poor fracture toughness. It was found that OC/DDM elastic moduli increased from 2.2 to 3.3 GPa as the DDM content increased two-fold above the maximum cross-link density into a high defect density region, while the fracture toughness remained unchanged. OC/DDM, as a "single phase" hybrid NC, was used as a matrix for ~ 100 nm diameter core–shell rubber reinforcing particles (CSR). Characterization of OC/DDM/CSR NCs showed that fracture toughness improved significantly on inclusion of CSR particles, with little effect on the elastic moduli and thermostability. Stress and strain at failure also improved, indicating better fracture toughness. SEM studies revealed that shear yielding and CSR pull-out were the likely sources of toughening [279, 281].

The octa(dimethylsiloxypropylglycidylether)octasiloxane (OG), prepared from octa(hydrodimethylsiloxy)octasilsesquioxane $[HMe_2SiOSiO_{1.5}]_8$ which was fully epoxidized in reaction with allyl glycidyl ether (Figure 2.12). The relationship between the dielectric properties (dielectric constant, ε', and loss factor, ε''; activation energy, E_a) and the ratio of epoxy resin (OG) to hardener of the epoxy resin thermosetting polymers was studied by N.L.D. Filho et al. [342].

The amplitude of the ε'' peak decreased with increasing OG content until about 73 wt% and slightly increased at higher OG loading. The temperature of the position of the ε'' peak increased with increase in OG content and reached maximum values for compositions in the range of 67–73 wt%, and then it decreased sharply at higher OG content. Further increase in concentration of OG up to 83 wt% reduced E_a. The changes in tensile modulus and fracture toughness mechanical properties as a function of OG content were similar [342].

Poly(silsesquioxanes) prepared by the sol–gel process of methyl(triethoxy)silane (MTES), 3-glycidoxypropyl(trimethoxy)silane (GPTMS), and TEOS with phytic acid as the catalyst were applied as anticorrosion coatings and showed a very good performance [437].

A novel mercaptopropyl-functionalized POSS (MPOSS)] was prepared *via* the hydrolytic condensation of (3-mercaptopropyl)triethoxysilane in ethanol solution catalyzed by he concentrated hydrochloric acid. It was applied for modification of epoxy resins based on DGEBA. The miscibility between epoxide and POSS occurred at a relatively high POSS loading. The impact resistance of DGEBA/POSS NCs showed distinct improvement. TGA–DTA analysis indicated that the MPOSS/DGEBA epoxy hybrids exhibited a lower thermal stability at a lower temperature but higher thermostability and higher efficiency in char formation at an elevated temperature. The T_g of the MPOSS/epoxy hybrids were lower when compared to the neat epoxy [438].

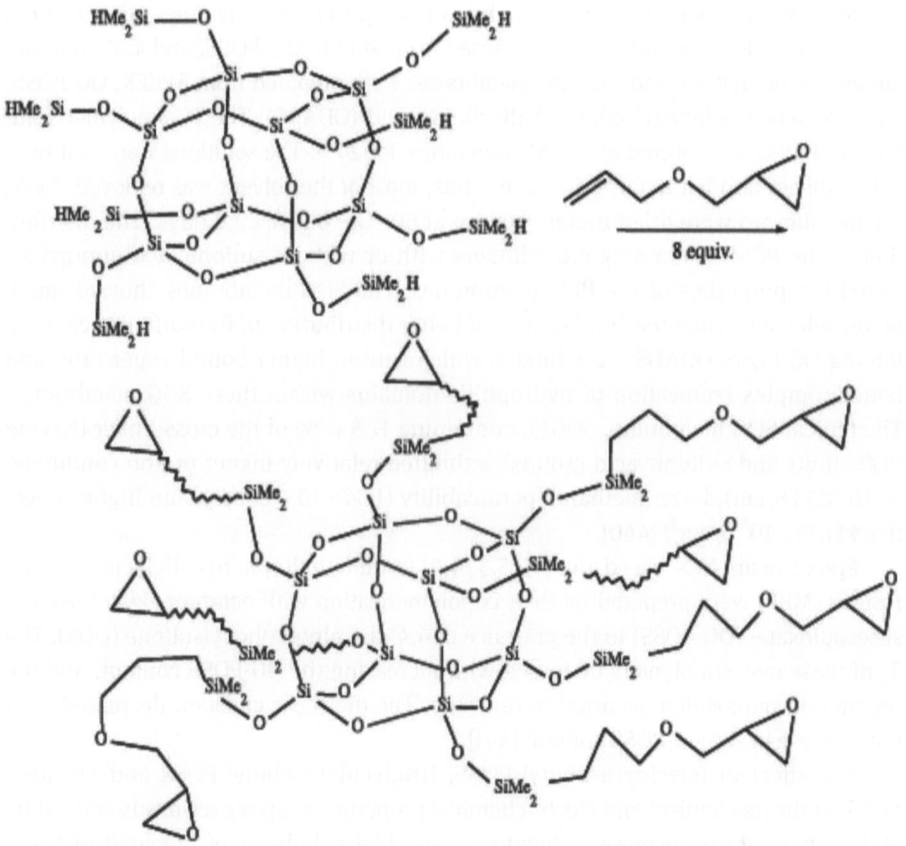

Figure 2.12: Reaction scheme of preparation of the octa[dimethylsiloxypropylglycidylether] octasiloxane (OG) [342]. Copyright Sociedade Brasileira de Química.

Epoxy-POSS NCs were prepared from epoxide functionalized octakis(glycidylsiloxy)octasilsesquioxane (OG) and DGEBA by photopolymerization. DSC and dynamic mechanical analysis (DMA) revealed that the NCs had higher T_gs than the neat epoxy resin. Moreover, DMA indicated that all of the NCs exhibited increased storage moduli in the rubbery state, due to both the nanoreinforcement effect of the POSS cages and the additional degree of cross-linking that resulted from the reactions between the epoxy and OG units. TGA revealed that the thermostability of the NCs was better than the pristine epoxy resin [439].

Octakis(dimethylsilyloxypropylglycidyl ether)octasilsesquioxane (OG-POSS) macromer was prepared as a colorless viscous liquid by the addition reaction of allyl glycidyl ether (AGE) with $Q_8M^H{}_8$, catalyzed by the Karstedt's Pt catalyst, which was carried out in dry toluene at 80 °C under an argon atmosphere within 8 h, followed by work-up. Two kinds of new cross-linked proton exchange membranes (PEM) were prepared by incorporation of POSS hybrids into sulfonated poly(ether ether ketone) (SPEEK). The OG membranes were obtained from SPEEK, OG-POSS, and 4,4′-diaminodiphenyl ether (ODA), and the SOG membranes were prepared from SPEEK, OG-POSS, and 4,4′-diaminodiphenyl ether-2,2′-disulfonic acid (ODADS). The 15% polymer solutions in DMSO were stirred at room temperature for 24 h. The solutions were cast onto Teflon dishes and left at 60 °C for 48 h. Thus, most of the solvent was removed. Next, the membranes were dried under vacuum at 60 °C/~ 0.2 Tr for 2 days. The distribution of the POSS containing cross-linkers with or without sulfonic acid groups affected the properties of the PEM membranes. The SOG membranes showed much better selectivity than the OG, because of better distribution of the sulfonic acid containing OG-POSS/ODADS cross-linkers which caused higher bound water ratio and better complex connection of hydrophilic domains within these SOG membranes. The typical SOG membrane, SOG15, containing 17.5 wt% of the cross-linker (having POSS units and sulfonic acid groups), exhibited relatively higher proton conductivity (0.0153 S/cm), lower methanol permeability (1.34×10^{-7} cm^2/s), and higher selectivity (1.14×10^5 S/cm^3) [440].

Epoxy resin NCs based on 3,3′,5,5′-4,4′-tetramethylbiphenyl diglycidyl epoxy resins (TMBP) were prepared by their copolymerization with octapropylglycidylether silsesquioxane (OG-POSS) in the presence of 4,4′-diaminodiphenylsulfone (DDS). The T_g of these new copolymers increased with increasing the OG-POSS content, and the thermal decomposition occurred in one step. The dielectric constant decreased with the increase in the OG-POSS content [441].

The effect of (glycidyl)isobutyl-POSS, (triglycidyl)isobutyl-POSS and glycidyl-POSS on the mechanical and electrochemical properties of epoxy resin was studied by M. Longhi et al. An increase in roughness and hydrophobicity of prepared NCs was observed. The sample containing glycidylisobutyl-POSS showed the best corrosion resistance. Thus, modified epoxy resin was used as coating on low alloy steel [442].

Copolysilsesquioxane oligomers prepared from 3-aminopropyl(triethoxy)silane (APTES) and phenyl(triethoxy)silane (PTES) were copolymerized with the epoxy resin (DGEBA) and filled with multiwalled carbon nanotubes (0.25 wt%). Hybrid NCs showed improved thermal stability and onset degradation temperatures higher than 340 °C under N_2. Young's modulus and TS values up to 2.9 GPa and 47 MPa were determined, respectively, and were slightly higher than for the neat epoxy resin [443].

Methylacryloylpropyl POSS (MAP-POSS) was thermally polymerized with bisphenol-A epoxy resin using diphenyliodonium fluoride-borate as a cationic initiator and benzoyl peroxide (BPO) as co-initiator. The T_g of MAP-POSS-epoxy NC containing 8 wt% MAP-POSS was 175.6 °C and was 6.1 °C higher than the pure epoxy resin [444].

A novel octakis-2{[3-(trimethoxysilyl)propyl]thio}ethyl(octasilsesqioxane) was prepared by a radical thiol-ene addition of 3-mercaptopropyltrimethoxysilane (MPTMS) to octavinyloctasilsesquioxane (POSS-Vi8), in the presence of 2,2′-azobisisobutyronitryl under mild conditions (60 °C, toluene solution). The partial addition of MPTMS to the vinyl monomer POSS-Vi8 afforded POSS derivatives with both types of functional groups: vinyl and 2-[3-(trimethoxysilyl)propylthio]ethyl, pendant to the POSS core. The chemical structures of the thiol-ene addition products were confirmed by the ^{29}Si-NMR and mass spectroscopy [445].

A thiol-ene click reaction was also used for a preparation of organic–silica hybrid monolith (with a high porosity) from oligomeric polyhedral methacryl-substituted silsesquioxane (POSS-MA) and 1,4-bis(mercaptoacetoxy)butane (BMAB), with addition of toluene and dodecanol as porogenic additives. The POSS–BMAB hybrid monolith exhibited good permeability, mechanical properties and high thermostability. The prepared POSS–BMAB hybrid monolith exhibited typical reversed-phase chromatographic behavior with respect to a series of test compounds (phenols, anilines, thiourea, benzene alkylbenzenes, and polycyclic aromatic hydrocarbon) and can be used in a capillary liquid chromatography [446].

Novel hydroxyl functionalized cubic silsesquioxanes such as octakis-2[(6-hydroxyhexyl)thio]ethyloctasilsesquioxane (POSS-S-OH) and heptaisobutyl-2[(6-hydroxyhexyl)thio]ethyloctasilsesquioxane (iBu-POSS-S-OH) were prepared by K. Rózga-Wijas et al. POSS structures, having 2[(6-hydroxyhexyl)thio]ethyl and the vinyl groups, pendant from the octahedral cage were also described. They were synthesized through thiol-ene click reaction of 6-mercapto-1-hexanol (MCH) to octavinyloctasilsesquioxane (POSS-Vi), and heptaisobutyl(vinyl)octasilsesquioxane (iBu-POSS-Vi), in the presence of 2,2′-AIBN. These functionalized POSS cages served as initiators for ring-opening polymerization (ROP) of L,L-dilactide, catalyzed by tin(II) 2-ethyl-hexanoate. The polymerization gave biodegradable hybrid star shape and linear systems with the octasilsesquioxane cage as the core and polylactide (PLA) arms [447].

A novel organic-mineral hybrid material containing sulfur (S-Vi-POSS) was prepared from sulfur with octavinyl-substituted octasilsesquioxane by using intensive homogenization techniques or developing a method of solution reaction. The S-Vi-POSS may be a very useful material for modification of properties of rubber vulcanizates [448].

Two kinds of POSS hybrids containing mono and polymeric double-decker silsesquioxane units (mono-DDSQ and poly-DDSQ) and two or many epoxy groups, respectively, were prepared by the hydrosilylation of two different allyl-substituted epoxy derivatives with Si–H difunctional DDSQ monomer by J. Cao et al. [449]. The curing kinetics, thermal, mechanical, and surface properties of epoxy resin containing different content of DDSQ moieties were studied. Both mono and poly DDSQ significantly improved thermal and mechanical properties of the epoxy thermosets. The mono DDSQ led to more flexible structure and toughened the epoxy resin better, while the branched polymeric DDSQ provided better thermostability. Moreover, the DDSQ changed surface properties of the epoxy resin from hydrophilic to hydrophobic. The addition of DDSQ had little effect on the initial decomposition temperature (~320 °C). The char residue at 750 °C increased with increasing DDSQ content. The decomposition rate decreased with increase in the DDSQ ratio [449].

Nanocomposites (NCs) of polystyrene (PS) containing five polyhedral oligomeric silsesquioxanes (POSS) of various structures (with different inorganic cage, functionalized with various organic pendant groups) were obtained by melt processing. The POSS monomers included octa-isobutyl and isooctyl derivatives and incompletely condensed silsesquioxane triols with *i*-Bu, *i*-Oct, and Ph substituents. The structure-properties relationships of the PS-POSS NCs were compared to the properties of neat PS. The rheological studies showed that the POSS molecules acted as a plasticizer and the processing aid. Theoretical calculations revealed small differences between the solubility parameters of POSS and the PS matrix suggesting better compatibility of components and no aggregation tendency. The addition of 5 wt% POSS to PS caused a decrease in the rigidity, T_g, and damping factor of the NC systems. The stronger effects on properties of NCs were observed in the case of POSS triols with open cage than in the POSS molecules with closed cage. The melt properties of NCs were significantly affected by the structure of POSS, by the type of the pendant organic groups, and by the interaction between the POSS organic groups and the host PS matrix, while the solid state properties were mostly dependent on the kind of POSS cages [450].

Liquid POSS derivatives: octakis[dimethyl(phenethyl)siloxy]silsesquioxane (OS-POSS) and octakis[dimethyl(4-acetoxy-phenethyl)siloxy]silsesquioxane (OA-POSS), were synthesized via the hydrosilylation of styrene and 4-acetoxystyrene, respectively, with octakis(dimethylsiloxy)silsesquioxane ($Q_8M^H{}_8$). Octakis[dimethyl(4-hydroxyphenethyl)siloxy]silsesquioxane (OP-POSS) was prepared from OA-POSS and hydrazine monohydrate. These octakis-functionalized POSS hybrids behaved as amorphous glasses and had low T_gs, which were dependent on the intermolecular interactions of their outer organic groups. NCs of OS-POSS, OA-POSS, and OP-POSS with PS, poly(4-acetoxystyrene), and poly(4-vinylpyridine), respectively, were obtained by the blending method. The effects of the resultant intermolecular aromatic hydrophobic, dipole–dipole, and hydrogen-bonding interactions on their properties were studied. The best dispersion formed OA-POSS in poly(4-acetoxystyrene) [451].

OA-POSS, OS-POSS, and OP-POSS were blended with the poly(ethylene oxide) (PEO). The thermal properties of POSS-PEO blends were analyzed by TGA, DSC, and dynamic mechanical analysis (DMA) methods. The 20 wt% loss (T_d) during heating rate: 20 °C/min was high: 442.8, 461.7, and 551.0 °C – for neat OA-POSS, OS-POSS, and OP-POSS, respectively, and was gradually increased with increasing loading of POSS hybrids. For 10% content of POSS derivatives, T_d decreased in the order: 378.8, 316.3, and 372.6 °C, respectively, for the above silsesquioxanes, while for pure PEO, it was only 308.1 °C. The OP-POSS/PEO system showed the highest thermal stability and lowest crystallization rate due to hydrogen bonding interactions (between its hydroxyl and ether units) that were stronger than between carbonyl and methylene groups in the OA-POSS/PEO NC [452].

Newer NCs of PS with POSS hybrids were prepared by in situ polymerization of styrene, in the presence of POSS fillers of the structure $RR'_7(SiO_{1.5})_8$ (R = 4-methoxyphenyl or 2,4-difluorophenyl and R' = cyclopentyl). Their thermostability was dependent on the effects of the electron-donor or electron-withdrawing substituents of the phenyl group [453].

A hydroxypropyl POSS hybrid, functionalized with eight hydroxyl groups (POSS-OH), was prepared by the hydrosilylation reaction of allyl alcohol with octakis [(hydro)dimethylsiloxy]octasilsesquioxane ($Q_8M^H{}_8$) in the presence of Karstedt complex. It was used for the preparation of polyimide-POSS NCs. POSS-OH was cured with polyamic acid. Next, the imidization step was carried out at 40–300 °C within 10 h. The formation of homogeneous dispersion of POSS nanoparticles was documented by SEM, AFM, and TEM. The incorporation of POSS-OH into PI matrix resulted in low dielectric constant (low-k) of PI-POSS NC, without worsening mechanical properties. The modulus of NC increased with increase in the weight fraction of POSS-OH in PI [454].

POSS-siloxane-modified polyimide (PI) hybrid NCs were obtained by the reaction of siloxane-modified polyamic acid (PAA) with octa(aminophenyl)silsesquioxane (OAPS). First, PAA was synthesized by the reaction of 4,4-diaminodiphenylsulfone with pyromellitic dianhydride in *N*-methylpyrrolidone (NMP), and was then reacted with bis(3-aminopropyl) polydimethylsiloxane. The siloxane-modified PI was mixed with different amounts of OAPS in NMP solution. The incorporation of siloxane units into PI-POSS NCs led to heterogeneous morphology. Data from thermal studies showed that the incorporation of POSS appreciably increased the T_g, the thermostability, and the char yield of hybrid NCs as compared to the neat PI [455].

NC films were prepared by blending of polyimide (PI), composed of oxydianiline (ODA) and pyromellitic dianhydryde (PMDA) units, with incompletely condensed trisilanolphenyl (TSP)-POSS. The formation of POSS nanoaggregates in the bulk and on the surface of NCs was observed. Their size and distribution was dependent on POSS content and affected volume electrical resistivity of NCs. Radial cracks and a brittle fracture were formed, as compared to ductile fracture in the case of PI [456].

Polyamide NCs containing 0.5, 2, or 4 wt% of octakis[3-glycidoxypropyl(dimethylsiloxy)]octasilsesquioxane (POSS, in a form of viscous oil) were obtained by melt blending. SEM results showed that POSS molecules were uniformly dispersed in a polyamide 6 matrix. FTIR and DSC analyses indicated that the addition of POSS promoted modifications of the polyamide crystal structure, and significantly increased the T_g (from 28 to 53 °C), crystallization temperature (from 163 to 189 °C), and the crystallization rate of the PA6/POSS composites. TS and elongation at break were distinctly improved by the addition of POSS. The impact strength of PA6/POSS composites was also increased, as compared with the neat polyamide 6. DMTA tests showed that both the storage modulus (G') and relaxation temperatures, and especially, T_g of the PA6/POSS composites were significantly enhanced [457]. Grafting cyclopentadienyl on silsesquioxanes gave the $(C_5H_5SiO_{3/2})_n$ structure and totally changed its electronic structure and its chemistry; cyclopentadienyl became a reactive site where a transition metal atom (e.g., Sc) was doped and thus served as an effective adsorption site for hydrogen molecules. This nanocomplex showed the following advantages: (1) the storage capacity in the fully grafted case was 5 wt% where hydrogen was bound molecularly with a binding energy of about 0.6 eV per H_2 molecule; (2) the POSS structure itself was stable. Based on theoretical considerations it was concluded that the functionalized POSS complexes can be an effective perspective material for hydrogen storage [458].

Fluorinated POSS (F-POSS) compounds were prepared (by S.T. Iacono et al.) and tested for potential applications such as wetting agents of materials in automotive and aerospace industries. It was found that fluorodecyl POSS showed higher water contact angles than those of PTFE. It was predicted that F-POSS may have the lowest surface energy among known solids. The combination of low surface energy with surface roughness and curvature, similar to that of the Lotus leaf, can create new superoleophobic materials [459].

Properties of cycloolefin copolymer (COC)-(allyl)hepta-iso-butyl octasilsesquioxane (POSS) NCs, prepared by melt compounding were studied by A. Dorigato et al. Composites retained their optical transparency of the matrix up to a nanofiller content of 5 wt%. At higher POSS contents, a crystalline phase was formed, due to the presence of agglomerated nanoparticles. A significant decrease in the T_g, thermal decomposition rate, elastic modulus, loss and storage shear modulus, ultimate elongation, and viscosity were observed with the increasing POSS content. The formation of a soft interphase at the nanoparticles/polymer boundary was proposed. Isobutyl groups on POSS nanoparticles limited the stress transfer process and acted as a molecular lubricating agent in the COC matrix [460].

A novel polypropylene/silsesquioxane (PP/POSS) composite nonwoven fabric with permanent electret was prepared through the melt-blown technique with corona charging. Field-emission scanning electron microscope (FESEM) and wide-angle X-ray diffraction (WAXRD) were used to study the distribution of octavinyl POSS nanoparticles on the surface of the fibers and the morphology of the composite fibers, respectively. POSS acted as a nucleating agent and accelerated the crystallization process during the nonisothermal cooling stage. Improved TS and elongation at break of the PP/POSS melt-blown nonwoven fabrics were observed when compared with PP nonwoven fabric. The utmost stable charge density of the PP/POSS melt-blown nonwoven fabric improved by 78.4% in comparison to the neat PP nonwoven fabric, while the maximum collection efficiency of monodisperse polystyrene aerosol (PSL) (particle size: 0.3 μm) reached 97.4% for PP/POSS composite melt-blown nonwoven fabric, which was higher by 9% as compared to the neat PP melt-blown nonwoven fabric [461].

The NCs of poly(oxymethylene) (POM) containing 0.5 and 1 wt% of octakis(dimethylsiloxy, ethyl epoxycyclohexyl) octasilsesquioxanes (POSS) were obtained by melt-blending. It was found that the addition of POSS led to an increase of both the TS and stiffness, especially for the POM/POSS NCs with 0.5 wt% POSS. The obtained polymer composites POM/POSS are promising construction materials [462].

The PVC was modified with 0.5 and 5 wt% of 3-chloropropyl-substituted octasilsesquioxane (ClPr-POSS). POSS was incorporated into PVC by direct mixing of solid powders or by addition of POSS to a solution of PVC in THF. Finally, the samples were made by melt mixing in the Brabender mixer. It was found (by DSC and DMA methods) that ClPr-POSS acted as a plasticizer, leading to a decrease of the T_g of poly(vinyl chloride) [463].

POSS molecules also exhibit interesting properties for space technology applications, mainly as materials that are resistant to atomic oxygen in low earth orbit and for use solar power systems in space [464].

Thermostable NCs, prepared from cyanate ester resins (CER, based on bisphenol E), were densely cross-linked and modified with different amounts (0.01–10 wt%) of epoxycyclohexyl-functionalized silsesquioxane (ECH-POSS). These NCs were characterized by many comprehensive techniques (FTIR, far-IR, TEM, DSC, DMA, TGA, SAXS, EDXS, and creep rate spectroscopy). Addition of ultra-low POSS amounts (≪1 wt%), covalently bound into CER network, caused substantial change to their nanostructure and to their properties – presumably due to molecular POSS dispersion and their spatial distribution in the matrix. The T_gs were increased by 20–50 °C, high-temperature elasticity and creep resistance properties were enhanced, and the thermostability under inert atmosphere was increased at $T < 400$ °C. At ultra-low POSS contents, not only did the molecular POSS dispersion and their quasi-periodic spatial distribution in the matrix took place in the chemically modified CER network, but possible enhanced long-range "constrained dynamics" effect was also observed. At higher contents of ECH-POSS (especially from 2 to 10 wt%), no effects or negative effects were observed due to the formation of the nanorod- or platelet-like structures and POSS-enriched nanodomains [465].

An octaphenyl POSS (Ph-POSS) was cross-linked in solution (via Friedel–Crafts reaction) giving a macroscopic polymer network that was carbonized at 900 °C, and monodispersed silica domains were subsequently etched with HF. Thus, a well-defined microporous carbon nanosphere (MCNS) material was obtained, which showed very large surface area (e.g., 2264 m^2/g) with excellent adsorption and super-capacitance properties [466].

The functionalized POSS-containing hybrid polymeric materials were prepared by different methods: polycondensation, polyaddition, free-radical polymerization, both conventional and controlled (ATRP), and ring-opening metathesis (including *click chemistry* in some cases). Linear copolymers with silsesquioxane units in the main or side chains, and star copolymers were obtained. The incorporation of POSS into the copolymer structure resulted in improved mechanical properties, increased glass transition and decomposition temperatures, and better resistance to oxidation, in comparison to the conventional polymeric systems [295].

The intumescent flame retardant (FR) PLA-POSS composites (IFR-PLA/POSS) were obtained via a melt-blending. The incorporation of POSS increased the mechanical properties of the PLA matrix. The limiting oxygen index (LOI) values increased significantly from 20.0% for neat PLA to 40.0% for PLA with 22.5 wt% phosphorus containing IFR and 7.5 wt% POSS; the latter passed the UL-94 V0 test. The thermostability of the residual char at the later stages of the thermal degradation improved significantly. The thermally stable poly(phosphoric acid) and polysiloxane derivatives were formed during the thermal degradation process of the PLA/POSS composite. The residual char of the FR PLA/POSS composite was composed of cross-linked phosphorous oxynitride and carbonized aromatic networks [467].

Blends of biodegradable (stiff and brittle at room temperature) PLA with POSS hybrids grafted with longer (P-1) and shorter (P-2) chains of poly(ethylene glycol) were prepared. PLA/P-1 blends were homogeneous despite P-1 having a relatively high M_w of 9,500 g/mol. The blend containing 20 wt% P-1 was a transparent and flexible elastomer-like material with T_g at 16 °C, exhibiting excellent drawability. It retained homogeneity, good drawability as well as flexibility within 6 months of aging at room temperature: a 2% modulus of elasticity below 100 MPa and a large elongation at break (800%). PLA/P-2 blends were only partially miscible. However, the blend with 15 wt% P-2 was transparent and ductile, with T_g at 49 °C, a relatively high yield strength of 29 MPa, and a large elongation at break (230%). These blends showed local plasticization effect [468].

Properties of poly(ε-caprolactone)/clay NCs containing 1, 5, or 10 wt% montmorillonite (MMT) were modified by hydroxyl functionalized, quaternized POSS surfactants. The octa(3-chloropropyl) POSS was quaternized with 2-dimethylaminoethanol. Modification of clay with quaternized POSS surfactants (in POSS:MMT weight ratio = 1:2) gave stable clay separation for its 3-D structure and facilitated the miscibility of polymer with clay in the NCs due to the star architecture. An improvement in the thermostability of PCL was observed even at 1 wt% clay content. A 25% mass loss of all NCs was observed at 373–381 °C [469].

Biodegradable hybrid organic–inorganic copolymers of poly(ε-caprolactone) (PCL) with 5.3–21.3 wt% POSS were prepared via ROP of ε-caprolactone. Chemical structures of the polymers were characterized by ^{1}H-NMR, FTIR, and GPC. XRD analysis showed that both POSS and PCL segment in POSS/PCL hybrids crystallized and formed two well-separated crystalline phases (except for a low content of POSS, 5.3 wt%). The POSS segment suppressed the partial crystallization of the PCL segment. POSS/PCL hybrids with the highest POSS content (21.3 wt%) formed "snowflake" shape crystals, while at low POSS content, classic spherulites were observed [470].

A novel hybrid silicon/graphite-based nanomaterial was prepared by grafting (aminopropyl)silsesquioxane (POSS-NH_2) on graphene oxide sheets (GOs). The combination of the amino functionalized POSS molecules and GO sheets provided a new hybrid silicon/graphite-based nanomaterial (GRA-POSS). The electrical conductivity of the reduced GO was restored in GRA-POSS, which opened practical application possibilities as polymer NCs. The surface grafting of the GO sheets with POSS-NH_2 was confirmed by FTIR and XPS. Its morphology was studied by field emission electron microscopy and AFM. The GRA-POSS hybrid, exposed to visible light, induced a photoconductivity response [471, 472].

Graphene nanosheets functionalized with POSS hybrids through amide groups (POSS-graphene, Figure 2.13) were prepared from a reaction of amine-substituted POSS with functional groups (e.g., epoxy and carboxyl groups) in GO. The POSS-graphene hybrids were highly soluble in organic solvents, thus being attractive for multifunctional applications. Thin films from a solution casting of the POSS-graphene showed superhydrophobic properties with a water/air contact angle of ~157°. The superhydrophobic POSS-graphene powder can be used for fabrication of a liquid marble. The POSS-graphene hybrids were also used as novel nanofillers to increase the T_g and thermal stability of polymers [473].

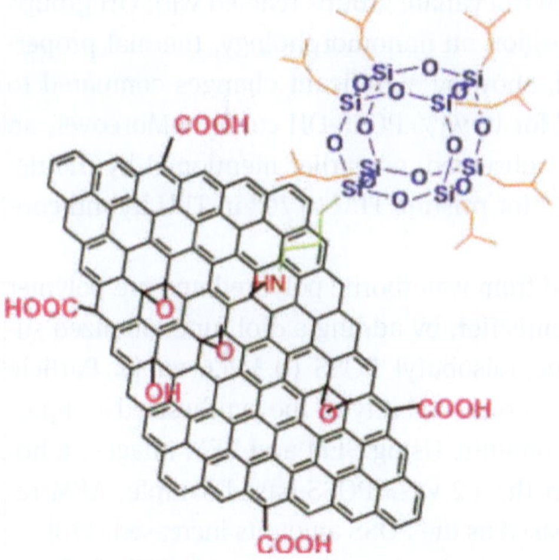

Figure 2.13: An example of one possible chemical structures of POSS-graphene hybrid. Reprinted (adapted) with permission from [473]. Copyright 2012, American Chemical Society.

The physical and chemical functionalization of carbon nanomaterials (carbon nanotubes, fullerenes, graphene, and graphene oxide) with POSS derivatives provided novel hybrid nanostructure materials, which could find numerous technological applications [474].

Single-walled carbon nanotubes (SWNTs), modified with POSS molecules through amide linkages, were prepared by reacting the acid-treated SWNTs with amine-functionalized POSS. The SWNTs-POSS hybrids were characterized by FTIR, XPS, and UV-vis spectra. They were dispersed in water or in organic solutions. The SWNTs-POSS exhibited extremely low toxicity in cytotoxicity tests and find biomedical applications [475].

A series of thermoplastic polyurethanes (POSS TPUs), composed of alternating multiblock structures formed by nanostructured hard segments of POSS diol and biodegradable soft segments of a PLA/caprolactone copolymer, P(DLLA-*co*-CL), and containing segments of PEG or other polyols were prepared by Q. Guo et al. They differed in hydrophilicity, degradability, MW, and T_g, while the hard-to-soft segment feed ratio (mol%) was constant. POSS-aggregates formed crystals and served as physical cross-links on the nanometer scale, while the soft segments enabled the controlled drug release from the POSS TPU stent coatings in phosphate-buffered saline (PBS) solution. Thus, 90% of the drug was released from a biodegradable stent coating between half-a-day and about 90 days [476].

Hybrids of TPU with dihydroxyl-functionalized POSS were fabricated by melt-reactive blending of both ingredients. (*Trans*-cyclohexanediol)isobutyl POSS (POSS-OH) (2, 5, and 10 wt%) was used as a chain modifier. A PU chain-scission-equilibrium reaction took place, leading to the formation of highly reactive isocyanate and hydroxyl chain ends. In the molten state, the isocyanate groups reacted with OH groups of POSS-OH. The effect of POSS concentration on nanomorphology, thermal properties, and surface properties was studied, showing significant changes compared to pristine TPU. The T_g increased by ~10 °C for 10 wt% POSS-OH content. Moreover, an increase in surface water wettability was observed, as earlier mentioned by the decrease of the water contact angle from 95° for pristine TPU to 70° in TPU hybrid containing 10 wt% of the reactive POSS [477].

Different kind of PU NCs was prepared from waterborne polyurethane pre-polymer in the presence of the anionic internal emulsifier, by adding a diol functionalized silsesquioxane: (1,2-propanediol)propoxy heptaisobutyl POSS (0.3–4.6 wt%). Particle size, viscosity, TS, modulus, T_g, and the thermal stability of the synthesized samples were improved by increasing the POSS contents. Using SEM and TEM images, a homogeneous morphology was observed in the 1.2 wt% POSS-based sample. AFM results showed that surface roughness increased as the POSS amounts increased [478].

The thermal stability and flammability of rigid polyurethane foams (RPUF) chemically modified with (1,2-propanediol)isobutyl POSS (PHI-POSS) as a pendant group and octa(3-hydroxy-3-methylbutyldimethylsiloxy) POSS (octa-POSS) as a cross-linking agent was studied. The RPUF were prepared in a three-step process from a sorbitol-based polyether polyol, polymeric 4,4′-diphenylmethane diisocyanate, and dimethyl propane phosphonate as an FR. The addition of the POSS modifier affected the PU cellular structure and resulted in an improvement of their thermal properties. Based on TGA data and flammability results, it was suggested that a char was formed at the surface and this layer acted as an insulating barrier, limiting heat and mass transfer with FR, thus leading to decreased heat release rate, especially for systems containing OCTA-POSS [479].

Multifunctional octasilsesquioxanes containing different reactive groups [EtO-Si, Si–H and (2-hydroxyethoxyethyl)dimethylsiloxy] were used in the fabrication of high-resolution secondary nanoimprint lithography (NIL) molds, which showed tunable chemical and mechanical properties when used as replication materials. They exhibited Young's moduli exceeding 4.5 GPa and excellent thermostability for high-temperature imprinting. Their low surface energies facilitated the separation of the mold from the imprint and were used for preparation of high-resolution replicas of SiO_2

imprint masters. These replicas were directly used to form high-quality imprints through both UV and thermal NIL, with a patterning resolution of a few nanometers [480].

Organosilicate hard films were obtained through acid hydrolysis, spin casting, and vitrification of a series of cubic silsequioxane monomers having different organic tethers, which were functionalized in their corners with triethoxysilane groups:

(a) octa(triethoxysilylethyl)(octadimethylsiloxy)octasilsesquioxane;
(b) [tetra(triethoxysilylethyl)(dimethylsiloxy)tetradimethylsiloxy] octasilsesquioxane;
(c) [tetra(triethoxysilylethyl)dimethylsiloxy][tetra(2-hydroxyethoxyethyl)dimethylsiloxy]octasilsesquioxane.

The degree of cross-linking and the cubic silsequioxane content of the vitrified films were dependent on the length of the organic ligands and the number of triethoxysilane groups. Thus, a series of high-quality optical-grade coatings with excellent properties was prepared. Improved mechanical properties, lower thermal expansion, higher density, lower porosity, and very hydrophilic surface properties resulted from higher concentrations of the cubic silsequioxane units in the network and higher cross-linking densities [481].

A double-decker-shaped silsesquioxane dianhydride (DDSQDA) (**3**) was prepared by the hydrosilylation reaction of the double-decker-shaped silsesquioxane (DDSQ) (**1**) with *cis*-5-norbornene-*endo*-2,3-dicarboxylic anhydride (**2**) [482].

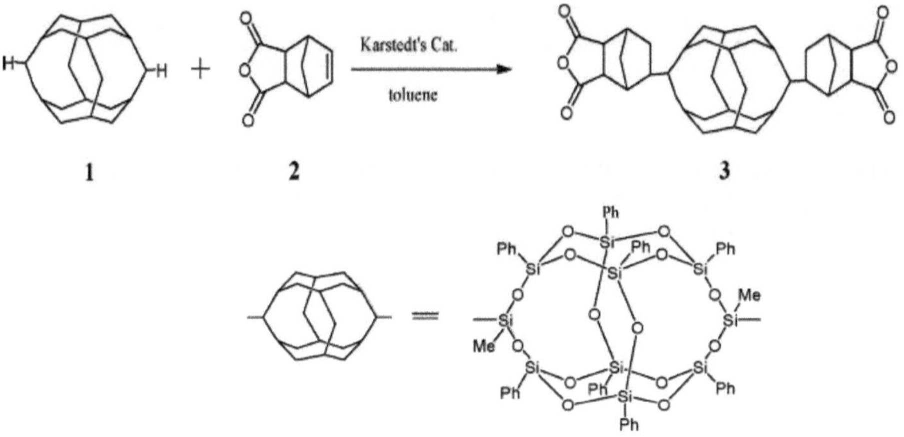

Scheme 2.14: Synthesis of a double-decker-shaped silsesquioxane dianhydride (DDSQDA) (**3**). Reprinted (adapted) with permission from [482]. Copyright 2007, American Chemical Society.

DDSQDA (**3**) was reacted with 4,4′-oxydianiline (ODA), giving the DDSQ-diamine (Scheme 2.15). Next, a series of semiaromatic polyimides containing POSS structures

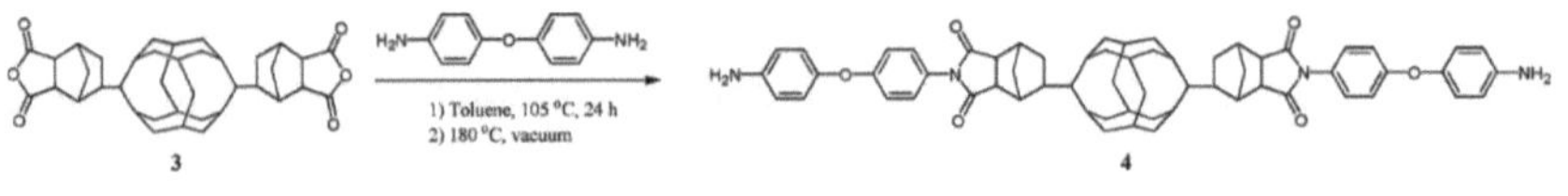

Scheme 2.15: Synthesis of the DDSQ-diamine (**4**) in reaction of DDSQDA (**3**) with 4,4′-oxydianiline (ODA). Reprinted (adapted) with permission from [482]. Copyright 2007, American Chemical Society.

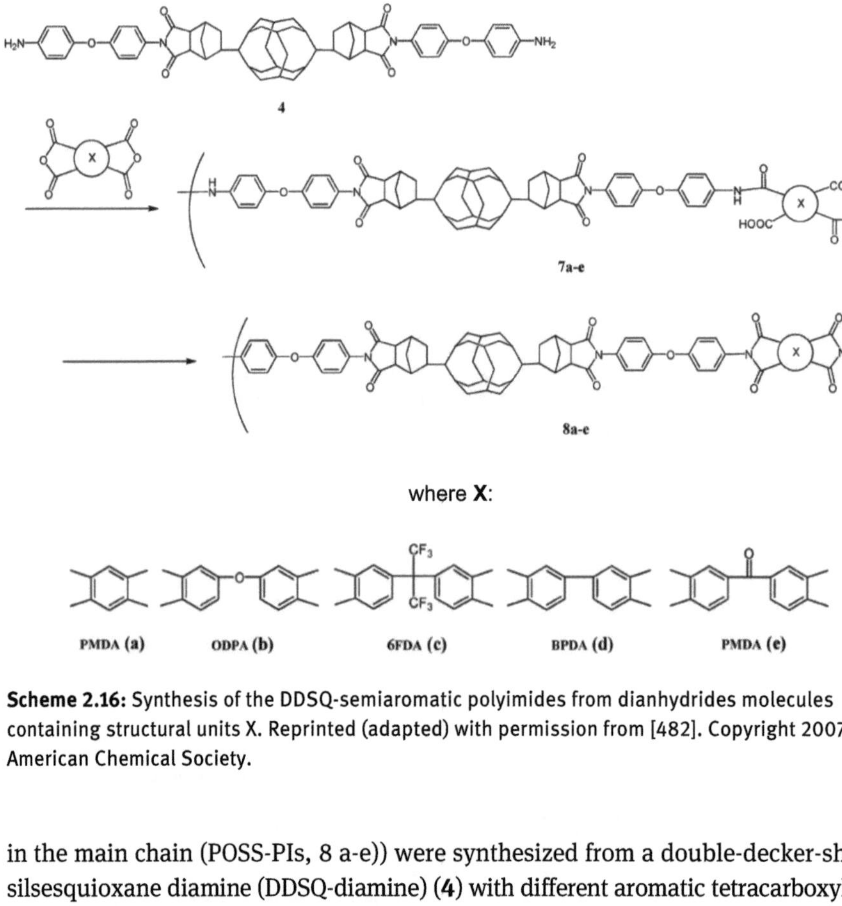

Scheme 2.16: Synthesis of the DDSQ-semiaromatic polyimides from dianhydrides molecules containing structural units X. Reprinted (adapted) with permission from [482]. Copyright 2007, American Chemical Society.

in the main chain (POSS-PIs, 8 a-e)) were synthesized from a double-decker-shaped silsesquioxane diamine (DDSQ-diamine) (**4**) with different aromatic tetracarboxylic dianhydrides (Scheme 2.16). The POSS-PIs showed good mechanical properties, very good thermostability, low water absorption, alkali resistance, and a low dielectric constant. The temperatures corresponding to 5% weight loss (T_{d5}) in air were observed at 495–514 °C. The polymer films had good mechanical properties: TS of 42–74 MPa and a small relative elongation at break of 2.9–6.0%. The highest elongation (6.0%) was exhibited by POSS-PI **8c** obtained from 4,4′-oxydiphthalic anhydride (ODPA). The water absorption of POSS-PI **8b** (<1%) was much lower than that of PI from pyromellitic dianhydride (PMDA)/ODA (6.0%). POSS-PIs **8** showed excellent acid and alkaline resistance, for example, POSS-PI **8b** saved its flexibility when immersed in 5% NaOH solution at 40 °C for 72 h. The dielectric constant of POSS-PI **8c** was 2.36 [482].

Synthesis and thermal, rheology, and mechanical properties of polyimides, epoxy resins, PMMA, polyurethanes, and other polymer NCs containing nanoparticles of POSS hybrids were reviewed by D. Gnanasekaran et al. [483]. Synthesis of different kinds of POSS-PDMS, POSS-propylene glycol (PPG), and polyurethane-POSS hybrid NC membranes, their characterization, properties, and applications for separation of gases (O_2/N_2, CO_2/N_2) was reviewed by B.R.S. Reddy and D. Gnanasekaran [484].

Functionalized polysilsesquioxane-based hybrid silica materials were prepared by a condensation of vinyl(triethoxy)silane with 3-aminopropyl(triethoxy)silane. They were used as solid amine sorbents for direct CO_2 capture from air. This amine-functionalized material showed a selective CO_2 capturing capacity from atmospheric air. Desorption began at a temperature of 60 °C and was completed at 80 °C. This sorbent revealed high recycling ability, and 50 cycles in ambient air and 100 cycles of adsorption/desorption in pure CO_2 were demonstrated without any loss in efficiency [485].

Quaternary ammonium-functionalized silsesquioxanes (Q-POSS) with different kinds and levels of quaternization (with alkyl halides) were prepared by P. Majumdar et al. – by the hydrosilylation of allyldimethylamine with $Q_8M^H{}_8$. Compositions of Q-POSS showed antimicrobial activity in solution against the gram-negative bacterium *Escherichia coli* and the gram-positive bacterium *Staphylococcus aureus*. Their antimicrobial properties in polysiloxane coatings [based on polydimethylsiloxane-α, ω-diols HO-PDMS-OH cross-linked with methyl(triacetoxy)silane] were strongly dependent on Q-POSS composition, similar to coating surface energy and surface morphology. The Q-POSS, having a relatively low quaternization and longer alkyl chains, exhibited the highest antimicrobial activity in solution, while coatings obtained using Q-POSS with the highest level of quaternization did not show antimicrobial activity, which was assigned to agglomeration of Q-POSS molecules through the formation of intermolecular interactions involving the quaternary ammonium groups. Probably, agglomeration reduced diffusion and inhibited interaction of the quaternary ammonium POSS halides with microbial cells [486].

POSS mimic a nanoscale silica particle and may find promising application, particularly in cardiovascular interventional devices, as a new generation of NCs. POSS acts as an amphiphile at the water–air interface [487] and it is used in stents [488, 489]. A novel type of NC, single functional side-group of POSS molecules in the form of a pendant side chain attached to poly(carbonate-urea)urethane (PCU) [491–493] – similar to that obtained by Lichtenhan et al. [490] – was prepared by R.Y. Kannan et al. [494]. Unlike carbon nanotubes [495], POSS nanocomposites were cytocompatible and can be used for tissue engineering. The inorganic POSS matrix exhibited a viscoelastic effect on the bulk of the NC [496], which was useful in vascular prostheses [497]. No significant difference in cell adhesion, viability, and proliferation between POSS NCs and standard cell culture plates was observed [498, 499]. The combination of the unique POSS molecules and polymers offers great promise for the future of biomedical devices, especially in cardiovascular bypass grafts [495].

The aggregation or crystallization processes of POSS hybrids, depending on their chemical structures, determine the interactions between the POSS moiety and their physical properties such as viscosity and melt-elasticity. The POSS cages are resistant to photooxidation and may find numerous applications, for example, in high-resolution lithography and special coatings. There is a growing interest in the use of POSS hybrids and POSS-based polymers in the field of biological and medical applications, due to their nontoxicity to biostability [500]. The POSS NCs show profitable mechanical properties, good hydrolytic and oxidative resistance, biocompatibility and biostability, resistance to calcification and fatigue, endothelialization properties, reduced in vitro inflammatory response, and anti-thrombogenic potential. The NCs containing POSS molecules are considered for many biomedical applications, for instance, as cardiovascular and breast implants, dental materials, biosensors, coating materials for quantum dot nanocrystals, in drug delivery systems, and in tissue engineering [501]. The POSS polymers containing silver nanoparticle exhibited improved wound healing [501] and a strong anti-inflammatory effect [502, 503]. POSS-modified materials are also used as effective sun protection agents [504] and as coatings for greenhouse covers [505].

A water-soluble macromonomer, based on octavinyl silsesquioxane, containing vinyl-terminated PEG 400 segments in each of the eight arms, promoting water solubility was prepared by the multistep method. TEM images showed spherical 22 nm aggregates in aqueous solution above the critical micelle concentration (CMC). The macromonomer was copolymerized with a vinyl-terminated, triblock copolymer of lactide–PEG–lactide, in a 1:4 v/v acetone/water mixture, providing a library of cross-linked hydrogels. The obtained copolymer networks contained a range of nm-µm sized pores and their swelling properties in water and buffered saline solution (PBS) at pH 7.4 were analyzed. At pH 7.4, the hydrogel networks underwent a slow hydrolysis with the release of PEG and lactic acid fragments. The hydrogels were noncytotoxic toward fibroblast cultures (at pH 7.4, within 23–28 days) and could be used as scaffolds for alveolar bone repair [506].

Trisilanol isooctyl-POSS was metallized with titanium atom, by reacting with titanium tetra(isobutoxide), in order to end-cap the POSS molecule. The prepared polyhedral oligomeric metallized silsesquioxane (POMS), in a monomeric form, was characterized by ^{29}Si-NMR, FTIR, RAMAN and UV–visible spectroscopy, and MALDI-TOF techniques. Pristine POSS and POMS were used in the synthesis of a polycarbonate (PC)–urethane copolymer, which showed better thermal stability and mechanical properties as compared to starting POSS and also controlled the surface properties. Its hydrophobicity and biocompatibility caused improved cell viability in appropriate biological tests, which is important for potential biomedical applications [507].

The surface properties of intraocular lenses (IOLs) were modified with PEG, POSS, heparin, F-heparin, 2-methacryloyloxyethyl phosphorylcholine, TiO_2, titanium, titanium nitride, vinyl pyrrolidone, and inhibitors of cytokines. It gave different kind of lenses: more hydrophobic or more hydrophilic, or lenses with a hydrophilic anterior and hydrophobic posterior surface, exhibiting a better biocompatibility in both in vitro and animal experiments [508].

An impact of polyhedral oligosilsesquioxanes (POSS) on mammalian cells was checked by W.A. Stańczyk et al. Toxicity of octaammonium chloride salt of octaaminopropyl POSS toward two cell lines – *mouse neuroblastoma (N2a)* and *embryonic mouse hippocampal cells (mHippoE-18)* – was studied in vitro. No harmful effect of this POSS on both cells was observed. Therefore, a noncytotoxic property of POSS suggested possibility to use it as a promising drug carrier [509]. Attempts to use, including synthetic methods, silsesquioxane NCs as potential nanocarriers in contemporary teranostics were reviewed, leading to the use of functionalized POSS molecules and dendrimers as well as organo-inorganic copolymers as covalent composites and systems for encapsulation of bioactive substances [510]. Two different synthetic routes were used in the preparation of the conjugates of functional POSS structures: octa(3-chloroammoniumpropyl)silsesquioxane and octa(carboxydecylthioethyl) silsesquioxane with anticancer drugs – anthracyclines [511]. Octa(3-chloroammoniumpropyl)octasilsesquioxane (T_8-POSS) was studied as a potential nanocarrier in co-delivery systems with doxorubicin (DOX). The toxicity of doxorubicin and T_8-POSS:DOX complexes at four different molar ratios (1:1; 1:2, 1:4, 1:8) toward microvascular endothelial cells (HMEC-1), breast cancer cells (MCF-7), and human cervical cancer endothelial cells (HeLa) was determined. Cytotoxic properties of T_8-POSS:DOX complexes indicated that the T_8-POSS:DOX system at the molar ratio of 1:8 was more effective than free DOX. It was found that the T_8-POSS is a promising, complex nanocarrier for doxorubicin delivery [512].

New kinds of hydrogels (OP-PD gels) containing POSS hybrid were prepared by the incorporation of cationic octaammonium (OA-POSS) particles into chemically cross-linked cationic poly(dimethylaminoethyl methacrylate) (PDMAEMA) hydrogels via in situ radical freezing polymerization. They showed considerably improved properties: better mechanical properties with higher TS and compressive strength. Owing to versatile swelling properties, OP-PD gels may find applications in controlled drug release [513].

The surface properties of SiC nanoparticles (NPs) were modified by a sonochemical coating with octa-*iso*-butyl (OI) POSS to form a compatible interface between the particle and a thermoset polymer. The structure of the OI-POSS coated SiC nanoparticles were analyzed by means of XPS, FTIR, and XRD techniques. The formation of a covalent bonding between SiC and OI-POSS was observed. The OI-POSS-coated

SiC NPs were used for reinforcement of thermoset epoxy resin to evaluate the mechanical and thermal properties of NCs. Their flexural strength, modulus, and T_g increased in comparison to the properties of neat epoxy resin [514].

A precipitated silica was grafted with silane coupling agents (SCA) [(3-isocyanatopropyl)triethoxysilane or *N*-2-(aminoethyl)-3-aminopropyl(trimethoxy)silane] by hydrolytic condensation reactions, carried out in methanol-water system (4:1, v/v). After vacuum evaporation of solvents, different amounts of methacryl-substituted POSS (3, 5, or 10 wt%) were deposited onto the surface of silanized silica in a similar way, in toluene, giving a new hybrid material (SiO_2/silane/POSS). The chemical immobilization of SCA and methacryl POSS onto silica surface significantly increased its hydrophobic character, which is important for the practical applications of this filler, e.g. in elastomers, providing good dispersion, high adhesion, and chemisorption to a polymer. Thermostability of SiO_2/silane/POSS hybrids was enhanced when compared to neat methacryl POSS modifier [515].

The surface of an inorganic TiO_2-SiO_2 composite oxide was functionalized with trisilanol- and phenyl-substituted POSS (Ph-POSS) and phenyl(trimethoxy)silane. Physicochemical properties of TiO_2-SiO_2/Ph-POSS hybrid were analyzed by means of various methods (FTIR, TEM, TGA, zeta potential). The total volume and mean size of pores for a specific surface area and a hydrophilic-hydrophobic character of the surface were also determined [516].

Three series of hyperbranched polysiloxanes (HB-PSOX) and poly(carbosiloxanes) (HB-PCSOX) were synthesized from octafunctional polyhedral oligosilsesquioxane monomers and various difunctional silanes by bimolecular nonlinear A_8-B_2 polymerization (BMNLP), and characterized by FTIR, ^{1}H- and ^{29}Si-NMR, SEC, and DSC, for refractive index [517].

Series 1 was prepared from T_8Vi_8 (Vi = CH_2=CH) and disilane monomers, and series 2 was prepared from $T_8(OSiMe_2OH)_8$ and divinylsilane monomers (where T denotes an $SiO_{3/2}$ unit), both through platinum-catalyzed hydrosilylation polyaddition. Series 3 was prepared from $T_8(OSiMe_2OH)_8$ and dichlorosilane monomers via the hydrolytic polycondensation. Of course, their properties were dependent on their structure, that is, content of methyl and phenyl groups.

These hyperbranched POSS polymers were further functionalized with alkoxysilane-end groups (useful in further curing) and mixed with linear polysiloxanes to obtain transparent and solid nanostructured POSS-containing coatings that can be used in high-performance space and solar applications [517].

Similar hyperbranched (HB) poly(carbosiloxanes) and polysiloxanes containing octafunctional silsesquioxane (POSS) branch points containing curable (triethoxy)-silane or silanol end-groups were formulated with linear polysiloxanes to prepare transparent and solid nanostructured materials containing POSS units for numerous high performance space and solar applications. The effect of the methyl and phenyl groups content, molecular architecture and linear polysiloxane MW on thermal,

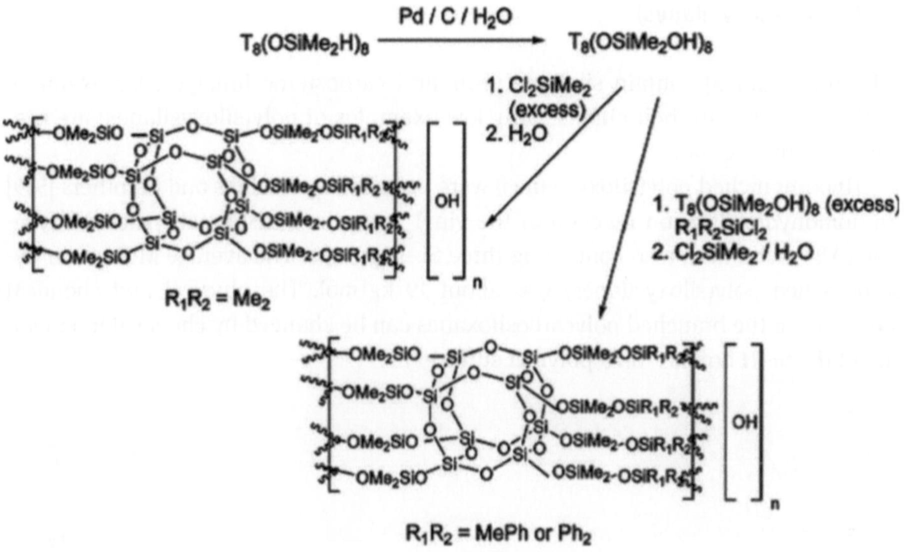

Scheme 2.17: Synthesis of HB-PSOX from $T_8(OSiMe_2OH)_8$ monomer. Reprinted (adapted) with permission from [517]. Copyright 2007, American Chemical Society.

physical, transparency, and proton, electron and UV radiation resistance properties was determined, and the physical properties of the nanomaterials were tailored for manufacture of adhesives to be used as flexible or rigid coatings. The methyl compositions exhibited an excellent electron resistance in comparison to a commercial space control material and to a POSS-free HB polymer control material, even when directly exposed to radiation in coating form, while the phenyl formulations showed poor UV and electron resistance [518].

Some other examples of copolymers containing silsesquioxane cubes included in their structures are described in Section 2.6. Over the last 30 years, there is an intensive increase in the number of practical applications of POSS hybrids and further progress in this field can be intriguing, exciting, and very optimistic.

2.6 Copolymers containing siloxane linkages

Many kinds of copolymers containing siloxane or silsesquioxane and other segments in their chemical structures are described in the following sections. Their chemical and physical properties are significantly modified this way.

2.6.1 Poly(siloxysilanes)

Poly(siloxysilanes) contain siloxane units and carbosilane linkages, for example, Si–C–C moieties, in their chains. Only few examples of poly(siloxysilanes) are presented in this section.

Hyperbranched poly(siloxysilanes) were prepared by Mathias and Carothers [519] via homohydrosilylation reaction of the vinyl group in a tris(dimethysiloxy)vinylsilane (AB_3 monomer), also containing three Si–H groups. The average MW of the hyperbranched poly(siloxysilanes) was about 19 kg/mol. The physical and chemical properties of the branched polycarbosiloxanes can be changed by chemical modification of the Si–H bonds ("on a polymer surface").

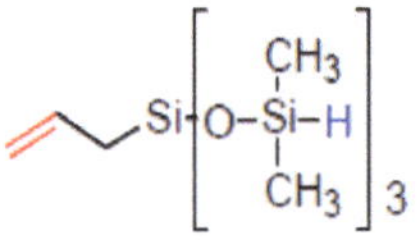

Figure 2.14: Tris(dimethysiloxy)vinylsilane (AB_3 monomer) [519].

Many other dendritic polymers containing silicon atoms in their chains belong to a group of poly(siloxysilanes). Polysiloxane copolymers have block or graft structures. Branched and comb-like structures exist if $RSiO_{3/2}$ units are included in their structures [see p.137 in ref. 8].

This is a very important group of siloxane–polyether copolymers, often used in the manufacture of polyurethane foams [520–522]. Most often, the distribution of the grafted siloxane units in the latter copolymers is random. Linear copolymers have different structure and composition: alkoxy-siloxane-alkoxy (A-B-A), siloxane-alkoxy-siloxane (B-A-B), and (siloxane-alkoxy)$_n$ (A-B)$_n$. Copolymers with Si–O moieties are synthesized via condensation or ester exchange reactions, while the hydrosilylation reactions lead to Si–C linkages. Ester exchange proceeds under milder conditions (than condensation of Si–OH groups or Si–Cl groups with oligo(oxyalkylene) glycols) in the presence of acidic or basic catalysts, which must be neutralized and the evolved alcohol is usually distilled off. Poly(dimethylsiloxane-*b*-1,2-oxypropylene) copolymers show better hydrolytic resistance than poly(dimethylsiloxane-*b*-oxyethylene) derivatives [523–525]. Alternatively, poly(oxyalkylene)-PDMS copolymers are prepared by the condensation of Si–H functional siloxanes with HO-terminated oligo(oxyalkylenes), in the presence of alkaline or tin catalysts [526] or by reacting polysiloxanes terminated with epoxy functional oxyalkylene groups [527].

The Si–C-linked copolymers can be prepared by the condensation of carbofunctional PDMS with telechelic (dihydroxy)oligo(oxyalkylenes) or the catalytic hydrosilylation of oligo(oxyalkylenes), with terminal unsaturated groups (mostly: vinyl or allyl) with Si–H functional α,ω-dihydrooligo(dimethylsiloxanes), or polysiloxanes containing pendant Si–H groups [528–530]. The branched copolymers may contain T units with two or three polyoxyalkylene segments and are more resistant to hydrolysis as compared to linear ones [528, 531].

Block PDMS–oxyalkylene copolymers are soluble in water and form micelles above the critical concentration, reducing the surface tension of water from 72 to 20–30 mN/m at 1 wt% concentration, depending on their structure [532, 533]. Due to their hydrophilic properties, polysiloxane-polyoxyalkylene copolymers find medical applications, for instance, in contact lenses (after cross-linking) [534].

Block polysiloxane–polyalkylene copolymers are prepared by anionic ROP of isoprene, styrene or α-methyl styrene followed by the addition of cyclosiloxane monomers (mainly D_4) [535]. Coupling of block copolymer dianion with Me_2SiCl_2 leads to multiblock copolymer with $(ABA)_n$ structure [536–539]. PDMS–polystyrene block copolymers with $M_n > 10^5$ g/mol and a silicone content of 80% exhibited elastomeric properties, similar to pure SEs, that is, quite poor TS, but high elongation at break (E_b). At 50% silicone content they behave like PS [540]. Block polysiloxane–polystyrene copolymers have two T_gs: ~–120 °C for siloxane segments and ~50 °C, depending on the PS block. Block PDMS copolymers with α-methyl styrene containing 73% siloxane units with $M_n \leq 2 \times 10^4$ g/mol show outstanding mechanical properties: TS = 1,300 psi and E_b = 935% [529].

New types of PDMS–polyene copolymers (with a relatively high T_g of 132–138 °C, good moisture resistance, low dielectric constant and low coefficient of thermal expansion, and flexural strength 28 kpsi) were prepared by the hydrosilylation of the C=C bonds in polyenes with PMHS in the presence of Pt catalysts [8, 541].

2.6.2 Copolymers containing polysiloxane, silsesquioxane, and carbosiloxane segments

An important group are copolymers of polysiloxanes with polycarbonates (PC), polyphthalate, poly(alkyl)esters, and PMMA.

Polysiloxane–PC copolymers based on bisphenol A or cyclobutylene carbonate segments [529] were prepared by phosgenation of bisphenol A and α,ω-dichloropoly(dimethylsiloxanes) in the presence of pyridine [542, 543] or with α,ω-bis-(dimethylamino)poly(dimethylsiloxanes) [544]. Phosgenation of mixtures of polysiloxanes, containing alkyl or aryl substituents of silicon and hydroxyaryloxy- or hydroxy alkyl-end groups with other bisphenols gave new block polysiloxane–PC copolymers [545–549].

Polysiloxane–PC block and graft copolymers were also prepared in bulk by polycondensation of carboxyalkyl- or carboxyaryl-functionalized PDMS with organic diols at 300–400 °C [550] or the by an interfacial method [551]. These copolymers formed two-phase systems (with two T_gs: a ~–110 °C for PDMS block and between 60 and 140 °C for PC block) [529, 552] and showed lower TS and improved elongation at break, in comparison to PC [553].

Polysiloxane–PC block copolymers with tetramethyl-1,3-cyclobutylene segments were prepared by the reaction of cyclobutylene carbonate prepolymer with α,ω-bis-(dimethylamino)poly(dimethylsiloxane) [229, 554, 555]. Cross-linking of acetoxy- or allyl-terminated block polysiloxane–PC copolymers or polysiloxane copolymers with side mercaptoalkyl groups provided elastomers with high mechanical strength even at high temperatures [556, 557].

Polysiloxane–polyphthalate copolymers were prepared by the polycondensation of α,ω-bis-(dimethylamino)poly(dimethylsiloxanes) or carbofunctional PDMS containing aminoalkylene end groups with appropriate polyesters or with phthaloyl or isophthaloyl dichlorides [229, 558]. Copolymers with amide linkages were hydrolytically resistant, as compared to those with Si–O–C moieties.

Polycondensation of 1,3-bis-(*p*-hydroxyphenyl) disiloxanes with terephthalic acid or its acid chloride gave disiloxane–terephthalic copolymers with a strictly alternating sequence, which showed outstanding thermal stability ($T_5 = 436$ °C) [559] and two-phase morphology with different T_gs: ~−110 to −120 °C for disiloxane segment and ~140–210 °C for polyester segment [229, 560, 558]. Graft copolymers with pendant siloxane segments were prepared by the hydrosilylation of allyl-functionalized polyester obtained by the polycondensation of terephthalic dichloride with allyl trimethylol propane [561].

Newer examples of polysiloxane copolymers with organic polymers have also been described. cyclolinear (CL) poly(organocarbosiloxanes) with ethylene groups instead of oxygen atoms in disiloxane MeSi–O–SiMe bridges between rings were prepared by the polyaddition reaction of dihydrocycloorganotetra(hexa, octa)siloxanes with divinylcycloorganotetra(hexa)siloxanes bearing methyl and ethyl substituents, in the presence of different platinum catalysts according to Scheme 2.18 [562].

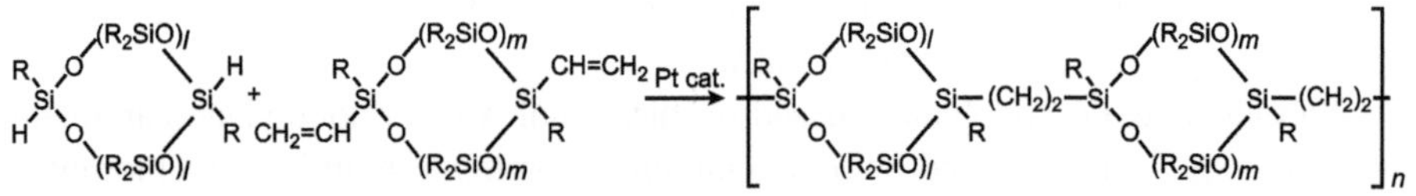

Scheme 2.18: Synthesis of CL PMCS **III–V**. Copyright 2013 by the authors; licensee MDPI, Basel, Switzerland.
l = m = 1, R = Me (**III-Me**); l = m = 1, R = Et (**III-Et**); l = m = 2, R = Me (**IV-Me**);
l = m = 2, R = Et (**IV-Et**); l = 2, m = 3, R = Me (**V-Me**).

CL poly(carbosiloxanes) with varying content and location of $(CH_2)_n$ groups in the monomer unit were prepared by heterofunctional polycondensation and polyaddition reactions of difunctional organocyclosiloxanes and organocyclocarbosiloxanes. Incorporation of CH_2 groups into the methylcyclohexasiloxane unit did not affected the ability of the polymer to self-organize with the formation of a mesomorphic state in a wide temperature range, while on incorporation of $(CH_2)_2$ segments into a cyclotetrasiloxane unit or in a bridge connecting two methylcyclotetra(hexa)siloxane units, changed copolymer properties. Comparison of the X-ray data of dihydroxy derivatives of decamethylcyclohexasiloxane and decamethyl-5-carbocyclohexasiloxane with packing of CL organosilicon polymers in bulk showed that the polymer retained the layered type of crystalline structure, which is typical for monomers. All CL poly (methylcarbosiloxanes) with different monomer units formed monomolecular Langmuir films at the air/water interface, excluding those having longer hydrophobic fragment than hydrophilic ones. The ability to form multilayers was dependent on the substituents at Si atom in the bridge between the cycles [563].

Two novel chiral LC monomers and a series of their corresponding chiral side-chain polymers based on poly(methylhydro)siloxane were also prepared. All polymers showed a high thermal stability with wide mesophase temperature ranges. The isotropic phase transition and T_gs of all polymers increased and were dependent on the molecular composition of LC polysiloxanes [564].

Side-chain discotic polysiloxane with 2,3,6,7-tetrakis(hexyloxy)-10-methoxytriphenylene-11-undecanoate moieties was obtained by the hydrosilylation reaction, using PMHS. The phase behavior and thermooptical properties of the polysiloxane and the side chain precursor 2,3,6,7-tetrakis(hexyloxy)-10-methoxytriphenylene-11-undecanoate were examined by polarizing optical microscopy, thermooptical analysis, DSC, and wide-angle X-ray scattering (WAXS) studies. A columnar planar alignment of LC in the layers was determined. This polymer could be a promising material for use in optoelectronic devices. The differences in phase transitions and morphology between the triphenylene precursor and the discotic polysiloxane were discussed [565].

The monovinylsilsesquioxane 1-(9-decenyl)-3,5,7,9,11,13,15-heptaethylpentacyclo-[9.5.1.13,9.15,15.17,13]octasiloxane was homopolymerized or copolymerized with ethylene and propylene, to obtain metallocene catalysts activated with methylalumoxane (MAO). Thus, 17–25 wt% of this new POSS monomer, as pendant T_8 cages, were incorporated into the structure of polyolefin copolymers, which caused the melting temperature to decrease up to 18 K with respect to polyethylene (PE) and increase the thermostability in air [231].

Copolymers with pendant silsesquioxane (POSS) groups were synthesized by a radical copolymerization of monomethacryloxypropyl-POSS monomers (heptaethyl or heptacyclopentyl substituted) with methacrylic acid, 2-(trifluoromethyl) acrylic acid, *tert*-butyl methacrylate (TBMA), *tert*-butyl trifluoro methacrylate (TBTFMA), and itaconic anhydride, in various compositions, to obtain AIBN as an initiator. Carboxylic acid groups in TBMA (or TBTFMA) mers were deprotected with a perfluorooctylsulfonate-based photoacid generator by heating, after exposure. Thus, the POSS-based resists with thickness of 157 nm and with good film-forming properties were obtained for use in lithographic applications. They showed no silicon outgassing, high sensitivity

(<10 mJ/cm^2 under open-field exposure), sub-100-nm resolution capabilities, and high absorbance (~4 μm^{-1}) [566].

Silsesquioxane–siloxane copolymers were also prepared by T. Gunji et al. via the dehydrocoupling of (octahydro)silsesquioxane $(HSiO_{3/2})_8$ ($T^H{}_8$) with diphenylsilanediol, tetraphenyldisiloxane-1,3-diol or poly(dimethylsiloxane-α,ω-diols) in the presence of (diethyl)hydroxylamine, followed by trimethylsilylation. Silica gel films were obtained by spin-coating of the concentrated reaction solutions and they showed a pencil hardness up to 6H (increasing with the curing temperature) and a small surface area below 400 °C, with a maximum 290–450 m^2/g at 650 °C [567].

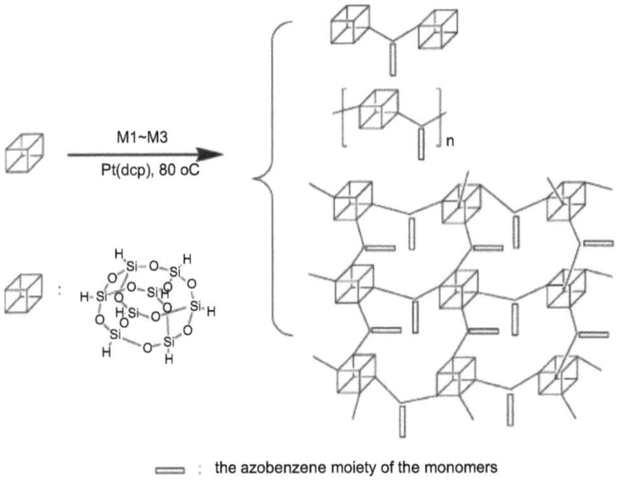

Scheme 2.19: Synthetic routes to different structures of the POSS hybrids containing azobenzene chromophores. Reprinted (adapted) with permission from [568]. Copyright 2007, American Chemical Society.

New POSS-based inorganic–organic hybrid materials with various architectures (bead-type, dumbbell-type, and network-type structure) were prepared by the hydrosilylation reaction of octahydrosilsesquioxane ($T^H{}_8$) with monomers containing different azobenzene chromophores. These hybrid composites were soluble in organic solvents (toluene, THF, and chloroform) and showed good film-forming and nonlinear optical properties. The structure of these hybrids was effectively tuned by changing the feed ratio and molecular structure of organic chromophore monomers. The incorporation of organic azobenzene chromophore into the inorganic POSS gave the hybrids with high thermostability and good optical limiting properties [568].

A polyfluorene–silsesquioxane copolymer (**PF-POSS**) with a star-like structure was prepared by the Ni(0)-catalyzed reaction of polydioctylfluorene (**PF**) with octa-(2-(4-bromo-phenyl)ethyl)octasilsesquioxane (**OBPE-POSS**). The incorporation of the silsesquioxane cage into the polyfluorene caused substantial reduction of the aggregation and enhanced the thermal stability. An elimination of the glass transition and crystallization as well as a significant reduction of the melting enthalpy in **PF-POSS** were observed in the DSC thermograms. The **PF-POSS** copolymer showed beneficial photoluminescent properties. The maximum luminescence intensity and quantum efficiency of **PF-POSS** were almost more than those of the **PF** electroluminescent device. This copolymer can be used for the manufacture of organic light-emitting diodes with improved thermal and optoelectronic characteristics [569].

Two kinds of inorganic–organic hybrid gels were prepared by a photohydrosilylation reaction of the 1,3,5,7,9,11,13,15-octakis(dimethylsilyloxy)pentacyclo-[9,5,1,1,1,1] octasilsesquioxane ($Q_8M^H{}_8$) or the cyclic siloxane, 1,3,5,7-tetramethylcyclotetrasiloxane ($D^H{}_4$), as cross-linking reagents with end-functionalized dienes, 1,5-hexadiene (HD) or 1,9-decadiene (DD), which were catalyzed with a bis(acetylacetonato) platinum in toluene. Their mesh size (1.4–1.6 nm), mesh size distribution, and network structures were quantitatively characterized by means of a novel scanning microscopic light scattering. Glass plates were coated with the $Q_8M^H{}_8$-DD gel by means of the photohydrosilylation reaction in a cast reaction medium [570].

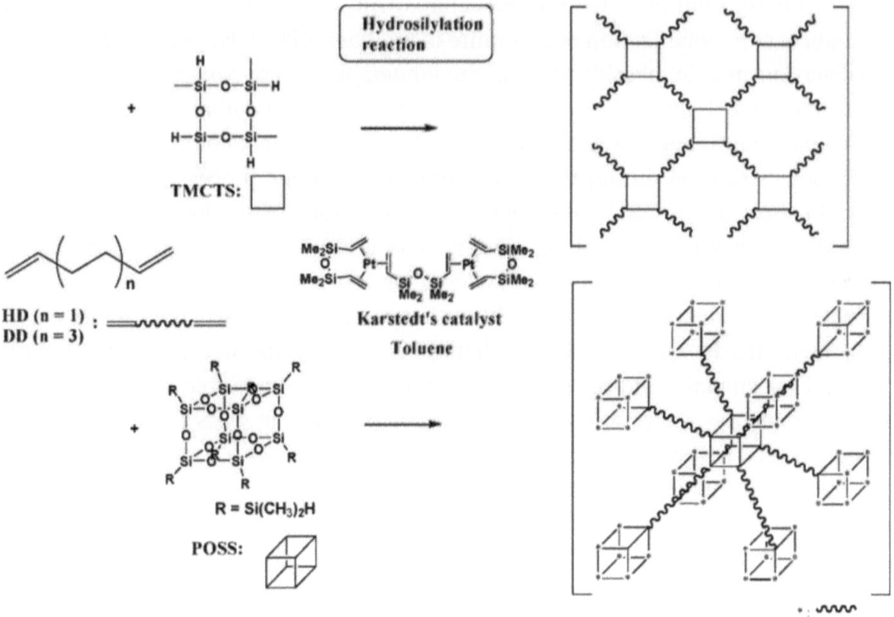

Scheme 2.20: Structure of organic–inorganic hybrid gels prepared by hydrosilylation reaction of HD and DD with $D^H{}_4$ and $Q_8M^H{}_8$ in the presence of Karstedt's catalyst. Reprinted (adapted) with permission from [570]. Copyright 2007, American Chemical Society.

NCs of monomethacryl heptacyclohexyl-substituted POSS [$(C_6H_{11})_7Si_8O_{12}C_7H_{11}O_2$, 1125.9 g/mol, POSS-1] copolymers with MMA, containing up to 51 mol% (92 wt%)

POSS-1 were prepared using radical copolymerization. The degree of polymerization decreased with increasing POSS-1 loading and the polydispersity of the copolymers was quite low. Addition of POSS-1 improved the thermostability of NCs and the ceramic yield of SiCO, especially for a high POSS-1 content [571].

High MW POSS-containing homopolymers (PHEMA-POSS) were synthesized via reversible addition–fragmentation chain transfer (RAFT) polymerization of HEMA-POSS monomer, derived from aminopropyl-substituted heptaisobutyloctasilsesquioxane and 2-hydroxyethyl methacrylate (HEMA). PHEMA-POSS was further modified with *N,N*-dimethylaminoethyl methacrylate in a RAFT process, providing a series of amphiphilic POSS-containing poly(*N,N*-dimethylaminoethyl methacrylate) diblock copolymers, PHEMA-POSS-*b*-PDMAEMA. They self assembled into different morphologies: irregular assembled aggregates or core–shell spheres and complex spheres (pearl-necklace-liked structure) or large compound vesicles. The results of complex studies (by dynamic laser scattering, UV spectroscopy, SEM, and TEM) revealed their thermo- and pH-responsive properties. The reversible transition of the assembled morphologies from spherical micelles to complex micelles was controlled through acid-base interactions [572].

POSS-based antireflective nanostructures (ARC) were prepared on glass by a free radical copolymerization of a mixture of methacryl POSS cage with 1,6-hexanediol diacrylate as a double-side nanoimprint lithography. These nanostructures showed excellent chemical, thermal, and antireflective properties. An optimum resist composition for nanoimprinting was achieved using a 1:12 molar ratio of both monomers. Thermal copolymerization during the nanoimprint lithographic process gave a uniform array of nanostructures on both sides of a glass substrate, with ~90% to 100% yields. The transmittance of the resulting glass increased to 98.2% (reflectance 1.26%) and quasi-omnidirectional transmittance was excellent (observed between −50° and +50° of angles of incidence). Moreover, ARC nanostructures showed strong adhesion to glass, better hardness, and better mechanical strength, thus suggesting their strong potential for commercial application. Owing to their incompatible mechanical and thermal properties, polymer-based antireflective coatings on glass could find potential commercial outdoor photovoltaic applications [573].

Polyhedral copolymers of silsesquioxane-derived methacrylate (PS-*co*-PMAPOSS) and styrene of various MAPOSS : styrene ratios with M_n over 100 kDa, were synthesized by concurrent ARGET ATRP–RAFT. Fibrous mats of the prepared copolymers were further developed through electrospinning process (in THF and DMF), using various POSS content. It was found that liquid–liquid phase separation followed by phase segregation of PS and POSS led to varying degrees of porosity of the electrospun fibers. Based on preliminary tests, it was suggested that highly porous PS-*co*-PMAPOSS fiber can be modified by silanization or with high-temperature or plasma treatment for further novel applications [574].

Poly(methyl methacrylate-*co*-*n*-butyl methacrylate-*co*-cyclopentyl methacryloxy propyl-polyhedral oligomeric silsesquioxane) [poly(MMA-*co*-BMA-*co*-POSSMA)] were tested in inverse gas chromatography (IGC) with polar and nonpolar solvents [575].

Linear polysiloxanes containing polyhedral oligomeric silsequioxane (POSS) side groups, soluble in various organic solvents, were prepared by the hydrosilylation reaction of poly(ethylhydrosiloxane) with different amounts of POSS and 1-octene using the Karstedt's catalyst. The graft copolymers containing at least 10 mol% POSS units formed free standing films, which were cast from solution [576].

The Si–H-monofunctionalized heptaisobutyl silsesquioxane T_8 derivative (*i*Bu_7POSS-SiH) was prepared from a reaction of trichlorosilane with 1,3,5,7,9,11,14-hepta(isobutyl) tricyclo-[7.3.3.15,11]-heptasiloxane-endo-3,7,14-triol (*i*Bu_7POSS-$(OH)_3$). It was used for side groups functionalization of (α,ω-divinyl)poly(dimethylsiloxane), α-vinyl-ω-*n*Bu-poly(dimethylsiloxane), α,ω-trimethylsilylpoly(dimethylsiloxane-*co*-methylvinylsiloxane), and poly(1,2-butadiene) by hydrosilylation reactions. Pendant side POSS moieties in amorphous copolymers indicated the formation of POSS aggregates at the nanometer scale. The type of the formed nanostructures during the self-association of *i*Bu_7POSS was dependent on the number of silsesquioxane units and their position with respect to the polymer chains. The process of self-organization through associative interactions was thermoreversible. The thermooxidative resistance of the studied samples seemed to be affected by arrangement of POSS units [577].

A great number of publications have been devoted to the chemical modifications of silsesquioxanes with oligo- and polyethers (PEO). Star-shaped cubic silsesquioxane–poly(ethylene oxide) (POSS–PEO) were synthesized by the hydrosilylation of $Q_8M^H{}_8$ with (allyl, methyl)-terminated oligomers of ethylene oxide ($CH_3(OCH_2CH_2)_n$ $OCH_2CH{=}CH_2$) [584]. The T_g and the crystallization temperature (T_c) of POSS–PEO were different from those of linear PEO. The overall crystallization growth rate was reduced due to the reduction of the mobility of the PEO crystallites in the presence of POSS and due to the star structure of the polymer. The presence of POSS led to the diffusion- and nucleation-controlled mechanisms in the crystallization process of POSS–PEO. In the case of linear PEO, only the nucleation-controlled mechanism was observed [578].

Hydrosilylation of unsaturated PEGs with both octa(hydrido)silsesquioxane ($T^H{}_8$) and octa(dimethylsiloxy)silsesquioxane ($Q_8M^H{}_8$) was used as a synthetic route to a homologous series of PEG (various chain length)-substituted octasilsesquioxanes. The PEG-substituted silsesquioxanes were prepared by the initial synthesis of a series of allyl-terminated PEG and the subsequent attachment to both $T^H{}_8$ and $Q_8M^H{}_8$. They were chemically characterized by spectroscopic methods: FTIR and NMR (^{1}H-, ^{13}C-, and ^{29}Si-NMR), and the properties of the POSS hybrids were thermally characterized by TGA and DSC. The functionalization of the POSS macromonomers ($Q_8M^H{}_8$ and $T^H{}_8$) with PEGs moieties caused a chain length-dependent increase in T_g and a concomitant suppression of crystallization [579].

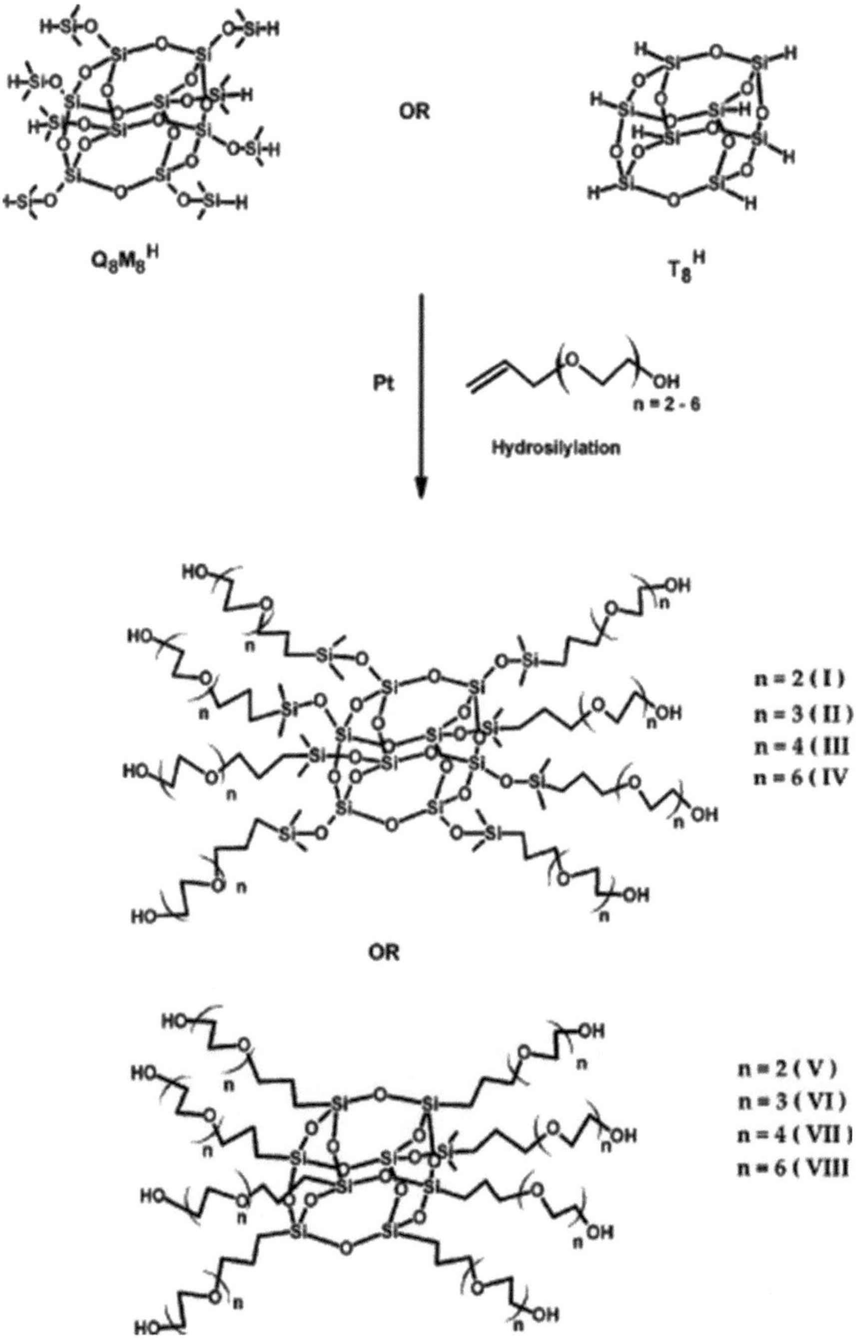

Scheme 2.21: Hydrosilylation of $Q_8M^H{}_8$ or $T^H{}_8$ with ethylene glycol monoallyl ethers. Reprinted (adapted) with permission from [579]. Copyright 2007, American Chemical Society.

The POSS derivatives of octakis[dimethyl(phenethyl)siloxy]silsesquioxane (OS-POSS), octakis[dimethyl(4-acetoxyphenethyl)siloxy]silsesquioxane (OA-POSS), and octakis[dimethyl-(4-hydroxyphenethyl)siloxy]silsesquioxane (OP-POSS) were prepared from octakis(dimethylsiloxy)silsesquioxane ($Q_8M^H{}_8$) via hydrosilylation reactions. These POSS hybrids were blended with the homopolymer poly(ethylene oxide) (PEO), giving a series of NCs. The OP-POSS/PEO NC showed the lowest crystallization rate and the highest thermal stability [452].

Comb-like polysiloxanes containing POSS and oligo(ethylene oxide) groups (EO) in the side chains were synthesized through the hydrosilylation grafting of poly(ethylhydrosiloxane) with 1-allyl-3,5,7,9,11,13,15-heptacyclopentylpentacyclo-[9.5.1.13,9.15,15.17,13]octasiloxane (allyl-CyPOSS) and allyl-functionalized oligo (ethylene glycol) monomethyl ethers. The T_g of the polymer increased with increasing content of POSS side groups, whereas the T_m and heat of fusion of the oligo(ethylene oxide) side group decreased. Moreover, the T_g and T_m increased when the length of the EO group was increased, while the decomposition temperature decreased [580].

POSS-embedded hyperbranched (HB) inorganic–organic hybrid silsesquioxane polymers (hbr-POSS) with ester "joints" were prepared by X. Wang, D. Wu et al. from hydroxyl-functionalized AB_7 POSS monomer, using the "one-pot" approach (Scheme 2.22). Their MWs and molecular sizes (~5, 11, and 15 nm) were tailored by controlling the polycondensation time. Next, the hydrophobic cores were modified into the amphiphilic HB polymers with adjustable hydrophilicity and hydrophobicity. They have building units of amphiphilic polymers, enabling morphological transition of their structures from micelle to vesicle, with the increased size of the core in aqueous solution, and find potential application as high-performance materials [581].

Octavinyl polyhedral oligomeric silsesquioxane (POSS) was functionalized with eight chains of liquid PEG (MW 400 g/mol), followed by acrylation. Next, this copolymer was incorporated into a hydrogel network of triblock poly(lactide-*b*-ethylene glycol-*b*-lactide) diacrylates by a redox-initiated copolymerization. The prepared organic–inorganic hybrid hydrogels contained 1–28 wt% of the inorganic cross-linked POSS. The hydrolytic degradation of hydrogels formed from these copolymers was studied by different methods (weight loss, cryogenic SEM, and ATR-FTIR) in a pH 7.4 PBS solution at 37 °C. Copolymers of acrylated PEG-lactide (MW 10^3 g/mol) with increasing amounts of POSS formed a more porous network and more resistant to hydrolysis. An increased POSS content resulted in decreased rate of degradation due to its hydrophobic nature and resistance to hydrolysis. The copolymer with increased lactide content showed increased degradation rate due to the increased number of hydrolytically sensitive ester groups in the chemical structure [582].

Copolymerization of POSS with PEG gave injectable in situ cross-linkable hybrid cements. Their mechanical, rheological, and in vitro bioactive properties were determined. It was confirmed that hybridization improved the mechanical and bioactive properties of POSS and PEG. The strength of the hybrids was similar to that of

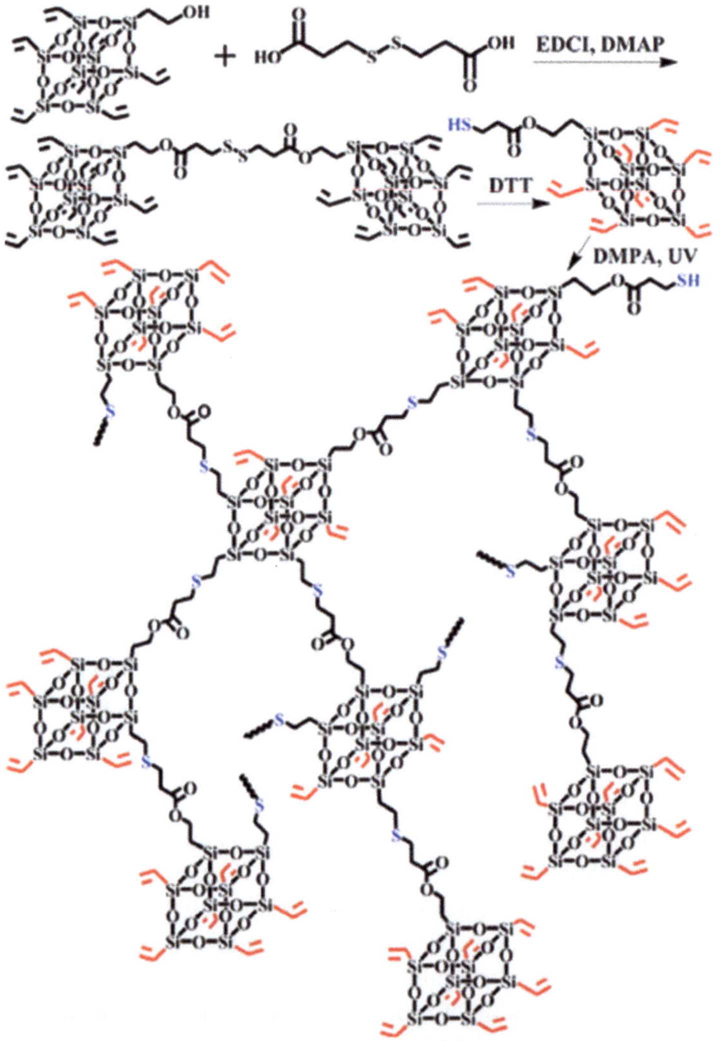

Scheme 2.22: Synthetic approach to an HB-POSS polymer from an AB_7POSS monomer with one thiol and seven vinyl groups [581]. Reprinted (adapted) with permission from [298]. Copyright 2015, Royal Chemical Society.

commercial cements. However, the Young's moduli of the hybrids were lower than those of commercial cements and more similar to those of the cancellous bone. The PEG–POSS-based hybrid materials are a promising alternative to commercial bone cements [583].

PEG-functionalized POSS derivatives (as a plasticizer), and lithium bis(trifluoromethanesulfonyl) imide (LiTFSI) were used to prepare a series of composite polymer electrolytes (CPEs), based on organic–inorganic hybrid star-shaped polymer (SSP).

The effects of PEG chain length and the composite compositions on the properties of CPEs were studied. Star-shaped polymer (SPP) was obtained through ATRP of poly(ethylene glycol) methyl ether methacrylate (PEGMA) with (methacryl)cyclohexyl-substituted POSS (MA-POSS), which was initiated by octafunctional substrate. The PEG-functionalized POSS derivatives were prepared by the hydrosilylation reaction of allyl-PEG with Si–H functional octakis(dimethylsilyloxy)silsesquioxane ($Q_8M^H{}_8$). The CPEs were found to be dimensionally stable enough to separate the electrodes in batteries, but they still showed high mobility of ion conducting P(PEG-MA) segments, as estimated by the low T_g. These solid-state CPEs showed about three times larger ionic conductivity (4.5×10^{-5} S/cm at 30 °C) than that of the matrix polymer (SPP) electrolyte (1.5×10^{-5} S/cm at 30 °C). An all-solid-state lithium battery prepared from the CPEs exhibited a larger discharge capacity than that prepared from the SPP electrolyte at 60 °C. The CPEs were electrochemically stable up to 4.2 V without the decomposition of electrolytes [584].

Polyhedral hydrophilic derivatives of oligosilsesquioxanes (POSS) with poly(ethylene oxide) (PEO), and derivatives with hydrophobic isobutyl groups (*i*-Bu) were formed in cylindrical structures in the presence of hydrated surfactant, Pluronic P105 (EO_{37}-PO_{56}-EO_{37}) – poly(ethylene oxide)-*b*-poly(propylene oxide)-*b*-poly(ethylene oxide). They showed different surface properties and different molecular structures in the presence of water and in water-in-xylene media [585].

A new POSS macromer functionalized with eight reactive isocyanate groups was synthesized by the hydrosilylation of octakis(hydridodimethylsiloxy)octasilsesquioxane with *m*-isopropenyl-α,α'-dimethylbenzyl isocyanate (*m*-TMI) in THF in the presence of Karstedt's catalyst (Scheme 2.23). This macromer was used as a nano-cross-linker in the preparation of the first cross-linked polyurethane with discrete POSS molecules dispersed within the polymer matrix [586].

PEG homopolymers terminated with POSS cages (POSS-PEG-POSS) having polyether soft segments and urethane units were prepared by reacting the hydroxyl end groups of PEG (with number-average molar weights of $M_n = 1–10$ kg/mol) with the isocyanate group of [(isocyanatopropyl)dimethylsilyloxy]heptacyclohexyloctasilsesquioxane macromers (Fig. 2.15).

These amphiphilic telechelic polymers obtained from nonsurface-active precursors exhibited surface activity at the air/water (A/W) interface. POSS moieties introduced the hydrophobic groups for hydrophilic PEG chains. For short PEG chains (1, 2, and 3.4 kg/mol), insoluble monolayers were formed, while POSS end groups were not sufficiently hydrophobic to keep higher molar mass hydrophilic PEG blocks (8 and 10 kg/mol) at the A/W interface. The POSS end groups were present at the A/W interface and the PEG chains were squeezed into the subphase, increasing the surface pressure. The X-ray reflectivity studies indicated on Y-type a double-layer structure of Langmuir-Blodgett multilayer films with a double-layer spacing of about 3.52 nm [587].

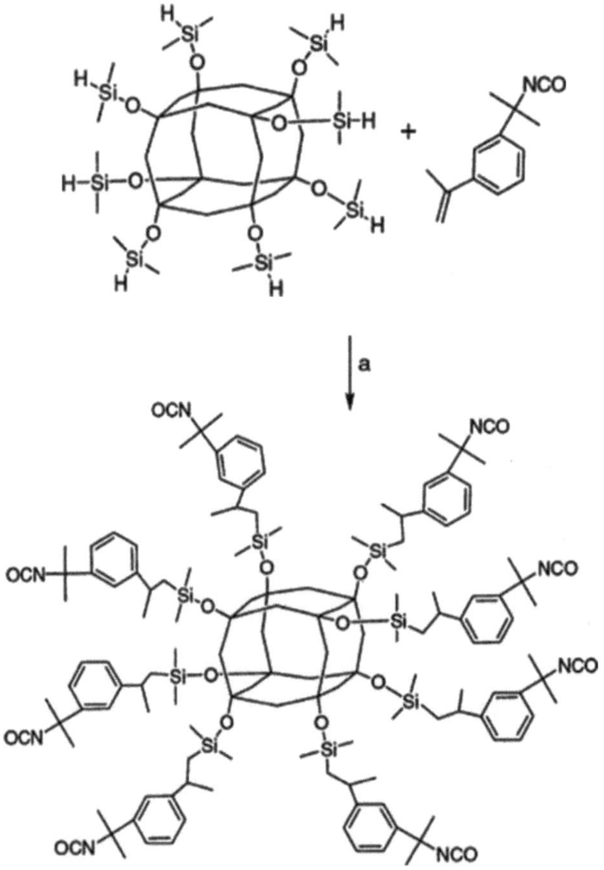

Scheme 2.23: Synthesis of $Q_8M_8{}^{TMI}$ (the core represents the Si_8O_{12} structure of the POSS). Conditions: (a) THF, Karstedt's catalyst, reflux for 2 days. Reprinted (adapted) with permission from [586]. Copyright 2007, American Chemical Society.

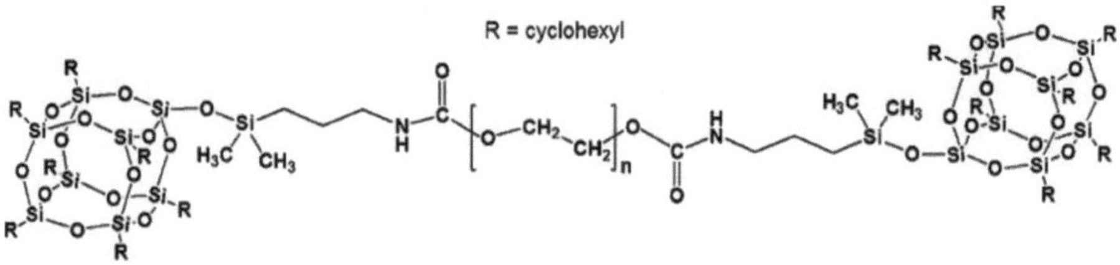

Figure 2.15: POSS-PEG-POSS telechelic polymers. Reprinted (adapted) with permission from [587]. Copyright 2007, American Chemical Society.

Reversible addition-fragmentation chain transfer (RAFT) polymerization allowed for a wider selectivity of the monomers containing carboxyl, amino and ionic groups, and it was used in different experimental conditions, including aqueous media. The POSS-containing RAFT reagent, prepared from (aminopropyl)isobutyl-substituted POSS was used in the RAFT polymerization of *N*-isopropylacrylamide (NIPAM) to give tadpole-shaped organic–inorganic hybrid poly(*N*-isopropylacrylamide) (PNIPAM). Similarly, from the same POSS-containing RAFT initiator in the polymerization of styrene, the organic–inorganic hybrid homopolymers and block copolymers were prepared [588].

A living/controlled polymerization method was combined with "click chemistry," which was based on the copper-catalyzed 1,3-dipolar cycloaddition between alkynes and azides. So-called Huisgen reaction was carried out under mild conditions using various functional groups. Click chemistry was most often used for functionalization of the pendant groups or for end groups of polymers and the preparation of well-defined polymers. The click reaction occurred between a polymer and small organic substrates or between two different polymer chains. The copper-catalyzed Huisgen click reaction between mono-, di-, and penta-functional azido-terminal polymers made by ATRP with alkyne-functional POSS molecules proceeded smoothly with formation of ditelechelic (dumbbell-shaped) and monochelic (tadpole-shaped), linear hybrid polymers as well as pentatelechelic, star-shaped hybrids. The inverse procedure was also used, that is, the reaction of azido-functional POSS molecule was "clicked" to alkyne-terminated polymers. Most of the POSS molecules behave as intrinsically hydrophobic. The polar groups attached to POSS molecules enable their compatibility with many other polymers. For instance, oligosilsesquioxanes containing oligo(ethylene glycol)s substituents (POSS-PEG monomers), having hydrophilic chemical structure, are soluble in water and easily disperse in polar monomers. The POSS molecules containing hydroxyl groups serve as an initiator in the ROP of lactide in the presence of a stannous catalyst [589].

A series of multiblock thermoplastic polyurethane (TPU) hydrogels with PEG soft segment and isobutyl-substituted POSS unit, incorporated into TMP diol (as the hard segment), was prepared using 4,4′-methylene-bis(phenyl isocyanate) (MDI). The MWs of PEG were 10, 20, or 35 kg/mol, while the molar ratio of the chain extender diol containing POSS to PEG (having constant MW 10 kg/mol) changed from 3:1 to 8:1. WAXD and SAXS experiments revealed that both hydrophilic soft segments (PEG) and hydrophobic hard segments (POSS) formed crystalline structures.

Owing to the thermodynamic incompatibility (of PEG and POSS units), microphase separation occured. The nanocrystals of inorganic–organic hybrid POSS networks served as physical cross-linking points. These polymers were swollen in water with equilibrium swelling ratio increasing from ~70% to ~600%, with growing PEG content. The water diffusion in the hydrogel was strongly dependent on PEG crystallinity. The shear modulus of the hybrid hydrogels ($0.3\ \text{MPa} < G < 4.0\ \text{MPa}$) were in the range higher than for most previously studied hydrogels. The high stiffness of such

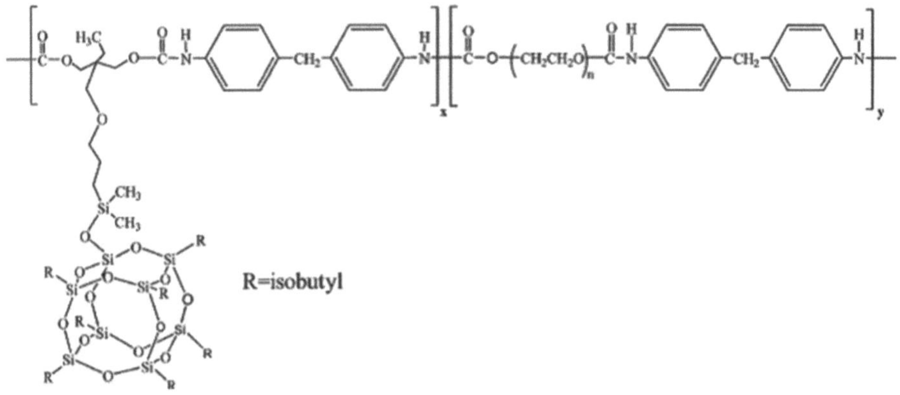

Figure 2.16: Chemical structure of the hybrid thermoplastic polyurethanes obtained through polycondensation reaction of 4,4′-methylenebis(phenyl isocyanate) (MDI) both with PEG and isobutyl-functionalized POSS-diol. Reprinted (adapted) with permission from [590]. Copyright 2010, American Chemical Society.

thermoplastic hydrogels was controlled by the POSS: PEG mole ratio but was independent of the MW of PEG segment. These thermoplastic PU-PEG copolymers modified with POSS side groups can be melt-processed into complex shapes and coatings and followed by hydration to form hydrogels. Multiblock TPU hydrogels with PEG soft segment and isobutyl-substituted POSS units can be used for coatings [590].

POSS derivatives are very attractive inorganic–organic nanofillers, also for polyurethane matrices. Due to their differential structure and functionality, they are easily soluble in many monomers and organic solvents, enabling incorporation into polyurethanes (and other polymers), even at the stage of synthesis. Binding POSS with polyurethane by chemical reactions led to improvement of many basic properties of PU and eliminated their disadvantages. The molecules of POSS have the ability to nucleate aggregation and crystallization in the polymer matrix, forming stable morphological structures. In the case of large POSS substituents, segmental vibrations can be reduced. Profitable second-order interactions between polymer chains containing POSS units may also exist.

POSS-modified nanohybrid polyurethanes may find numerous applications, for example, as attractive materials in material engineering, microelectronics, optics, and also as medical implants of improved biological resistance [591, 592].

TPU–POSS copolymers (TPU–POSS) were prepared from PU prepolymer, different amounts of octa(*N*-phenyl)aminopropyl-substituted polyhedral octasilsesquioxane (POSS) as a chain modifier, and a slight mole excess of hydroquinone bis(2-hydroxyethyl)ether, in a one-step process in MEK solution. The TPU-POSS, containing 0.2, 0.4, and 0.6 wt% POSS, was characterized by FTIR, TGA, DSC, and capillary rheometry.

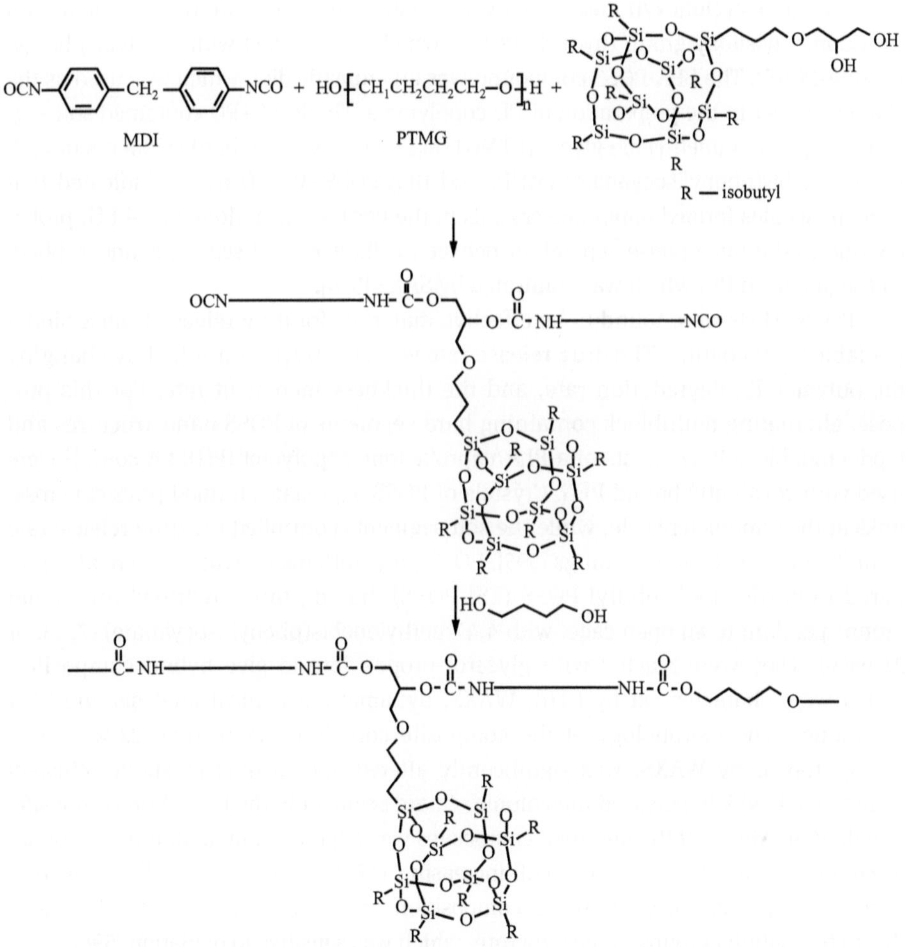

Scheme 2.24: Synthesis of polyurethane containing oligosilsesquioxane pendant groups. Reprinted (adapted) with permission from [592]. Copyright 2011, ZW CHEMPRESS – SITPChem (Poland).

The FTIR confirmed a complete conversion of the NCO terminal groups due to the absence of the band at ~2,253 cm^{-1}. The TGA showed that incorporation of POSS improved the thermal properties of TPU–POSS. The incorporation of 0.4 wt% of POSS increased the T_g of the flexible phase. Addition of POSS increased melting enthalpy of TPU–POSS, in comparison to neat TPU, due to the formation of larger crystals. The flexible phase exhibited an enhanced thermostability, probably due to increased phase separation between the rigid and flexible domains. The intrinsic viscosities increased with increasing amounts of POSS [593].

A PU elastomer containing 21 or 34 wt% POSS additives as molecular reinforcement in the hard segment was studied by B.X. Fu et al. The POSS molecules included

$(RSiO_{3/2})_8$ (R = cyclohexyl) and its derivative with one corner group substituted with hydridomethylsiloxy group (hydrido-POSS), which was reacted with 3-(allybisphenol-A) (BPA-POSS). The BPA-POSS compound was an aromatic diol, which was used as the chain extender in the preparation of PU copolymers. The POSS–PU contained soft segments of polytetramethylene glycol (PTMG) (M_w = 2,000 g/mol), hard segments of 4,4′-methylene-bis(phenylisocyanate) (MDI), and BPA-POSS. WAXD results indicated that POSS molecules formed nanoscale crystals in the hard segment domains of PU, probably due to the microphase separation between solid-like hard segments and rubbery soft segments in PU, which was confirmed by SAXS [594].

POSS–TPUs were found to be effective materials for drug release from a biodegradable stent coating. The drug release rate was effectively controlled by changing the polymer T_g, degradation rate, and the thickness increment rate. For this purpose, alternating multiblock containing hard segments of POSS nanostructures and biodegradable soft segments of a PLA/caprolactone copolymer [P(DLLA-*co*-CL)] were used with covalently bound PEG. Crystals of POSS aggregates formed physical cross-links at the nanometer scale, while the soft segments controlled the drug release rate from the POSS TPU stent coatings [595]. POSS-polyurethane derivatives were also prepared from trisilanolisobutyl POSS (TSI-POSS), having three hydroxyl functional groups pendant to an open cage, with 4,4′-methylenebis(phenyl isocyanate) (7, 13, or 22 wt%). They were reacted with glycerol propoxylate to give hybrid composites, which were characterized by FTIR, WAXS, dynamic mechanical analysis and TGA techniques. The morphology of the composite containing more than 22 wt% TSI-POSS, studied by WAXS, was significantly altered and distinct crystallite clusters were formed, which increased the volume of hard segment in the PU hybrid composite. The thermostability of the designed composites was dependent on a kind of functional group in the TSI-POSS structure. With increasing TSI-POSS content, the T_g of the composites was enhanced, while the decomposition temperature was slightly decreased due to the isobutyl groups in the structure, which was sensitive to oxidation [596].

Stable aqueous and homogenous hybrid PU/POSS dispersions were obtained from linear PU by the incorporation of various amounts (~2–10 wt%) of a secondary amino-terminated monofunctional POSS into the polymer chains built from hexamethylene diisocyanate, hexamethylene diisocyanate trimer (HDI-T), poly(tetramethylene oxide), and 2,2′-bis(hydroxymethyl)propionic acid. The GPC, FTIR, TEM, and XRD results confirmed that the POSS units were incorporated into the PU matrices based on HDI-T. Thus, the physical properties of the PU/POSS hybrid films were modified – the TS and surface hydrophobicity were increased [597].

POSS-capped PEG–bisurea copolymer and an octyl-capped copolymer were prepared according to Scheme 2.25. The physical association of POSS end groups caused reinforcement of shear-thinning hydrogels – the POSS-capped copolymer tended to gel more easily. Hydrogels made of the POSS-capped copolymer showed significantly enhanced mechanical strength and the thermal stability compared to octyl-capped copolymer [598].

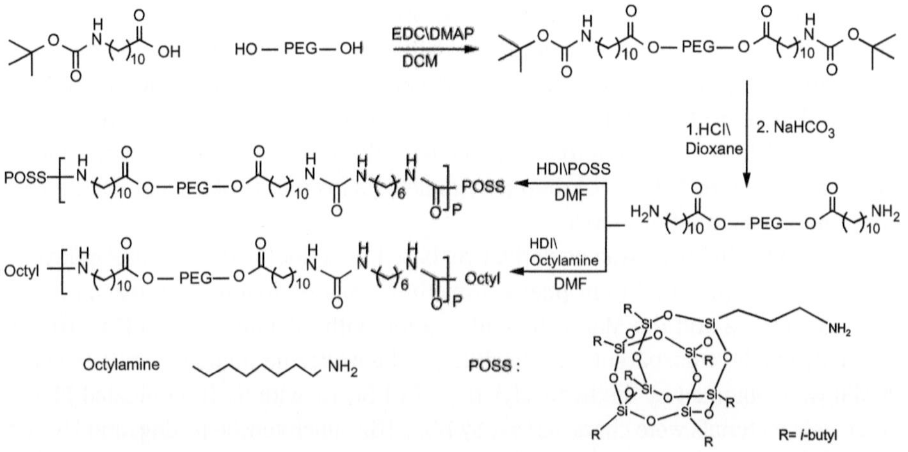

Scheme 2.25: Synthesis of the POSS-capped and octyl-capped copolymers. Reprinted (adapted) with permission from [598]. Copyright 2017, Royal Chemical Society.

Star-shaped poly-L-lactide-PEG POSS hybrid membranes (SPPS) were prepared by the photo-cross-linking of (methacryl)urethane functionalized star poly-L-lactide with poly(ethylene glycol) diacrylate (PEGDA) and methacrylate POSS. SPPS showed tunable mechanical properties (5.6–153 MPa in tensile modulus, 24–157% in elongation), biodegradation, biomineralization activity, and good osteoblast biocompatibility. The membrane of SPPS hybrids may find promising use in guiding bone regeneration [599].

Star-shaped copolymers (with number average MWs: $M_n = (3.1–7.9) \times 10^4$ g/mol) were prepared by grafting amphiphilic poly(L-aspartate)-*b*-poly(ethylene glycol) block copolymers on carboxy-functionalized POSS. 1-(3-Aminopropyl) imidazole was conjugated to the pendant groups of poly(L-aspartate) segments of these copolymers to form pH-sensitive micelles, which were used for trapping the anticancer drug doxorubicin (DOX) in the micelles. The mean diameter of the micelles containing DOX molecules were in the range of 50–60 nm and were much smaller than those of blank micelles (100–200 nm). The drug content and encapsulation efficiency of the micelles decreased with the substitution sequence of the side groups, carboxyl, benzyl and imidazole. The release of the drug micelles with imidazole groups was dependent on pH and more than 90% of the DOX could be released within 48 h in a weak acidic medium (pH 5.0). Most importantly, the pH-sensitive micelles exhibited better anticancer activity [600].

Organic–inorganic polymer hybrids were synthesized from octasilsesquioxanes containing OH groups and organic polymers: poly(*N*-vinylpyrrolidone), poly(2-methyl-2-oxazoline), and poly(*N*,*N*-dimethylacrylamide) through hydrogen bonding. They formed homogeneous, transparent films [601].

A high yield of highly transparent hydrophobic poly(siloxane, double-decker silsesquioxane) containing chloropropyl substituents was prepared by the dehydrocarbonative polycondensation reaction between di[(3-chloropropyl)isopropoxysilyl]-bridged double-decker octaphenylsilsesquioxane (DDSQ) and hexamethyltrisiloxane $H(Me_2SiO)_2SiMe_2H$ at ambient room temperature. Both halogenated polymers showed significantly higher T_g (~15 °C) than poly(dimethylsiloxane) (PDMS) and can be cast from solution as thin films [602].

A new bis-allyl benzoxazine derivative, based on a double-decker silsesquioxane (DDSQ-BZ) was prepared from phenyl(trimethoxy)lsilane, followed by reaction with 4-acetoxystyrene and the Mannich condensation with allylamine and CH_2O. Highly thermally stable, transparent, and flexible polybenzoxazine prepolymers (DDSQ-BZ-PDMS) were obtained by the hydrosilylation of DDSQ-BZ with Si–H-terminated PDMS. These new materials were characterized by DSC, TGA, microtensile testing, and UV–vis spectroscopy. The char yield (73 wt%) of DDSQ-BZ-PDMS was much higher than that of a typical polybenzoxazine [603].

Random and triblock copolymers containing double decker silsesquioxane units (DDSQ) and fluorinated poly aryl ether sulfone segments showed increased water contact angles and T_g values, but decreased dielectric constants (k values) with increasing contents of DDSQ units. The block copolymer had higher T_g values (up to 187 °C) than random copolymers (173 °C) with the same POSS molar content, due to their different sequence distribution. Their char yields at 800 °C in air ranged from 41.9 to 44.8 wt%. The dielectric constants of the triblock copolymers were drastically reduced (2.71 at 1 MHz) owing to the presence of DDSQ units and fluorine atoms [604].

POSS-poly(carbonate-urea)urethane copolymer (POSS-PCU) was prepared from *trans*-cyclohexanediol heptaisobutylsilsesquioxane (Figure 2.17), polycarbonate polyol, methylene diisocyanate (MDI), ethylenediamine, and diethylamine in *N,N*-dimethylacetamide (DMAC) solution. Antithrombogenic properties of POSS-PCU were compared to that of conventional polymers used in vascular bypass grafts [605].

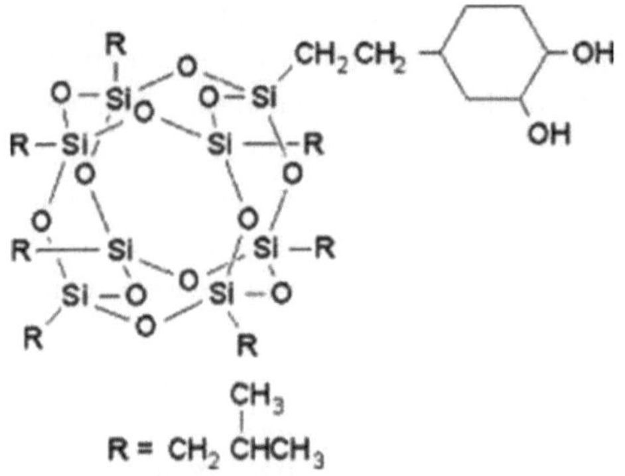

Figure 2.17: Molecular structure of *trans*-cyclohexanediol isobutyl-POSS. Reprinted (adapted) with permission from [605]. Copyright 2006, American Chemical Society.

PLA NCs were prepared by solution blending of commercial poly(L-lactide) (PLLA) and biodegradable core–shell particles, which were obtained via ring-opening copolymerization of a mixture of ε-caprolactone and L-lactide, initiated by polyhedral octa (3-hydroxy-3-methylbutyldimethylsiloxy)silsesquioxane (octa-POSS) in the presence of tin(II)2-ethyl-hexanoate. This gave random poly(ε-caprolactone-*co*-lactide) (PCL-PLA) a rubbery core, which was followed by the polymerization of D-lactide to form poly(D-lactide) (PDLA) as the outer shell. The rubbery properties of PCL-PLA was confirmed by DSC, which showed a T_g of ~−7 °C. The outer PDLA layer facilitated strong interactions between the core–shell rubber particles and the PLLA matrix, and led to the formation of a strong matrix/rubber stereocomplex. This was confirmed by FTIR, DSC, and XRD. The resulting biodegradable NCs showed a 10-fold increase in elongation at break and retained some mechanical properties (TS and Young's modulus) [606].

Copolymers of poly(L-lactide)s (PLLA) with 0.02–1.00 mol% of octa(glycidylether) POSS (OPOSS) cages were synthesized by the solution ROP of L-lactide in the presence of the $Sn(Oct)_2$ catalyst. FTIR and ^1H-NMR spectroscopic techniques confirmed the formation of secondary hydroxyl groups due to the reaction between PLLA main chains and the OPOSS cages. PLLA/PLLA-OPOSS NCs containing 1–30 wt% PLLA-OPOSS were prepared by solution-mixing of the neat PLLA polymer with various contents of the PLLA-OPOSS hybrid containing 0.50 mol% OPOSS. The T_g of the PLLA/PLLA-OPOSS NCs remained unchanged and the thermooxidative stability of the NCs was improved with PLLA-OPOSS content of 1–20 wt%, in comparison to the neat PLLA polymer. The presence of OPOSS enhanced the crystallinity of PLLA hybrids. Most of the PLLA/PLLA-OPOSS NCs had a higher crystallization rate at 120 °C. The nucleation density increased with the increase of PLLA-OPOSS loading [607].

A series of hybrid star PLLA (sPLLA) was prepared through the hydrosilylation between octakis(dimethylsiloxy)silsesquioxane ($Q_8M^H{}_8$) and telechelic PLLA macromolecules, functionalized with vinyl groups (mPLLA). The obtained sPLLA had fairly low polydispersity (M_w/M_n ~ 1.30) and five to seven arms of sPLLA, which decreased with increase in the chain length of mPLLA. Both the T_gs and T_ms of sPLLAs were higher than those of the mPLLA arms. The incorporation of POSS cages did not change the crystalline structure of PLLA, while the crystallinity of sPLLA was increased because the POSS core acted as a heterogeneous nucleating agent in the matrix and promoted the crystallization ability of PLLA. High-resolution TEM observation indicated that POSS was dispersed in the crystalline PLLA matrix as 5–20 nm aggregates. Microspheres of sPLLA with mean diameter 1 to 2 μm, prepared via emulsion solvent evaporation method, showed higher loading capacity, higher encapsulation efficiency, and lower drug release rate than mPLLA microspheres [608].

Polylactide (PLLA) scaffolds containing 0–5% POSS, functionalized with PEG, were prepared by electrospinning. The cytocompatibility [with human mesenchymal stem cells (hMSCs)] and hydrolytic degradation of this new PLLA/PEG-POSS NC was evaluated as a potential material for cartilage regeneration. The PEG-POSS nanoadditive did not exhibit any cytotoxicity. PEG-POSS was well dispersed in PLLA, but caused decrease in mean fiber diameter, increasing the specific surface area of the scaffold. The hydrolytic degradation rate of the PLLA/PEG-POSS NC was decreased and showed higher durability with time. The addition of PEG-POSS to the PLLA scaffolds did not affect its cytocompatibility to get hyaline cartilage from hMSCs [609].

Trans-(cyclohexanediol)isobutyl POSS (POSS-OH) (see Fig. 2.17) and (aminopropyl) heptaisobutyl POSS (POSS-NH_2) were grafted onto the PLA matrix, preliminarily functionalized with maleic anhydride groups. Thermal properties of the prepared hybrids were studied by DSC measurements. The POSS cages grafted onto the macromolecular chain led to an improvement of PLA crystallinity and a decrease of surface water wettability in the films made of PLA-g-MA/POSS-NH_2 [610].

A series of poly(octanediol citrate)-silsesquioxane (POC-POSS) hybrids with highly tunable elastomeric properties (in a hydrated state), biodegradation, and osteoblasts biocompatibility were prepared by the one-pot thermal polymerization. The POC–POSS hybrids exhibited significantly improved stiffness and ductility in both dry or hydrated conditions, as well as good antibiodegradation ability (20–50% weight loss within 3 months). The POC–POSS hybrids exhibited significantly enhanced osteogenic differentiation, expression of osteogenic markers, and calcium deposition. The high mechanical stability in the hydrated state and for enhanced osteogenic activity seem to favor future applications of POC–POSS hybrid elastomers for drug delivery, and as scaffolds and nanoscale vehicles for bone tissue regeneration [611].

Monomethacrylate-functionalized POSS molecules were polymerized and copolymerized (by Lichtenhan et al.) with other acrylic comonomers into novel linear silsesquioxane-based polymers [349]. The POSS-containing hybrid polymers with the novel architecture were prepared by living and controlled-living polymerization methods, for example, the synthesis of the homopolymers and triblock copolymers from a POSS-based methacrylate monomer using atom transfer radical polymerization (ATRP) was carried out by Pyun et al. [612]. The star-shaped hybrid polymers were prepared via ATRP of MMA using the octafunctional POSS molecule as an initiator[275].

Acrylic-polysiloxane coatings for building materials were prepared from acrylic acid-based resin emulsions, silicones, film-forming agents, water-proofing agents, thickeners, dispersants, defoaming agents, preservatives, and fillers (40–60 wt% of TiO_2, 30–40 wt% of mica powders, and 10–20 wt% of talc or $CaCO_3$) [613]. Silicone acrylic emulsion coatings were prepared by emulsion polymerization of the mixtures: methyl methacrylate (MMA), butyl acrylate (BA), acrylic acid (AA), and vinylsilicone monomer (VPS), under the initiation of $K_2S_2O_8$ in the presence of complex emulsifiers [614]. Acrylic polymers containing reactive epoxy groups were cured with epoxysilane oligomers, methylphenyl silicone resin, and $MeSi(OEt)_3$ in the presence of dibutyltin dilaurate (DBTL). Titanium white (TiO_2) and $BaSO_4$ were used as fillers. The obtained composites were used as weather-resistant coatings for blast steel [615]. PMMA-*g*-PDMS graft copolymers were synthesized by miniemulsion polymerization of PDMS macromonomer with MMA in the presence of AIBN as an initiator. Sodium dodecylsulfate (SDS) and PMMA-*b*-PEO block copolymer were used as surfactants, with methanol as agent reducing the interfacial tension [616]. UV-curable silicones were prepared from poly(dimethylsiloxane-α,ω-diol) and (acryloxymethyl)dimethylacryloxysilane by the hydrolytic polycondensation [617].

The ATRP method was used in the preparation of well-defined amphiphilic diblock copolymers of PEG and poly(pentafluorostyrene) (PPFS), using modified PEG as a macro-initiator. (Aminopropyl)isobutyl POSS was grafted to the PEG-*b*-PPFS via aromatic nucleophilic substitution reactions of F atoms in *para* position of the aromatic rings. The DSC data revealed that the grafted POSS acted as a plasticizer; the T_g of the PPFS block in the copolymer decreased significantly with increasing POSS content [618].

A novel water-soluble macromonomer, based on octavinyl silsesquioxane, contained vinyl-terminated PEG 400 attached to each of the eight arms. It was copolymerized in a 20:80 v/v acetone/water mixture with a vinyl-terminated, triblock copolymer of lactide-PEG-lactide to form a library of cross-linked hydrogels that were designed for use as scaffolds for bone repair. The cross-linked copolymer networks contained a range of nm-μm sized pores and showed swelling properties at pH 7.4 when the hydrogel networks underwent a slow hydrolysis with the release of PEG and lactic acid fragments. These hydrogels were noncytotoxic toward fibroblast cultures at pH 7.4 [619].

A novel POSS poly(carbonate-urea) urethane (POSS-PCU) NC polymer with covalently attached anti-CD34 antibodies showed enhanced capability to capture circulating endothelial progenitor cells. The POSS-PCU was used as a bypass graft, lacrimal duct, and as a bioartificial trachea. Its superior biocompatibility and unique biophysical properties make it amenable to be used for coating of medical devices, for example, for stents used after balloon angioplasty to improve re-endothelialization [620].

A very interesting group of silicon polymers are branched regular structures, so-called dendrimers [204, 621, 622]. Molecular brushes with PDMS side chains were prepared by atom transfer radical copolymerization (ATRP) via grafting through method using well-defined macromonomers with methacrylate functionality: monomethacryloxypropyl poly (dimethylsiloxane) and PEG methyl ether methacrylate. The obtained molecular brushes, containing amorphous PDMS fraction and crystallizable PEO segments, behaved as soft gels (with real shear modulus $G' \ll 10^4$ Pa) after annealing at high temperature [623].

A highly hydrophilic, antifouling dendronized poly(3,4,5-tris(2(2-(2-hydroxylethoxy)ethoxy)ethoxy)benzyl methacrylate) (PolyPEG) brush was prepared by surface-initiated ATRP (SI-ATRP) on PDMS substrates, which were first oxidized in H_2SO_4/H_2O_2 solution to transform the Si–CH_3 groups on their surfaces into Si–OH groups. Next, the PDMS surface was functionalized with ATRP initiator and the PolyPEG was grafted onto the PDMS surface via Cu-mediated ATRP. Antiadhesion properties were investigated via protein adsorption as well as by bacterial and cell adhesion studies. The PDMS–PolyPEG surface exhibited durable wettability and stability, and significant antiadhesion and antifouling properties. The PDMS–PolyPEG copolymer brushes may find biomedical applications [624].

Many other copolymers containing siloxane linkages have been described in the literature.

2.7 Antiadhesive, hydrophobic, and superhydrophobic properties of silicones

Hydrophobic properties of PDMS and other silicone polymers are known for a long time. In recent years, superhydrophobic modification of different surfaces was achieved, for example, through applications of functional silanes –fluorinated silanes or polysiloxanes [625]. A surface roughness and application of thin coatings of chemicals led to an unusual water repellent surface, causing water to bounce and roll off the surface [626].

The superhydrophobic surface was prepared from a supramolecular organosilane, 2-(3-(triethoxysilyl)propylaminocarbonylamino)-6-methyl-4[1H]pyrimidinone, and low MW PDMS (M_n = 583 g/mol). The supramolecular organosilane had quadruple hydrogen bonds and an intrinsically phase-separated structure in solution, which was retained even in the gel state, while a water droplet (diameter = 3 mm) on its surface formed the contact angle >150° [627].

Superhydrophobic coatings of PMSQ nanofilaments were obtained from equimolar amounts of liquid trichloromethylsilane (TCMS) and water vapor in the reaction vessel [628]. These filaments of several micrometers in length had diameters ranging from 20 to 150 nm. A surface of silicone-nanofilament coating was not made wet by water; droplets easily slid off and the rinsed sample remained virtually dry. This well-known Lotus effect [629] was caused by high surface roughness and the hydrophobic nature of silicone [630] and at pH 3–7, exhibited water contact angles ~150–160° [628].

Superhydrophobic PDMS surfaces were prepared by the surface-initiated ATRP polymerization (SIP) method, either flat or as the wall of microfluidic channels. First, oligo (ethylene glycol) methacrylate was incorporated into a structure of additive-SE-based vinyl-PDMS, which was cross-linked with prepolymer dimethylsiloxane containing vinyl groups and Si–H groups in the presence of a Pt catalyst. The obtained initiator-integrated PDMS substrates (iPDMS) were then treated with bromoisobutyryl bromide as the initiation moiety, followed by SIP with a second monomer, 1*H*,1*H*,2*H*,2*H*-perfluorodecyl methacrylate (FMA). Highly branched copolymers with poly(FMA) branches and poly [oligo(ethylene glycol) methacrylate] [poly(OEGMA)] cores were formed on the surface. The superhydrophobic surface with a contact angle of water at 155° was obtained from the iPDMS of ultralow initiator density [631].

Superhydrophobic surfaces on wood were obtained from potassium methyl siliconate (PMS) through a solution-immersion method. The silanol was formed by reacting PMS aqueous solution with CO_2, which was assembled via hydrogen bonds with the hydroxyl groups (–OH) present on the wood surface. The polymethylsilsesquioxane coating was formed through the polycondensation reaction between the hydroxyl and silanol groups. The morphology of products were characterized by FTIR, SEM, and the surface chemical composition was analyzed using energy-dispersive X-ray analysis and contact angle (CA) measurement. The properties of wood surface changed from hydrophilic to superhydrophobic, and the water CA was about 153° and the sliding angle was 4.6° [632].

Optically transparent superhydrophobic thin films of micro- and nanosilica hierarchical structures were prepared by spin-coating from suspensions of silica particles, obtained from a precursor solution of silane, H_2O, and ethanol with molar ratio of 1:4:4. These superhydrophobic films were characterized by SEM, optical transmission, and contact angle measurements [633].

The superhydrophobic coatings were prepared by a sol–gel method using hydrolysis and condensation reactions of poly(methylhydrosiloxane) (PMHS) with 3-aminopropyl(triethoxy)silane (KH550). The mass ratio of PMHS/KH550 of the sol–gel, made by spraying the sol–gel on microscope glass, affected the water contact angle (WCA) and the water sliding angle (WSA) of the film. With the mass ratio of PMHS/KH550 0.25 of the sol–gel, WCA of the corresponding film was 157° and the WSA was less than 1°. The diameter of the superhydrophobic film particle was about 40 μm. It was composed of micro-balls, whose diameter was about 2 μm. The microballs were composed of nanospheres with diameters less than 200 nm [634].

Pullulan-/poly(vinyl alcohol) (PULL/PVA)-blend membrane, exhibiting superhydrophobic properties (water contact angle larger than 150°) was prepared by modification with perfluorooctyl(triethoxy)silane (PFOTES) and the electrospinning method. The addition of a minor quantity of PVA resulted in an improvement in thermal stability and TS of the PULL membranes [635].

Durable and robust superhydrophobic thin coatings were created on textiles surfaces, in the absence of fluoro-free organosilanes, by dip-coating in a dispersion of silica nanoparticles modified with hexadecyltriethoxylsilane via a modified Stöber method [636]. Superhydrophobic textiles were prepared by dip-coating, using synthesis of nanosilica from TEOS solution (by the modified Stöber method), and its surface modification with hexadecyl(triethoxy)silane (HDTES), without the application of fluoroorganosilanes. The superhydrophobic coatings on textiles showed very good water repellency and excellent mechanical (especially abrasion and scratching resistance, adhesion of double side tape), chemical properties (e.g., exposure to acid and organic solvents and stability during laundering), and environmental stability (especially against external conditions and UV irradiation). Such superhydrophobic NC coatings also improved TS and elongation at break of the textiles [637].

Superhydrophobic surfaces were made from a copolymer of styrene with 1H,1H,2H,2H-perfluorooctyl methacrylate and vinyltriethoxysilane (PSFV) by the casting method from solution in THF, followed by the phase separation technique using addition of ethanol. The PSFV was synthesized by bulk polymerization of monomers. The superhydrophobic PSFV surface showed the water contact angle and sliding angle to be 162° and 4°, respectively. This simple method of preparation of superhydrophobic surfaces may find different practical applications [638]. Superhydrophobicity stimulates a variety of scientific and industrial applications in water repellency, friction reduction, self-cleaning, antifouling, and so on. Genzer and Efimenko deposited superhydrophobic monolayers on a stretched elastomeric PDMS substrate using $F(CF_2)_y(CH_2)_xSiCl_3$ molecules (from the vapor phase). After deposition, the strain was released and the film returned to its original size, while $F(CF_2)_y(CH_2)_xSiCl_3$ molecules were grafted on PDMS and formed a densely organized mechanically assembled monolayer. The water contact angle increased by up to 30° in comparison to the unstretched surface [639].

The effects of low surface energy and nanoroughness topography achieved, simultaneously or in sequence, are necessary to achieve superhydrophobicity of materials. Fluorine- and silicone-containing polymers as well as hydrocarbons exhibit low surface energy and can be used for the modification of cellulose surfaces by grafting, chemical vapor deposition, adsorption, and so on [640]. Surface properties of inorganic supports (silica or magnesium silicate) were modified (in an emulsion system) with selected alkylsilanes: 3-(2,2,3,3,4,4,5,5-octafluoropentyloxy)propyl(triethoxy)silane ($H(CF_2)_4CH_2O(CH_2)_3Si(OEt)_3$) (OPF), octadecylsilane,, and octyl(triethoxy) silane ($C_8H_{17}Si(OEt)_3$) (3, 5, or 10 wt%). Thus, hydrophobic powders were prepared. Fluoro-carbofunctional (triethoxy)silane (OPF) was synthesized by the hydrosilylation reaction of 1,1,2,2,3,3,4,4-octafluoropentyl allyl ether with $HSi(OEt)_3$ catalyzed by $[\{Rh(OSiMe_3)(cod)\}_2]$ [641]. Superhydrophobic films of hydroxide zinc carbonate (HZC), modified with fluoroalkyl(trimethoxy)silane $CH_3(CF_2)_6(CH_2)_3Si(OCH_3)_3$ (FAS), were prepared on aluminum substrate by a convenient in situ deposition process, immersing the aluminum substrates into an aqueous solution containing zinc nitrate hexahydrate and urea. Next, the obtained HZC films with different morphologies were modified with fluoroalkylsilane by immersing the $CH_3(CF_2)_6(CH_2)_3Si(OCH_3)_3$ (FAS) molecules in absolute ethanol solution. The HZC films displayed microporous, rose petal-like, or block-shaped structure, or pinecone-like structure depending on the deposition condition. The highest water contact angle of 156.2° was obtained after modification with FAS. The obtained the superhydrophobic surface considerably improved the corrosion resistance of aluminum [642].

Superhydrophobic polyimide (PI) films were modified on a the surface with 1H,1H,2H,2H-perfluorodecyl(triethoxy)silane $C_{10}F_{17}H_4Si(OCH_2CH_3)_3$ (FAS) under prolonged UV irradiation, providing unique micro/nanostructures. It was accompanied with an increase of the root mean square (RMS) surface roughness from 1.74 to 53.70 nm and a gradual increase of the water contact angle (WCA) from 105 to 159°. After 72 h of UV radiation exposure, the micro/nanostructured- and FAS-modified PI films exhibited super hydrophobicity with WCA larger than 150°, sliding angle (SA) less than 5°, and high thermal stability. The superhydrophobic properties continued to remain even after annealing at 350 °C [643].

The SEM images of the thin silica film prepared from TEOS by sol–gel and dip-pulling methods showed that its surface was covered with uniform particles of 20–30 nm and pores of 50–60 nm. A single-layer hydrophobic antireflective properties of silica were obtained by coating with 1H,1H,2H,2H-perfluorodecyl(triethoxy)silane, leading to a high transmittance (97%) and the large static water contact angle (130.6°). The film thickness was controlled by regulating the pulling speed for making the transmittance of the film, reaching a peak at a thickness of about 550 nm [644]. Optically transparent superhydrophobic silica thin films on glass surfaces with static contact angles 155–162° were prepared by spin-coating silica particles suspended in a precursor solution of methyl(triethoxy)silane (MTES) or its mixture with decyl(trimethoxy)silane, ethanol and H_2O with a molar ratio of 1:4:4. The surface transmittance ranged from 75% to 95%. The glass surfaces were modified with different kinds of silica particles: micro-, nano-, and hierarchical structures. The resulting superhydrophobic films were also characterized by SEM [645].

The superhydrophobic and optically transparent coatings based on methyl(triethoxy)silane were deposited on the glass substrates using the spray deposition method, followed by post surface modification with trimethyl(chloro)silane. A decrease in the surface free energy of coating was observed. The deposited coating surface with micro-nano scale hierarchical structures showed high optical transparency, improved thermostability, static water contact angle ~167°, low sliding angle ~1–3°, and stable superhydrophobic properties [646]. Superhydrophobic silica nanoparticles were prepared by microwave-assisted two-step sol–gel method. Silica particles of different sizes were obtained from TEOS. Next, they were hydrophobized with hexadecyl(trimethoxy)silane (HDTMS). Under microwave irradiation, high conversion degrees were obtained in relatively short reaction times. Silica NPs showed a bimodal size distribution, high water contact angles (≈ 150°), and low hysteresis values [647]. Homogeneous organic–inorganic hybrid hydrophobic xerogel nanocoatings prepared from TEOS and HDTMS in the presence of *n*-octylamine were used in stone restoration and showed improved resistance against acid rains [648]. Superhydrophobic polysiloxane coatings on nanosilica were prepared by acid-catalyzed sol–gel hydrolytic polycondensation of silanes mixture (triethoxymethylsilane, 0.015 mmol, and diethoxydimethylsilane, 0.005 mmol) in isopropanol and water (v/v = 2:1). The samples were dried overnight and thermally stabilized at 160 °C for 2 h. The superhydrophobicity of the polysiloxane surfaces was slightly improved by the surface abrasion. These properties exhibited stable superhydrophobicity even after long distance friction of the surface [649].

A biomimetic superhydrophobic surface was formed on SiO_2-containing PDMS with various micro- and nanostructured surfaces prepared using grass leaf, sand paper, and PET sheet with parallel groove geometry as templates via PDMS replica molding. Next, the PDMS surfaces were treated by a flame for 20 s. The effect of surface microstructures on the transparency of PDMS was also studied. The flame-treated surfaces of PDMS replicas and a flat PDMS sheet showed superhydrophobicity with CA values slightly exceeding 150° in both directions (perpendicular and parallel to the grooves) and ultra-low sliding angle reached a limiting value of 1° [650].

Superhydrophobic NC fiber membranes were prepared by electro-spinning of poly(vinylidene fluoride) (PVDF)-blend with silica nanoparticles modified with silane coupling agent (SCA), which were obtained by the sol–gel method. The effects of the SCA [vinyl(trimethoxy)silane, vinyl(triethoxy)silane or *n*-octyl(triethoxy)silane] and the mass ratio of the modified silica particles and PVDF on the hydrophobic properties of the composite membranes were studied. The incorporation of SCA-modified silica particles into the PVDF membrane increased the surface roughness, super hydrophobicity, and self-cleaning property of the NC membranes. The water contact angle (CA) ranged from 150° to 160° when the mass ratio of modified nanosilica with PVDF matrix increased from 0.5:1 to 5:1 [651].

The superhydrophobic coatings composed of silica nanoparticles and perfluorooctylated quaternary ammonium silane coupling agent (PFSC) were prepared on a cotton fabric. The silica sol was prepared by alkaline hydrolysis of TEOS in an aqueous ethanol and was catalyzed by ammonia. Then, the fabric was immersed in the silica sol and dip-coated twice. After drying, the fabric was immersed in the methanol solution containing PFSC, and again dip-coated twice and cured at 160 °C. The modified fabric showed a high hydrophobicity and oleophobicity, and the water and oil contact angles were about 145° and 131°, respectively [652]. Superhydrophobic surfaces showing a dual-sized surface roughness were prepared by a complex coating of silica nanoparticles, functionalized with amino groups, onto cotton textiles, followed by hydrophobization with 1H,1H,2H,2H-perfluorodecyl(trichloro)silane, stearic acid, or their mixtures. The wettability and morphology of the fabricated surfaces were characterized by contact angle measurements (142–168°), FTIR, SEM, TEM, and TGA [653]. Superhydrophobic cotton fabrics were obtained by the one-step chemical modification in solutions of bifunctional polysiloxanes, which contained in their structure trialkoxysilyl or glycidyl functional groups (capable of bonding to cotton), and fluoroalkyl groups showing surface activity. No stiffening, any decrease in mechanical properties and color changes of modified fabrics were observed. The superhydrophobic properties of cotton fabrics were saved even after multiple washings. They were determined by measuring the water contact angle by drop profile tensiometry [654]. Strongly hydrophobic and fire-retarded cotton fabrics were prepared in a two-step modification process. First, guanidine carbonate and/or ammonium dihydrogen phosphate were immobilized in aminopropyl(triethoxy)silane, using the sol–gel process. This was followed by impregnation of the modified fabrics with fluorofunctional silane or polysiloxane. The water contact angle reached 141–143°. As a result of the performed modifications, multifunctional fabrics with a substantially reduced flammability (and outstanding values of the limiting oxygen index (LOI) 46–71%) improved the thermostability, and high hydrophobicity was obtained. The modified fabrics were also characterized using SEM-EDS and FTIR techniques [655].

Multifunctional fabrics with super anti-wetting (superhydrophobic) properties and controllable adhesion (e.g., oil–water separation, self-cleaning, asymmetric/anisotropic wetting for microfluidic manipulation, air/liquid directional gating, and micro-template for patterning) have attracted a lot of interest in both fundamental research and special industry applications. For these, very important are coating durability, air permeability, UV-shielding, photocatalytic self-cleaning, self-healing, and patterned antiwetting properties [656].

Colorful superhydrophobic poly(ethylene terephthalate) (PET) textile surfaces with self-cleaning property were prepared by Xue et al. by chemical etching with sodium hydroxide solution and coating with PDMS [657]. The original textiles must be first cleaned with deionized water to remove the impurities. The cleaned textiles were dipped in NaOH solution for 10 min. Then, the soaked textiles are coated doubled-side by a polyethylene film and heated at 120 °C. Finally, the textiles were rinsed by neutral water (pH 7) and dried in an oven. After etching the solid surfaces by PET fibers, the contact area between the water droplets and textile surface was not only decreased, but air was also trapped for improving the fiber hydrophobicity, thus allowing the water to easily roll off from the textiles. The PDMS-modified textiles have remarkable durability against different pH solutions and exhibit excellent resistance to washing, abrasion, and exposure to UV light. The superhydrophobic surfaces can be colored by conventional dyeing or by thermal transfer printing [657].

Reinforced ethyl cellulose (SEC) sponges were prepared by cross-linking ethyl cellulose (EC) with epichlorohydrin (ECH) and complexing with silanized carbon nanotubes (Si-CNTs), followed by coating nanosilica on the surface of porous sponges, and subsequently modifying with hexadecyl(trimethoxy)silane (HDTMS). The SEC sponges exhibit superhydrophobic and superoleophilic properties (water contact angle: 158°, oil contact angle 0°, and sliding angle: 3°), outstanding mechanical strength, and absorption ability of oils and organic solvents. The absorption capacity was 64 times their own weight, which slightly decreased to 86.4% of its initial value after 50 separation cycles. The SEC sponges were used as an ideal absorbent in oil spillage cleanup [658].

Superhydrophobic coatings (water contact angle > 150°) and good mechanical properties are also achieved from high solid-fractal-waterborne dispersions, not containing polysiloxanes. These coatings exhibit oleophobic properties also (olive oil contact angle 120°) [659].

2.8 Physiological properties of organosilicon compounds and silicones

Silicon is present in plants, animals, and humans. The presence of silicon is necessary for the growth of hair and nails. Silicon is consumed each day with food (~0.5 g), but only 20–30 mg is introduced into the human blood stream. Ortho- and oligosilicic acids are water soluble and are easily removed from the human body. A constant concentration of silicon in the blood is regulated by the kidneys. Chlorosilanes attack tissues of the skin and the eyes, while tetra-substituted organosilanes are nontoxic if they do not contain any functional groups. Amino- and alkoxy-substituted silanes can be extremely toxic, for example, (4-aminobutyl)(diethoxy)methylsilane [660]. Ethoxysilanes are toxic for the kidneys and the liver [661]. (3-Iodopropyl)silatrane is toxic (median lethal dose $LD_{50} > 29$ mg/kg), and aryl and mercaptomethyl silatranes are very toxic to warm-blooded animals, while they do not poison bacteria, fungi, and frogs [662].

On the other hand, methyl, ethyl, ethynyl, chloromethyl, and (1-chloroethyl)silatranes are nontoxic ($LD_{50} > 2{,}000$ mg/kg). Highly toxic are organosilyl phosphines, arsines, stibines, bismuthines [663], hexaalkyl stannosiloxanes, and plumbosiloxanes [664].

Liquid and solid PDMS have found numerous technical applications in medicine, for example in face-lifting and in construction of artificial limbs, cardiac valves, gullets, trachea, etc. Silicone breast implants are based on silicone elastomers (SEs) and silicone rubbers (SRs) plasticized with silicone oils of relatively low MWs. SRs serve as catheters, drainage, and tracheotomy tubes. Synthetic arteries impregnated with silicones are well tolerated immunologically; they are elastic, do not break, and avoid coagulation. Silicones are useful in heart-lung medical equipment for blood circulation, as they prevent foaming. High MW silicones are physiologically inert, but low MW siloxane (e.g., as plasticizers of breast implants) are active. 2,6-*cis*-Diphenylhexamethylcyclotetrasiloxane is used in the treatment of prostate disease and as a drug for decreasing sexual activity [662].

Silicone oils are considered as physiologically inert liquids. According to the FDA and other institutions, very small amounts of silicone oils (called as *dimethicones* or *cyclomethicones*) are considered as ingredients of drugs, cosmetics, and safe food additives. A low toxicity ($LD_{50} > 15$ g/kg) was determined in tests of the digestive tract after inhalation of *cyclomethicone* vapors (median lethal concentration $LC_{50} > 1{,}200$ ppm). However, occasionally irritations of the epithelial and eyeball is noticed and increase in liver weight in rats that were fed with 5% and 10% of *dimethicone* or *cyclomethicone* as food additive for 90 days. They were also subjected to inhalation of 5% and 10% of *dimethicone* or *cyclomethicone* vapors. One crucial problem exists in the use of SEs, plasticized with low MW silicone oils, for breast implants [665]. Low MW polysiloxanes, especially slightly soluble in water, may penetrate the cell membranes into blood and undergo partial enzymatic biodegradation. Thus, greater caution is recommended, especially in medical applications of silicones [665–667].

PDMS, used as functional additive to food (E-900), should have viscosities in the range 200–300 cSt and appropriate MWs ~10,000 g/mol, which corresponds to an average degree of polymerization (DP) ~ 120 [670, 671]. In pharmaceutical applications and in food industry, PDMS is allowed to be used only with linear structures. Daily consumption of PDMS with the mentioned viscosity with food cannot exceed 1.5 mg/kg of the body weight [668–671].

According to [672, 673] pharmaceutical preparations contain PDMS (named *Simeticone* or *Dimeticone*) with DP 20–400 and viscosity 20–1,300 cSt, while PDMS with viscosity <50 cSt can be used only for external applications. PDMS was detected in the liver, lymph nodes, and spleen, approximately 3–4 years after the implantation of breast implants [674]. Silicone polymers may also affect the conformation change in some proteins (e.g., myoglobin) [675].

Silicones biodegradation in the body was observed in vivo [676, 677] and in vitro. After a few months in the blood and liver of experimental animals, emerged a variety of low MW biodegradable silicone products. The toxicity of the cyclosiloxanes was evaluated. Administration of D_3 (hexamethylcyclotrisiloxane), D_4 (octamethylcyclotetrasiloxane), D_5 (decamethylcyclopentasiloxane), and D_6 (dodecamethylcyclohexasiloxane) intraperitoneally to animals caused pneumonia, followed by the liver and hepatocellular necrosis [666, 667].

Analysis of the reaction body after the administration of a lethal dose of cyclic variations of PDMS showed that the toxicity of D_4 can be compared to the toxicity of carbon tetrachloride or trichlorethylene. Cyclosiloxanes migrating with breast implants may accumulate in other organs and increase their weight [678]. Similar studies were conducted by the Dow Corning Corporation in 1990. After oral administration of D_3, D_4, or D_5 (with food) to rats, hepatomegaly was observed [679]. Other studies have shown that cyclosiloxanes accumulate in the liver, lungs, blood, and fat, causing kidney and adrenal hyperplasia in rats of both sexes [680].

Studies of immunotoxic properties showed that cyclosiloxanes strongly stimulate the immune response, manifested by acute inflammation, denaturation, and a change in the conformation of the two proteins – fibrinogen and fibronectin. Recently, attention was drawn to the disease of the immune system in children nursed by mothers with breast implants. This problem still remains in the realm of scientific discussion. Experimental studies in monkeys [681] and rats [682, 683] indicated that the PDMS and cyclosiloxanes (D_n), after oral administration, were absorbed and appeared in the peripheral blood and in urine. They recognized the negative impact of D_n in reproduction, stating their teratogenic and oncogenic effects [684].

Currently, it is believed that polysiloxanes of lower MWs and especially those that are soluble in water can penetrate cell membranes and the blood by enzymes that are prone to biodegradability. In the light of the controversial research, it seems to be a common view with total indifference to physiological silicone oils and their derivatives [665]. Therefore, companies producing silicones and interested in the effects of their actions on living organisms, contribute greatly to the development of studies on the toxicity of these compounds.

According to industry producers, there is compelling evidence that siloxanes are nontoxic to the environment [685].

2.9 Biomedical and cosmetic applications of silicones and modified silica

Functional silicones, especially modified with amine or polyether moieties are ingredients of many well-known hair and skin formulations [686]. PDMS are also used as sun tan lotions in blends with UV-B additives such as *p*-aminobenzoic acid derivatives or (*p*-methoxy)cinnamates.

Carbofunctional silicones find many applications in the manufacture of personal care and cosmetics products, as also in versatile aesthetics formulations [687–691].

General structure of oligosiloxanes and polysiloxanes:

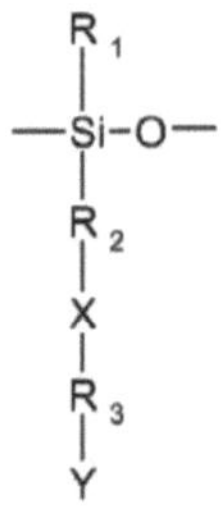

where R_1, R_2, and R_3 are alkyl or alkylene groups containing 1–10 or 1–18 C atoms, X and Y – alkyl or aryl quaternary amido or imido groups, containing 1–20 C atoms - they showed good antimicrobial properties [692].

In recent years, silicone gels and silicone gel sheetings are coming into prominence in scar therapy. Different types and structures of silicone polymers have been used in dermatological products. Mechanisms of their action on scars are discussed [693]. Vinyl functionalized PDMS and cross-linked with H-siloxanes are used as ingredients of dental composites [694]. A polysiloxane copolymer grafted with quaternary ammonium group of *tert*-butyl-aminoethyl methacrylate and covalently attached fluorescein was compatible with SEs and showed antibacterial properties against gram-negative bacteria *Escherichia coli*. This polymer probably formed a stable link with bacteria and showed a concentration-dependent damaging effect on *Escherichia coli* cells, as a result of binding to the inner bacterial membrane [695].

A polysiloxane with 5,5-dimethylhydantoin-based *N*-halamine pendants was prepared by alcoholysis of *N*-halamine precursor 3-(3-hydroxypropyl)-5,5-dimethylhydantoin (HPDMH), followed by chlorination of hydantoin ring with *tert*-butyl hypochlorite. The polyethylene (PE) fibers impregnated with *N*-halamine-grafted polysiloxane formed a 68 nm coating layer in supercritical carbon dioxide and exhibited biocidal properties against colonies of bacteria (*Staphylococcus aureus* and *Escherichia coli*). The coating layer on PE fibers was stable for washing cycles, storage, and UV irradiation [696].

New star-shaped poly-L-lactide-PEG-polyhedral oligomeric silsesquioxane (SPPS) hybrid membranes were prepared and used for bone regeneration purposes by a photocross-linking method using inositol as a core. SPPS showed tunable mechanical properties (5.6–6.0, ~107–153 MPa in tensile modulus, 24–36%, ~131–157% in elongation, more than 90% recovery), biodegradation, biomineralization activity, and good osteoblast biocompatibility. It appears that these hybrid membranes could find potential applications in bone tissue regeneration [599].

The following properties of silicones favor their biomedical applications [697]:
- physiological indifference,
- moderate biocompatibility,
- good elasticity over a wide temperature range,

- excellent resistance to biodegradation,
- excellent resistance to oxidation and ultraviolet light,
- outstanding ageing resistance,
- excellent dielectric properties over a wide range of temperatures.

Different types of medical-grade SRs are usually compounded with a very pure, finely divided amorphous silica filler to increase TS and tear strength. Incorporation of fillers affects the blood compatibility of silicones. Biocompatible SRs, reinforced with fillers, have been prepared from two types of modified amorphous silica: (1) the Si–H-functionalized silica, by reaction with $MeHSiCl_2$; and (2) silica with a covalently immobilized antithrombogenic heparinoid agent. Hydroxylated- and hexadecanoyl-modified silicones showed enhanced albumin binding and biocompatibility.

A SR based on poly(phenylsiloxane) was found to be a very biostable material with high dielectric properties and can be used as a coating for pacemakers. Several silicone-polyurethane thermoplastic interpenetrating polymer network (IPN) copolymers (Cardiothane^{TM}, Rimplast^{TM}) were based on vinyl functional SE which was mixed with a polymer matrix (e.g., polyurethane), followed by the addition of a crosslinker. They showed exceptional blood contacting and tissue biocompatibility, high strength, and other good physical properties. Graft polymerization of hydrophilic vinyl monomers onto silicone surface improved the hemocompatibility of silicones.

Silicones also show good blend-compatibility with various block copolymer compositions such as styrene-*b*-butadiene-*b*-styrene copolymers (SBS) and styrene-*b*-ethylene-*b*-butylene-styrene copolymers (SEBS). These blends had good surface and processing properties. Polypropylene (PP) also showed good compatibility when added to silicone-modified block copolymers. The resulting silicone/PP/SEBS compositions have a uniquely favorable balance of properties for biomedical use. They are free of phthalate plasticizers, exhibit hydro-phobic properties, and improved elastomer strength. Biomaterials based on silicones and hydrogels show surface wettability, which increased markedly with increasing content of the hydrogel phase. An antithrombogenicity of silicones can be improved by an ion implantation technique. Plasma treatment affects hydrophobicity of silicones, but the hydrophilicity of surfaces modified this way often disappears with time.

The manufacturing process of catheters with balloons and some wound dressings are dependent on dispersing the silicones in solvents (fluorocarbon or trichloroethane). They are later dipped or cast to form uniformly smooth surfaces.

Most important biomedical applications of silicones include [697]:
- tubings for carrying blood, drugs, or nutrients;
- implants and encapsulants for plastic, metallic, orthopedic and other surgical specialties;
- special goods: rods, rings, sheets, seals, tubes, valves, and so on;
- membranes of blood oxygenators in heart-lung circulation;
- drug delivery systems and aids for pharmaceutical processing.

Soft SR-polyacrylamide (PAA) gel composites suitable for biological and medical use are prepared by a radical grafting of their blend. The swelling of the composite materials in water and in water solutions is fast and pH-independent.

SRs are useful for treating embolisms in kidney arteries and as a filler for balloon-catheters. They are used for occlusion of aneurisms or pathological vessels. SRs are sterilized by ethylene oxide, steam autoclave, dry heat, or radiation, leading to an indefinite shelf-life [698].

PDMS-based biomaterial containing covalently bonded heparin was prepared by coupling the terminal aldehyde groups of diazotized heparin to the aminopropyl groups. Primary amine groups on the surface of PDMS elastomer were obtained by hydrosilylation grafting of aminopropyl vinyl ether (APVE) to Si–H groups, which were formed by argon plasma treatment (see Scheme 2.1) [119].

Monofunctional poly(ethylene oxide) (MPEO) of MW 350, 750, and 2,000, respectively, functionalized with $Si(OEt)_3$ groups were used as cross-linkers for poly (dimethylsiloxane)diols, together with $Si(OEt)_4$, giving copolymers with poly(ethylene oxide) (PEO)-rich surfaces. Their hydrophobicity was considerably higher in air. Silicone copolymers having low MW PEO chains (350 Da) reduced the adsorption of proteins: buffer (fibrinogen) (90%) or plasma (85%). They were cast as films, or molded into complex shapes [699].

A reverse microemulsion synthesis was used to prepare amine-functionalized silica nanoparticles of three different average diameters (50, 100, and 200 nm) with similar amine content [700]. Silica nanoparticles and nanorods (SNRs), functionalized with aminoalkoxysilanes, were modified with *N*-diazeniumdiolate nitric oxide (NO) donors and subsequently used in the manufacture of a new class of NO-releasing materials, which show antibacterial activity. The resulting hybrid nanoparticles, consisting of *N*-(6-aminohexyl)aminopropyl(trimethoxy)silane and TEOS, were highly monodisperse in size. *N*-Diazenium-diolate derivatives of secondary amines were used as NO donors. They belong to a new class of NO-releasing materials. The aspect ratio and size of the SNRs were tuned by changing the silane concentration, pH, and the temperature, during the surfactant-templated synthesis. The bactericidal efficacy of the NO-releasing nanoparticles against *Pseudomonas aeruginosa* increased with decreasing particle size. Smaller diameter NPs associate with the bacteria at a faster rate and to a greater extent than larger particles [701, 702].

Mesoporous silica nanoparticle (MSN), modified with chitosan were used as biodegradable pH-responsive drug nanocarriers, which were linked by hydrogen bond between $-NH_2$ on chitosan and –OH on MSN [702]. The adsorption and release of Naproxen was studied from the high quality mesoporous silica (MCM-41) as a model adsorbent. The MCM-41 silica easily formed tablets or pellets; it had hexagonally arranged pores with a regular structure, high pore volume (~1–2 cm^3/g), and high adsorption capacity (high specific surface area ~1000 m^2/g). Both ends of the cylindrical pores were likely open. The regularity of a pore shape and the uniformity of pore dimension were very important for new drug delivery systems [703].

2.10 Applications of silane-modified thermoplastics in medicine

Silane-modified thermoplastics and silica find new applications in medicine as well. A carbofunctional silane (CFS) was used in the manufacture of contact lenses. These special membranes pass oxygen and this ability is measured by the so-called *equivalent oxygen percentage* (EPO). The minimum value of EPO needed is 5–7%. For soft lenses made of PDMS, with a thickness of 0.2 mm, EPO is ~20%. Because of the hydrophobic properties of PDMS, which can lead to a damage of the cornea, lenses made of pure PDMS are not used. They are often made of carbofunctional silanes or silicone-methacrylic copolymers, that is, (methacryloxypropyl)(trimethylsiloxy)silane (TRIS) or (methacryloxypropyl)pentamethyldisiloxane. Lenses made from these materials attain EPO values only of ~10% [27].

In biomedical and pharmaceutical applications, hydrogels have been used for a long time, for example, a biocompatible hydrogel of poly(2-hydroxyethyl methacrylate) (PHEMA) absorbs a large amount of water and is used in the manufacture of ophthalmic intraocular or contact lenses, vascular prostheses, drug delivery, and soft-tissue replacement. Bioactive PHEMA-silica hybrids are manufactured by the addition of silica nanoparticles to a solution of HEMA monomer with PHEMA or the solution of HEMA and TEOS (as a silica precursor) using in situ sol–gel processes. A hybrid monomer of HEMA and 3-aminopropyltriethoxysilane (APTES) and its polymers were prepared and composite materials were synthesized using hydrolytic condensation reactions of TEOS, followed by in situ radical polymerization of HEMA. The manufactured hybrid materials were used as bioactive scaffolds for bone engineering [704].

Composites of SR with hydrogels are often used as drug release materials. A SR-polyacrylamide (PAAm) composite hydrogel showed high specific surface area, pH-independent swelling degree, and nonionogenic character [121, 705].

Owing to their biocompatibility and biodegradability, PLA finds numerous biomedical applications. In triblock copolymers of poly(lactide-*b*-methylvinylsiloxane-*b*-lactide), in which the vinyl groups were substituted with carboxylic acids at neutral pH, the carboxylate functional polysiloxane central block binds to the surface of magnetic nanoparticles, while the PLA tail blocks provide dispensability in PLA solvents through interparticle steric repulsive forces. These magnetic materials were used in magnetic field-directed drug delivery, magnetic cell separations, and magnetic hypertherma therapy in the treatment of tumors [706].

Polysiloxanes, functionalized with quaternary ammonium salt (QAS) groups, show permanent bactericidal and fungicidal properties. Newer low surface energy antimicrobial coatings were based on hybrid siloxane-epoxy coatings containing QAS moieties, which are capable of self-decontaminating in a variety of environments. These coatings reduced pathogenic bacteria on the surface by up to 99.9% of [707].

Since the 1970s, a wide range of polyurethane elastomers find biomedical applications due to a combination of good hydrolytic and oxidative resistance, excellent mechanical properties, good processability, and good biocompatibility. Commercial functionalized copolymers (polyurethanes or polyureas) with ~10 wt% PDMS contain PC or polyether units as soft segments. They are used in the manufacture of vascular grafts, intra-aortic balloons, in artificial heart grafts, implants, and other medical applications [708].

Synthetic polymers with pendant sugar moieties play the role of a simplified model of biopolymers bearing oligosaccharides. They also find application as artificial glycoconjugates in biochemistry and medicine. New polysiloxane copolymers with pendant glucosylthioureylene groups were prepared by a reaction of amino-functionalized PDMS with glucosyl isothiocyanate. They showed optical activity and can be used as biomedical and biotechnological materials and also as a model of amphiphilic polymers. Saccharide-based polysiloxanes, that is, PDMS-grafted amylase or PDMS-grafted saccharose were also prepared and characterized [709].

Siloxane bonds play an important role in the manufacture of medical implants. The surfaces of gold electrodes of thickness-shear mode acoustic wave sensors were coated with two organosiloxanes: octaphenylcyclotetrasiloxane and poly(mercaptopropylmethylsiloxane). The flow-through adsorption of some proteins – human serum albumin, fibrinogen, hemoglobin, immunoglobulin G, cytochrome C, and others – to the two siloxane surfaces and the gold electrode were detected by acoustic network analysis [710].

An organic–inorganic hybrid material (epoxy-SiO_2) was formed by incorporating epoxy structure units covalently into a silica glass network. The precursor was prepared by the reaction of DGEBA with 3-aminopropyl(trimethoxy)silane (APTES) via the sol–gel method. The hydrolytic polycondensation of the precursor with TEOS in THF at room temperature gave epoxy-SiO_2 hybrid sol–gel material containing 50 wt% SiO_2. Thermal properties of the hybrid material were characterized by DSC and TGA. The solid hybrid epoxy-SiO_2 material was used as a powder component of bone cement. The liquid ingredient contained bis-phenol-A glycidyl methacrylate (Bis-GMA), triethyleneglycol dimethacrylate (TEGDMA), and 20, 25, and 55 vol% of MMA, respectively. Mechanical properties (Young's modulus, compressive strength, hardness, and impact strength) of the new epoxy–SiO_2 bone cement exceeded those of commercial bone cement, while the TS and bending strengths of both cements were similar. The new epoxy–SiO_2 bone cement exhibited very low cytotoxicity as compared to the commercial bone cement [711].

PDMS-TEOS ormosils with different Ti and Ca contents showed in vitro bioactivity, which was dependent on the polar surface characteristics of these materials, the presence of a highly polar surface with intermediate base/acid ratio, and specific roughness [712].

Hydrolytic polycondensation of PDMS and TEOS with (tetraisopropyl)titanate (TiPT) at a molar ratio TiPT/(TiPT + TEOS) of 0.15, with the addition of calcium nitrate gave a pore- and crack-free transparent monolithic organic–inorganic hybrid material composed of a Si–O–Ti–O network modified with methyl groups, and a Ca^{2+}

ion ionically bonded to the network. This hybrid material showed a high bending strength, Young's modulus, and strain at failure, similar to those of the human cancellous bone and a high bioactivity. It can be useful as a new type of bioactive bone-repairing material [713].

A high MW poly(dimethyl-*co*-methylvinylsiloxane) ($M_v = 450{,}000$ Da) was used as a matrix for a composite with silica and a mixture of Fe_2O_3/TiO_2 as a color modifier. It was cross-linked with 2,4-dichlorobenzoyl peroxide as an initiator. This heat curable silicone-based formulation was used for relining removable dental prosthesis, which were stable even after 700,000 mastication cycles [714].

Novel dental resin composites containing various amounts of (octamethacryl) octasilsesquioxane (POSS), different methacryl functionalized resins and nanosilica, were cured by the UV light method. Addition of POSS decreased the volume shrinkage and initially increased their mechanical properties [715].

Silicon-based interpenetrating polymer networks (IPNs) were prepared from a telechelic commercial ladder silsesquioxane oligomer (LDS), a 1,4-bis(dimethylvinylsilyl)benzene (PCS) and $D^H{}_4$ by the condensation and hydrosilylation reactions. The obtained IPN were cured with Ti and Pt catalysts, followed by processing using melt-and-compression-molding method. The Izod impact strength and flexural modulus of LDS/PCS (8:2 w/w) were larger than those of each component due to a positive effect of IPN formation in the LDS/PCS combined system [716].

An aqueous solution containing 8.0–12.5 wt% of orthosilicic acid oligomers of the general formula $[(OSi{\equiv})_n]Si(OH)_{4-n}(OH)_{3n}$ ($n = 2$ or 3) – which were analyzed by ^{29}Si-NMR – and 30–40 wt% of aminoalcohol hydrochloride or choline chloride and alkali metal chloride showed bioactive properties [717].

Pressure-sensitive adhesives based on two-component PDMS rubber composition have shown great potential as adhesives in healthcare products, especially for gentle skin adhesives. They showed excellent biocompatibility in in-vitro cell culture experiments [718].

Polysiloxanes and silica nanoparticles are often used in the field of controlled drug delivery owing to their biocompatibility, versatility, physical and chemical resistance, and their ability for functionalization [719]. For example, the nanoporous SiO_2-CaO and PDMS-modified SiO_2-CaO xerogel materials with different amounts of PDMS were used as delivery systems for a drug to a bone – metronidazole (MT). The release of MT from xerogels was strongly dependent on the composition of the matrix [720].

Homogeneous PDMS/PVA microsphere composite hydrogels were prepared in an emulsion containing PVA aqueous solution, which was used as both emulgator/dispersant and raw material. These hydrogels showed relatively high equilibrium water content and swelling ratio, as well as enhanced water retention and compressive modulus, and could find applications as dressings in wound healing [721].

In the recent years, growing applications have been observed for POSS-poly (carbonate-urea) urethane (POSS-PCU) as a bioscaffold for tissue engineering. The

POSS-PCU is a versatile NC biomaterial. An integration of synthetic implants with host tissue can be improved by its topographical modifications. Optimization of POSS-PCU properties by dispersion of porogens (sodium bicarbonate ($NaHCO_3$) and sodium chloride (NaCl) and sucrose) onto the material surface was elaborated by C. Crowley et al., with the principle goal of increasing surface porosity. The effect of porogens on the material's mechanical strength, surface chemistry, wettability, and cytocompatibility was determined. Primary human bronchial epithelial cells (HBECs) were used for surface modifications of POSS-PCU scaffolds with a focus on in vitro cytocompatibility. Implantation of this scaffold in vivo showed that larger pore sizes favored cellular integration and vascular ingrowth [722].

A new silicone hydrogel material was prepared by the sequential interpenetrating network method from siloxane oligomers having Si–H and Si–vinyl functional groups. A hydrophilic network was obtained by a radical copolymerization of hydrophilic monomers (*N*-vinyl pyrrolidone, *N,N*-dimethylacrylamide) and a cross-linking agent (ethylene glycol dimethacrylate). In the hydrated state, the obtained transparent material exhibited hydrophilicity, high mechanical properties, oxygen and water permeability, and good wettability without chemical or plasma surface treatment [723].

References

[1] J.W. White, *Chapter: Recent Achievements in Industrial Silicon Polymer Science*, In: *Organosilicon Chemistry, Proceedings of the Xth International Symposium, Poznań, 1993*, B. Marciniec and J. Chojnowski, Eds., Gordon & Breach Sci. Publ., 1995, 363–386.
[2] B. Arkles, *Silicon Compounds, Silicon Esters*, Kirk-Othmer Encyclopedia of Chemical Technology, Fourth, Vol. 22, 1997, 69–81.
[3] J. Chruściel, Silikony – materiały polimerowe dla najnowszych technologii, *Rynek Chemik*, 1997, 7-8, 16–18. [In Polish].
[4] W. Noll, *Chemistry and Technology of Silicones*, Academic Press, New York, 1968.
[5] S. Pawlenko, *Organosilicon Chemistry*, Walter de Gruyter, Berlin – New York, 1986.
[6] G. Koerner, *Silicones: Chemistry and Technology*, Boca Raton, CRC Press. ISBN 0849377404, 1991.
[7] S.J. Clarson and J.A. Semlyen, *Siloxane Polymers*, Prentice Hall, New Jersey, 1993.
[8] H.R. Kricheldorf, *Silicon in Polymer Synthesis*, Springer-Verlag, Berlin, Heidelberg, 1996.
[9] I. Manners, Polymers and the periodic table: recent developments in inorganic polymer science, *Angewandte Chemie (International Ed. In English)*, 1996, 35, 1602–1621.
[10] M.A. Brook, Chapter 9: *Silicones*, In: *Silicon in Organic, Organometallic and Polymer Chemistry*, J. Wiley & Sons, Inc., New York-Chichester-Weinheim-Brisbane-Singapore–Toronto, 2000, 274–277.
[11] P. Rościszewski and M. Zielecka, *Silikony – właściwości i zastosowanie* (*Silicones - properties and application*), WNT, Warsaw, 2002. [In Polish].
[12] S.J. Clarson, J.J. Fitzgerald, M.J. Owen, S.D. Smith and M.E. Van Dyke, *Synthesis and Properties of Silicones and Silicone-Modified Materials*, ACS Symposium Series No. 838, 15 May 2003.
[13] S.J. Clarson, J.J. Fitzgerald, M.J. Owen, S.D. Smith and M.E. Van Dyke, Eds., *Science and Technology of Silicones and Silicone-Modified Materials*. ACS Symposium Series, Vol. 964, Washington, DC, American Chemical Society, 2007, ISBN 9780841274372.

[14] D.E. Leyden, *Silanes, Surfaces and Interfaces*, Gordon & Breach, 1985.

[15] M. Mazurek, Chapter 23: *Polimery krzemoorganiczne (Organosilicon polymers)*, In: *Chemia polimerów (Chemistry of Polymers), vol. 2*, Z. Florjańczyk and S. Penczek, Eds., Oficyna Wydawnicza Politechniki Warszawskiej, Warsaw (Poland), 1997, 322–347, [In Polish].

[16] A. Colas, Silicones: Preparation, Properties and Performance. https://www.dowcorning.com/content/publishedlit/01-3077.pdf

[17] G.J.J. Out, A.A. Turetskii, M. Möller and D. Oelfin, *Macromolecules*, 1994, 27, 3310.

[18] J. Chojnowski, *Journal of Inorganic and Organometallic Polymers and Materials*, 1991, 1, 299.

[19] A. De Gunzbourg, J.-C. Favier and P. Hémery, Anionic polymerization of octamethylcyclotetrasiloxane in aqueous emulsion I: preliminary results and kinetic study, *Polymer International*, 1994, 35(2), 179–188.

[20] M. Barrère, F. Ganachaud, D. Bendejacq, M.-A. Dourges, C. Maitre and P. Hémery, Anionic polymerization of octamethylcyclotetrasiloxane in miniemulsion II. Molar mass analyses and mechanism scheme, *Polymer*, 42(17), 7239–7246.

[21] G. Palaprat, F. Ganachaud, M. Mauzac and P. Hémery, Cationic polymerization of 2,4,6,8-tetramethylcyclotetrasiloxane processed by tuning the pH of the miniemulsion, *Polymer*, 2005, 46, 11213–11218.

[22] Y. Liu, Ed., *Silicone Dispersions*, CRC Press, 2016, 369 pages. ISBN 9780367872564 - CAT# K460629.

[23] O.N. Dolgov, M.G. Voronkov and M.P. Grinblat, *Kremneorganitcheskie zhidkie kautchuki i materialy na ikh osnove*, Izd. "Khimia", Leningrad, 1975, 54.

[24] B. Hardman and A. Torkelson, Silicones, In: *Encyclopedia of Polymer Science and Engineering*, Vol. 15, 2nd, H. Mark, N.M. Bikales, C.G. Overberger and G. Menges, Eds., Wiley, New York, 1989.

[25] D.P. Chattopathyay and D.D. Vyas, *Indian Journal of Fibre & Textile Research*, 2010, 35(1), 68.

[26] P.S. Purohit and P. Somasundaran, *Journal of Colloid and Interface Science*, 2014, 426, 235.

[27] B. Arkles, Look What You Can Make Out of Silicones (Biomedical Applications of Silicones), *Chemtech*, 1983, 13, 542–555. ISSN 0009-2703.

[28] H.R. Allcock and F.W. Lampe, *Contemporary Polymer Chemistry*, 2nd, Prentice Hall, 1990, 202.

[29] G. Odian, *Principles of Polymerization*, Third, Wiley Intersci. Publ., New York, Chichester, Brisbane, Toronto, Singapore, 1991, 138.

[30] N. Bosq, N. Guigo, J. Persello and N. Sbirrazzuolo, Melt and glass crystallization of PDMS and PDMS silica nanocomposites, *Physical Chemistry Chemical Physics*, 2014, 16, 7830–7840.

[31] M. Handke, B. Handke, A. Kowalewska and W. Jastrzębski, New polysilsesquioxane materials of ladder-like structure, *Journal of Molecular Structure*, 2009, 924–926, 254–263.

[32] P. Rościszewski, *Zastosowanie silikonów (The use of silicones)*, WNT, Warsaw, Poland, 1964, [In Polish].

[33] J. Maciejewski, *Silikony w przemyśle spożywczym (Silicones in the food industry)*, WNT, Warszawa, 1969, [In Polish].

[34] O.N. Dolgov, M.G. Voronkov and N.P. Grinblat, *Kremniyorganicheskie zhidkie kauchuki i materialy na ich osnove*, Leningrad, 1975.

[34a] A.V. Karlin, V.O. Reikhsfeld, E.G. Kagan, V.D. Lobkov, Y.A. Yużelevskiy and Grintsevich I.G., In: *Khimia i Technologia Kriemniejorganicheskich elastomerov, Ed., V.O.* Reiksfeld, Khimia, Leningrad, 1973.

[35] M. Schätz, *Silikonovy kauchuk*, Leningrad, 1975.

[36] J.M. Klosowski, Chapter 6: *Sealants in Construction*, In: *Silicone Sealants*, Marcel Dekker, New York, 1989, *265–297*.

[37] L.D. Lower and J.M. Klosowski, Chapter 32: *Silicone Adhesives and Sealants*, *Handbook of Adhesive Technology*, Ed, A. Pizzi and K.L. Mittal, Marcel Dekker, New York, Basel, Hong Kong, 1994, 521–529.

[38] S.L. Bontems, J. Stein and M.A. Zumbrum, *Journal of Polymer Science Part A: Polymer Chemistry*, 1994, 31, 2697.

[39] J.M. Pujol, J.M. Frances and M. Letoffe, *Organosilicon Chemistry, Proceedings of the Xth International Symposium, Poznan, 1993, Ed. Marciniec* B. *and Chojnowski J.*, Gordon & Breach Sci. Publ., 1995, 503.

[40] J. Chruściel, Silikonowe kauczuki i materiały uszczelniające (*Silicone rubbers and sealing materials), Rynek Chemik*, 1997, 9, 19. [In Polish].

[41] J. Chruściel, Crosslinking of polydimethylsiloxane-α,ω-diols with polymethylhydrosiloxanes and mechanical properties of resulting silicone rubbers, *Polimery*, 1999, 44(9), 586–596. [In Polish].

[42] M. Fejdyś, J. Chruściel and A. Miazga, Mechanical properties of silicone vulcanizates crosslinked with polymethylhydrosiloxanes, containing silica and calcium carbonate, *Polimery*, 2006, 51, 48–57. [In Polish].

[43] S.J. Clarson, Chapter 5: Preparation and properties of silicone elastomers, *R.G. Jones, et al., eds.*, In: *Silicon-Containing Polymers, Kluwer Academic Publishers*, 2000, *139–155.*

[44] A.C.M. Kuo, Poly(dimethylsiloxane), Polymer Data Handbook, Oxford University Press, Inc, 1999.

[45] Silastic Liquid Silicone Rubber Product Selector Guide. Dow Corning Corp., Midland, Mich. Form No.45-115-96.

[46] D. Corning 1732, Dow Corning 1737, and Dow Corning 1739. Dow Corning Products for High-Performance Sealing Application, Dow Corning Corp., Midland, Mich. Form No. 10-336B-90.

[47] H. Saechtling and W. Zebrowski, *Tworzywa sztuczne, Poradnik* (*Plastics, Handbook*), 4th, WNT, Warsaw, 1978, [In Polish].

[48] J.J. Pielichowski and A.A. Puszyński, *Technologia tworzyw sztucznych* (*Technology of plastics*), WNT, Warszawa, 1992. [In Polish]

[49] D. Żuchowska, *Polimery konstrukcyjne* (), WNT, Warszawa, 1995. [In Polish]

[50] https://en.wikipedia.org/wiki/Silly_Putty

[51] A. Thayer, What's That Stuff? Silly Putty, *Chemical & Engineering News*, Vol. 78, Nr 48, November 27, 2000.

[52] J. Maciejewski, Apparatus for Absorbing Mechanical Energy in Particular Energy of Very Strong Shocks, *US Pat.* 4,235,317 (1980).

[53] J. Maciejewski, Apparatus for Absorbing Mechanical Energy, in Particular Energy of Very Strong Shocks, *US Pat.* 4,307,794 (1981).

[54] J. Maciejewski, Sposób otrzymywania lepko-sprężystych kompozycji polisiloksanowych (Method for obtaining viscoelastic polysiloxane compositions), *Pat. PL* 109 050 (1981).

[55] J. Maciejewski, Hydrostatically damping and shock absorbing non-vulcanizable polysiloxane and boron compound for mechanical energy absorption, *US Pat.* 4339339 (1982).

[56] J. Maciejewski, Kompozycja polisiloksanowa absorbująca energię mechaniczną (Polysiloxane composition absorbing mechanical energy), *Pat. PL* 126182 (1985).

[57] J. Maciejewski, M. Michalski and Z. Słówko, Kompozycja borosiloksanowa (Borosiloxane composition), Pat. PL 126276 (1985).

[58] J. Maciejewski, Urządzenie do absorpcji energii mechanicznej, zwłaszcza energii bardzo silnych uderzeń (A device for absorbing mechanical energy, especially the energy of very strong impacts), Pat. PL 127164 (1985).

[59] J. Maciejewski, The utilization and damping characteristics of silicones in shock-absorbing structures and technology, *Polimery*, 1992, 37(11-12), 504–507. (In Polish). DOI: dx.doi.org/10.14314/polimery.1992.504.

[60] J. Maciejewski and A. Kubicki, Hydrostatically damping shock and vibration energy absorbing non-vulcanizable silicone elastomer, *US Patent* 5580917 (1996).

[61] J. Chruściel, Postęp w dziedzinie chemii polimetylowodorosiloksanów (*Progress in a field of polymethylhydrosiloxanes*), *Polimery*, 1999, 44(7/8), 462–474.
[62] J. Chruściel, J. Kulpiński and Z. Romanowski, (a) Synthesis of Some Block Polymethylhydrosiloxanes, *Gliwickie Seminarium Polimerowe, Gliwice, 26.06.1997 r., Proceedings*, pp. 65-68 (1997); (b) New Block Polymethylhydrosiloxanes, *Zeszyty Naukowe Politechniki Slaskiej, Seria: Chemia*, 1999, 140, 109–114.
[63] J.E. Mark, *Accounts of Chemical Research*, 1985, 18, 202.
[64] J.E. Mark and M.Y. Tang, *Journal of Polymer Science. Part B, Polymer Physics*, 1984, 22, 1849.
[65] T.L. Smith, B. Haidav and J.L. Hedrick, *Rubber Chemistry and Technology*, 1990, 63, 256.
[66] M.Y. Tang and J.E. Mark, *Macromolecules*, 1984, 17, 616.
[67] J.E. Mark and B. Erman, *Rubberlike Elasticity: A Molecular Primer*, Wiley-Interscience, New York, 1988.
[68] B.D. Viers and J.E. Mark, *Polymer Preprints*, 1995, 36, 354.
[69] J.E. Mark, *Advances in Polymer Science* = Fortschritte Der Hochpolymeren-Forschung, 1982, 44, 4.
[70] E.E. Hamurcu and B.M. Baysal, *Polymer*, 1993, 34, 5163.
[71] B. Erman and J.E. Mark, *Structure and Properties of Rubberlike Networks*, Oxford Univ. Press, New York, 1997.
[72] Z. Lasocki, J. Chruściel and J. Maciejewski, *27th International Symposium on Macromolecules, IUPAC, Strasbourg (France), 1981, Abstracts of Communications, vol. 1*, p. 410–414.
[73] S.J. Clarson and A.M.S. Al-ghandi, *Polymer Communications*, 1990, 31, 322.
[74] B. Arkles, J. Goff, S. Sulaiman and A. Sikorsky, Ultra-High Elongation Silicone Elastomers, *Rubber World*, 29–34, June 2016.
[75] https://www.simtec-silicone.com/capabilities/properties-of-lsr/
[76] https://www.tss.trelleborg.com/pl-pl/products-and-solutions/liquid-silicone-rubber
[77] M. Zielecka, K. Cyruchin, P. Rościszewski, H. Bęza, T. Romańczuk and T. Jakubas, Emulsja elastomerów silikonowych, zwłaszcza do impregnacji wyrobów włókienniczych (Emulsion of silicone elastomers, in particular for the impregnation of textiles), *Pat. PL*165436 (1994). In Polish
[78] H. Jiang, S. Manolache, A.C. Wong and F.S. Denes, Synthesis of dendrimer-type poly(ethylene glycol) structures from plasma-functionalized silicone rubber surfaces, *Journal of Applied Polymer Science*, 2006, 102(3), 2324–2337.
[79] L. Chen, G. Chen and L. Lu, Piezoresistive behavior study on finger-sensing silicone rubber/ graphite nanosheet nanocomposites, *Advanced Functional Materials*, 2007, 17, 898–904.
[80] B. Shentu, T. Gan and Z. Weng, Modification of Fe_2O_3 and its effect on the heat-resistance of silicone rubber, *Journal of Applied Polymer Science*, 2009, 113(5), 3202–3206.
[81] Y.B. Kim, D. Cho and W.H. Park, Fabrication and characterization of TiO_2/poly (dimethylsiloxane) composite fibers with thermal and mechanical stability, *Journal of Applied Polymer Science*, 2010, 116(1), 449–454.
[82] S. Mishra, N.G. Shimpi and A.D. Mali, Influence of stearic acid treated nano-$CaCO_3$ on the properties of silicone nanocomposites, *Journal of Polymer Research*, 2011, 18, 1715–1724.
[83] S.-J. Choi, T.-H. Kwon, H. Im, D.-I. Moon, D.J. Baek, M.-L. Seol, J.P. Duarte, Y.-K. Choi and A. Polydimethylsiloxane, (PDMS) sponge for the selective absorption of oil from water, *ACS Applied Materials & Interfaces*, 2011, 3, 4552–4556. DOI: dx.doi.org/10.1021/am201352w.
[84] M. Łubkowska, J. Gąsior, A. Miazga, R. Muskus, I. Frąckowiak and A. Idziak, Spoiwo silikonowe sieciowane w systemie addycyjnym oraz sposób jego otrzymywania (*Silicone binder crosslinked in the addition system and the method of its preparation*), Pat. PL 215086 (2013).
[85] A. Bele, G. Stiubianu, C.-D. Varganici, M. Ignat and M. Cazacu, Silicone dielectric elastomers based on radical crosslinked high molecular weight polydimethylsiloxane co-filled with silica and barium titanate, *Journal of Materials Science*, 2015, 50, 6822–6832.

[86] C. Tugui, G. Stiubianu, M. Iacob, C. Ursu, A. Bele, S. Vlad and M. Cazacu, Bimodal silicone interpenetrating networks sequentially built as electroactive dielectric elastomers, *Journal of Materials Chemistry. C*, 2015, 3, 8963–8969.
[87] C. Racles, M. Cazacu, B. Fischer and D.M. Opris, Synthesis and characterization of silicones containing cyanopropyl groups and their use in dielectric elastomer actuators, *Smart Materials and Structures*, 2013, 22, 104004.
[88] C. Racles, M. Ignat, A. Bele, M. Dascalu, D. Lipcinski and M. Cazacu, Silicone-based elastic composites able to generate energy on micromechanical impulse, *Smart Materials and Structures*, 2016, 25, 085024.
[89] S.J. Dünki, M. Dascalu, F.A. Nüesch and D.M. Opris, Silicones with enhanced permittivity for dielectric elastomer actuators, *Proc. SPIE* 9798, Electroactive Polymer Actuators and Devices (EAPAD) 2016, 97982K, April 20, 2016. http://http://dx.doi.org/10.1117/12.2218881 doi:10.1117/12.2218881
[90] L.-H. Cai, T.E. Kodger, R.E. Guerra, A.F. Pegoraro, M. Rubinstein and D.A. Weitz, Soft poly (dimethyl-siloxane) elastomers from architecture driven entanglement free design, *Advanced Materials*, 2015, 27, 5132–5140.
[91] Y. Xia and G.M. Whitesides, Soft Lithography, *Angewandte Chemie International Edition*, 1998, 37, 550–575.
[92] M.A. Wenshi, J. Li, B. Deng, X. Lin and X. Zhao, Temperature vulcanized silicone rubber composites prepared by an *in-situ* reduction method, *Journal of Wuhan University of Technology. Materials Science Edition*, 2013, 28(1), 127–131.
[93] Z. Tao, H. Wang, X. Li, Z. Liu and Q. Guo, Expanded graphite/polydimethylsiloxane composites with high thermal conductivity, *Journal of Applied Polymer Science*, 2017, 134, 44843.
[94] C.L. Mangun, A.C. Mader, N.R. Sottos and S.R. White, Self-healing of a high temperature cured epoxy using poly(dimethylsiloxane) chemistry, *Polymer*, 2010, 51, 4063–4068.
[95] S.H. Cho, H.M. Andersson, S.R. White, N.R. Sottos and P.V. Braun, Polydimethylsiloxane-based self-healing materials, *Advanced Materials*, 2006, 18, 997–1000.
[96] S.H. Cho, S.R. White and N.R. Sottos, Self-healing polymer coatings, *Advanced Materials*, 2009, 21, 645–649.
[97] M.W. Keller, S.R. White and N.R. Sottos, A self-healing poly(dimethyl siloxane) elastomer, *Advanced Functional Materials*, 2007, 17, 2399–2404.
[98] M.W. Keller, S.R. White and N.R. Sottos, Torsion fatigue response of self-healing poly (dimethylsiloxane) elastomers, *Polymer*, 2008, 49, 3136–3145.
[99] B.A. Beiermann, M.W. Keller and N.R. Sottos, Self-healing flexible laminates for resealing of puncture damage, *Smart Materials & Structures*, 2009, 18, 085001/1–7.
[100] X.K.D. Hillewaere and F.E. Du Prez, Fifteen chemistries for autonomous external self-healing polymers and composites, *Progress in Polymer Science*, 2015, 49–50, 121–153.
[101] P. Zheng and T.J. McCarthy, A surprise from 1954: Siloxane equilibration is a simple, robust, and obvious polymer self-healing mechanism, *Journal of the American Chemical Society*, 2012, 134(4), 2024–2027. DOI: https://doi.org/10.1021/ja2113257.
[102] H. Lu and S. Feng, Supramolecular silicone elastomers with healable and hydrophobic properties crosslinked by "salt-forming vulcanization", *Journal of Polymer Science. Part A, Polymer Chemistry*, 2017, 55, 903–911.
[103] X. Li, R. Yu, T. Zhaoa, Y. Zhang, X. Yang, X. Zhao and W. Huang, A self-healing polysiloxane elastomer based on siloxane equilibration synthesized through amino-ene Michael addition reaction, *European Polymer Journal*, 2018, 108, 399–405.
[104] Z. Liu, P. Hong, Z. Huang, T. Zhang, R. Xu, L. Chen, H. Xiang and X. Liu, Self-healing, reprocessing and 3D printing of transparent and hydrolysis resistant silicone elastomers, *Chemical Engineering Journal*, 2020, 387(124142).

[105] S. Dinkar, A. Dhanabalan and B.H.S. Thimmappa, Preparation and properties of poly(dimethyl siloxane)-colloidal silica/functionalized colloidal silica nanocomposites, *Journal of Applied Polymer Science*, 2012, 126, 1585–1592.
[106] C.S. O'Bryan, T. Bhattacharjee, S. Hart, C.P. Kabb, K.D. Schulze, I. Chilakala, B.S. Sumerlin, W.G. Sawyer and T.E. Angelini, Self-assembled micro-organogels for 3D printing silicone structures, *Science Advances*, 2017, 3, e1602800.
[107] X. Han, L. Wang, J. Li, X. Zhan, J. Chen and J. Yang, Separation of ethanol from ethanol/water mixtures by pervaporation with silicone rubber membranes: effect of silicone rubbers, *Journal of Applied Polymer Science*, 2011, 119, 3413–3421.
[108] Y. Shirazi, A. Ghadimi and T. Mohammadi, Recovery of alcohols from water using polydimethyl-siloxane–silica nanocomposite membranes: characterization and pervaporation performance, *Journal of Applied Polymer Science*, 2012, 124, 2871–2882.
[109 A. Rozicka, W. Kujawski, V. Guarnieri, L. Lorenzelli, A. Vasiliev and V. Filippov, Hydrophobic membranes for system monitoring underwater gas pipelines, *Architecture, Civil Engineering, Environment*, 2012, 4, 99–106.
[110] M. Imiela, R. Anyszka and D.M. Bieliński, Ceramizable polymer composites, *Elastomery*, 2014, 18(4), 16–24. In Polish.
[111] C. Racles, M. Dascalu, A. Bele and M. Cazacu, Chapter 11: Reactive and Functional Silicones for Special Applications, Springer Nature Switzerland AG, T.J. Gutiérrez, ed., In: *Reactive and Functional Polymers Volume One*, 2020, 235–292. DOI: https://doi.org/10.1007/978-3-030-43403-8_11.
[112] J.H. Kim, M.J. Kim, B. Lee, J.M. Chun, V. Patil and Y.-S. Kim, Durable ice-lubricating surfaces based on polydimethylsiloxane embedded silicone oil infused silica aerogel, *Applied Surface Science*, 2020, 512, 145728.
[113] R. Bui and M.A. Brook, Catalyst free silicone sealants that cure underwater, *Advanced Functional Materials*, 2020, 2000737.
[114] S. Zheng and M.A. Brook, Reversible redox crosslinking of thiopropylsilicones, *Macromolecular Rapid Communications*, 2020, 2000375.
[115] R. Yoda, Elastomers for biomedical applications, *Journal of Biomaterials Science. Polymer Edition*, 1998, 9(6), 561–626.
[116] P. Cevik, Chapter 12: Maxillofacial Silicone Elastomers in Dentistry, T.J. Gutiérrez, ed., In: *Reactive and Functional Polymers Volume One*, Springer Nature, Switzerland AG, 293–300, 2020. DOI: https://doi.org/10.1007/978-3-030-43403-8_12.
[117] S. Sikdar and S. Majumdar, Chapter 10: Reactive Silicones as Multifacetic Materials, T.J. Gutiérrez, ed., In: *Reactive and Functional Polymers Volume One*, Springer Nature, Switzerland AG, 207–234, 2020. DOI: https://doi.org/10.1007/978-3-030-43403-8_10.
[118] I. Gill, E. Pastor and A. Ballesteros, Lipase-silicone biocomposites: efficient and versatile immobilized biocatalysts, *Journal of the American Chemical Society*, 1999, 121(41), 9487–9496.
[119] B. Olander, A. Wirsen and A.-C. Albertsson, Silicone elastomer surface functionalized with primary amines and subsequently coupled with heparin, *Biomacromolecules*, 2003, 4(1), 145–148.
[120] O. Girshevitz, Y. Nitzan and C.N. Sukenik, Solution-deposited amorphous titanium dioxide on silicone rubber: a conformal, crack-free antibacterial coating, *Chemistry of Materials: A Publication of the American Chemical Society*, 2008, 20, 1390–1396.
[121] A. Mashak and A. Rahimi, Silicone polymers in controlled drug delivery systems: a review, *Iranian Polymer Journal*, 2009, 18(4), 279–295.
[122] M. Naeimi, A. Karkhaneh, J. Barzin, M.T. Khorasani and A. Ghaffarieh, Novel PDMS-based membranes: sodium chloride and glucose permeability, *Journal of Applied Polymer Science*, 2013, 127, 3940–3947. DOI: 10.1002/app.37709.

[123] M. Itoh, F. Oka, M. Suto, S.D. Cook and N. Auner, Characterization and some insights into the reaction chemistry of polymethylsilsesquioxane or methyl silicone resins, *International Journal of Polymer Science*, 2012, ID 526795.
[124] W.R. Huang, *Silicon Mater Appl.*, 1998, 2, 10.
[125] G.J. Wang and J. Xie, *Journal of Building Materials*, 2004, 7, 299.
[126] H.Z. Wang, *New Chemistry of Materials*, 1992, 3, 35.
[127] F. Bao, Z.G. Sun and S.Q. Huang, *Silicone Material*, 2002, 16, 9.
[128] H.M. Pan, B.E. Tan, S.J. Huang and L.Z. Zhang, *Chin. Adhes.*, 1998, 8, 1.
[129] W.H. Daudt and L.J. Tyler, *U.S. Pat.* 2,676,182 (1954).
[130] Y.H. Liu, *Pat. WO* 2005,040,250.
[131] D.R. Juen and B.X. Zhong, *U.S. Pat.*, 6,072,012 (2000).
[132] L.E. Carpenter, J.R. Cassiday, L.N. Cook, L.M. Madore, T.A. Oberhellman, R.G. Schmidt and H.X. Zhang, U.S. Pat. 5,741,876 (1998).
[133] G.A. Vincent and J.S. Tonge, *Eur. Pat.* 1,083,195 (2001).
[134] G.A. Vincent, *Eur. Pat.* 0,812,870 (1997).
[135] V.L. Berry, L.N. Cook, T.M. Leaym and R.G. Schmidt, *Pat. WO* 2007,145,765,
[136] S. Venkatraman and R. Gale, *Biomaterials*, 1998, 19, 1119.
[137] Z.M. Zhou, A.A. Sherman, W.K. Dunshee and W.J. Winkler, *Pat. WO* 2005,065,736.
[138] L.F. Hanneman, V. Krukonis and J.J. Watkins, *U.S. Pat.* 5,436,061 (1995).
[139] J.R. Hahn and J.A. Vallender, *U.S. Pat.* 3,983,298 (1976).
[140] J. Kosal, *Pat. WO* 2003,029,381.
[141] A. Nakamura, *Pat. WO* 2006,003,853.
[142] W. Huang, Y. Huang and Y.Z. Yu, *Journal of Applied Polymer Science*, 1998, 70, 1753.
[143] W. Huang, Y. Huang and Y.Z. Yu, *Chinese Journal of Polymer Science*, 1999, 17, 429.
[144] J.T. Woodard and V.L. Metevia, *U.S. Pat.* 4,655,767 (1987).
[145] P. Wei, C.X. Xiong and L.P. Liu, *Silicone Material*, 2007, 21, 76.
[146] I.N. Florian, *Journal of Materials Science Letters*, 2003, 22, 1237.
[147] C.C. Currie and J.W. Kell, *U.S. Pat.* 2,814,601 (1957).
[148] J.T. Goodwin, *U.S. Pat.* 2,857,356 (1958).
[149] H. Kobayashi and W. Nishiumi, *U.S. Pat.* 5,527,873 (1996).
[150] T. Jermouni, M. Smaihi and N. Hovnanian, Hydrolysis and initial polycondensation of phenyltrimethoxysilane and diphenyldimethoxysilane, *Journal of Materials Chemistry*, 1995, 5, 1203–1208.
[151] S. Özgümüş, T.B. Ĺyim, I. Acar and E. Küçükoğlu, Synthesis of novel silicone modified acrylic resins and their film properties, *Polymers for Advanced Technologies*, 2007, 18, 213–219.
[152] Z. Shen, Z. Xia and Y. Zhang, Modification, characterization and properties of epoxy resin (E-20) with silicone intermediate (RSN-6018), *Progress in Organic Coatings*, 2018, 114, 115–122.
[153] Y. Li, S. Yang, H. Liu, X. Li, Z. Dai, N. Zhao and J. Xu, Polymethylsilsesquioxane and hydroxyl-terminated polydimethylsiloxane composite: vapor incubation before thermal curing, *Journal of Applied Polymer Science*, 2009, 111, 1454–1461.
[154] B. Li, S. Zhang, Q. Xu and B. Wang, Preparation of composite polyacrylate latex particles with in situ-formed methylsilsesquioxane cores, *Polymers for Advanced Technologies*, 2009, 20, 1190–1194.
[155] M. Jia, C. Wu, W. Li and D. Gao, Synthesis and characterization of a silicone resin with silphenylene units in Si-O-Si backbones, *Journal of Applied Polymer Science*, 2009, 114, 971–977.
[156] S. Ma, W. Liu, N. Gao, Z. Yan and Y. Zhao, Synthesis and properties of LED-packaging epoxy resin toughened by a novel polysiloxane from hydrolysis and condensation, *Macromolecular Research*, 2011, 19(9), 972–979.

[157] Y. Qu, G. Huang, X. Wang and J. Li, Study on the mechanism of the formation of polyhedral oligomeric silsesquioxanes by the 2D correlation infrared spectra, *Journal of Applied Polymer Science*, 2012, 125, 3658–3665.

[158] J. Kurjata, K. Rózga-Wijas and W. Stańczyk, Investigation of hydrolysis and condensation of methyltriethoxysilane in aqueous systems, *European Journal Chemistry*, 2013, 4(4), 343.

[159] F. Sun, Y. Hu and H.-G. Du, Synthesis and characterization of MQ silicone resins, *Journal of Applied Polymer Science*, 2012, 125, 3532–3536.

[160] C.-H. Lee, W.-C. Chen, J.-Y. Hsu and H.-L. Chen, Effect of molecular architecture of copolymer template on the morphology of mesoporous methylsilsesquioxane, *Polymer*, 2007, 48, 3546–3554.

[161] Y. Mizuta, Y. Daiko, A. Mineshige and T. Yazawa, Solvent effect on distribution of phenyl groups for $C_6H_5SiO_{3/2}$–SiO_2 coatings prepared on polycarbonate substrate, *Journal of Sol-gel Science and Technology*, 2012, 62, 92–97.

[162] T. Gunji, Y. Hayashi, A. Komatsubara, K. Arimitsua and Y. Abe, Preparation and properties of flexible freestanding films via polyalkoxysiloxanes by acid-catalyzed controlled hydrolytic polycondensation of tetraethoxysilane and tetramethoxysilane, *Applied Organometallic Chemistry*, 2012, 26, 32–36.

[163] Z. Zhang, J. Huang, Y. Lin and H. Rao, Oxygen-enriching properties of a new hybrid membrane cross-linked from PDMS and fluorine-containing silicone resin, *Polymer Engineering and Science*, 2013, 53, 2328–2335.

[164] K.-H. Haas, S. Amberg-Schwab, K. Rose and G. Schottner, Functionalized coatings based on inorganic organic polymers (ORMOCER®s) and their combination with vapor deposited inorganic thin films, *Surface & Coatings Technology*, 1999, 111, 72–79.

[165] O. Foussaier, M. Menetrier, -J.-J. Videau and E. Duguet, Polydimethylsiloxane-based ORMOSIL microstructure: correlation with compressive behavior, *Materials Letters*, 2000, 42, 305–310.

[166] A. Tamayo, L. Téllez, M. Rodriguez-Reyes, M.A. Mazo, F. Rubio and J. Rubio, Surface properties of bioactive TEOS–PDMS–TiO_2–CaO ormosils, *Journal of Materials Science*, 2014, 49, 4656–4669.

[167] S.-R. Zhai, L. Zhang, B. Zhai and Q.-D. An, Facile sol–gel synthesis of thiol-functionalized materials from TEOS-MPTMS-PMHS system, *Journal of Sol-gel Science and Technology*, 2012, 61, 23–33.

[168] M. Du, N. Mao and S.J. Russell, Control of porous structure in flexible silicone aerogels produced from methyltrimethoxysilane (MTMS): the effect of precursor concentration in sol-gel solutions, *Journal of Materials Science*, 2016, 51, 719–731.

[169] Z. Li, X. Cheng, S. He, X. Shi, H. Yang and H. Zhang, Tailoring thermal properties of ambient pressure dried MTMS/TEOS co-precursor aerogels, *Materials Letters*, 2016, 171, 91–94.

[170] D. Zhang, W. Huo, J. Wang, T. Li, X. Cheng, J. Li and A. Hang, Synthesis of allyl-ended hyper-branched organic silicone resin by halloysite-supported platinum catalyst, *Journal of Applied Polymer Science*, 2012, 126, 1580–1584.

[171] J.-S. Kim, S.-C. Yang and B.-S. Bae, Thermally stable transparent sol-gel based siloxane hybrid material with high refractive index for light emitting diode (LED) encapsulation, *Chemistry of Materials* : A Publication of the American Chemical Society, 2010, 22, 3549–3555.

[172] J.-S. Kim, S.C. Yang, S.-Y. Kwak, Y. Choi, K.-W. Paik and B.-S. Bae, High performance encapsulant for light-emitting diodes (LEDs) by a sol–gel derived hydrogen siloxane hybrid, *Journal of Materials Chemistry*, 2012, 22, 7954–7960.

[173] K.H. Jung, J.Y. Bae, S.J. Park, S. Yoo and B.-S. Bae, High performance organic-inorganic hybrid barrier coating for encapsulation of OLEDs, *Journal of Materials Chemistry*, 2011, 21, 1977–1983.

[174] H.Y. Kim, J.-Y. Bae, Y.H. Kim, Y.B. Kim and B.-S. Bae, Low temperature curable epoxy siloxane hybrid materials for LED encapsulant, *Journal of Applied Polymer Science*, 2014, 131, 39968.

[175] T.C. Loh, C.M. Ng, R.N. Kumar, H. Ismail and Z. Ahmad, Improvement of thermal ageing and transparency of methacrylate based poly(siloxane-silsesquioxane) for optoelectronic application, *Journal of Applied Polymer Science*, 2017, 134, 45285.
[176] Y. Liu, X. Luan, Y. Feng, X. Tan, Y. Han and X. Sun, Self-adhesive epoxy modified silicone materials for light emitting diode encapsulation, *Polymers for Advanced Technologies*, 2017, 28, 1473–1479.
[177] J. Dul, Z. Lasocki and M. Jarosz, Sposób wytwarzania termoplastycznych i termotopliwych żywic silikonowych (*A method of producing thermoplastic and hot-melt silicone resins*), Pat. PL, 1980, 97290.
[178] W. Wojnowski, J. Dul and Z. Lasocki, Sposób otrzymywania kopolimerów organiczno-krzemo-organicznych (*The method of obtaining organic-organosilicon copolymers*), Pat. PL, 1983, 118111.
[179] J. Dul, Z. Lasocki, E. Leśniak, M. Szczepanowska and W. Żuraw, Sposób wytwarzania żywic silikonowych (*Method for producing silicone resins*), Pat. PL 160224B1 (1993).
[180] E. Leśniak, Z.M. Michalska and J. Chojnowski, One-step synthesis of thermoplastic phenyl-silsesquioxane polymer and its copolymers with diphenylsiloxanes, *Journal of Inorganic and Organometallic Polymers*, 1998, 8(1), 1–21.
[181] M.K. Lee and D.J. Meier, Synthesis and properties of diarylsiloxane and (aryl/methyl)siloxane polymers: 1. Thermal properties, *Polymer*, 1993, 34(23), 4882–4892.
[182] J. Dul, Z. Lasocki and E. Leśniak, Sposób wytwarzania silikonowych środków przeciw pienieniu (*Method of obtaining silicone antifoaming agents*), Pat. PL, 1994, 164031.
[183] Y. Zhang, Y. Huang, X. Liu and Y. Yu, Studies on the silicone resins cured with polymethyl-silazanes at ambient temperature, *Journal of Applied Polymer Science*, 2003, 89, 1702–1707.
[184] B. Gardelle, S. Duquesne, V. Rerat and S. Bourbigot, Thermal degradation and fire performance of intumescent silicone-based coatings, *Polymers for Advanced Technologies*, 2013, 24, 62–69.
[185] J. Ji, X. Ge, W. Liang, X. Pang, R. Liu, S. Wen, J. Sun, X. Chen and J. Ge, Synthesis of high molecular weight vinylphenyl-containing MQ silicone resin via hydrosilylation reaction, *Coatings*, 2019, 9, 605. DOI: 10.3390/coatings9100605.
[186] P. Rościszewski, J. Sołtysiak, M. Łubkowska and Z. Walkowiak, New reactive silicone resins, *Polimery*, 2016, 61(1), 3–9.
[187] B. Ceccaroli, E. Ovrelid and S. Pizzini, Solar Silicon Processes: Technologies, Challenges, and Opportunities, CRC Press, 2016, 258 pages.
[188] H.H. Wang, X.R. Li, G.Q. Fei and J. Mou, Synthesis, morphology and rheology of core-shell silicone acrylic emulsion stabilized with polymerisable surfactant, *EXPRESS Polymer Letter*, 2010, 4, 670–680.
[189] B. Li, S. Zhang, Q. Xu and B. Wang, Preparation of composite polyacrylate latex particles with in situ-formed methylsilsesquioxane cores, *Polymers for Advanced Technologies*, 2009, 20, 1190–1194.
[190] J. Kurjata, W. Fortuniak, S. Rubinsztajn and J. Chojnowski, $B(C_6F_5)_3$ catalyzed dehydrocarbon polycondensation of $PhSiH_3$ with $(MeO)_4Si$ as model polyfunctional comonomers in new route to hydrophobic silicone TQ resins, *European Polymer Journal*, 2009, 45, 3372–3379.
[191] W. Fortuniak, J. Chojnowski, S. Słomkowski, P. Pośpiech and J. Kuriata, Route to hydrophilic, hydrophobic and functionalized cross-linked polysiloxane microspheres, *Polymer*, 2013, 54, 3156–3165.
[192] S. Słomkowski, W. Fortuniak, J. Chojnowski, P. Pośpiech and U. Mizerska, Polysiloxane microcapsules, microspheres and their derivatives, *Polimery*, 2017, 62, 499–508.
[193] P. Pośpiech, J. Chojnowski, U. Mizerska, T. Makowski, K. Strzelec and N. Sienkiewicz, Polysiloxane microspheres functionalized with imidazole groups as a palladium catalyst support, *Applied Organometallic Chemistry*, 2016, 30, 399–407.

[194] X. Zang, L. Cai, Y. Yuan and Z. Li, Fluoroalkylation of silsesquioxanes-modified polysiloxane and its surface property, *Journal of Inorganic and Organometallic Polymers and Materials*, 2015, 25, 975–981.

[195] D. Devapal, S. Packirisamy, K.J. Sreejith, P.V. Ravindran, B.K. George and C. Synthesis, Ceramic conversion studies of borosiloxane oligomers from phenyltrialkoxysilanes, *Journal of Inorganic and Organometallic Polymers and Materials*, 2010, 20, 666–674.

[196] P. Jia, H. Liu, Y. Wang and X. Cai, A novel epoxy-functionalized hyperbranched polysiloxane (HPSi) endowing methyl phenyl silicone resin (Si603)/epoxy systems with enhanced compatibility and fire retardancy performance, *RSC Advances*, 2015, 5, 97413–97421.

[197] https://termoorganika.pl/farba-dyspersyjno-krzemianowa-aqua-pro

[198] http://cedar.com.pl/pl/produkty/farby/farby-termoodporne

[199] https://www.noxan.pl/spray/farba-zaroodporna-spray-czarny-srebrny-750st-c-500ml.html

[200] https://www.wacker.com/cms/en/industries/coatings/industrial/metal-coating/high-temp/mufflers.jsp?country=PL&language=pl

[201] https://www.dowcorning.com/content/publishedlit/Chapter7.pdf

[202] G. Nishino, S. Kanda, H. Sugimoto, K. Inomata and E. Nakanishi, Preparation and coating properties of an acrylic melamine resin containing silicone segments, *Polymer Bulletin*, 2012, 68, 2049–2060.

[203] F. Wang, Z. Huang, G. Zhang and X. Li, Preparation and thermal stability of heat-resistant phenolic resin system constructed by multiple heat-resistant compositions containing boron and silicon, *High Performance Polymers*, 2017, 29(4), 493–498. DOI: https://doi.org/10.1177/0954008316644971.

[204] D.A. Loy and K. Rahimian, Chapter 3: Building Hybrid Organic-Inorganic Materials Using Silsesquioxanes, In: Hanbook of Organic-Inorganic Hybrid Materials and nanocomposites, Vol. 1, Hybrid Materials, H.S. Nalwa (Ed.), Am. Sci. Publ, 2003, 125–265.

[205] A. Provatas and J.G. Matisons, Silsesquioxanes: synthesis and applications, *Trends in Polymer Science*, 1997, 5, 327–332.

[206] E. Leśniak, Polimery, Silsesquioxanes. Part I, 2001, 46(7-8), 516–521.

[207] E. Leśniak, Silsesquioxanes. Part II, *Polimery*, 2001, 46(9), 582–589.

[208] J.F. Brown Jr., Double chain polymers and nonrandom crosslinking, *Journal of Polymer Science Part C*: Polymer Letters, 1963, 1(1), 83–97.

[209] R.H. Baney, M. Itoh, A. Sakakibara and T. Suzuki, Silsesquioxanes, *Chemical Reviews*, 1995, 95, 1409–1430. ISSN 0009-2665.

[210] T. Li, L. Shi and M. Ye, *Chinese Journal of Polymer Science*, 1995, 13(4), 366.

[211] H. Adachi, E. Adachi and S. Yamamoto, *Materials Research Society Symposia Proceedings*, 1991, 227, 95.

[212] J.J. Chruściel and E. Leśniak, Chapter 9: Modification of thermoplastics with reactive silanes and siloxanes, A.Z. El-Sonbati, Ed., In: *Thermoplastic Elastomers*, ISBN 979-953-307-353-7, InTech - Open Science, Rijeka, Croatia. DOI: 10.5772/34694, 155–192.

[213] S. Yamamoto, N. Yasuda, A. Ueyama, H. Adachi and M. Ishikawa, Mechanism for the formation of poly(phenylsilsesquioxane), *Macromolecules*, 2004, 37(8), 2775–2778.

[214] K.G. Williams, S.P. Gido and E.B. Coughlin, Chapter 4: *Polymers and Copolymers Containing Covalently Bonded Polyhedral Oligomeric Silsesquioxanes Moieties*, C. Hartmann-Thompson, Ed., In: *Applications of Polyhedral Oligomeric Silsesquioxanes, Advances in Silicon Science*, Vol. 3, DOI 10.1007/978-90-481-3787-9_4, © Springer Science + Business Media B.V., 2011, 167–207.

[215] R.M. Laine, C. Zhang, A. Sellinger and L. Viculis, Polyfunctional cubic silsesquioxanes as building blocks for organic/inorganic hybrids, *Applied Organometallic Chemistry*, 1998, 12, 715–723.

[216] R.M. Laine, M. Roll, M. Asuncion, S. Sulaiman, V. Popova, D. Bartz, D.J. Krug and P.H. Mutin, Perfect and nearly perfect silsesquioxane (SQs) nanoconstruction sites and Janus SQs, *Journal of Sol-gel Science and Technology*, 2008, 46, 335–347.
[217] D.B. Cordes, P.D. Lickiss and F. Rataboul, Recent developments in the chemistry of cubic polyhedral oligosilsesquioxanes, *Chemical Reviews*, 2010, 110, 2081–2173.
[218] D.B. Cordes and P.D. Lickiss, Chapter 2: *Preparation and Characterization of Polyhedral Oligo-silsesquioxanes*, C. Hartmann-Thompson, Ed., In: *Applications of Polyhedral Oligomeric Silsesquioxanes*, Advances in Silicon Science, Vol. 3, DOI 10.1007/978-90-481-3787-9_2, © Springer Science+Business Media B.V., 2011, 135–166.
[219] E. Markovic, K. Constantopolous and J.G. Matisons, Chapter 1: *Polyhedral Oligomeric Silsesquioxanes: From Early and Strategic Development through to Materials Application*, C. Hartmann-Thompson, Ed., In: *Applications of Polyhedral Oligomeric Silsesquioxanes*, Advances in Silicon Science, Vol. 3, DOI 10.1007/978-90-481-3787-9_1, © Springer Science +Business Media B.V., 2011, 1–46.
[220] A.J. Ward, A.F. Masters and T. Maschmeyer, Chapter 3: *Metallasilsesquioxanes: Molecular Analogues of Heterogeneous Catalysts*, C. Hartmann-Thompson, Ed., In: *Applications of Polyhedral Oligomeric Silsesquioxanes*, Advances in Silicon Science, Vol. 3, DOI 10.1007/ 978-90-481-3787-9_3, © Springer Science+Business Media B.V, 2011, 47–133.
[221] C. DeArmitt, Chapter 5: *Polyhedral Oligomeric Silsesquioxanes in Plastics*, C. Hartmann-Thompson, Ed., In: *Applications of Polyhedral Oligomeric Silsesquioxanes*, Advances in Silicon Science, Vol. 3, DOI 10.1007/978-90-481-3787-9_5, © Springer Science+Business Media B.V., 2011, 209–228.
[222] S.T. Iacono, A.J. Peloquin, D.W. Smith Jr. and J.M. Mabry, Chapter 6: *Fluorinated Polyhedral Oligo-silsesquioxane Surfaces and Superhydrophobicity*, C. Hartmann-Thompson, Ed., In: *Applications of Polyhedral Oligomeric Silsesquioxanes*, Advances in Silicon Science, Vol. 3, DOI 10.1007/978-90-481-3787-9_6, © Springer Science+Business Media B.V, 2011, 229–246.
[223] F.J. Feher and R.L. Blanski, Olefin polymerization by vanadium-containing silsesquioxanes: synthesis of a dialkyl-oxo-vanadium(V) complex that initiates ethylene polymerization, *Journal of the American Chemical Society*, 1992, 114(14), 5886–5887.
[224] F.J. Feher, D. Soulivong, A.G. Eklud and K.D. Wyndham, Cross-metathesis of alkenes with vinyl-substituted silsesquioxanes and spherosilicates: a new method for synthesizing highly-functionalized Si/O frameworks, *Chemical Communications*, 1997, 1185–1186.
[225] J.R. Severn, R. Duchateau, R.A. Van Santen, D.D. Ellis and A.L. Spek, Homogeneous models for chemically tethered silica-supported olefin polymerization catalysts, *Organometallics*, 2002, 21(1), 4–6.
[226] R. Duchateau, H.C.L. Abbenhuis, R.A. Van Santen, A. Meetsma, S.K.H. Thiele and M.F.H. Van Tol, Half-sandwich titanium complexes stabilized by a novel silsesquioxane ligand: soluble model systems for silica-grafted olefin polymerization catalysts, *Organometallics*, 1998, 17(24), 5222–5224.
[227] N. Maxim, P.C.M.M. Magusin, P.J. Kooyman, J.H.M.C. Van Wolput, R.A. Van Santen and H.C.L. Abbenhuis, Synthesis and characterization of microporous Fe–Si–O materials with tailored iron content from silsesquioxane precursors, *The Journal of Physical Chemistry*. B, 2002, 106(9), 2203–2209.
[228] C. Bonhomme, P. Toledano, J. Maquet, J. Livage and L. Bonhomme-Coury, *Journal of the Chemical Society. Dalton Transactions*, 1997, 1617.
[229] A.R. Bassindale, M. Pourny, P.G. Taylor, M.B. Hursthouse and M.E. Light, Fluoride-ion encapsulation within a silsesquioxane cage, *Angewandte Chemie International Edition*, 2003, 42(30), 3488–3490.

[230] A.R. Bassindale, D.J. Parker, M. Pourny, P.G. Taylor, P.N. Horton and M.B. Hursthouse, Fluoride ion entrapment in octasilsesquioxane cages as models for ion entrapment in zeolites. Further examples, X-ray crystal structure studies, and investigations into how and why they may be formed, *Organometallics*, 2004, 23(19), 4400–4405.

[231] A. Tsuchida, C. Bolin, F.G. Sernetz, H. Frey and R. Mülhaupt, Ethene and propene copolymers containing silsesquioxane side groups, *Macromolecules*, 1997, 30, 2818–2824.

[232] M.G. Voronkov and V.I. Lavrent'yev, Polyhedral oligosilsesquioxanes and their homo derivatives, *Topics in Current Chemistry*, 1982, 102, 199.

[233] D.A. Loy, K.J. Shea and B. Polysilsesquioxanes, Highly porous hybrid organic-inorganic materials, *Chemical Reviews*, 1995, 95(5), 1431–1442.

[234] J. Lichtenhan and J.C. Salmone, ed., Polymeric Materials Encycyclopedia, Vol. 10, CRC Press, NY, 1996, 7768.

[235] R.M. Laine, Nanobuilding blocks based on the $[OSiO_{1.5}]_x$ ($x = 6, 8, 10$) octasilsesquioxanes, *Journal of Materials Chemistry*, 2005, 15, 3725–3744.

[236] J.D. Lichtenhan, H.Q. Vu, J.A. Carter, J.W. Gilman and F.J. Feher, Silsesquioxane-siloxane copolymers from polyhedral silsesquioxanes, *Macromolecules*, 1993, 26(8), 2141–2142.

[237] J.W. Gilman, D.S. Schlitzer and J.D. Lichtenhan, Low earth orbit resistant siloxane copolymers, *Journal of Applied Polymer Science*, 1996, 60(4), 591.

[238] R.I. Gonzalez, S.H. Phillips and G.B. Hoflund, Situ oxygen-atom erosion study of polyhedral oligomeric silsesquioxane-siloxane copolymer, *Journal of Spacecraft and Rockets*, 2000, 37, 463–467.

[239] R. Weidner, N. Zeller, B. Deubzer and V. Frey, *US Patent*, 1991, 5,047,492.

[240] S. Dathe, E. Popowski, G. Sonnek, T. Feiher, H. Jancke and U. Scheim, *Euro Patent* 0,348,705 A1 (1989).

[241] C. Freyer, J. Wolferseder and U. Peetz, *Patent EP* 0,624, 691 A1 (1993).

[242] Y. Zhang, S.H. Lee, K. Liang, H. Toghiani and C.U. Pittman, Phenolic resin–trisilanolphenyl polyhedral oligomeric silsesquioxane (POSS) hybrid nanocomposites: Structure and properties, *Polymer*, 2006, 47(9), 2984–2996.

[243] M. Ikeda, H. Saito, Improvement of polymer performance by cubic-oligosilsesquioxane, *Reactive & Functional Polymers*, **67**, 1148–1156 (2007).

[244] M.D. Stewart and C.G. Willson, Imprint materials for nanoscale devices, *MRS Bulletin*, 2005, 30(12), 947–951.

[245] M.D. Stewart, J.T. Wetzel, G.M. Schmid, F. Palmieri, E. Thompson, E.K. Kim, D. Wang, K. Sotodoe, K. Jen, S.C. Johnson, J. Hao, M.D. Dickey, Y. Nishimura, R.M. Laine, D.J. Resnick and C.G. Willson, *Proceedings of SPIE – The International Society for Optical Engineering*, 2005, 219, 5751.

[246] I. Hasegawa, S. Sakka, Y. Sugahara, K. Kuroda and C. Kato, Silicate anions formed in tetramethyl-ammonium silicate methanolic solutions as studied by ^{29}Si nuclear magnetic resonance, *Journal of the Chemical Society. Chemical Communications*, 1989, 208–210.

[247] I. Hasegawa and S. Moto-Jima, Dimethylvinylsilylation of $Si_8O_{20}{}^{8-}$ silicate anion in methanol solutions of tetramethylammonium silicate, *Journal of Organometallic Chemistry*, 1992, 441 (3), 373–380.

[248] I. Hasegawa and S. Sakka, *Chem. Lett.*, Rapid solidification of (2-hydroxyethyl)tetrammonium Silicate, 1988, 8, 1319–1322.

[249] P.A. Agaskar, New synthetic route to the hydridospherosiloxanes Oh-H8Si8O12 and D5h-H10Si10O15, *Inorganic Chemistry*, 1991, 30(13), 2707–2708.

[250] D. Hoebbel, K. Endres, T. Reinert and I. Pitsch, Inorganic-organic polymers derived from functional silicic acid derivatives by additive reaction, *Journal of Non-crystalline Solids*, 1994, 176(2-3), 179–188.

[251] D. Hoebbel, I. Pitsch and D. Heidmann, Eurogel '91, Elsevier Science Publication, 1992, 467.
[252] M. Pach and R. Stösser, Scavenger assisted trapping of atomic hydrogen in Si_8O_{12}-cages, *The Journal of Physical Chemistry*. A, 1997, 101(44), 8360–8365.
[253] B. Hong, T.P.S. Thoms, H.J. Murfee and H.J. Lebrun, Highly branched dendritic macromolecules with core polyhedral silsesquioxane functionalities, *Inorganic Chemistry*, 1997, 36(27), 6146–6147.
[254] F.J. Feher, K.D. Wyndham and D.J. Knauer, Synthesis, characterization and lectin binding study of carbohydrate functionalized silsesquioxanes, *Chemical Communications*, 1998, 2393–2394.
[255] P.R. Dvornic, C. Hartmann-Thompson, S.E. Keinath and E.J. Hill, Organic–inorganic polyamidoamine (PAMAM) dendrimer– polyhedral oligosilsesquioxane (POSS) nanohybrids, *Macromolecules*, 2004, 37(20), 7818–7831.
[256] G.M. Kim, H. Qin, X. Fang, F.C. Sun and P.T. Mather, Hybrid epoxy-based thermosets based on polyhedral oligosilsesquioxane: Cure behavior and toughening mechanisms, *Journal of Polymer Science. Part B, Polymer Physics*, 2003, 41(24), 3299–3313.
[257] B.S. Kim and P.T. Mather, Amphiphilic telechelics incorporating polyhedral oligosilsesquioxane: 1. synthesis and characterization, *Macromolecules*, 2002, 35(22), 8378–8384.
[258] A.J. Waddon and E.B. Coughlin, Crystal structure of polyhedral oligomericsilsequioxane (POSS) nano-materials: a study by x-raydiffraction and electron microscopy, *Chemistry of Materials: A Publication of the American Chemical Society*, 2003, 15(24), 4555–4561.
[259] G. Cardoen and E.B. Coughlin, Hemi-telechelic polystyrene-POSS copolymers as model systems for the study of well-defined inorganic/organic hybrid materials, *Macromolecules*, 2004, 37(13), 5123–5126.
[260] B.X. Fu, B.S. Hsiao, H. White, M. Rafailovich, P. Mather, H.G. Joen, S. Phillips, J. Lichtenhan and J. Schwab, Nanoscale reinforcement of polyhedral oligomeric silsesquioxane (POSS) in polyurethane elastomer, *Polymer International*, 2000, 49(5), 437–440.
[261] B.X. Fu, W. Zhang, B.S. Hsiao, M. Rafailovich, J. Sokolov, G. Johansson, B.B. Sauer, S. Phillips and R. Blanski, Synthesis and characterization of segmented polyurethanes containing polyhedral oligomeric silsesquioxanes nanostructured molecules, *High Performance Polymers*, 2000, 12, 565–571.
[262] H.C. Lin, S.W. Kuo, C.F. Huang and F.C. Chang, Thermal and surface properties of phenolic nanocomposites containing octaphenol polyhedral oligomeric silsesquioxane, *Macromolecular Rapid Communications*, 2006, 27(7), 537–541.
[263] E.S. Baker, J. Gidden, S.E. Anderson, T.S. Haddad and M.T. Bowers, Isomeric structural characterization of polyhedral oligomeric silsesquioxanes (POSS) with styryl and epoxy phenyl capping agents, *Nano Letters*, 2004, 4(5), 779–785.
[264] B.X. Fu, A. Lee and T.S. Haddad, Styrene–butadiene–styrene triblock copolymers modified with polyhedral oligomeric silsesquioxanes, *Macromolecules*, 2004, 37(14), 5211–5218.
[265] E.T. Kopesky, T.S. Haddad, R.E. Cohen and G.H. McKinley, Thermomechanical properties of poly(methyl methacrylate)s containing tethered and untethered polyhedral oligomeric silsesquioxanes, *Macromolecules*, 2004, 37(24), 8992–9004.
[266] J.F. Brown, L.H. Vogt and P.I. Perescott, Preparation and characterization of the lower equilibrated phenylsilsesquioxanes, *Journal of the American Chemical Society*, 1964, 86(6), 1120–1125.
[267] C. Zhang, F. Babonneau, C. Bonhomme, R.M. Laine, C.L. Soles, H.A. Hristov and A.F. Yee, Highly porous polyhedral silsesquioxane polymers. synthesis and characterization, *Journal of the American Chemical Society*, 1998, 120(33), 8380–8391.
[268] A. Sellinger and R.M. Laine, Thermally curable and photocurable inorganic/organic hybrids, *Macromolecules*, 1996, 29, 2327–2330.

[269] J. Choi, J. Harcup, A.F. Yee, Q. Zhu and R.M. Laine, Organic/inorganic hybrid composites from cubic silsesquioxanes, *Journal of the American Chemical Society*, 2001, 123(46), 11420–11430.

[270] A. Sellinger and R.M. Laine, Silsesquioxanes as synthetic platforms. 3. Photocurable, liquid epoxides as inorganic/organic hybrid precursors, *Chemistry of Materials: A Publication of the American Chemical Society*, 1996, 8(8), 1592–1593.

[271] C. Zhang and R.M. Laine, Hydrosilylation of allyl alcohol with $[HSiMe_2OSiO_{1.5}]_8$: Octa(3-hydroxypropyldimethylsiloxy)octasilsesquioxane and its octamethacrylate derivative as potential precursors to hybrid nanocomposites, *Journal of the American Chemical Society*, 2000, 122(29), 6979–6988.

[272] R.M. Laine, J. Choi and I. Lee, Organic–Inorganic Nanocomposites with Completely Defined Interfacial Interactions, *Advanced Materials*, 2001, 13, 800–803.

[273] C. Zhang and R.M. Laine, Silsesquioxanes as synthetic platforms. II. Epoxy-functionalized inorganic-organic hybrid species, *Journal of Organometallic Chemistry*, 1996, 521(1-2), 199–201.

[274] R. Tamaki, Y. Tanaka, M.Z. Asuncion, J. Choi and R.M. Laine, Octa(aminophenyl) silsesquioxane as a Nanoconstruction Site, *Journal of the American Chemical Society*, 2001, 123(49), 12416–12417.

[275] R.O.R. Costa, W.L. Vasconcelos, R. Tamaki and R.M. Laine, Organic/Inorganic Nanocomposite Star Polymers via Atom Transfer Radical Polymerization of Methyl Methacrylate Using Octafunctional Silsesquioxane Cores, *Macromolecules*, 2001, 34, 5398–5407.

[276] R.M. Laine, C. Zhang, A. Sellinger and L. Viculis, Polyfunctional cubic silsesquioxanes as building blocks for organic/inorganic hybrids, *Journal of Applied Organometallic Chemistry*, 1998, 12, 715–723.

[277] R. Tamaki, J. Choi, R.M. Laine and A. Polyimide, Nanocomposite from octa(aminophenyl)-silsesquioxane, *Chemistry of Materials: A Publication of the American Chemical Society*, 2003, 15(3), 793–797.

[278] J. Choi, R. Tamaki, S.G. Kim and R.M. Laine, Organic/inorganic imide nanocomposites from aminophenylsilsesquioxanes, *Chemistry of Materials: A Publication of the American Chemical Society*, 2003, 15(17), 3365–3375.

[279] J. Choi, A.F. Yee and R.M. Laine, Organic/inorganic hybrid composites from cubic silsesquioxanes. epoxy resins of octa(dimethylsiloxyethylcyclohexylepoxide) silsesquioxane, *Macromolecules*, 2003, 36(15), 5666–5682.

[280] J. Choi, S.G. Kim and R.M. Laine, Organic/inorganic hybrid epoxy nanocomposites from aminophenylsilsesquioxanes, *Macromolecules*, 2004, 37(1), 99–109.

[281] J. Choi, A.F. Yee and R.M. Laine, Toughening of cubic silsesquioxane epoxy nanocomposites using core-shell rubber particles: a three-component hybrid system, *Macromolecules*, 2004, 37(9), 3267–3276.

[282] C.M. Brick, R. Tamaki, S.G. Kim, M.Z. Asuncion, M. Roll, T. Nemoto and R.M. Laine, Spherical, polyfunctional molecules using poly (bromophenylsilsesquioxane)s as nanoconstruction sites, *Macromolecules*, 2005, 38(11), 4655–4660.

[283] C.M. Brick, Y. Ouchi, Y. Chujo and R.M. Laine, Robust polyaromatic octasilsesquioxanes from polybromophenylsilsesquioxanes, Br_xOPS, via Suzuki coupling, *Macromolecules*, 2005, 38(11), 4661–4665.

[284] Z.L. Zhang, M.A. Horsch, M.H. Lamm and S.C. Glotzer, Tethered nano building blocks: toward a conceptual framework for nanoparticle self-assembly, *Nano Letters*, 2003, 3(10), 1341–3146.

[285] X. Zhang, E.R. Chan and S.C. Glotzer, Self-assembled morphologies of monotethered polyhedral oligomeric silsesquioxane nanocubes from computer simulation, *The Journal of Chemical Physics*, 2005, 123, 184718.

[286] D. Neumann, M. Fisher, L. Tran and J.G. Matisons, Synthesis and characterization of an isocyanate functionalized polyhedral oligosilsesquioxane and the subsequent formation of

an organic–inorganic hybrid polyurethane, *Journal of the American Chemical Society*, 2002, 124(47), 13998–13999.
[287] P. Maitra and S.L. Wunder, Oligomeric poly(ethylene oxide)-functionalized silsesquioxanes: interfacial effects on T_g, T_m, and ΔH_m, *Chemistry of Materials: A Publication of the American Chemical Society*, 2002, 14(11), 4494–4497.
[288] G.H. Mehl and J.W. Goodby, Liquid-crystalline, substituted octakis(dimethylsiloxy) octasilsesqui-oxanes: oligomeric supermolecular materials with defined topology, *Angewandte Chemie*, 1996, 35(22), 2641–2643.
[289] K. Takahashi, S. Sulaiman, J.M. Katzenstein, S. Snoblen and R.M. Laine, New aminophenyl-silsesquioxanes – synthesis, properties, and epoxy nanocomposites, *Australian Journal of Chemistry*, 2006, 59(8), 564–570.
[290] S. Sulaiman, C. Brick, C.M. De Sana, J.M. Katzenstein, R.M. Laine and R.A. Basheer, Tailoring the global properties of nanocomposites. Epoxy resins with very low coefficients of thermal expansion, *Macromolecules*, 2006, 39(16), 5167–5169.
[291] M.Z. Asuncion and R.M. Laine, Silsesquioxane barrier materials, *Macromolecules*, 2007, 40(3), 555–562.
[292] C. Brick, E.R. Chan, S.C. Glotzer, D.C. Martin and R.M. Laine, Self-lubricating nano-ball-bearings, *Advanced Materials*, 2007, 19(1), 82–86.
[293] H. Mori, Design and synthesis of functional silsesquioxane-based hybrids by hydrolytic condensation of bulky triethoxysilanes, *International Journal of Polymer Science*, 2012, Article ID 173624, 17 pages.
[294] F.J. Feher, R. Terroba and J.W. Ziller, A new route to incompletely-condensed silsesquioxanes: base-mediated cleavage of polyhedral oligosilsesquioxanes, *Chemical Communications*, 1999, 2309–2310.
[295] P. Groch, K. Dziubek and K. Czaja, Silsesquioxanes and their application in the synthesis of polymeric materials, *Polimery*, 2015, 60, 219–231.
[296] H.V. Tran, R.J. Hung, D.A. Loy, D.R. Wheeler, J. Byer, W. Conley and C.G. Willson, Synthesis of siloxanes and silsesquioxanes for 157 nm lithography, *Abstracts of Papers of the American Chemical Society*, 2001, 221, 283-POLY Part 2.
[297] B.D. Ghosh, K.F. Lott and J.E. Ritchie, Structural characterization of a sol-gel-prepared anhydrous proton-conducting electrolyte, *Chemistry of Materials: A Publication of the American Chemical Society*, 2006, 18(2), 504–509.
[298] A.S. Lee, -S.-S. Choi, S.Y. Oh, H.S. Lee, B. Kim, S.S. Hwang and K.-Y. Baek, Incompletely condensed POSS-based spin-on-glass networks for impeccable ultra low-k integration, *Journal of Materials Chemistry. C*, 2015, 3, 11605–11611.
[299] P.A. Agaskar, The higher hydridospherosiloxanes: synthesis and structures of $H_nSi_nO_{1.5n}$ (n = 12, 14, 16, 18), *Inorganic Chemistry*, 1990, 29(9), 1603.
[300] A.R. Bassindale and T.E. Gentle, Siloxane and hydrocarbon octopus molecules with silsesquioxane cores, *Journal of Materials Chemistry*, 1993, 3(12), 1319–1325.
[301] J.V. Crivello and R. Malik, Synthesis and photoinitiated cationic polymerization of monomers with the silsesquioxane core, *Journal of Polymer Science. Part A, Polymer Chemistry*, 1997, 35(3), 407–425.
[302] C. Frye and W.T. Collins, Oligomeric silsesquioxanes, (HSiO3/2)n, *Journal of the American Chemical Society*, 1970, 92(19), 5586–5588.
[303] P.A. Agaskar, V.W. Day and W.G. Klemperer, A new route to trimethylsilylated spherosilicates. Synthesis and structure of [Si12O18](OSiMe3)12, D3h-[Si14O21](OSiMe3)14, and C2v-[Si14O21](OSiMe3)14, *Journal of the American Chemical Society*, 1987, 109(5554).
[304] T. Shioda, T. Gunji, N. Abe and Y. Abe, Preparation and properties of polyhedral oligomeric silsesquioxane polymers, *Applied Organometallic Chemistry*, 2011, 25, 661–664.

[305] K. Constantopoulos, D. Clarke, E. Markovic, D. Uhrig, S. Clarke, J.G. Matisons and G. Simon, New family of POSS monomers suitable for forming urethane polymerizable nanocomposite coatings, *Polymer Preprints*, 2004, 45(1), 668–669.
[306] H. Chen and R.E. Tecklenburg, Characterization of low and intermediate molecular weight hydrogen silsesquioxanes by mass spectrometry, *Journal of the American Society for Mass Spectrometry*, 2006, 17, 1437–1441.
[307] V. Belot, R. Corriu, D. Leclercq, P.H. Mutin and A. Vioux, Thermal reactivity of hydrogenosilsesquioxane gels, *Chemistry of Materials: A Publication of the American Chemical Society*, 1991, 3, 127–131.
[308] Z.S. Xie, D.R. Dai and R.B. Zhang, *Chinese Journal of Polymer Science*, 1992, 10(4), 361–365.
[309] R.B. Zhang, Z. Li, Z.S. Xie, P. Xie, D.R. Dai and D.S. Liu, *CN Pat.* ZL 94 1 00507.0
[310] Z. Li, X. Cao, H. Xu, P. Xie, M. Cao and R. Zhang, Synthesis and characterization of reactive ladderlike polyallylsilsesquioxane and polyvinylsilsesquioxane, *Reactive and Functional Polymers*, 1999, 39, 1–7.
[311] Y. Zhang, Z. Li, C.-Q. Liu, R.B. Zhang, C.-F. Zhu and C. Wang, The hydrosilylation reaction of ladderlike polyallylsilsesquioxane (allyl-T) and polyhydrosilsesquioxane (H-T), *Chinese Journal of Polymer Science*, 2002, 20, 31–38.
[312] -C.-C. Yang, W.-C. Chen, L.-M. Chen and C.-J. Wang, Characterization of poly-(silsesquioxane) by thermal curing, *Proceedings of the National Science Council, Republic Of China / A*, 2001, 25(6), 339–343.
[313] S.Z. Chen, J.F. Liu, H.J.H. Chen and F.S. Huang, Study of nanoimprint pattern transfer on hydrogen silsesquioxane, *Journal of Vacuum Science & Technology B*, 2006, 24, 1934–1940.
[314] C.K. Yang, G. Pandraud, K. Babaei Gavan, E. Van Der Drift and P.J. French, Further Investigation of Hydrogen Silsesquioxane E-Beam Resist as Etching Mask for Cryogenic Silicon, Proceedings of SAFE, pp. 527–530 (2008).
[315] P.T. Liu, J.T. Sheu and T.C. Chang, Direct Patterning of Low-Permittivity Hydrogen Silsesquioxane, NSRRC Activity Report 2002/2003.
[316] S.K. Singh, A.A. Kumbhar and R.O. Dusane, Resisting oxygen plasma damage in low-k hydrogen silsesquioxane films by hydrogen plasma treatment, *Materials Letters*, 2006, 60, 1579–1581.
[317] L. O'Faolain, M.V. Kotlyar, N. Tripathi, R. Wilson and T.F. Krauss, Fabrication of photonic crystals using a spin-coated hydrogen silsesquioxane hard mask, *Journal of Vacuum Science & Technology B*, 2006, 24, 336–339.
[318] A.E. Grigorescu, M.C. Van Der Krogt, C.W. Hagen and P. Kruit, 10 nm lines and spaces written in HSQ, using electron beam lithography, *Microelectronic Engineering*, 2007, 84, 822–824.
[319] H. Namatsu, T. Yamaguchi, M. Nagase, K. Yamazaki and K. Kurihara, *Microelectronic Engineering*, 1998, 41/42, 331–334.
[320] I. Junarsa, M.P. Stoykovich, P.F. Nealey, Y. Ma, F. Cerrina and H.H. Solak, *Journal of Vacuum Science & Technology B*, 2005, 23(1), 138–143.
[321] T.-S. Chang, T.-C. Chang, P.-T. Liu, T.-S. Chang and F.-S. Yeh, Integration issues for siloxane-based hydrogen silsesquioxane (HSQ) applied on TFT-LCDs, *Thin Solid Films*, 2006, 498, 70–74.
[322] S.-W. Hwang, G.-R. Lee, J.-H. Min and S.H. Moon, Angular dependence of etch rate of a silsesquioxane-based low dielectric constant material in fluorocarbon plasmas, *Electrochemical and Solid-State Letters*, 2003, 6(1), G12–G15.
[323] D. Park, T.H. Stievater, W.S. Rabinovich, N. Green, S. Kanakaraju and L.C. Calhoun, Characterization of hydrogen silsesquioxane as a Cl_2/BCl_3 inductively coupled plasma etch mask for air-clad InP-based quantum well waveguide fabrication, *Journal of Vacuum Science & Technology B*, 2006, 24, 3152-.

[324] M.J. Word, I. Adesida and P.R. Berger, Nanometer-period gratings in hydrogen silsesquioxane fabricated by electron beam lithography, *Journal of Vacuum Science & Technology B*, 2003, 21, L12–L15.
[325] G.M. Schmid and L.E. Carpenter, II, J.A. Liddle, Nonaqueous development of silsesquioxane electron beam resist, *Journal of Vacuum Science & Technology B*, 2004, 22, 3497–3502.
[326] J.K.W. Yang, B. Cord, K.K. Berggren, J. Klingfus, S.-W. Nam, K.-B. Kim and M.J. Rooks, Understanding of hydrogen silsesquioxane electron resist for sub-5-nm-half-pitch lithography, *Journal of Vacuum Science & Technology B*, 2009, 27, 2622–2627.
[327] S.-I. Lee, J.H. Sim, H.-J. Lee, R. Kasica, H.-M. Kim, C.L. Soles, K.-B. Kim and D.Y. Yoon, Organosilicate polymer e-beam resists with high resolution, sensitivity and stability, *Applied Organometallic Chemistry*, 2013, 27, 644–651.
[328] M. Guillorn, J. Chang, N. Fuller, J. Patel, M. Darnon, A. Pyzyna, E. Joseph, S. Engelmann, J. Ott, J. Newbury, D. Klaus, J. Bucchignano, P. Joshi, C. Scerbo, E. Kratschmer, W. Graham, B. To, J. Parisi, Y. Zhang and W. Haensch, Hydrogen silsesquioxane-based hybrid electron beam and optical lithography for high density circuit prototyping, *Journal of Vacuum Science & Technology B*, 2009, 27, 2588–2592.
[329] D. Morecroft, J.K.W. Yang, S. Schuster, K.K. Berggren, Q. Xia, W. Wu and R.S. William, Sub-15 nm nanoimprint molds and pattern transfer, *Journal of Vacuum Science & Technology B*, 2009, 27(6), 2837–2840.
[330] S. Schuster, D. Morecroft, J.K.W. Yang and K.K. Berggren, Lift-off of Sub-15-nm Hydrogen Silsesquioxane (HSQ) Structures, manuscript in review (2009).
[331] S. Strobel, K.J. Harry, H. Duan, J.K.W. Yang, V.R. Manfrinato and K.K. Berggren, Electrochemical development of hydrogensilsesquioxane by applying an electrical potential, *Nanotechnology*, 2011, 22(375301), 7. DOI: 10.1088/0957-4484/22/37/375301.
[332] N. Clark, A. Vanderslice and R. Grove, III, R.R. Krchnavek, Time-dependent exposure dose of hydrogen silsesquioxane when used as a negative electron-beam resist, *Journal of Vacuum Science & Technology B*, 2006, 24, 3073–3076.
[333] H.-S. Lee, J.-S. Wi, S.-W. Nam, H.-M. Kim and K.-B. Kim, Two-step resist-development process of hydrogen silsesquioxane for high-density electron-beam nanopatterning, *Journal of Vacuum Science and Technology. B*, 2009, 27, 188–192.
[334] J.N. Bremmer, Y. Liu, K.G. Gruszynski and F.C. Dall, Int. Dielectrics for ULSI Multilevel Interconnection Conference, 1997, p. 333.
[335] S. Maghsoodi, S. Wang, G.S. Becker, J.D. Albaugh, C.R. Yeakle, D.K. Choi, R.R. Warner, G.A. Cerny, J.E. Hamon, D. Ha and E.S. Moyer, *SID 03 Digest*, 2003, p. 1512.
[336] J.N. Bremmer, Y. Liu, K.G. Gruszynski and F.C. Dail, Cure of Hydrogen silsesquioxane for intermetal dielectric applications, *Materials Research Society Symposia Proceedings*, 1997, 476(37), Low-dielectric constant materials III: April 1-4, 1997, San Francisco, California, U.S.A.
[337] H.J.H. Chen, J.F. Liu, Y.J. Hsu, J.C.C. Syu and F.S. Huang, Fabrication of Au nanowires on hydrogen silsesquioxane by nanoimprint transfer, *Nanotechnology*, 2005, 16, 2913–2918.
[338] S.-W. Ryu, H.-J. Choi, S. Choo, C.-H. Kim and H. Lee, Fabrication of rigid stamp on a cylindrical substrate using hydrogen silsesquioxane/ZrO_2 nanoparticle composite material for roll-to-roll nanoimprinting process, *Journal of Sol-gel Science and Technology*, 2015, 73, 628–633.
[339] I. Majoros, T.M. Marsylko and J.P. Kennedy, Synthesis and characterization of novel well-defined stars consisting of eight polyisobutylene arms emanating from an octa (dimethylsiloxy)octasilsesquioxane core, *Polymer Bulletin*, 1997, 38, 15.
[340] A. Provatas, M. Luft, J.C. Mu, A.H. White, J.G. Matisons and B.W. Skelton, Silsesquioxanes: Part I: A key intermediate in the building of molecular composite materials., *Journal of Organometallic Chemistry*, 1998, 565, 159–164.

[341] N.L.D. Filho, H.A. De Aquino, G. Pires and L. Caetano, Relationship between the dielectric and mechanical properties and the ratio of epoxy resin to hardener of the hybrid thermosetting polymers, *Journal of the Brazilian Chemical Society*, 2006, 17, 533–541.

[342] D. Hoebel and M. Wicker, Die Konstitution des Tetramethylammoniumsilicats der Zusammensetzung 1,0 $N(CH_3)_4OH$ 1,0 $SiO_2 \cdot 8{,}0–8{,}3\ H_2O$, *Zeitschrift Für Anorganische Und Allgemeine Chemie*, 1971, 384(1), 43–52.

[343] I. Hasegawa, Co-hydrolysis products of tetraethoxysilane and methyltriethoxysilane in the presence of tetramethylammonium ions, *Journal of Sol-gel Science and Technology*, 1993, 1(1), 57–63.

[344] M. Moran, C.M. Casado, I. Cuadrado and J. Losada, Ferrocenyl substituted octakis (dimethylsiloxy) octasilsesquioxanes: a new class of supramolecular organometallic compounds. Synthesis, characterization, and electrochemistry, *Organometallics*, 1993, 12(11), 4327–4333.

[345] I. Hasegawa, K. Ino and H. Onishi, An improved procedure for syntheses of silyl derivatives of the cubeoctameric silicate anion, *Applied Organometallic Chemistry*, 2003, 17(5), 287–290.

[346] M.S. Soh, A.U.J. Yap and A. Sellinger, Methacrylate and epoxy functionalized nanocomposites based on silsesquioxane cores for use in dental applications, *European Polymer Journal*, 2007, 43(2), 315–327.

[347] C.M. Hessel, E.J. Henderson and J.G.C. Veinot, Hydrogen silsesquioxane: a molecular precursor for nanocrystalline Si–SiO_2 composites and freestanding hydride-surface-terminated silicon nanoparticles, *Chemistry of Materials: A Publication of the American Chemical Society*, 2006, 18(26), 6139–6146.

[348] J.D. Lichtenhan, Y.A. Otonari and M. Cam, Linear hybrid polymer building blocks: methacrylate-functionalized polyhedral oligomeric silsesquioxane monomers and polymers, *Macromolecules*, 1995, 28(24), 8435–8437.

[349] T.F. Baumann, T.V. Jones, T. Wilson, A.P. Saab and R.S. Maxwell, Synthesis and characterization of novel PDMS nanocomposites using POSS derivatives as cross-linking filler, *Journal of Polymer Science. Part A, Polymer Chemistry*, 2009, 47, 2589–2596.

[350] M. Morh, C.M. Casado and I. Cuadrado, Ferrocenyl substituted octakis(dimethylsiloxy)-octasilsesquioxanes: a new class of supramolecular organometallic compounds. Synthesis, characterization, and electrochemistry, *Organometallics*, 1993, 12(11), 4327–4333.

[351] P. Maitra, H. Zhang, J. Zhang and S.L. Wunder, Anionic/PEO Functionalized Silsesquioxanes as Single Ion Conductors for Lithium Batteries, *Electrochemical Society*, 2002, 195.

[352] P.R. Chinnam and S.L. Wunder, Polyoctahedral Silsesquioxane-Nanoparticle Electrolytes for Lithium Batteries: POSS-Lithium Salts and POSS-PEGs, *Chemistry of Materials: A Publication of the American Chemical Society*, 2011, 23(23), 5111–5121.

[353] P.R. Chinnam, H. Zhang and S.L. Wunder, Blends of pegylated polyoctahedralsilsesquioxanes (POSS-PEG) and methyl cellulose as solid polymer electrolytes for lithium batteries, *Electrochimica Acta*, 2015, 170, 191–201.

[354] A.R. Polu and H.-W. Rhee, Nanocomposite solid polymer electrolytes based on poly(ethylene oxide) /POSS-PEG ($n = 13.3$) hybrid nanoparticles for lithium ion batteries, *Journal of Industrial and Engineering Chemistry*, 2015, 31, 323–329.

[355] J. Zhang, C. Maa, J. Liu, L. Chen, A. Pan and W. Wei, Solid polymer electrolyte membranes based on organic/inorganic nanocomposites with star-shaped structure for high performance lithium ion battery, *Journal of Membrane Science & Technology*, 2016, 509, 138–148.

[356] A.R. Polu, H.-W. Rhee, M.J.K. Reddy, A.M. Shanmugharaj, S.H. Ryu and D.K. Kim, Effect of POSS-PEG hybrid nanoparticles on cycling performance of polyether-LiDFOB based solid polymer electrolytes for all solid-state Li-ion battery applications, *Journal of Industrial and Engineering Chemistry*, 2017, 45, 68–77.

[357] F. Eckstorff, Y. Zhu, R. Maurer, T.E. Müller, S. Scholz and J.A. Lercher, Materials with tunable low-k dielectric constant derived from functionalized octahedral silsesquioxanes and spherosilicates, *Polymer*, 2011, 52, 2492–2498.
[358] M. Seino, T. Hayakawa, Y. Ishida and M. Kakimoto, Hydrosilylation polymerization of double-decker-shaped silsesquioxane having hydrosilane with diynes, *Macromolecules*, 2006, 39(10), 3473–3475.
[359] H.W. Su, W.C. Chen, W.C. Lee and J.S. King, New photocurable acrylic/silsesquioxane hybrid optical materials: synthesis, properties, and patterning, *Macromolecular Materials and Engineering*, 2007, 292, 666–673.
[360] A. Kowalewska, K. Rózga-Wijas and M. Handke, Alkoxymethylcyclosiloxanes - new efficient precursors of crystalline $(CH_3SiO_{3/2})_8$ silsesquioxane and polymethyl silsesquioxanes, *e-Polymers*, 2008, (no. 150).
[361] M. Handke, A. Kowalewska and W. Mozgawa, Spectroscopic study of ceramic precursors obtained by hydrolytic condensation of ethoxycyclotetrasiloxane, *Journal of Molecular Structure*, 2008, 887, 152–158.
[362] H. Li, J. Zhang, R. Xu and D. Yu, Direct synthesis and characterization of crosslinked polysiloxanes via anionic ring-opening copolymerization with octaisobutyl-polyhedral oligomeric silsesquioxane and octamethylcyclotetrasiloxane, *Journal of Applied Polymer Science*, 2006, 102(4), 3848–3856.
[363] M. Handke and A. Kowalewska, Siloxane and silsesquioxane molecules - precursors for silicate materials, *Spectrochimica Acta. Part A: Molecular Spectroscopy*, 2011, 79(4), 749–757.
[364] C. Zhang and R.M. Laine, Silsesquioxanes as synthetic platforms. II. Epoxy-functionalized inorganic-organic hybrid species, *Applied Organometallic Chemistry*, 1996, 521, 199–201.
[365] P. Żak, B. Marciniec, M. Majchrzak and C. Pietraszuk, Highly effective synthesis of vinyl-functionalised cubic silsesquioxanes, *Applied Organometallic Chemistry*, 2011, 696, 887–891.
[366] B. Marciniec, M. Dutkiewicz, H. Maciejewski and M. Kubicki, New, Effective Method of Synthesis and Structural Characterization of Octakis(3-chloropropyl)octasilsesquioxane, *Organometallics*, 2008, 27(4), 793–794.
[367] M. Dutkiewicz, H. Maciejewski and B. Marciniec, Functionalization of polyhedral oligomeric silsesquioxane (POSS) via nucleophilic substitution, *Synthesis*, 2009, 12, 2019–2024.
[368] D.R. Do Carmo, M. De, S. Magossi, U. De, O. Bicalho and D.R. Silvestrini, Synthesis, characterization and thermal properties of silsesquioxane organically modified with 4,5-diphenyl-2-imidazolethiol, *International Journal of Chemistry*, 2014, 6, 22–33.
[369] S. Chimjarn, R. Kunthom, P. Chancharone, R. Sodkhomkhum, P. Sangtrirutnugula and V. Ervithayasuporn, Synthesis of aromatic functionalized cagerearranged silsesquioxanes (T_8, T_{10}, and T_{12}) via nucleophilic substitution reactions, *Dalton Transactions*, 2015, 44, 916–919.
[370] Y.E. Aziz, A.R. Bassindale, P.G. Taylor, P.N. Horton, R.A. Stephenson and M.B. Hursthouse, Facile synthesis of novel functionalized silsesquioxane nanostructures containing an encapsulated fluoride anion, *Organometallics*, 2012, 31, 6032–6040.
[371] R.M. Laine, M. Roll, M. Asuncion, S. Sulaiman, V. Popova, D. Bartz, D.J. Krug and P.H. Mutin, Perfect and nearly perfect silsesquioxane (SQs) nanoconstruction sites and Janus SQs, *Journal of Sol-gel Science and Technology*, 2008, 46, 335–347.
[372] M.E. Pérez-Ojeda, B. Trastoy, Á. Rol, M.D. Chiara, I. Garcia-Moreno and J.L. Chiara, Controlled click-assembly of well-defined hetero-bifunctional cubic silsesquioxanes and their application in targeted bioimaging, *Chemistry – A European Journal*, 2013, 19, 6630–6640.
[373] H. Toyodome, Y. Kaneko, K. Shikinaka and N. Iyi, Preparation of carboxylate group-containing rod-like polysilsesquioxane with hexagonally stacked structure by sol-gel reaction of 2-cyanoethyl-triethoxysilane, *Polymer*, 2012, 53, 6021–6026.

[374] M.K. Kolel-Veetil, D.D. Dominguez and T.M. Keller, Dendritic networks containing polyhedral oligomeric silsesquioxane (poss) and carborane clusters, *Journal of Polymer Science. Part A, Polymer Chemistry*, 2008, 46, 2581–2587.

[375] Z. Shen, J. Kim, J. Shen, C.M. Downing, S. Lee, H.H. Kung and M.C. Kung, Spherosilicates with peripheral malonic acid and vinyl end groups, *Chemical Communications*, 2013, 49, 3357.

[376] G. Zhou, T. Simerly, L. Golovko, I. Tychinin, V. Trachevsky, Y. Gomza and A. Vasiliev, Highly functionalized bridged silsesquioxanes, *Journal of Sol-Gel Science and Technology*, 2012, 62, 470–482.

[377] J.H. Jung, K. Chou, J.C. Furgal and R.M. Laine, Synthesis of acetoxyphenyl- and hydroxyphenyl-terminated polyfunctional T_8, T_{10}, T_{12} silsesquioxanes and initial studies on their use in the formation of highly crosslinked polyesters, *Applied Organometallic Chemistry*, 2013, 27, 666–672.

[378] L.A. Bivona, O. Fichera, L. Fusaro, F. Giacalone, M. Buaki-Sogo, M. Gruttadauria and C. Aprile, A polyhedral oligomeric silsesquioxane-based catalyst for the efficient synthesis of cyclic carbonates, *Catalysis Science and Technology*, 2015, 5, 5000.

[379] Z.-X. Zhang, J. Hao, P. Xie, X. Zhang, C.C. Han, R. Zhang and A well-defined ladder, polyphenyl-silsesquioxane (Ph-LPSQ) synthesized via a new three-step approach: monomer self-organization-lyophilization – surface-confined polycondensation, *Chemistry of Materials: A Publication of the American Chemical Society*, 2008, 20, 1322–1330.

[380] S. Chang, T. Matsumoto, H. Matsumoto and M. Unno, Synthesis and characterization of hepta-cyclic laddersiloxanes and ladder polysilsesquioxane, *Applied Organometallic Chemistry*, 2010, 24, 241–246.

[381] S.-S. Choi, A.S. Lee, H.S. Lee, H.Y. Jeon, K.-Y. Baek, D.H. Choi and S.S. Hwang, Synthesis and characterization of UV-curable ladder-like polysilsesquioxane, *Journal of Polymer Science. Part A, Polymer Chemistry*, 2011, 49, 5012–5018.

[382] A. Koleżyński, W. Jastrzębski, W. Szczypka, A. Kowalewska, M. Nowacka and M. Sitarz, The structure and bonding properties of chosen phenyl ladder-like silsesquioxane clusters, *Journal of Molecular Structure*, 2013, 1044, 314–322.

[383] H.W. Ro, V. Popova, D.J. Krug, A.M. Forster, R.M. Laine and C.L. Soles, Cubic silsesqui-oxanes as tunable high-performance coating materials, *Applied Organometallic Chemistry*, 2013, 27, 652–659.

[384] M. Nowacka, C. Fischer, A. Kowalewska, M. Hebda and K. Hodor, Thermally induced phenomena leading to degradation of poly (silsesquioxane) materials, *European Polymer Journal*, 2017, 86, 17–28.

[385] N. Takamura, L. Viculis, C. Zhang and R.M. Laine, Completely discontinuous organic/inorganic hybrid nanocomposites by self-curing of nanobuilding blocks constructed from reactions of $[HMe_2SiOSiO_{1.5}]_8$ with vinylcyclohexene, *Polymer International*, 2007, 56, 1378–1391.

[386] J. Han and S. Zheng, Highly porous polysilsesquioxane networks via hydrosilylative polymerization of macrocyclic oligomeric silsesquioxanes, *Macromolecules*, 2008, 41, 4561–4564.

[387] M. Ikeda and H. Saito, Improvement of polymer performance by cubic-oligosilsesquioxane, *Reactive and Functional Polymers*, 2007, 67, 1148–1156.

[388] N. Hao, M. Böhning, S. Wohlrab and A. Schönhals, Polymer based nanocomposites with phenethylPOSS as nanofiller investigated by dielectric spectroscopy and gas transport measurements, *Macromolecular Symposia*, 2008, 267(1), 69–73.

[389] E.T. Kopesky, T.S. Haddad, G.H. McKinley and R.E. Cohere, Miscibility and viscoelastic properties of acrylic polyhedral oligomericsilsesquioxane–poly(methyl methacrylate) blends, *Polymer*, 2005, 46(13), 4743–4752.

[390] M.Z. Asuncion, I. Hasegawa, J.W. Kampf and R.M. Laine, The selective dissolution of rice hull ash to form $[OSiO_{1.5}]_8[R_4N]_8$ (R = Me, CH_2CH_2OH) octasilicates. Basic nanobuilding blocks and possible models of intermediates formed during biosilicification processes, *Journal of Materials Chemistry*, 2005, 15, 2114–2121.
[391] R.M. Laine, V.N. Popova-Guerguieva, D.W. Bartz, M.Z. Asuncion and D.J. Krug, Multi-functional silsesquioxanes for novel coating applications, Pat. WO WO2009002660.
[392] C. Hartmann-Thompson, Chapter 7: Polyhedral Oligomeric Silsesquioxanes in Electronics and Energy Applications, C. Hartmann-Thompson, Ed., In: *Applications of Polyhedral Oligomeric Silsesquioxanes*, Advances in Silicon Science, Vol. 3, DOI 10.1007/978-90-481-3787-9_7, © Springer Science+Business Media B.V., 2011.
[393] M. Joshi and B.S. Butola, Polymeric nanocomposites - polyhedral oligomeric silsesquioxanes (POSS) as hybrid nanofiller, *Journal of Macromolecular Science Part C- Polymer Reviews*, 2004, 44, 389–410.
[394] L. Zheng, A.J. Waddon, R.J. Farris and E.B. Coughlin, X-ray Characterizations of polyethylene polyhedral oligomeric silsesquioxane copolymers, *Macromolecules*, 2002, 35(6), 2375.
[395] Y. Liu, Z. Shi, H. Xu, J. Fang, X. Ma and J. Yin, Preparation, characterization, and properties of novel polyhedral oligomeric silsesquioxane–polybenzimidazole nanocomposites by Friedel –Crafts reaction, *Macromolecules*, 2010, 43(16), 6731.
[396] E. Ayandele, B. Sarkar and P. Alexandridis, Polyhedral Oligomeric Silsesquioxane (POSS)-Containing Polymer Nanocomposites, *Nanomaterials*, 2012, 2, 445–475. DOI: 10.3390/nano2040445.
[397] W. Zhang and A.H.E. Müller, Architecture, self-assembly and properties of well-defined hybrid polymers based on polyhedral oligomeric silsequioxane (POSS), *Progress in Polymer Science*, 2013, 38, 1121.
[398] Y.J. Lee, S.W. Kuo, W.J. Huang, H.Y. Lee and F.C. Chang, Miscibility, specific interactions, and self-assembly behavior of phenolic/polyhedral oligomeric silsesquioxane hybrids, *Journal of Polymer Science. Part B, Polymer Physics*, 2004, 42(6), 1127–1136.
[399] C. DeArmitt, POSS® User's Guide, Hybrid Plastics, USA.
[400] F.J. Feher, D.A. Newman and J.F. Walzer, Silsesquioxanes as models for silica surfaces, *Journal of the American Chemical Society*, 1989, 111(5), 1741–1748.
[401] F.M. Capaldi, M.C. Boyce and G.C. Rutledge, The mechanical properties of crystalline cyclopentyl polyhedral oligomeric silsesquioxane, *The Journal of Chemical Physics*, 2006, 124, 214709.
[402] G.A. Gadodia, Synthesis and Study of Hybrid Organic - Inorganic Polyhedral Oligomeric Silsesquioxanes (POSS) Based Polymers, University of Massachusetts-, Amherst, 2009.
[403] B.X. Fu, C. DeArmitt and J. Schwab, Tailor-Made Polyhedral Oligomeric Silsesquioxane (POSS) Molecules for Thermosets, Hybrid Plastics, Inc., Sept. 2008. https://www.trfa.org/erc/docretrieval/uploadedfiles/Technical%20Papers/2008%20Meeting/Fu-HybridPlastics_ppt-POSS_Molecules.pdf
[404] M. Zaborski, A. Strąkowska, A. Kosmalska, H. Maciejewski and M. Dutkiewicz, POSS compounds as modifiers and additives for elastomeric composites, *Polimery*, 2013, 58, 772–781.
[405] A. Strąkowska, A. Kosmalska, M. Masłowski, T. Szmechtyk, K. Strzelec and M. Zaborski, POSS as promoters of self-healing process in silicone composites, *Polymer Bulletin*, 2019, 76(3387-3402). DOI: https://doi.org/10.1007/s00289-018-2522-8.
[406] Y.Y. Jo, A.S. Lee, K.-Y. Baek, H. Lee and S.S. Hwang, Multi-crosslinkable self-healing poly-silsesqui-oxanes for the smart recovery of anti-scratch properties, *Polymer*, 2017, 124, 78–87.
[407] W. Zhang and A.H.E. Müller, Architecture, self-assembly and properties of well-defined hybrid polymers based on polyhedral oligomeric silsequioxane (POSS), *Progress in Polymer Science*, 2013, 38(1121). DOI: http://dx.doi.org/10.1016/j.progpolymsci.2013.03.002.

[408] R.K. Bharadwaj, R.J. Berry and B.L. Farmer, Molecular dynamics simulation study of norbornene-POSS polymers, *Polymer*, 2000, 41, 7209–7221.
[409] F.J. Feher and K.J. Weller, Synthesis and characterization of labile spherosilicates: $[(Me_3SnO)_8Si_8O_{12}]$ and $[(Me_4SbO)_8Si_8O_{12}]$, *Inorganic Chemistry*, 1991, 30(5), 880–882.
[410] M.W. Ellsworth and D.L. Gin, Recent advances in the design and synthesis of polymer-inorganic nanocomposites, *Polymer News*, 1999(24), 331–341.
[411] T.S. Haddad and J.D. Lichtenhan, Hybrid organic-inorganic thermoplastics: Styryl-based polyhedral oligomeric silsesquioxane polymers, 1996, 29(22), 7302–7304.
[412] T.S. Haddad, P.T. Mather, H.G. Jeon, A. Romo-Uribe, A.R. Farris and J.D. Lichtenhan, Thermoplastics modified with nanoscale inorganicmacromers, *Materials Research Society Symposia Proceedings*, 1998, 519, 381–386.
[413] J. Wu and P.T. Mather, POSS polymers: physical properties and biomaterials applications, *Journal of Macromolecular Science, Part C: Polymer Reviews*, 2009, 49, 25–63.
[414] H. Bergenudd, P. Eriksson, C. DeArmitt, B. Stenberg and J.E. Malmström, Synthesis and evaluation of hyperbranched phenolic antioxidants of three different generations, *Polymer Degradation and Stability*, 2002, 76(3), 503–509.
[415] C. DeArmitt, P. Wheeler and S. Hait, unpublished work (2009), Hybrid Plastics, Hattiesburg, MS, USA.
[416] S. Iyer and D.A. Schiraldi, Role of specific interactions and solubility in the reinforcement of bisphenol A polymers with polyhedral oligomeric silsesquioxanes, *Macromolecules*, 2007, 40 (14), 4942–4952.
[417] S.Y. Soong, R.E. Cohen, M.C. Boyce and W. Chen, The effects of thermomechanical history and strain rate on antiplasticization of PVC, *Polymer*, 2008, 49(6), 1440–1443.
[418] R. Misra, B.X. Fu and S.E. Morgan, Surface energetics, dispersion, and nanotribomechanical behavior of POSS/PP hybrid nanocomposites, *Journal of Polymer Science. Part B, Polymer Physics*, 2007, 45(17), 2441–2455.
[419] R. Misra, B.X. Fu and S.E. Morgan, *65th Annual Technical Conference Society of Plastics Engineers*, 62–66 (2007).
[420] S.Y. Soong, R.E. Cohen and M.C. Boyce, Polyhedral oligomeric silsesquioxane as a novel plasticizer for poly(vinyl chloride), *Polymer*, 2007, 48(5), 1410–1418.
[421] E.G. Shockey, A.G. Bolf, P.F. Jones, J.J. Schwab, K.P. Chaffee, T.S. Haddad and J.D. Lichtenhan, Functionalized polyhedral oligosilsesquioxane (POSS) macromers: new graftable POSS hydride, POSS α-olefin, POSS epoxy, and POSS chlorosilane macromers and POSS–siloxane triblocks, *Applied Organometallic Chemistry*, 1999, 13(4), 311–327.
[422] A. Lee and J.D. Lichtenhan, Thermal and viscoelastic property of epoxy–clay and hybrid inorganic–organic epoxy nanocomposites, *Journal of Applied Polymer Science*, 1999, 73(10), 1993–2001.
[423] H. Kamo, S. Kuga, T. Ono and M. Ikeda, *US Pat.* 193533 (2002).
[424] C. DeArmitt and P. Wheeler, POSS keeps high temperature plastics flowing, *Plastics Additives & Compounding*, 2008, 10(4), 36–39.
[425] H. Araki and K. Naka, Synthesis of dumpbell-shaped trifluoropropyl-substituted POSS derivatives linked by simple aliphatic chains and their optical transparent thermoplastic films, *Macromolecules*, 2011, 45, 6039–6045, ISSN 0024-9297.
[426] P.A. Wheeler, R. Misra, R.D. Cook and S.E. Morgan, Polyhedral oligomeric silsesquioxane trisilanols as dispersants for titanium oxide nanopowder, *Journal of Applied Polymer Science*, 2008, 108(4), 2503–2508.
[427] J.D. Lichtenhan, B.X. Fu, M. Blue, P. Wheeler, R. Misra and S.E. Morgan, *World Pat.* 054418 A2 (2008).

[428] Y. Li, H. Su, X. Feng, K. Yue, Z. Wang, Z. Lin, X. Zhu, Q. Fu, Z. Zhang, S.Z.D. Cheng and W.-B. Zhang, Precision synthesis of macrocyclic giant surfactants tethered with two different polyhedral oligomeric silsesquioxanes at distinct ring locations via four consecutive "click" reactions, *Polymer Chemistry*, 2015, 6(5), 827–837. DOI: 10.1039/c4py01360c.
[429] Z. Lin, P. Lu, X. Yu, W.-B. Zhang, M. Huang, K. Wu, K. Guo, C. Wesdemiotis, X. Zhu, Z. Zhang, K. Yue and S.Z.D. Cheng, Sequential "Click" Synthesis of "Nano-Diamond-Ring-like" Giant Surfactants Based on Functionalized Hydrophilic POSS/C60 Tethered with Cyclic Polystyrenes, *Macromolecules*, 2014, 47, 4160–4168.
[430] H. Saito, M. Ikeda, *Japan Kokai Tokkyo Koho* 2002-302,599 (2002)
[431] I.S. Isayeva and J.P. Kennedy, Amphiphilic membranes crosslinked and reinforced by POSS, *Journal of Polymer Science. Part A, Polymer Chemistry*, 2004, 42, 4337–4352.
[432] C.H. Worthley, K.T. Constantopoulos, M. Ginic-Markovic, E. Markovic and S. Clarke, A study into the effect of POSS nanoparticles on cellulose acetate membranes, *Journal of Membrane Science*, 2013, 431, 62–71.
[433] S.V. Parambath, D. Ponnamma, K.K. Sadasivuni, S. Thomas and R. Stephen, Effect of nanostructured polyhedral oligomeric silsesquioxane on the physical properties of poly(vinyl alcohol), *Journal of Applied Polymer Science*, 2017, 134, 45447.
[434] B. Balanuca, A. Lungu, A.-M. Hanganu, L.R. Stan, E. Vasile and H. Iovu, Hybrid nanocomposites based on POSS and networks of methacrylated camelina oil and various PEG derivatives, *European Journal of Lipid Science and Technology* : EJLST, 2014, 116, 458–469.
[435] K.P. Pramoda, W. Chen and T.S. Chung, Study of curing kinetics of ladder-like polyepoxysiloxane, *Polymer Bulletin*, 2001, 47(1), 55–63.
[436] R.M. Laine, J. Choi and I. Lee, Organic-inorganic nanocomposites with completely defined interfacial interactions, *Advanced Materials*, 2001, 13(11), 800–803.
[437] W. Xing, B. You and L. Wu, The microstructure and anticorrosion performance of phytic acid-catalyzed polysilsesquioxane coatings, *Journal of Sol-Gel Science and Technology*, 2007, 42, 187–195.
[438] J. Fu, L. Shi, Y. Chen, S. Yuan, J. Wu, X. Liang and Q. Zhong, Epoxy nanocomposites containing mercaptopropyl polyhedral oligomeric silsesquioxane: morphology, thermal properties, and toughening mechanism, *Journal of Applied Polymer Science*, 2008, 109, 340–349.
[439] J.-M. Huang, H.-J. Huang, Y.-X. Wang, W.-Y. Chen and F.-C. Chang, Preparation and characterization of epoxy/polyhedral oligomeric silsesquioxane hybrid nanocomposites, *Journal of Polymer Science. Part B, Polymer Physics*, 2009, 47, 1927–1934.
[440] Y.-C. Yen, Y.-S. Ye, -C.-C. Cheng, C.-H. Lu, L.-D. Tsai, J.-M. Huang and F.-C. Chang, The effect of sulfonic acid groups within a polyhedral oligomeric silsesquioxane containing cross-linked proton exchange membrane, *Polymer*, 2010, 51, 84–91.
[441] M. Pan, C. Zhang, B. Liu and J. Mu, Dielectric and thermal properties of epoxy resin nano-composites containing polyhedral oligomeric silsesquioxane, *Journal of Materials Science Research*, 2013, 2(1), 153–162.
[442] M. Longhi, L.P. Zini, V. Pistor, S.R. Kunst and A.J. Zattera, Evaluation of the mechanic and electrochemical properties of an epoxy coating with addition of different polyhedral oligomeric silsesquioxanes (POSS) applied on substrate of low alloy steel, *Materials Research*, 2017, 20(5), 1388–1401. DOI: http://dx.doi.org/10.1590/1980-5373-MR-2016-0892.
[443] M.A. De Farias, L.A.F. Coelho and S.H. Pezzin, Hybrid nanocomposites based on epoxy/silsesquioxanes matrices reinforced with multi-walled carbon nanotubes, *Materials Research*, 2015, 18(6), 1304–1312.

[444] G.-X. Hou, J.-G. Gao and C. Tian, Hybrid free radical-cationic thermal polymerization of methylacryloylpropyl-POSS/epoxy resins nanocomposites, *Journal of Polymer Research*, 2013, 20(221). DOI: 10.1007/s10965-013-0221-6.

[445] K. Rózga-Wijas and J. Chojnowski, Synthesis of new polyfunctional cage oligosilsesquioxanes and cyclic siloxanes by thiol-ene addition, *Journal of Inorganic and Organometallic Polymers* and Materials, 2012, 22, 588–594.

[446] S. Shen, F. Ye, C. Zhang, Y. Xiong, L. Su and S. Zhao, Preparation of polyhedral oligomeric silsesquioxane based hybrid monoliths by thiol-ene click chemistry for capillary liquid chromatography, *Analyst*, 2015, 140, 265–271.

[447] K. Rózga-Wijas, W.A. Stańczyk, J. Kurjata and S. Kaźmierski, Star-shaped and linear POSS-polylactide hybrid copolymers, *Materials*, 2015, 8, 4400–4420.

[448] R. Anyszka, M. Kozanecki, A. Czaderna, M. Olejniczak, J. Sielski, M. Siciński, D.M. Bieliński, M. Imiela, T. Gozdek, D. Pietrzak and J. Wręczycki, The 2nd Intern. Conference on Material Technology and Environ-mental Materials (MTEM 2016), Oct. 21-23, 2016, Beijing, China.

[449] J. Cao, H. Fan, B.-G. Li and S. Zhu, Synthesis and evaluation of double-decker silsesquioxanes as modifying agent for epoxy resin, *Polymer*, 2017, 124, 157–167.

[450] N.T. Dintcheva, E. Morici, R. Arrigo, F.P. La Mantia, V. Malatesta and J.J. Schwab, Structure-properties relationships of polyhedral oligomeric silsesquioxane (POSS) filled PS nanocomposites, *eXPRESS Polymer Letters*, 2012, 6, 561–571.

[451] Y.-C. Sheen, C.-H. Lu, C.-F. Huang, S.-W. Kuo and F.-C. Chang, Synthesis and characterization of amorphous octakis-functionalized polyhedral oligomeric silsesquioxanes for polymer nano-composites, *Polymer*, 2008, 49, 4017–4024.

[452] K.-W. Huang, L.-W. Tsai and S.-W. Kuo, Influence of octakis-functionalized polyhedral oligomeric silsesquioxanes on the physical properties of their polymer nanocomposites, *Polymer*, 2009, 50, 4876.

[453] I. Blanco, L. Abate, M.L. Antonelli, F.A. Bottino and P. Bottino, Phenyl hepta cyclopentyl – polyhedral oligomeric silsesquioxane (ph,hcp-POSS)/polystyrene (PS) nanocomposites: t he influence of substituents in the phenyl group on the thermal stability, *eXPRESS Polymer Letters*, 2012, 6, 997–1006.

[454] M.A. Wahab, K.Y. Mya and C. He, Synthesis, morphology, and properties of hydroxyl terminated-POSS/polyimide low-k nanocomposite films, *Journal of Polymer Science. Part A, Polymer Chemistry*, 2008, 46, 5887–5896.

[455] A. Chandramohan, S. Nagendiran and M. Alagar, Synthesis and characterization of polyhedral oligomeric silsesquioxane–siloxane-modified polyimide hybrid nanocomposites, *Journal of Composite Materials*, 2011, 46(7), 773–781.

[456] R. Verker, E. Grossman and N. Eliaz, Effect of the POSS–polyimide nanostructure on its mechanical and electrical properties, *Composites Science and Technology*, 2012, 72, 1408–1415.

[457] R. Jeziórska, B. Świerz-Motysia, A. Szadkowska, B. Marciniec, H. Maciejewski, M. Dutkiewicz and I. Leszczyńska, Effect of POSS on morphology, thermal and mechanical properties of polyamide 6, *Polimery*, 2011, 56(809).

[458] Q. Sun, Q. Wang, P. Jena, B.V. Reddy and M. Marquez, Hydrogen storage in organometallic structures grafted on silsesquioxanes, *Chemistry of Materials: A Publication of the American Chemical Society*, 2007, 19(12), 3074–3078.

[459] S.T. Iacono, A. Vij, A.J. Peloquin, G.Y. Dennis, W. Smith Jr. and J.M. Mabry, Fluorinated Polyhedral Oligomeric Silsesquioxanes (F-POSS), OMB Report No. 0704-0188, AFSR Lab. (2010).

[460] A. Dorigato, A. Pegoretti and C. Migliaresi, Physical properties of polyhedral oligomeric silsesquioxanes–cycloolefin copolymer nanocomposites, *Journal of Applied Polymer Science*, 2009, 114, 2270–2279.

[461] X. Song, S. Zhou, Y. Wang, W. Kang and B. Cheng, Mechanical and electret properties of poly-propylene unwoven fabrics reinforced with POSS for electret filter materials, *Journal of Polymer Research*, 2012, 19, 9812.
[462] D. Czarnecka-Komorowska, T. Sterzyński, H. Maciejewski and M. Dutkiewicz, The effect of polyhedral oligomeric silsesquioxane (POSS) on morphology and mechanical properties of polyoxymethylene (POM), *Composites Theory and Practice*, 2012, 12, 232–236.
[463] T. Sterzyński, J. Tomaszewska, J. Andrzejewski and K. Skórczewska, Evaluation of glass transition temperature of PVC/POSS nanocomposites, *Composites Science and Technology*, 2015, 117, 398–403.
[464] H.W. Brandhorst Jr., Chapter 8: Polyhedral Oligomeric Silsesquioxanes in Space Applications, C. Hartmann-Thompson, Ed., In: *Applications of Polyhedral Oligomeric Silsesquioxanes*, Advances in Silicon Science 3, DOI 10.1007/978-90-481-3787-9_8, © Springer Science +Business Media B.V., 2011.
[465] V. Bershtein, A. Fainleib, L. Egorova, O. Grigoryeva, D. Kirilenko, S. Konnikov, V. Ryzhov, O. Starostenko, P. Yakushev, M. Yagovkina and J.-M. Saiter, The impact of ultra-low amounts of introduced reactive POSS nanoparticles on structure, dynamics and properties of densely cross-linked cyanate ester resins, *European Polymer Journal*, 2015, 67, 128–142.
[466] Z. Li, D. Wu, Y. Liang, R. Fu and K. Matyjaszewski, Synthesis of well-defined microporous carbons by molecular-scale templating with polyhedral oligomeric silsesquioxane moieties, *Journal of the American Chemical Society*, 2014, 136, 4805–4808.
[467] X. Wang, S. Xuan, L. Song, H. Yang, H. Lu and Y. Hu, synergistic effect of poss on mechanical properties, flammability, and thermal degradation of intumescent flame retardant polylactide composites, *journal of macromolecular science part b physics*, 2012, 51, 255–268.
[468] A. Zubrowska, E. Piórkowska, A. Kowalewska and M. Cichorek, Novel blends of polylactide with ethylene glycol derivatives of POSS, *Colloid and Polymer Science*, 2015, 293, 23–33.
[469] M.D. Hossain, Y. Yoo and K.T. Lim, Synthesis of poly(ε-caprolactone)/clay nanocomposites using polyhedral oligomeric silsesquioxane surfactants as organic modifier and initiator, *Journal of Applied Polymer Science*, 2011, 119, 936–943.
[470] S. Xia, Q. Liu, Y. Shen, P. Yao and B. Deng, Poly(ε-caprolactone)/polyhedral oligomeric silsesquioxane hybrids: crystallization behavior and thermal degradation, *Journal of Applied Polymer Science*, 2016, 133, 44113.
[471] L. Valentini, M. Cardinali, J.M. Kenny, M. Prato and O. Monticelli, On a novel nanomaterial based on graphene and POSS: GRAPOSS; https://arxiv.org/ftp/arxiv/papers/1112/1112.3459.pdf
[472] L. Valentini, M. Cardinali, J.M. Kenny, M. Prato and O. Monticelli, A photoresponsive hybrid nano-material based on graphene and polyhedral oligomeric silsesquioxanes, *European Journal of Inorganic Chemistry*, 2012, 5282–5287.
[473] Y. Xue, Y. Liu, F. Lu, J. Qu, H. Chen and L. Dai, Functionalization of graphene oxide with polyhedral oligomeric silsesquioxane (POSS) for multifunctional applications, *The Journal of Physical Chemistry Letters*, 2012, 3, 1607–1612.
[474] G. Potsi, A. Rossos, A. Kouloumpis, M.K. Antoniou, K. Spyrou, M.A. Karakassides, D. Gournis and P. Rudolf, Carbon nanostructures containing polyhedral oligomeric silsesquioxanes (POSS), *Current Organic Chemistry*, 2016, 20(6), 662–673.
[475] Y. Xue and H. Chen, Synthesis and cytotoxicity of POSS modified single walled carbon nanotubes, *Journal of Nanomaterials*, 2015, Article ID 407437.
[476] Q. Guo, P.T. Knight and P.T. Mather, Tailored drug release from biodegradable stent coatings based on hybrid polyurethanes, *Journal of Controlled Release*, 2009, 137, 224–233.
[477] O. Monticelli, A. Fina, D. Cavallo, E. Gioffredi and G. Delprato, On a novel method to synthesize POSS-based hybrids: an example of the preparation of TPU based system, *eXPRESS Polymer Letters*, 2013, 7(12), 966–973.

[478] H. Honarkar, M. Barmar, M. Barikani and P. Shokrollahi, Synthesis and characterization of polyhedral oligomeric silsesquioxane-based waterborne polyurethane nanocomposites, *The Korean Journal of Chemical Engineering*, 2016, 33(1), 319–329. DOI: 10.1007/s11814-015-0102-2.
[479] S. Michałowski, E. Hebda and K. Pielichowski, Thermal stability and flammability of polyurethane foams chemically reinforced with POSS, *Journal of Thermal Analysis and Calorimetry*, 2017, 130, 155–163. DOI: 10.1007/s10973-017-6391-4.
[480] H.W. Ro, V. Popova, L. Chen, A.M. Forster, Y. Ding, K.J. Alvine, D.J. Krug, R.M. Laine and C.L. Sol, Cubic silsesquioxanes as a green, high-performance mold material for nanoimprint lithography, *Advanced Materials*, 2011, 23, 414–420.
[481] H.W. Ro, V. Popova, D.J. Krug, A.M. Forster, R.M. Laine and C.L. Soles, Cubic silsesquioxanes as tunable high-performance coating materials, *Applied Organometallic Chemistry*, 2013, 27, 652–659.
[482] S. Wu, T. Hayakawa, R. Kikuchi, S.J. Grunzinger, M. Kakimoto and H. Oikawa, Synthesis and characterization of semiaromatic polyimides containing POSS in main chain derived from double-decker-shaped silsesquioxane, *Macromolecules*, 2007, 40(16), 5698–5705.
[483] D. Gnanasekaran, K. Madhavan and B.S.R. Reddy, Developments of polyhedral oligomeric silsesquioxanes (POSS), POSS nanocomposites and their applications: a review, *Journal of Scientific and Industrial Research*, 2009, 68, 437–464.
[484] B.R.S. Reddy and D. Gnanasekaran, Chapter 10: Structure-Gas Transport property Relationship of Poly(dimethylsiloxane-urethane Nanocomposite Membranes, In: *Advances in Nano-composites – Synthesis, Characterization, and Industrial Applications*, B.R.S. Reddy, Ed., InTech Open Science, Croatia, 2011, 195–226.
[485] K.A.S. Abhilash, T. Deepthi, R.A. Sadhana and K.G. Benny, Functionalized polysilsesquioxane-based hybrid silica solid amine sorbents for the regenerative removal of CO_2 from air, *ACS Applied Materials & Interfaces*, 2015, 7, 17969–17976.
[486] P. Majumdar, E. Lee, N. Gubbins, S.J. Stafslien, J. Daniels, C.J. Thorson and B.J. Chisholm, Synthesis and antimicrobial activity of quaternary ammonium-functionalized POSS (Q-POSS) and polysiloxane coatings containing Q-POSS, *Polymer*, 2009, 50, 1124.
[487] J. Deng, J.T. Polidan, J.R. Hottle, C.E. Farmer-Creely, B.D. Viers and A.R. Esker, Polyhedral oligomeric silsesquioxanes: a new class of amphiphiles at the air/water interface, *Journal of the American Chemical Society*, 2002, 124(15194-15195).
[488] R.A. Sahatjian, Implantable medical devices. *Patent WO* 2004/032799.
[489] P.J. Santierre, Bioactive surface modifiers for polymers and articles made therefrom. *Patent WO* 2002/098477, 2002.
[490] B.X. Fu, B.S. Hsiao, S. Pagola, P. Stephens, H. White, M. Rafailovich, J. Sokolov, P.T. Mather, H.G. Jeon, S. Phillips, J. Lichtenhan and J. Schwab, Structural development during deformation of polyurethane containing polyhedral oligomeric silsesquioxanes (POSS) molecules, *Polymer*, 2001, 42, 599–611.
[491] A.M. Seifalian, H.J. Salacinski, A. Tiwari, A. Edwards, S. Bowald and G. Hamilton, In vivo biostability of a poly(carbonate-urea)-urethane graft, *Biomaterials*, 2003, 24, 2549–2557.
[492] H.J. Salacinski, S. Hancock and A.M. Seifalian, Polymer for use in conduits and medical devices (silsequioxanes/pendant arms), Patent 0401202.7 (2004). (b) A. Seifalian, H. Salacinski, B. Ramesh, A. Darbyshire, S. Hancock, Polymer for use in conduits, medical devices and biomedical surface modification, *US Pat.* 7820769B2 (2010).
[493] H.J. Salacinski, G. Hamilton and A.M. Seifalian, Surface functionalization and grafting of heparin and/or RGD by an aqueous-based process to a poly(carbonate-urea)urethane cardiovascular graft for cellular engineering applications, *Journal of Biomedical Materials Research*, 2003, 66A, 688–697.

[494] R.Y. Kannan, H.J. Salacinski, P.E. Butler and A.M. Seifalian, Polyhedral oligomeric silsesquioxane nano-composites: the next generation material for biomedical applications, *Accounts of Chemical Research*, 2005, 38, 879–884.
[495] D. Cui, F. Tian, C.S. Ozkan, M. Wang and H. Gao, Effect of single wall carbon nanotubes on human HEK293 Cells, *Toxicology Letters*, 2005, 155, 73–85.
[496] N.R. Tai, H.J. Salacinski, A. Edwards, G. Hamilton and A.M. Seifalian, Compliance properties of conduits used in vascular reconstruction, *British Journal of Surgery*, 2000, 87, 1516–1524.
[497] R.Y. Kannan, H.J. Salacinski, P.E. Butler, G. Hamilton and A.M. Seifalian, Current status of prosthetic bypass grafts: a review, *Journal of Biomedical Materials Research. Part B, Applied Biomaterials*, 2005, 74, 570–581.
[498] G. Punshon, D.S. Vara, K.M. Sales, A.G. Kidane, H.J. Salacinski and A.M. Seifalian, Interactions between endothelial cells and a poly(carbonate-silsesquioxane-bridge-urea) urethane, *Biomaterials*, 2005, 26, 6271–6279.
[499] R.Y. Kannan, K.M. Sales, H.J. Salacinski, P.E. Butler and A.M. Seifalian, Endothelialisation of poly (carbonate-siloxane-urea) urethane, *Medical Journal of Malaysia*, 2004, 59(Suppl. B), 107–108.
[500] H. Ghanbari, S.M. Marashi, Y. Rafiei, K. Chaloupka and A.M. Seifalian, Chapter 9: Biomedical Application of Polyhedral Oligomeric Silsesquioxane Nanoparticles, C. Hartmann-Thompson, Ed., In: Applications of Polyhedral Oligomeric Silsesquioxanes, Advances in Silicon Science, Vol. 3. DOI 10.1007/978-90-481-3787-9_9, © Springer Science+Business Media B.V., 2011.
[501] J. Tian, K.K. Wong, C.M. Ho, C.N. Lok, W.Y. Yu and C.M. Che, Topical delivery of silver nanoparticles promotes wound healing, *ChemMedChem*, 2007, 2(1), 129–136.
[502] P.L. Nadworny, J. Wang, E.E. Tredget and R.E. Burrell, Anti-inflammatory activity of nanocrystalline silver in a porcine contact dermatitis model, *Nanomedicine*, 2008, 4(3), 241–251.
[503] R.G. Sibbald, J. Contreras-Ruiz, P. Coutts, M. Fierheller, A. Rothman and K. Woo, *Advances in Skin & Wound Care*, 2007, 20(10), 549–558.
[504] X. Chen, D. Wang, X. Li, Y. Liu and L. Hu, *Materials Science Forum*, 2009, 610-613(2), 1039–1043.
[505] L. Hu, X. Chen, D. Wang and Y. Liu, *Polymer Materials Science & Engineering*, 2009, 100, 271–272.
[506] D.K. Wang, S. Varanasi, E. Strounina, D.J.T. Hill, A.L. Symons, A.K. Whittaker and F. Rasoul, Synthesis and characterization of a POSS-PEG macromonomer and POSS-PEG-PLA hydrogels for periodontal applications, *Biomacromolecules*, 2014, 15, 666–679.
[507] H. Yahyaei, M. Mohseni, H. Ghanbari and M. Messori, Synthesis and characterization of polyhedral oligomeric titanized silsesquioxane: A new biocompatible cage like molecule for biomedical application, *Materials Science and Engineering: C*, 2016, 61, 293–300.
[508] Q. Huang, G.P.-M. Cheng, K. Chiu and G.-Q. Wang, Surface modification of intraocular lenses, *Chinese Medical Journal*, 2016, 129, 206–214.
[509] A. Janaszewska, K. Gradzinska, M. Marcinkowska, B. Klajnert-Maculewicz and W.A. Stańczyk, In vitro studies of polyhedral oligo silsesquioxanes: evidence for their low cytotoxicity, *Materials*, 2015, 8(6062-6070). DOI: 10.3390/ma8095291.
[510] K. Gradzińska, K. Łabęcka, A. Kowalewska and W.A. Stańczyk, Silsesquioxane nanocarriers in diagnostics and biomedicine, *Polimery*, 2016, 61(4), 231–138.
[511] K. Piorecka, E. Radzikowska, J. Kurjata, K. Rozga-Wijas, W.A. Stańczyk and E. Wielgus, Synthesis of the first POSS cage–anthracycline conjugates via amide bonds, *New Journal of Chemistry*, 2016, 40, 5997–6000.
[512] E. Sobierajska, M. Konopka, A. Janaszewska, K. Piorecka, A. Blauz, B. Klajnert-Maculewicz, M. Stańczyk and W.A. Stańczyk, Unusual enhancement of doxorubicin activity on co-delivery with polyhedral oligomeric silsesquioxane (POSS), *Materials*, 2017, 10(559). DOI: 10.3390/ma10050559.
[513] Y. Chen, G. Zeng and W. Liu, Robust and stimuli-responsive POSS hybrid PDMAEMA hydrogels for controlled drug release, *Journal of Biomedical Materials Research. Part A*, 2016, 104A, 2345–2355.

[514] M. Reza-E-Rabby, S. Jeelani and V.K. Rangari, Structural analysis of polyhedral oligomeric silsesquioxane coated SiC nanoparticles and their applications in thermoset polymers, *Journal of Nanomaterials*, 2015, Article ID 894856, 13 pages.
[515] K. Szwarc-Rzepka, F. Ciesielczyk and T. Jesionowski, Preparation and physicochemical properties of functionalized silica/octamethacryl-silsesquioxane hybrid systems, *Journal of Nanomaterials*, 2013, Article ID 674237, 15 pages.
[516] M. Nowacka, D. Ambrożewicz and T. Jesionowski, TiO_2-SiO_2/Ph-POSS functional hybrids: preparation and characterisation, *Journal of Nanomaterials*, 2013, Article ID 680821, 10 pages.
[517] N. Xu, E.J. Stark, P.R. Dvornic, D.J. Meier, J. Hu and C. Hartmann-Thompson, Hyperbranched polycarbo-siloxanes and polysiloxanes with octafunctional polyhedral oligomeric silsesquioxane (POSS) branch points, *Macromolecules*, 2012, 45(11), 4730–4739.
[518] N. Xu, E.J. Stark, P.I. Carver, P. Sharps, J. Hu and C. Hartmann-Thompson, Hyperbranched polyhedral oligomeric silsesquioxane (HB-POSS) nanomaterials for high transmission and radiation-resistant space and solar applications, *Journal of Applied Polymer Science*, 2013, 130, 3849–3861.
[519] L.J. Mathias and T.W. Carothers, Hyperbranched poly(siloxysilanes), *Journal of the American Chemical Society*, 1991, 113(10), 4043–4044.
[520] G. Riess, G. Hurtvez and P. Bahadur, Encyklopedia of Polymer Science and Engineering, 2nd Ed., Vol. 2, J. Wiley, New York, 1985, 324.
[521] J.H. Saunders and K.H. Frisch, Polyurethanes, Chemistry and Technology Part II, Intersci. Publ., J. Wiley, New York, 1964, 67.
[522] W. Becker and D. Braun, Kunstoffe Handbuch Polyurethane Neuherausgabe, Hanser Verlag Műnchen, 1973, 101.
[523] D.L. Bailey and F.M.O. Connor, *Brit. Pat.* 880022 (1958).
[524] H. Koepnick and D. Delfs, *Pat. DE* 1108917 (1961).
[525] D.L. Bailey and F.M.O. Connor, *Pat. DE* 1012602 (1957).
[526] J.C.A. Delaval and P.A.E. Guinet, *Brit. Pat.* 1060057 (1965).
[527] E.P. Pluedemann, *Pat. DE* 1445347 (1971).
[528] J.E. Plumb and J.H. Atherton, Block Copolymers, D.C. Allport and W.H. James, Eds.)., James Appl. Sci. Publ., London, 1973, 305.
[529] A. Noshay and J.E. McGrath, Block Copolymers, Overview and Critical Survey, Academic Press, New York, 1977, 400.
[530] C.H. Blevins, C.J. Murphy, P.L. Mlatoch and C.H. Greene, *US Pat.* 5045571 (1991).
[531] H. Koepnick and D. Delfs, *Pat. DE* 1257433 (1967).
[532] H. Haman and G. Tschernko, The theory of cell opening in the formation of elastic polyurethane foams, *Plaste Und Kautschuk*, 1979, 26(11), 619–624.
[533] J. Ritter, J.G. Dubjaga and A.B. Komarowan, *Plaste Und Kautschuk*, 1979, 26(11), 624.
[534] P.R. Dvornic and R.W. Lenz, High Temperature Siloxane Elastomers, Hűttig & Wepf Verlag, 1990, 136.
[535] J.W. Dean, *Journal of Polymer Science. Part B, Polymer Physics*, 1970, 8(10), 677. ???.
[536] J.C. Saam, A.H. Ward and F.W. Fearon, *Advances in Chemistry Series*, 1973, 129, 239.
[537] G. Clouet, *Pat. EP* 421894 (1990).
[538] Y. Kawakami, H. Karasawa, T. Aoki, Y. Yamamura and Y. Yamashita, *Polymer Journal*, 1985, 17, 1159.
[539] Y. Kawakami, H. Karasawa, T. Aoki, Y. Yamamura, H. Hisada and Y. Yamashita, *Polymer Communications*, 1985, 26, 133.
[540] J.C. Saam, A.H. Ward and F.W. Fearon, *Polymer Preprint. ACS Division of Polymer Chemisty*, 1972, 13(1), 524. (1973).
[541] R.T. Leibfried, *Pat. EP* 423412 (1990).

[542] (a) H.A. Vaughn, *Brit. Pat.* 989379 (1965); (b) H.A. Vaughn, *Brit. Pat.* 140465 (1969).
[543] H.A. Vaughn, *Am. Chem. Soc. Div. Org. Coat. Plast. Chem.*, 1969, Abstracts, 29, 133.
[544] A. Noshay, M. Matzner and T.C. Wiliams, *Industrial & Engineering Chemistry Product Research and Development*, 1973, 12(4), 268.
[545] P. Horlacher et al., *Pat. DE-OS* 3924992 (1989).
[546] V. Serini et al., *Pat. DE-OS* 3929401 (1989).
[547] G. Freitag et al., *Pat. EP* 374635 (1989).
[548] P. Horlacher et al., *Pat. EP* 411370 (1990).
[549] H.A. Vaughn, *U.S. Pat.* 3419634 (1968).
[550] C.M. Hawkins and R.R. Galucci, *Pat. EP* 402674 (1989).
[551] J.S. Riffle, J.E. McGrath, R. Freelin and A.K. Banthia, *Journal of Macromolecular Science. Chemistry*, 1081, 15(5), 967.
[552] R.P. Kambour, Chapter: Structure and Properties of Alternating Block Polymers of Dimethylsiloxane and Bisphenol-A Carbonate, In: Block Polymers, S.L. Agarwall, Ed., Springer, Boston, MA, 263–276. DOI: https://doi.org/10.1007/978-1-4684-1842-2_19.
[553] T.L. Maglia and D.G. LeGrand, *Polymer Engineering and Science*, 1970, 10(6), 349.
[554] I. Goodman (Ed.), Developments in Block copolymers, p. 164, *Appl. Sci Publ.* (1982).
[555] M. Matzner, A. Noshay and J.E. McGrath, *Polymer Preprint. ACS Division of Polymer Chemisty*, 1973, 14(1), 68.
[556] H.A. Vaughn, *U.S. Pat.* 3419635 (1968).
[557] R.V. Viventi, *U.S. Pat.* 3600288 (1971).
[558] I. Yilgör and J.E. McGrath, *Polym. Prepr., Advances in Polymer Science* = Fortschritte Der Hochpolymeren-Forschung, 1988, 86, 1.
[559] K. Matssuwaka and H. Inoue, *J. Polym. Sci., Polymer Chemistry Edition*, 1990, 28, 13.
[560] J.S. Riffle, J.E. McGrath, I. Yilgör, C. Iran, C.I. Wilker and A.K. Banthia, *Polymer Preparation. Advanced Polymer Science*, 1983, 24(1), 81.
[561] R. Mikami and T. Okawa, *Pat. EP* 400613 (1990).
[562] T.V. Astapova, E.V. Matukhina, P.V. Petrovskii, I.V. Blagodatskikh, N.N. Makarova and K. Yu., Godovsky, Synthesis of cyclolinear polyorganocarbosiloxanes capable of self-organization, *Polymer Science*, 1999, A41, 377–383.
[563] N.N. Makarova, T.V. Astapova, A.I. Buzin, A.P. Polishchuk, N.V. Chizhova and I.M. Petrova, Self-organizing cyclolinear organosilicon polymers in bulk and on the surface of water, *International Journal of Molecular Sciences*, 2013, 14, 18215–18238.
[564] J.-W. Wang and B.-Y. Zhang, Synthesis and properties of novel chiral side-chain liquid crystalline polysiloxanes containing different chiral pendant groups, *Journal of Materials Science*, 2014, 49, 604–613.
[565] T. Makowski, T. Ganicz, W. Zajączkowski, W. Pisula, W.A. Stańczyk and A. Tracz, Synthesis and thermotropic behavior of side chain polysiloxane bearing triphenylene moiety, *EXPRESS Polymer Letters*, 2015, 9, 636–646.
[566] E. Tegou, V. Bellas, E. Gogolides, P. Argitis, D. Eon, G. Cartry and C. Cardinaud, Polyhedral oligomeric silsesquioxane (POSS) based resists: material design challenges and lithographic evaluation at 157 nm, *Chemistry of Materials: A Publication of the American Chemical Society*, 2004, 16, 2567–2577.
[567] T. Gunji, T. Shioda, K. Tsuchihira, H. Sekia, T. Kajiwara and Y. Abe, Preparation and properties of polyhedral oligomeric silsesquioxane–polysiloxane copolymers, *Applied Organometallic Chemistry*, 2010, 24, 545–550.
[568] X. Su, S. Guang, H. Xu, X. Liu, S. Li, X. Wang, Y. Deng and P. Wang, Controllable preparation and optical limiting properties of POSS-based functional hybrid nanocomposites with different molecular architectures, *Macromolecules*, 2009, 42(22), 8969–8976.

[569] W.-J. Lin, W.-C. Chen, W.-C. Wu, Y.-H. Niu and A.K.-Y. Jen, Synthesis and optoelectronic properties of starlike polyfluorenes with a silsesquioxane core, *Macromolecules*, 2004, 37, 2335–2341.
[570] N. Naga, Y. Kihara, T. Miyanaga and H. Furukawa, Synthesis of organic-inorganic hybrid gels from siloxane or silsesquioxane and α,ω-nonconjugated dienes by means of a photo hydrosilylation reaction, *Macromolecules*, 2009, 42(10), 3454–3462.
[571] N. Amir, A. Levina and M.S. Silverstein, Nanocomposites through copolymerization of a polyhedral oligomeric silsesquioxane and methyl methacrylate, *Journal of Polymer Science. Part A, Polymer Chemistry*, 2007, 45, 4264–4275.
[572] L. Hong, Z. Zhang, Y. Zhang and W. Zhang, Synthesis and self-assembly of stimuli-responsive amphiphilic block copolymers based on polyhedral oligomeric silsesquioxane, *Journal of Polymer Science. Part A, Polymer Chemistry*, 2014, 52, 2669–2683.
[573] H.K. Raut, S.S. Dinachali, A.Y. He, V.A. Ganesh, M.S.M. Saifullah, J. Lawc and S. Ramakrishna, Robust and durable polyhedral oligomeric silsesquioxane-based anti-reflective nanostructures with broadband quasi-omnidirectional properties, *Energy & Environmental Science*, 2013, 6, 1929–1937.
[574] T. Pisuchpen, V. Intasanta and V.P. Hoven, Highly porous organic–inorganic hybrid fiber from copolymers of styrene and polyhedral oligomeric silsesquioxane-derived methacrylate: Syntheses, fiber formation and potential modification, *European Polymer Journal*, 2014, 60, 38–48.
[575] Q.-C. Zou, H. Yu and H. Chen, Investigation of the surface properties of POSS-modified polyacrylate by inverse gas chromatography, *e-Polymers*, 2012, no. 057.
[576] H.-S. Ryu, D.-G. Kim and J.-C. Lee, Polysiloxanes containing polyhedral oligomeric silsesquioxane groups in the side chains; synthesis and properties, *Polymer*, 2010, 51, 2296–2304.
[577] A. Kowalewska, W. Fortuniak, J. Chojnowski, A. Pawlak, K. Gadzinowska and M. Zaród, Polymer nano-materials through self-assembly of polymeric POSS systems, *Silicon*, 2012, 4, 95–107.
[578] K.Y. Mya, K.P. Pramoda and C.B. He, Crystallization behavior of star-shaped poly(ethylene oxide) with cubic silsesquioxane (CSSQ) core, *Polymer*, 2006, 47, 5035–5043.
[579] E. Markovic, M. Ginic-Markovic, S. Clarke, J. Matisons, M. Hussain and G.P. Simon, Poly (ethylene glycol)-octafunctionalized polyhedral oligomeric silsesquioxane: synthesis and thermal analysis, *Macromolecules*, 2007, 40(8), 2694–2701.
[580] H.-S. Ryu, D.-G. Kim and J.-C. Lee, Synthesis and properties of polysiloxanes containing polyhedral oligomeric silsesquioxane (POSS) and oligo (ethylene oxide) groups in the side chains, *Macromolecular Research*, 2010, 18, 1021–1029.
[581] D. Li, Y. Niu, Y. Yang, X. Wang, F. Yang, H. Shena and D. Wu, Synthesis and self-assembly behavior of POSS-embedded hyperbranched polymers, *Chemical Communications*, 2015, 51, 8296–8299.
[582] D. Wang, P.M. Fredericks, A. Haddad, D.J.T. Hill, F. Rasoul and A.K. Whittaker, Hydrolytic degradation of POSS-PEG-lactide hybrid hydrogels, *Polymer Degradation and Stability*, 2011, 96, 123–130.
[583] J. Engstrand, A. López, H. Engqvist and C. Persson, Polyhedral oligomeric silsesquioxane (POSS)-poly(ethylene glycol) (PEG) hybrids as injectable biomaterials, *Biomedical Materials*, 2012, 7(3), 035013.
[584] D.-G. Kim, J. Shim, J.H. Lee, S.-J. Kwon, J.-H. Baik and J.-C. Lee, Preparation of solid-state composite electrolytes based on organic/inorganic hybrid star-shaped polymer and PEG-functionalized POSS for all-solid-state lithium battery applications, *Polymer*, 2013, 54, 5812–5820.
[585] B. Sarkar, E. Ayandele, V. Venugopal and P. Alexandridis, Polyhedral oligosilsesquioxane (POSS) nanoparticle localization in ordered structures formed by solvated block copolymers, *Macromolecular Chemistry and Physics*, 2013, 214, 2716–2724.

[586] D. Neumann, M. Fisher, L. Tran and J.G. Matisons, Synthesis and characterization of an isocyanate functionalized polyhedral oligosilsesquioxane and the subsequent formation of an organic-inorganic hybrid polyurethane, *Journal of the American Chemical Society*, 2002, 124(47), 13998–13999.
[587] W. Lee, S. Ni, J. Deng, B.-S. Kim, S.K. Satija, P.T. Mather and A.R. Esker, Telechelic poly(ethylene glycol)-POSS amphiphiles at the air/water interface, *Macromolecules*, 2007, 40(3), 682–688.
[588] W. Zhang, X. Zhung, X. Li, Y. lin, J. Bai and Y. Chen, Preparation and characterization of organic-inorganic hybrid polymers containing polyhedral oligomeric silsesquioxane via RAFT polymerization, *Reactive & Functional Polymers*, 2009, 69(124-129).
[589] W. Zhang and A.H.E. Müller, Synthesis of tadpole-shaped POSS-containing hybrid polymers via "click chemistry", *Polymer*, 2010, 51, 2133–2139.
[590] J. Wu, Q. Ge and P.T. Mather, PEG-POSS Multiblock polyurethanes: synthesis, characterization, and hydrogel formation, *Macromolecules*, 2010, 43(18), 7637–7649.
[591] B. Janowski and K. Pielichowski, Polimery nanohybrydowe zawierające poliedryczne oligo-silseskwioksany (*Nano-hybrid polymers containing polyhedral oligosilsesquioxanes*), *Polimery*, 2008, 53(2), 85–98. In Polish.
[592] M. Jancia and K. Pielichowski, Modern polyurethane elastomers modified with polyhedral silsesquioxanes, *Chemik*, 2011, 65(10), 1035–1040.
[593] G.H. Lopes, J. Junges, R. Fiorio, M. Zeni and A.J. Zattera, Thermoplastic polyurethane synthesis using POSS as a chain modifier, *Materials Research*, 2012, 15, 698–704.
[594] B.X. Fu, B.S. Hsiao, H. White, M. Rafailovich, P.T. Mather, H.G. Jeon, S. Phillips, J. Lichtenhan and J. Schwab, Nanoscale reinforcement of polyhedral oligomeric silsesquioxane (POSS) in polyurethane elastomer, *Polymer International*, 2000, 49, 437–440. – spr.
[595] Q. Guo, P.T. Knight and P.T. Mather, Tailored drug release from biodegradable stent coatings based on hybrid polyurethanes, *Journal of Controlled Release: Official Journal of the Controlled Release Society*, 2009, 137, 224–233.
[596] R. Pan, R. Shanks, I. Kong and L. Wang, Trisilanolisobutyl POSS/polyurethane hybrid composites: preparation, WAXS and thermal properties, *Polymer Bulletin*, 2014, 71, 2453–2464.
[597] T. Hao, X. Liu, G.-H. Hu, T. Jiang and Q. Zhang, Preparation and characterization of polyurethane/POSS hybrid aqueous dispersions from mono-amino substituted POSS, *Polymer Bulletin*, 2017, 74, 517–529.
[598] Z. Qin, B. Qu, L. Yuan, X. Yu, J. Li, J. Wang, H. Lv and X. Yang, Injectable shear-thinning hydrogels with enhanced strength and temperature stability based on polyhedral oligomeric silsesquioxane end-group aggregation, *Polymer Chemistry*, 2017, 8, 1607–1610.
[599] M. Xie, J. Ge, B. Lei, Q. Zhang, X. Chen and P.X. Ma, Star-Shaped, Biodegradable, and Elastomeric PLLA-PEG-POSS Hybrid Membrane With Biomineralization Activity for Guiding BoneTissue Regeneration, *Macromolecular Bioscience*, 2015, 15, 1656–1662.
[600] Y. Pu, L. Zhang, H. Zheng, B. He and Z. Gu, Drug release of pH-sensitive poly(L-aspartate)-*b*-poly(ethylene glycol) micelles with POSS cores, *Polymer Chemistry*, 2014, 5, 463–470.
[601] K.M. Kim, T. Inakura and Y. Chujo, Organic-inorganic polymer hybrids using octasilsesquioxanes with hydroxyl groups, *Polymer Bulletin*, 2001, 46(5), 351–356.
[602] R. Sodkhomkhum and V. Ervithayasuporn, Synthesis of poly(siloxane/double-decker silsesqui-oxane) via dehydrocarbonative condensation reaction and its functionalization, *Polymer*, 2016, 86, 113–119.
[603] Y.-T. Liao, Y.-C. Lin and S.-W. Kuo, Highly thermally stable, transparent, and flexible polybenzoxazine nanocomposites by combination of double-decker-shaped polyhedral silsesquioxanes and poly-dimethylsiloxane, *Macromolecules*, 2017, 50, 5739–5747.
[604] Q. Jiang, J. Hao, Y. Wei and J. Mu, Synthesis and characterization of block and random POSS/fluorinated PAES tricopolymers, *Polymers for Advanced Technologies*, 2017, 28, 658–664.

[605] R.Y. Kannan, H.J. Salacinski, J. De Groot, I. Clatworthy, L. Bozec, M. Horton, P.E. Butler and A.M. Seifalian, The antithrombogenic potential of a polyhedral oligomeric silsesquioxane (POSS) nanocomposite, *Biomacromolecules*, 2006, 7(1), 215–223.

[606] Y. Sun and C. He, Biodegradable "core–shell" rubber nanoparticles and their toughening of poly(lactides), *Macromolecules*, 2013, 46, 9625–9633.

[607] J. Zou, X. Chen, X.B. Jiang, J. Zhang, Y.B. Guo and F.R. Huang, Poly(L-lactide) nanocomposites containing octaglycidylether polyhedral oligomeric silsesquioxane: Preparation, structure and properties, *EXPRESS Polymer Letters*, 2011, 5, 662–673.

[608] X. Zhang, C. Wang, S. Fang, J. Sun, C. Li and Y. Hu, Synthesis and characterization of well-defined star PLLA with a POSS core and their microspheres for controlled release, *Colloid and Polymer Science*, 2013, 291, 789–803.

[609] C. Gomez-Sanchez, T. Kowalczyk, G. Ruiz De Eguino, A. Lopez-Arraiza, A. Infante, C.I. Rodriguez, T.A. Kowalewski, M. Sarrionandia and J. Aurrekoetxe, Electrospinning of poly(lactic acid)/ polyhedral oligomeric silsesquioxane nanocomposites and their potential in chondrogenic tissue regeneration, *Journal of Biomaterials Science. Polymer Edition*, 2014, 25, 802–825.

[610] L. Gardella, S. Colonna, A. Fina and O. Monticelli, On novel bio-hybrid system based on PLA and POSS, *Colloid and Polymer Science*, 2014, 292, 3271–3278.

[611] Y. Du, M. Yu, X. Chen, P.X. Ma and B. Lei, Development of Biodegradable Poly(citrate)-Polyhedral Oligomeric Silsesquioxanes Hybrid Elastomers with High Mechanical Properties and Osteogenic Differentiation Activity, *ACS Applied Materials & Interfaces*, 2016, 8, 3079–3091.

[612] J. Pyun, K. Matyjaszewski, J. Wu, G.M. Kim, S.B. Chun and P.T. Mather, ABA triblock copolymers containing polyhedral oligomeric silsesquioxane pendant groups: synthesis and unique properties, *Polymer*, 2003, 44, 2739–2750. ISSN 0032-3861.

[613] D. Liu, Z. Nie, W. Xu, L. Sun and G. Chen, *Chinese Pat. Appl.* CN 1997-105779 (15.04.1997).

[614] M. Yang, H. Liu, H. Zhang, D. Chen, J. Wang and C. Wang, *Tuliao Gongye*, 2000, 30(1), 12–15.

[615] S. Yasuaki, S. Susumu, Y. Tomeo, T. Tadashi and M. Koji, Jap. Pat. JP 2000063612 A2 (29 Feb. 2000).

[616] J.-F. Stumbé, F. Calderara and G. Riess, *Polymer Bulletin*, 2001, 47(3-4), 277–282.

[617] H.K. Chu, *ACS Symp. Series, vol. 729, Silicones and Silicone-Modified Materials*, pp. 170–179 (2000).

[618] H. Hussain, B.H. Tan, K.Y. Mya, Y. Liu, C.B. He and T.P. Davis, Synthesis, micelle formation, and bulk properties of poly(ethylene glycol)-*b*-poly(pentafluorostyrene)-*g*-polyhedral oligomeric silsesqui-oxane amphiphilic hybrid copolymers, *Journal of Polymer Science. Part A, Polymer Chemistry*, 2009, 55, 152–163.

[619] D.K. Wang, S. Varanasi, E. Strounina, D.J.T. Hill, A.L. Symons, A.K. Whittaker and F. Rasoul, Synthesis and characterization of a POSS-PEG macromonomer and POSS-PEG-PLA hydrogels for periodontal applications, *Biomacromolecules*, 2014, 15, 666–679.

[620] A. Tan, Y. Farhatnia, D. Goh, G. Natasha, A. De Mel, J. Lim, S.-H. Teoh, A.V. Malkovskiy, R. Chawla, J. Rajadas, B.G. Cousins, M.R. Hamblin, M.S. Alavijeh and A.M. Seifalian, Surface modification of a polyhedral oligomeric silsesquioxane poly(carbonate-urea) urethane (POSS-PCU) nanocomposite polymer as a stent coating for enhanced capture of endothelial progenitor cells, *Biointerphases*, 2013, 8, 23.

[621] G.R. Newkome, C.N. Moorefield and F. Vögtle, Dendrimers and Dendrons: Concepts, Syntheses, Applications, Wiley-VCH Verlag GmbH & Co, 2002, 104–138. 172-189, 298-299, 377-393.

[622] D.A. Tomalia and J.M.J. Frechet, Discovery of dendrimers and dendritic polymers: a brief historical perspective, Journal of Polymer Science. Part A, Polymer Chemistry, 2002, 40, 2719–2728.

[623] D. Neugebauer, Y. Zhang, T. Pakula and K. Matyjaszewski, PDMS-PEO densely grafted copolymers, *Macromolecules*, 2005, 38(21), 8687–8693.

[624] Z. Zhang, J. Wang, Q. Tu, N. Nie, J. Sha, W. Liu, R. Liu, Y. Zhang and J. Wang, Surface modification of PDMS by surface-initiated atom transfer radical polymerization of water-soluble dendronized PEG methacrylate, *Colloids and Surfaces. B, Biointerfaces*, 2011, 88, 85–92.
[625] M. Qu, J. He and J. Zhang, Chapter 9: Superhydrophobicity, Learn from the Lotus Leaf, InTech, Croatia (2010).
[626] S.S. Latthe, A.B. Gurav, C.S. Maruti and R.S. Vhatkar, Recent progress in preparation of superhydrophobic surfaces: a review, *Journal of Surface Engineered Materials and Advanced Technology*, 2012, 2, 76–94.
[627] J.T. Han, D.H. Lee, C.Y. Ryu and K. Cho, Fabrication of superhydrophobic surface from a supramolecular organosilane with quadruple hydrogen bonding, *Journal of the American Chemical Society*, 2004, 126, 4796–4797.
[628] G.R.J. Artus, S. Jung, J. Zimmermann, H.-P. Gautschi, K. Marquardt and S. Seeger, Silicone nanofilaments and their application as superhydrophobic coatings, *Advanced Materials*, 2006, 18, 2758–2762.
[629] W. Barthlott and C. Neinhuis, *Planta*, 1997, 202, 1.
[630] J. Bico, C. Marzolin and D. Quéré, *Europhysics Letters*, 1999, 47, 220.
[631] T. Qian, Y. Li, Y. Wu, B. Zheng and H. Ma, Superhydrophobic poly(dimethylsiloxane) via surface-initiated polymerization with ultralow initiator density, *Macromolecules*, 2008, 41, 6641–6645.
[632] C. Liu, S. Wang, J. Shib and C. Wang, Fabrication of superhydrophobic wood surfaces via a solution-immersion process, *Applied Surface Science*, 2011, 258, 761–765.
[633] P.-H. Chen, -C.-C. Hsu, P.-S. Lee and C.-S. Lin, Fabrication of semi-transparent superhydrophobic surface based on silica hierarchical structures, *Journal of Mechanical Science and Technology*, 2011, 25(1), 43–47.
[634] D. Su, C. Huang, Y. Hu, Q. Jiang, L. Zhang and Y. Zhu, Preparation of superhydrophobic surface with a novel sol–gel system, *Applied Surface Science*, 2011, 258, 928–934.
[635] M.R. Karim and M.S.L. Islam, Thermal behavior with mechanical property of fluorinated silane functionalized superhydrophobic pullulan/poly(vinyl alcohol) blends by electrospinning method, *Journal of Nanomaterials*, 2011, Article ID 979458.
[636] J.T. Han, S.Y. Kim, J.S. Woo and G.W. Lee, Superhydrophobic/superhydrophilic surfaces from a carbon nanotube based composite coating, *Advanced Materials*, 2008, 20, 3724.
[637] J. Zhang, B. Li, L. Wu and A. Wang, Facile preparation of durable and robust superhydrophobic textiles by dip coating in nanocomposite solution of organosilanes, *Chemical Communications*, 2013, 49, 11509–11511.
[638] Q. Shang, B. Fu, H. Liu, M. Wang and G. Xiao, Facile creation of superhydrophobic surface with fluorine–silicon polymer under ambient atmosphere, *Journal of Coatings Technology and Research*, 2012, 9(5), 589–595.
[639] J. Genzer and K. Efimenko, Recent developments in superhydrophobic surfaces and their relevance to marine fouling: a review, *Biofouling*, 2006, 22(5), 339–360.
[640] J. Song and O.J. Rojas, Approaching super-hydrophobicity from cellulosic materials: a review, *Nordic Pulp and Paper Research Journal*, **28**(2), 216–238 (2013).
[641] D. Ambrożewicz, F. Ciesielczyk, M. Nowacka, J. Karasiewicz, A. Piasecki, H. Maciejewski and T. Jesionowski, Fluoroalkylsilane versus alkylsilane as hydrophobic agents for silica and silicates, *Journal of Nanomaterials*, 2013, Article ID 631938.
[642] J. Liang, Y. Hu, Y. Wu and H. Chen, Fabrication and corrosion resistance of superhydrophobic hydroxide zinc carbonate film on aluminum substrates, *Journal of Nanomaterials*, 2013, Article ID 139768.
[643] H.Y. Gu, Z.Y. Qi, W. Wu, Y. Zeng and L.X. Song, Superhydrophobic polyimide films with high thermal endurance via UV photo-oxidation, *eXPRESS Polymer Letters*, 2014, 8, 588–595.

[644] W. Dou, P. Wang, D. Zhang and J. Yu, An efficient way to prepare hydrophobic antireflective SiO_2 film by sol-gel method, *Materials Letters*, 2016, 167, 69–72.
[645] P.-H. Chen, -C.-C. Hsu, P.-S. Lee and C.-S. Lin, Fabrication of semi-transparent superhydrophobic surface based on silica hierarchical structures, *Journal of Mechanical Science and Technology*, 2011, 25(1), 43–47.
[646] S.A. Mahadik, D.B. Mahadik, M.S. Kavale, V.G. Parale, P.B. Wagh, H.C. Barshilia, S.C. Gupta, N.D. Hegde and A.V. Rao, Thermally stable and transparent superhydrophobic sol–gel coatings by spray method, *Journal of Sol-gel Science and Technology*, 2012, 63, 580–586.
[647] A. Santiago, A. González, J.J. Iruin, M.J. Fernández-Berridi and L. Irusta, Preparation of superhydrophobic silica nanoparticles by microwave assisted sol–gel proces, *Journal of Sol-Gel Science and Technology*, 2012, 61, 8–13.
[648] F. Xu, D. Li, H. Zhang and W. Peng, TEOS/HDTMS inorganic–organic hybrid compound used for stone protection, *Journal of Sol-Gel Science and Technology*, 2012, 61, 429–435.
[649] M. Qu, J. He, S. Zhe, K. Li, X. Liu and C. Yu, Fabrication of mechanical durable polysiloxane superhydrophobic materials, *Journal of Nanomaterials*, 2015, Article ID 284685.
[650] G. Lia, D. Hea, Y. Lina, Z. Chena, Y. Liua and X. Peng, Fabrication of biomimetic superhydrophobic surfaces by a simple flame treatment method, *Polymers for Advanced Technologies*, 2016, 27, 1438–1445.
[651] H. Sun, Y. Xu, Y. Zhou, W. Gao, H. Zhao and W. Wang, Preparation of superhydrophobic nanocomposite fiber membranes by electrospinning poly(vinylidene fluoride)/silane coupling agent modified SiO_2 nanoparticles, *Journal of Applied Polymer Science*, 2017, 134, 44501.
[652] M. Yu, G. Gu, W.D. Meng and F.L. Qing, Superhydrophobic cotton fabric coating based on a complex layer of silica nanoparticles and perfluorooctylated quaternary ammonium silane coupling agent, *Applied Surface Science*, 2007, 253, 3669–3673.
[653] C.-H. Xue, S.-T. Jia, J. Zhang, L.-Q. Tian, H.-Z. Chen and M. Wang, Preparation of superhydrophobic surfaces on cotton textiles, *Science and Technology of Advanced Materials*, 2008, 9, 035008.
[654] M. Przybylak, H. Maciejewski, A. Dutkiewicz, I. Dąbek, M. Nowicki, Fabrication of superhydrophobic cotton fabrics by a simple chemical modification, *Cellulose*, **23**, 2185–2197 (2016); DOI 10.1007/s10570-016-0940-z
[655] M. Przybylak, H. Maciejewski, A. Dutkiewicz, D. Wesołek and M. Władyka-Przybylak, Multifunctional, strongly hydrophobic and flame-retarded cotton fabrics modified with flame retardant agents and silicon compounds, *Polymer Degradation and Stability*, 2016, 128, 55–64.
[656] H. Liu, S.-W. Gao, J.-S. Cai, C.-L. He, -J.-J. Mao, T.-X. Zhu, Z. Chen, J.-Y. Huang, K. Meng, K.-Q. Zhang, S.S. Al-Deyab and Y.-K. Lai, Recent progress in fabrication and applications of superhydrophobic coating on cellulose-based substrates, *Materials*, 2016, 9, 124.
[657] C.H. Xue, Y.R. Li, P. Zhang, J.Z. Ma and S.T. Jia, Washable and wear-resistant superhydrophobic surfaces with self-cleaning property by chemical etching of fibers and hydrophobization, *ACS Applied Materials & Interfaces*, 2014, 6, 10153–10161.
[658] Y. Lu and W. Yuan, Superhydrophobic/superoleophilic and reinforced ethyl cellulose sponges for oil/water separation: synergistic strategies of crosslinking, Carbon Nanotube Composite, and Nanosilica Modification, *ACS Applied Materials & Interfaces*, 2017, 9, 29167–29176.
[659] A.B. López, J.C. De La Cal and J.M. Asua, From fractal polymer dispersions to mechanically resistant waterborne superhydrophobic coatings, *Polymer*, 2017, 124, 12–19.
[660] V.K. Rowe, H.C. Spencer and S.L. Bass, Toxicological studies on certain commercial silicones and hydrolyzable silane intermediates, *The Journal of Industrial Hygiene and Toxicology*, 1948, 30, 332–352.

[661] U.C. Pozzani and C.P. Carpenter, Response of rodents to repeated inhalation of vapors of tetraethyl orthosilicate, *Archives of Industrial Hygiene and Occupational Medicine*, 1951, 4, 465–468.
[662] M.G. Voronkov, Biological activity of silatranes, *Topics in Current Chemistry*, 1979, 84, 77–135.
[663] G. Fritz and G. Poppenburg, *Zeitschrift Für Anorganische Und Allgemeine Chemie*, 1964, 331, 147.
[664] H. Schmidbaur and H. Husek, Hexaalkylplumbosiloxane, *Journal of Organometallic Chemistry*, 1964, 1, 257.
[665] J. Łukasiak, B. Falkiewicz, E. Dąbrowska and M. Stołyhwo, *Bio-mat. Chem. Toksykol.*, 1996, 29, 199.
[666] P. Rościszewski, J. Łukasiak, A. Dorosz, J. Galiński and M. Szponar, Biodegradation of polyorgano-siloxanes, *Macromolecular Symposia*, 1998, 130, 337.
[667] J. Łukasiak, A. Dorosz, M. Prokopowicz, P. Rościszewski and B. Falkiewicz, Chapter; Biodegradation of Silicones (Organosiloxanes), In: *Biopolymers*, Vol. 9, Miscellaneous Biopolymers and Biodegradation of Synthetic Polymers, S. Matsumura and A. Steinbüchel, 2002, 539–568. Eds. ISBN 978-3-527-30228-4.
[668] Eighteen Report of the Joint FAO/WHO Expert Committee on Food Additives (JECFA), World Health Organization, Genewa, 1974, 24.
[669] K. Mojsiejewicz-Pieńkowska and J. Łukasiak, Some directions of polydimethylsiloxanes applications and possibilities of these compounds analyses, *Polimery*, 2006, 51, 186–191.
[670] JECFA valuations, 1994, April, 9.
[671] A. Rutkowski, Dodatki funkcjonalne do żywności (Functional food additives), Wydawnictwo Agro & Food Technology, Katowice (1993). In Polish
[672] European Pharmacopeia, 2002, vol. 4.
[673] British Pharmacopeia, 2001, vol. 1.
[674] D.P. Gorczyca, The Augmented breast, In: Radiologic and Clinical Perspectives, Thieme, New York, 1997.
[675] A.B. Anderson and C.R. Robertson, *Biophysical Journal*, 1995, 68, 2091.
[676] B. Pfleiderer and L. Garrido, *Magnetic Resonance in Medicine*, 1995, 33, 8.
[677] M.L. Raimondi, C. Sassava and J.R. Bellobono, *Journal of Biomedical Materials Research*, 1995, 29(nr 1), 59.
[678] E. Libermann, E.D. Lykissa, R. Barrios, C. Nan Ou, G. Kala and S. Kala, *Environmental Health Perspectives*, 1999, 107(2), 161.
[679] Dow Corning Report, No. 1179-11 T. 37322-37, 409, 1/31/90, 1990.
[680] S. Varaprath, K.L. Salyers, K.P. Plotzke and S. Nonavati, *Analytical Biochemistry*, 1998, 256, 14.
[681] Final Report on the Safety Assessment of Cyclomethicone, *Journal of the American College of Toxicology*, 1991, 10, 9.
[682] J. Łukasiak, Z. Jamrógiewicz, W. Czarnowski, J. Krechniak and B. Falkiewicz, *Bromatologia I Chemia Toksykologiczna*, 1999, 32, 99.
[683] J. Łukasiak, Z. Jamrógiewicz, D. Jachowska, W. Czarnowski, M. Hrabowska, M. Prokopowicz and B. *Falkiewicz, Polimery*, 2001, 46, 546.
[684] Dow Corning Report, No. 1996-I 0000-41337, DCC 833-610001, 8/27/96, 1996.
[685] J. Lambert, Silicone Safety and the Cosmetic Industry, http://www.cosmeticsandtoiletries.com/research/chemistry/Silicone-Safety-and-the-Cosmetic-Industrypremium-231904601.html?prodrefresh=y&ajs_aid=006e6db6-c262-47cb-a9d5-7b74b8be5fd0&ajs_uid=1027F2629590A5U
[686] J. Roidl, *Perfum. Kosmet.*, 1986, 67, 232.
[687] H. Van Dort, R. Haller, P. Pretzer, A. Urrutia, G. Brissette, I. Van Reeth and V. Caprasse, New Formulation Opportunities with Carbinol-Functional Silicones, https://www.google.pl/search?q=New+Formula-tion+Opportunities+with+Carbinol-Functional+Silicones¡utf-8œutf-8&client=firefox-b&gfe_rd=cr&ei=4sWHWJOyAseCYJvNkVA

[688] A. Urrutia, H.M. Van Dort, G. Brissette, P. Pretzer, R. Haller, I. Van Reeth and V. Caprasse, Silicone Carbinol Fluid: Beyond Aesthetics for Color Cosmetics, https://www.dowcorning.com/content/publishedlit/27-1253-01.pdf
[689] Carbinol Functional Silicon-Based Technologies for Coatings, http://www.pcimag.com/articles/82951-carbinol-functional-silicon-based-technologies-for-coatings
[690] H.X. Kuah and X.J. Loh, Chapter 8: "*Silicones: The Future for Beauty and Everyday Care*, In: *Polymers for Personal Care Products and Cosmetics*, X.J. Loh, Ed., Royal Society of Chemistry, 2016, 135–153. DOI:10.1039/9781782623984-00135. https://books.google.pl/books?isbn=1782622950.
[691] Chapter 28: "Silicones: A Key Ingredient in Cosmetic and Toiletary Formulations", In: Handbook of Cosmetic Science and Technology, Fourth Edition, A.O. Barel, M. Paye and H.I. Maibach, Eds., CRC Press, 2014, 321–330.
[692] Pat. WO 2008006744A3 (2008).
[693] M. Jamrógiewicz, M. Żebrowska, J. Łukasiak and M. Sznitowska, Silikonowe preparaty do leczenia powierzchniowego blizn (Silicone therapy for topical scar treatment), *Farmacja Polska*, 2010, 66(6), 437–442. In Polish.
[694] P. Hatzi, M. Tzakis and G. Eliades, Setting characteristics of vinyl-polysiloxane interocclusal recording materials, *Dental Materials*, 2012, 28, 783–791.
[695] J. Jońca, C. Tukaj, W. Werel, U. Mizerska, W. Fortuniak and J. Chojnowski, Bacterial membranes are the target for antimicrobial polysiloxane-methacrylate copolymer, *Journal of Materials Science. Materials in Medicine*, 2016, 27(55). DOI: https://doi.org/10.1007/s10856-016-5669-6.
[696] Y. Chen, Q. Zhang, Q. Han, Y. Mi, S. Sun, C. Feng, H. Xiao, P. Yu and C. Yang, Synthesis of polysiloxane with 5,5-dimethylhydantoin-based N-halamine pendants for biocidal functionalization of polyethylene by supercritical impregnation, *Journal of Applied Polymer Science*, 2017, 134(44721).
[697] P. Hron, J. Slechtova, K. Smetana, B. Dvorankova and P. Lopour, Silicone rubber-hydrogel composites as polymeric biomaterials. IX. Composites containing powdery polyacrylamide hydrogel, *Biomaterials*, 1997, 16, 1069–1073.
[698] R. Yoda, Elastomers for biomedical applications, *Journal of Biomaterials Science. Polymer Edition*, 1998, 9(6), 561–626. and referencies cited therein. DOI: 10.1163/156856298X00046.
[699] H. Chen, M.A. Brook and H. Sheardown, Silicone elastomers for reduced protein adsorption, *Biomaterials*, 2004, 25, 2273–2282.
[700] A.W. Carpenter, D.L. Slomberg, K.S. Rao and M.H. Schoenfisch, Influence of scaffold size on bactericidal activity of nitric oxide-releasing silica nanoparticles, *ACS Nano*, 2011, 5(9), 7235–7244.
[701] Y. Lu, D.L. Slomberg, B. Sun and M.H. Schoenfisch, Shape- and nitric oxide flux-dependent bactericidal activity of nitric oxide-releasing silica nanorods, *Small*, 2013, 9, 2189–2198.
[702] W.T. Liu, Y. Yang, P.H. Shen, X.J. Gao, S.Q. He, H. Liu and C.S. Zhu, Facile and simple preparation of pH-sensitive chitosan-mesoporous silica nanoparticles for future breast cancer treatment, *EXPRESS Polymer Letters*, 2015, 9(12), 1068–1075.
[703] A.P. Krasucka and J. Goworek, MCM-41 silica as a host material for controlled drug delivery systems, *Annales Universitatis M. Curie-Skłodowska, Lublin – Polonia, LXX, 2 sectio AA*, 45–66 (2015).
[704] G. Luciani, A. Constantini, B. Silvestri, F. Tescione, F. Branda and A. Pezzella, Synthesis, structure and bioactivity of pHEMA/SiO_2 hybrids derived through in situ sol-gel process, *Journal of Sol-Gel Science and Technology*, 2008, 46, 166–175. ISSN 0928-0707.
[705] A. Mashak, In vitro drug release from silicone rubber-polyacrylamide composite, *Silicone Chemistry*, 2008, 3, 295–301. ISSN 1572-8994.

[706] R.T. Ragheb and J.S. Riffle, Synthesis and characterization of poly(lactide-*b*-siloxane-*b*-lactide) copolymers as magnetite nanoparticle dispersants, *Polymer*, 2008, 49, 5397–5404. ISSN 0032-3861.
[707] R.R. Pant, J.L. Buckley, P.A. Fulmer, J.H. Wynne, D.M. McCluskey and J.P. Philips, Hybrid siloxane epoxy coatings containing ouaternary ammonium moieties, *Journal of Applied Polymer Science*, 2008, 110, 3080–3086. ISSN 1097-4628.
[708] J.P. Sheth, A. Aneja, G.L. Wilkes, E. Yilgor, G.E. Atilla, I. Yilgor and F.L. Beyer, Influence of system variables on the morphological and dynamic mechanical behavior of PDMS based segmented polyurethane and polyurea copolymers: a comparative perspective, *Polymer*, 2004, 45, 6919–6932. ISSN 0032-3861.
[709] C.J. Zhou, R.F. Guan and S.Y. Feng, The preparation of a new polysiloxane copolymer with glucosylthioureylene groups on the side chains, *European Polymer Journal*, 2004, 40, 165–170. ISSN 0014-3057.
[710] B.A. Cavic and M. Thompson, Protein adsorption to organosiloxane surfaces studied by acoustic wave sensor, *Analyst*, 1998, 123, 2191–2196.
[711] J.M. Yang, C.H. Shih, C.-N. Chang, F.H. Lin, J.M. Jiang, Y.G. Hsu, W.Y. Su and L.C. See, Preparation of epoxy–SiO_2 hybrid sol–gel material for bone cement, *Journal of Biomedical Materials Research*, 2003, 64A, 138–146.
[712] A. Tamayo, L. Tellez, M. Rodriguez-Reyes, M.A. Mazo, F. Rubio and J. Rubio, Surface properties of bioactive TEOS–PDMS–TiO_2–CaO ormosils, *Journal of Materials Science*, 2014, 49, 4656–4669.
[713] Q. Chen, N. Miyata and T. Kokubo, Bioactivity and mechanical properties of poly(dimethylsiloxane)-modified calcia–silica hybrids with added titania, *Journal of the American Ceramic Society*, 2003, 86(5), 806–810.
[714] M. Cazacu, C. Racles, A. Vlad, M. Anthone and N. Forna, *Journal of Composite Materials*, Silicone-based composite for relining of removable dental prosthesis, 2009, 43(19), 2045–2055.
[715] J. Wang, Y. Liu, J. Yu, Y. Sun and W. Xie, Study of POSS on the properties of novel inorganic dental composite resin, *Polymers*, 2020, 12, 478.
[716] M. Tsumura and T. Iwahara, Silicon-based materials prepared by IPN formation and their properties, *Journal of Applied Polymer Science*, 2000, 78, 724–731.
[717] M. Zielecka, K. Cyruchin and A. Szulc, Biologically active silicon preparation and a method for obtaining a biologically active silicon preparation, *Pat. PL* 194223 (2007).
[718] S.C.L. Fischer, K. Kruttwig, V. Bandmann, R. Hensel and E. Arzt, Adhesion and cellular compatibility of silicone-based skin adhesives, *Macromolecular Materials and Engineering*, 2017, 302, 1600526.
[719] I. Blanco, Polysiloxanes in theranostics and drug delivery: a review, *Polymers*, 2018, 10, 755.
[720] K. Czarnobaj, The role of polydimethylsiloxane in the molecular structure of silica xerogels intended for drug carriers, *Scientia Pharmaceutica*, 2015, 83, 519–533.
[721] H. Wang, F. Wang, X. Li, X. Peng, Z. Cia and Z. Wang, Preparation and performance investigation of polydimethylsiloxane microsphere/polyvinyl alcohol composite hydrogel, *Materials Letters*, 2018, 228, 399–402.
[722] C. Crowley, P. Klanrit, C.R. Butler, A. Varanou, M. Plate, R.E. Hynds, R.C. Chambers, A.M. Seifalian, M.A. Birchall and S.M. Janes, Surface modification of a POSS-nanocomposite material to enhance cellular integration of a synthetic bioscaffold, *Biomaterials*, 2016, 83, 283–293.
[723] N.A. Chekina, V.N. Pavlyuchenko, V.F. Danilichev, N.A. Ushakov, S.A. Novikov and S.S. Ivanchev, A new polymeric silicone hydrogel for medicalapplications: synthesis and properties, *Polymers for Advanced Technologies*, 2006, 17, 872–877.

Chapter 3
Modification of organic polymers with silanes, silicones, silica, and silicates

Reactive silanes, silicones, silicon-containing copolymers, silsesquioxanes, silica or nanosilica, and various silicates are very often applied for chemical and physical modification of different polymers, for instance, thermoplastic polymers or chemo- and thermoreactive polymers, for example, polyurethanes or epoxy resins [1, 2].

The carbofunctional silanes (CFS) of a general formula $\mathbf{X_nSi(R'Y)_{4-n}}$ and carbofunctional polysiloxanes (CFPS) find numerous practical applications in many areas, owing to the presence of two kinds of the functional groups in their chemical structures. Functional groups X form chemical bonds with inorganic materials, while the functional groups Y react with organic polymers. Most often methoxy- or ethoxy-substituted CFS of the structure $(RO)_3Si(CH_2)_3Y$ are used, in which organic functional groups Y exist: amine ($-NH_2$), methacryloxy ($-O-CO-C(CH_3)=CH_2$), or glicydyloxy ($-O-CH_2-(CH-CH_2O)$) groups.

3.1 Modification of polymer properties by application of CFS

The CFS are often applied in sol–gel processes for a preparation of hybrid organic–inorganic materials [3–5]. The alkoxysilane groups of the CFS undergo hydrolysis and elementary condensation equilibrium reactions (3.1)–(3.3), which are catalyzed both by acids and bases:

$$\equiv Si-OR + H_2O \longrightarrow \equiv Si-OH + ROH \tag{3.1}$$

$$\equiv Si-OR + HO-Si\equiv \longrightarrow -O-Si\equiv + ROH \tag{3.2}$$

$$\equiv Si-OH + HO-Si\equiv \longrightarrow \equiv Si-O-Si\equiv + H_2O \tag{3.3}$$

The hydrolysis reactions proceed faster than condensation reactions in acidic media, but in basic media, condensation reactions are faster. Products of the sol–gel processes of (tetraethoxy)silane ($Si(OEt)_4$, TEOS) are silica gel, branched poly(ethoxysilicates), or silsesquioxanes, as presented in Scheme 3.1 [6]. Octakis(tetraalkylaminoxy)-substituted polyhedral octaphenylsilsesquioxane (POSS) was used as an intermediate substrate for the synthesis of octakis(dimethylsiloxy)octasilsesquioxane $QM^H{}_8$ ($M^H = HMe_2SiO$) [7].

Many kinds of organic polymers were modified with the CFS. For example, new derivatives of 2,6-di-*tert*-butylphenol antioxidants containing (trialkoxy)silane groups, prepared by the sol–gel method, were applied for a stabilization of isotactic polypropylene (PP) [8]. (Aminoalkyl)silanes are quite often used for the modification of polymer properties. Chlorinated polymers, for example, poly(vinyl

https://doi.org/10.1515/9783110643671-003

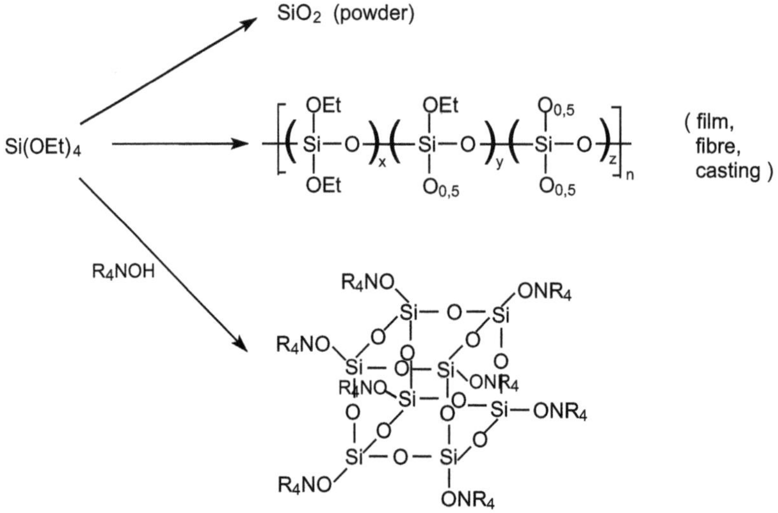

Scheme 3.1: Products formed in the sol–gel process of (tetraethoxy)silane. Reprinted (adapted) with permission from [1]. Copyright 2012, Intech–Open Sci.

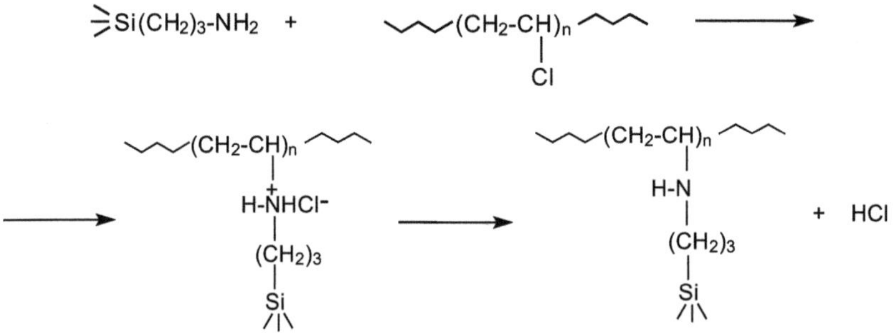

Scheme 3.2: Reactions of (aminopropyl)silane with poly(vinyl chloride). Reprinted (adapted) with permission from [1]. Copyright 2012, Intech–Open Sci.

chloride), easily formed quaternary salts with (aminoalkyl)silanes (Scheme 3.2) [9].

Silanes containing oxirane groups were applied for surface treatment of fillers that were used in epoxide composites. They are also utilized for a modification of novolac resins through reaction of hydroxyl groups of phenol with epoxy groups. Unsaturated polyesters (PES) modified with (methacryloxypropyl)silanes were utilized in a manufacture of glass-reinforced laminates. Polycarbonates (PC) and polysulfones (PSU) form amide or sulfonamide bonds, while polyurethanes form urea bonds, respectively. In this way, modified polyamides, PC and PSU, were easily reinforced with the fillers, giving composite materials exhibiting very good mechanical properties [9].

Silyl functional reagents, terminating polymer chains ("*endcappers*"), were prepared by adding the Michael reactions of 3-aminopropyl(trimethoxy)silane (APTMS) to acrylates. They were used for an incorporation of (trimethoxy)silane groups (-Si $(OCH_3)_3$) on chain ends of polyurethanes, enabling their cross-linking under action of atmospheric moisture. It established a base for elaboration of a technology of newer one-component, solventless adhesion materials (*solvent free*) [10].

CFS were also applied for the surface modification of the inactive fillers and polymers, for example, in polyolefin (PO) composites filled with calcium carbonate or mica [11–14]. PP filled with $CaCO_3$ is the fundamental plastic for production of garden furniture. Two amino-functional silanes were applied for the improvement of a tensile strength (TS) and decrease of deformability of the PP composites containing nonmodified $CaCO_3$ and modified with silanes [15–18]. The polycondensation of (triethoxy) silyl groups under process conditions took place quickly and led to reinforcement of interactions between ingredients of the composite PP/$CaCO_3$ [15].

In order to enhance the TS and softness properties of cotton, more flexible Si–O bonds between the silane coupling agents (SCA) and the cotton fiber were formed. The surface of the cotton fiber was modified by condensation reactions with 3-glycidoxypropyl(triethoxy)silane (GPTES) in an ethanol–water medium. The Fourier-transform infrared (FTIR) spectra showed two additional peaks at 860 cm^{-1} (Si–OH symmetric stretch) and 1,207 cm^{-1} (Si–O–C bending), respectively, as a result of silane absorption on the fiber surface. The surface morphology and the structural characteristics of the raw and modified fibers were studied by scanning electron microscopy (SEM) coupled with X-ray diffraction, respectively. The moisture absorption of modified cotton fiber was lower than for the raw cotton fiber, and the swelling of modified cotton fiber was decreased in the polar solvents but increased in nonpolar solvent. The absorption of *reactive brown 10* and *reactive orange 14* by the modified fiber was comparatively higher than for untreated cotton fiber [19].

3.1.1 Modification of polyolefin properties with silanes

A cross-linking is one of the modification methods of polyolefins (POs) properties, such as high temperature, chemical and stress cracking resistance, and also as shape memory. It can be done by different methods: irradiation, peroxide process, azo process, and moisture curing of silane-functionalized POs [20, 21]. The so-called SIOPLAS technology was developed in the late 1960s and modified in the late 1980s. Both irradiation and peroxide cross-linking processes have some disadvantages such as high costs for irradiation equipment, the possibility of precuring in peroxide cross-linking processes, the inability to cross-link PP through tertiary-bonded carbon atoms, and the formation of voids in cable insulations.

3.1.1.1 Silane-cross-linked polyolefins

The new cross-linking SIOPLAS technology is a two-step process based on the chemistry of organofunctional silanes, which were first grafted onto PO chains. Most often, vinyl(trimethoxy)silane (VTMS) was melt grafted onto POs in the presence of small amounts of peroxide initiators, usually in extruders. Small amounts of silanes (~2%) and 5–15 wt% of dicumyl peroxide were used. The silane grafting reaction was very fast, and it allowed to use reactive processing on the industrial scale. Within a few minutes in the extruder, very high grafting yields were obtained. At least 80% of the added silane was grafted onto polyethylene (PE). In a second postreactor step, the siloxane linkages (Si–O–Si) were formed (see Scheme 3.3). The possibility of silane homopolymerization was almost excluded under the conditions of the reactive processing. Vinyl(triethoxy)silane (VTES), and especially (3-methacryloxypropyl)silane, had limited application. The reactive processing techniques were applied for silane grafting onto any kind of olefin homopolymer (low-density PE (LDPE), linear LDPE (LLDPE), high-density PE (HDPE), very low density VLDPE, PE waxes, or all kinds of PP), PO blends or ethylene copolymers (EVA, EP, and EPDM elastomers), and polyisobutene [20].

The silane-grafted POs retained thermoplastic properties and were processed in the usual way, while final goods subsequently subjected to cross-linking in the presence of moisture. Before the moisture cross-linking step, a master batch containing an organotin catalyst was melt mixed with the silane-grafted PO in order to achieve cross-linking rates of practical importance. The alkoxysilyl groups were hydrolyzed to silanol groups (Si–OH), which further condensed forming the siloxane bonds and the three-dimensional network. Cross-linking occurred within few hours or days besides the production line in hot water tanks or steam chambers, most often at 60–90 °C. The silane cross-linking has many advantages among curing technologies. Most of those advantages are the consequence of separating the cross-linking step from the shaping step and the specific structure of the network as well, the latter feature was responsible for better thermomechanical properties in comparison with peroxide cross-linked PO networks.

In a modified one-step version of this technology called as MONOSIL process, PE, silane, grafting initiator, and cross-linking catalyst were added in a high-shear mixing extruder with a longer screw [22, 23].

An alternative method to the silane grafting is a one-step random copolymerization of ethylene with vinyl(trialkoxy)silanes at high pressure (150–350 MPa) and temperature (180–290 °C) in the presence of free radical initiators (usually peroxides), which was developed by the Union Carbide and Mitsubishi Petrochemicals. This process has some advantages since the silane units (<5 wt%) can be randomly and homogeneously distributed in the polymer chain, so it was possible to achieve a certain degree of cross-linking with a small amount of the silane incorporated in the polymer matrix [20, 24]. Alternatively, a copolymer of ethylene with 1,3-divinyl-1,1,3,3-tetramethyldisiloxane, a poly(ethylene-*co*-divinylsiloxane), was prepared in high-pressure reactors at 180–220 °C and p ~1,300 bar [25].

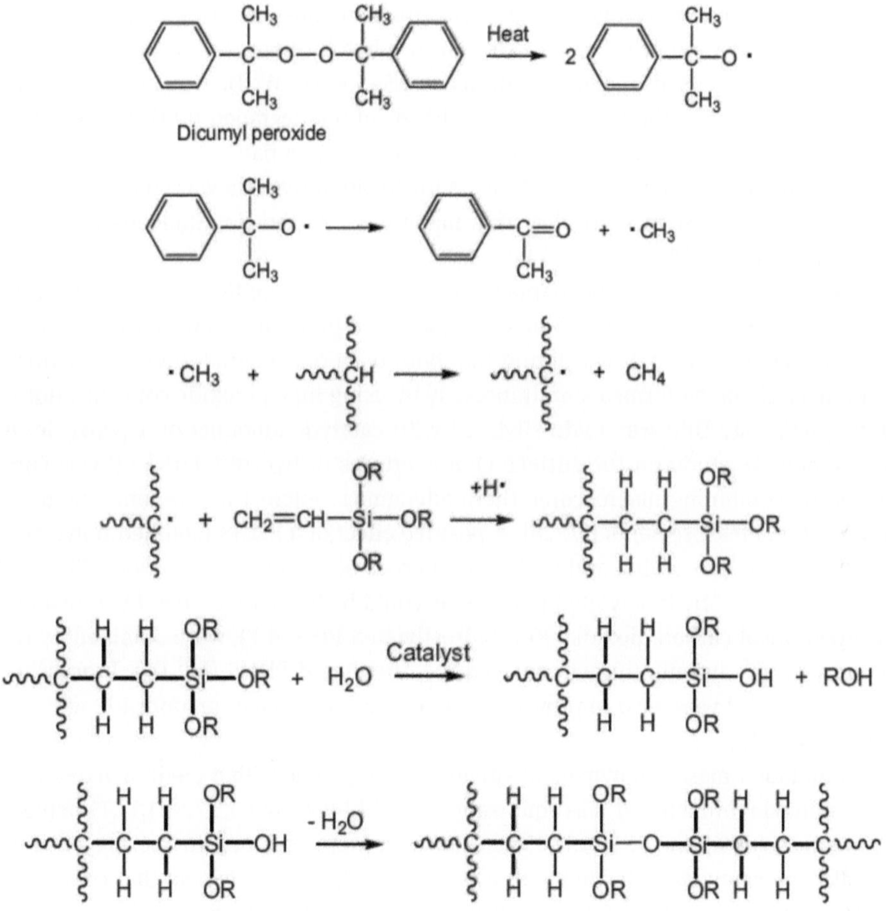

Scheme 3.3: Basic reactions leading to silane grafting onto PE and further cross-linking of PE [20]. Reprinted (adapted) with permission from [1]. Copyright 2012, Intech–Open Sci.

3.1.1.2 Hydrosilylation of polyolefins

A hydrosilylation of terminal double bonds in POs through reactive processing was found to be an alternative useful modification method of POs properties:

$$\equiv Si\text{-}CH{=}CH_2 \quad + \quad H\text{-}Si\equiv \quad \longrightarrow \quad \equiv Si\text{-}CH_2\text{-}CH_2\text{-}Si\equiv \; + \; \equiv Si\text{-}\underset{\displaystyle CH_3}{\underset{|}{CH}}\text{-}Si\equiv$$

PP-containing terminal double bonds were modified with a hydride-terminated polydimethylsiloxane (PDMS; $M^H{}_2D_n$, where M^H = $HMe_2SiO_{0.5}$, D = Me_2SiO) at three different temperatures through a hydrosilylation reaction, catalyzed in the melt phase by Karstedt's platinum catalyst [26–28]. The double bonds were formed by peroxide-initiated degradation of PP in an extruder or a batch mixer. An organic peroxide, *Lupersol 101*, was used in concentrations of 0.5–5 wt%. $M^H{}_2D_n$ was used as a substrate hydrosilylating the terminal double bonds of the degraded PP (DPP). Reactive processing experiments were carried out in a hot press, a batch mixer, and a single screw extruder. The reaction time of the hydrosilylation reaction was short enough to be completed in a screw extruder. This made possible the simultaneous extrusion and modification of PP.

Two different reaction mechanisms were used to initiate the hydrosilylation reaction: a radical chain addition mechanism and a platinum-catalyzed mechanism with Karstedt's catalyst. It was found that both reactions, degradation and hydrosilylation, could be performed simultaneously by using high peroxide concentrations, while previously DPP was hydrosilylated with catalytic amounts of a peroxide. A comprehensive study on the surface characteristics of hydrosilylated PP was conducted by combining macroscopic thermodynamics, microstructure, and chemical composition measurements [27, 28]. A positive effect of a PDMS-modified polyolefin additive on the processing and surface properties of LLDPE was observed [29]. Surface properties of hydrosilylated polyolefins could be further modified by annealing in supercritical carbon dioxide [30]. Hydrosilylated PP and PE were obtained by reacting with di- and multifunctional hydride-terminated PDMS ($M^H{}_2D_n$). Processing properties of PP were also improved by an addition of a small amount (~2 wt%) of silicone oil [31].

High molar mass copolymers of ethylene or propylene with α-olefin macromonomer, monovinyl-functional silsesquioxane cage 1-(9-decenyl)-3,5,7,9,11,13,15-heptaethylpentacyclo-[9.5.1.13,9.15,15.17,13]octasiloxane, were obtained in the presence of metallocene catalysts, activated with alumoxanes. These copolymers that contained up to 25 wt% (1.2 mol%) of spherosilicate pendant units were responsible for a decrease in the melting temperature by 18 K with respect to PE. A thermal stability under air was improved in the PE copolymer in comparison to PE [32].

Cross-linked styrene–isoprene–siloxane copolymers were prepared by cross-linking of elastomeric glassy styrene–isoprene prepolymer with oligo- or polyhydrosiloxanes. They were used for the preparation of membranes to pervaporative removal of volatile organic compounds from water [33, 34]. The hydrosilylation of polybutadiene (PB) with hydrosilanes was performed with rhodium or platinum catalysts, and Pt nanoclusters on a laboratory scale. Many of silyl derivatives of PB had functional group and could find useful applications [35–38].

A PDMS-grafted copolymer, which can be potentially applied as a processing agent or a modifier of surface properties, was synthesized via reactive melt mixing of ethylene–ethyl acrylate–maleic anhydride terpolymer and aminopropyl-terminated PDMS [39].

A large family of polystyrene-based block copolymers, modified with reactive silyl groups, may find wide potential applications [40–44]. For example, PTEPM (poly-3-

(triethoxysilyl)propylmethacrylate) was prepared by radical polymerization of a gellable monomer 3-(triethoxysilyl)propyl methacrylate (TEPM) mediated by a chain transfer agent (CTA), cumyl dithiobenzoate (CDB), that is, reversible addition-fragmentation chain transfer (RAFT) polymerization. A series of well-defined PTEPM-*block*-polystyrene

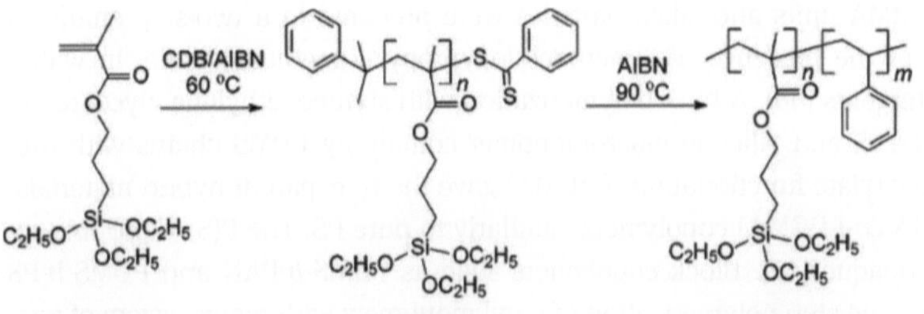

Scheme 3.4: Synthesis of PTEPM and PTEPM-*b*-PS diblock copolymer by RAFT polymerization. Reprinted (adapted) with permission from [40]. Copyright 2007, American Chemical Society.

(PTEPM-*b*-PS) diblock copolymers were obtained via sequential RAFT polymerization of styrene mediated by the PTEPM macromolecular CTA (Scheme 3.4).

The bulk microphase separation behaviors of PTEPM-*b*-PS diblock copolymers were studied by a combination of small-angle X-ray scattering (SAXS) and transmission electron microscopy. With change of the block composition, three different morphologies of microphase separation, that is, lamellae, hexagonally packed cylinders, and spheres, were obtained [40].

3.2 Applications of carbofunctional polysiloxanes (CFPS)

The CFPS are most often used for the syntheses of hybrid silicone–organic copolymers and for the modification of many polymers and plastics. By this way, numerous new polymeric materials with profitable properties were obtained.

3.2.1 Copolymers containing acrylate units and reactive silyl groups or siloxane segments

Acryl–siloxane copolymers, for example, the graft copolymers of polysiloxane with styrene, methyl methacrylate, and chloroprene were synthesized by a radical polymerization from the commercial CFPS containing side and terminal mercaptopropyl groups [45]. Some silyl derivatives of acrylates, methacrylates, are useful for the preparation of membranes showing the selectivity with reference to oxygen. Poly(silyl methacrylate) may be applied in contact lenses (CLs). The increase of polymethacrylate solubilities with an increase in the incorporation of silyl groups was observed. Polymers based on 2-hydroxyethyl methacrylate (HEMA) were also modified with some silyl groups [46]. Waterborne silicone-acrylic dispersions were obtained by miniemulsion copolymerization of divinyl-terminated PDMS and methyl methacrylate (MMA), butyl acrylate (BA), and acrylic acid [47]. Monodispersed, anisotropic polymers composed of styrene and/or MMA units and silica particles were prepared in a two-step emulsion polymerization in the presence of 3-methacryloxypropyl(trimethoxy)silane, in water, without any detergents [48]. A bulk polymerization with styrene, ethylene glycol dimethacrylate (EGDMA), and silicone macromonomer containing PDMS chains with urethane and methacrylate functionalities (SiUMA) gave the transparent hybrid materials: the P(S-*co*-SiUMA-*co*-EDGMA) copolymers, similarly to pure PS. The P(S-*co*-SiUMA) copolymers were opaque [49]. Block copolymers such as PDMS-*b*-PAN and PDMS-*b*-PS were prepared in one-step polymerization of vinyl monomers with redox system of α,ω-dihydroxypropyl-terminated PDMS with ceric ammonium nitrate at ambient temperature in water. These copolymers with a low silicone content (1–2 wt%) showed much higher contact angle values than the corresponding homopolymers and may act as compatibilizers in the blending application of vinyl polymers with polysiloxanes [50].

The precipitation copolymerization of poly(*N*-isopropylacrylamide) (PNIPAAm) with 3-methacryloxypropyl(triethoxy)silane gave thermoresponsive shape memory polymers (SMPs), which were soluble in water solution below 32 °C, due to the hydrogen bonding interactions between polymer chains and water molecules [51]. They may find biomedical applications. A self-assembly silane monolayer was formed by 3-aminopropyl(triethoxy)silane (APTES), followed by the graft polymerization of NIPAAm on the glass surface. These kinds of glasses bonded with NIPAAm films may be used as environmentally switchable materials [52]. Amphiphilic multiblock and triblock copolymers containing poly(*N*,*N*-dimethylacrylamide) (PDMAAm) and PDMS segments (PDMAAm-*b*-PDMS-*b*-PDMAAm) may also find a variety of biomedical applications [53]. A similar ABC triblock copolymer with a very low surface energy such as PDMS-*b*-PMMA (poly(methyl methacrylate))-*b*-PHFBMA (2,2,3,3,4,4,4-heptafluorobutyl methacrylate) synthesized via ATRP [54] could be applied for antifouling coatings. Transparent siloxane-fluoroacryl IPN networks with good hydrophobic and oleophobic properties were prepared in situ from α,ω-bis-(3-hydroxypropyl)polydimethylsiloxanes, isocyanurate derivatives of hexamethylene diisocyanates $R(NCO)_x$, 3,3,4,4,5,5,6,6,7,7,8,8,8-tridecafluoro-1-octanol, EGDMA, dicyclohexyl acrylate, and dipercarbonate (DCP), in the presence of dibutyltin dilaurate [55].

Amphiphilic copolymers of poly(ethylene glycol) (PEG)-fluoroalkyl acrylate and polysiloxane methacrylate units, blended with a PDMS matrix in different ratios were used as fouling-release coatings and showed significantly reduced levels of hard fouling compared to the control coatings of neat PDMS [56]. Polymerizations of amphiphilic fumarate monomers (containing both hydrophilic and hydrophobic substituents) gave copolymers with high oxygen permeability, hydrophilicity, and transparency required for CL materials. Dissymmetric fumarate monomers having both alkoxyethyl group [2-(2-methoxy-ethoxy)ethyl, 2-(2-(2-methoxyethoxy)ethoxy)ethyl] and a bulky 3-[tris(trimethylsiloxy)silyl]propyl group were prepared by the radical polymerization with styrene and *N*-vinylpyrrolidone (NVP). Various membranes, prepared by copolymerization of fumarate monomers with NVP or NVP and HEMA, showed much better transparency as compared to the membranes containing 3-[tris-(trimethylsilyloxy)silyl]-propyl methacrylate. Polyfumarates containing silyl substituents can be used as high performance materials, biodegradable polymers, low dielectric materials, oxygen permeability membranes, crystal liquid, contact lenses (CL) materials, and so on. The CL materials should have both the oxygen-permeable property for breathing of corneal cell and the hydrophilic property for not repelling a tear and to protect a lipid deposition [57].

Poly(vinyl alcohol) (PVA)-*graft*-PDMS and triblock PVA-*b*-PDMS-*b*-PVA copolymers with various contents of hydrophilic and hydrophobic segments were prepared in microsuspension by the reaction of an epoxy-terminated PDMS with some pendant alcohol groups in P(VA-*co*-Vac). They can be applied as dispersants and in the preparation of thermoplastic elastomers (TPEs) [58, 59].

PVDF-based copolymers may be used as membrane materials [60]. They have repellent properties with respect to boiling acetone, water, acids, and oil [61].

Silicone-containing particles with hybrid structure can be prepared by miniemulsion polymerization, polymerization in nonaqueous media, hybridization of organic polymer and polysiloxane, and emulsion polymerization of silicone monomers in silicone polymer dispersions. Alternatively, silicones can be incorporated into polymers through physical methods. For fabrication of hybrid aqueous polymer dispersions (APD) were used:

(1) emulsion polymerization of unsaturated organic monomers in aqueous dispersions of silicone polymers or copolymers,
(2) emulsion copolymerization of unsaturated organic monomers with (alkoxy)silanes or polysiloxanes with unsaturated functional groups, and
(3) emulsion polymerization of (alkoxy)silanes (in particular with unsaturated functional groups) and/or cyclic siloxanes in organic polymer dispersions.

Silicones form either the core or the shell in hybrid particles. A very large group of silicone-containing APD with hybrid particle structure has been applied mainly as: (1) coatings that have specific surface properties such as enhanced water repellency or antigraffiti or antisoiling properties owing to migration of silicone to the surface, and (2) impact modifiers for thermoplastics and thermosets [62].

3.2.2 Modification of thermoplastic polymers with silicones, CFS, and CFPS

Many kinds of step-growth and other thermoplastics (PES, polyamides, polyimides (PIs), polylactide, etc.) were modified with CFPS.

3.2.2.1 Modification of polyester properties with silicones

Aromatic silicone PES and polyamides (SAPAs) (which were liquids at room temperature (RT)) were synthesized under mild conditions using α,ω-dihydroxyalkyl-terminated PDMS. The SAPAs were also prepared by transesterification reaction of α,ω-diaminoalkyl-terminated PDMS with dimethyl terephthalate catalyzed by lipase in toluene at a temperature of 80–90 °C. Polysiloxane–PES copolymers were synthesized by polycondensation of a series of diacids and α,ω-bis(3-hydroxypropyl)-PDMS with *Novozyme-435* enzyme [63, 64].

Novel thermoplastic PES such as poly(ester-siloxane)s were composed of poly (butylene terephthalate) (PBT) as hard segments (HS) and PDMS-containing prepolymers as soft segments (SS). In order to increase the compatibility of PDMS copolymers of siloxanes with terminal aliphatic PES, a triblock copolymer polycaprolactone-*block*-PDMS-*block*-polycaprolactone (PCL-PDMS-PCL) was used. The thermal stability of the poly(ester-siloxane)s was higher than that of PBT homopolymer, and the reduced crystallinity of the HSs was observed [65].

Poly(ϵ-caprolactone) (PCL)-based shape memory polymers (SMPs) have received much attention due to their biocompatibility, biodegradability, and elasticity of PCL, which was useful as a crystalline switching segment for SMPs as its T_{trans} was a well-defined T_m in the range of 45–60 °C with increasing M_n. Such T_{trans} values can be useful for in vivo deployment as well as other applications that require low heating. To modify properties of PCL-based SMPs, AB polymer networks were prepared through incorporation of organic HS or SS. PDMS with extremely low T_g (−123 °C) was used as an effective SS ingredient. Schoener et al. prepared organic–inorganic SMPs comprised of inorganic PDMS segments, terminated with aminopropyl groups and organic PCL segments. The resulting SMP network exhibited excellent shape fixity and recovery. By changing the PDMS length, the thermal, mechanical, and surface properties were systematically alternated [66].

Carbon nanotubes (CNTs) are the ideal reinforcing agents for high-strength polymer composites because of their tremendous mechanical strength, their Young's modulus is ca. 1,000 GPa, and it is much higher than that of the conventional carbon fibers (200–800 GPa). For good reinforcement effect, the surface of CNTs should be strongly bonded to the polymer matrix. Sidewall functionalization of CNTs led to incorporation of mainly carboxylic groups and made possible to chemically attach these nanofillers with polymers. The nanocomposites (NCs) of poly(L-lactide) (PLLA) were reinforced by (aminopropyl)isooctyl POSS-modified multiwalled nanotubes (MWNTs) (MWNT-*g*-POSS). The homogeneous MWNT PLLA/MWNT-*g*-POSS composites without any aggregation were obtained. The fractal surface of the composites showed not only a uniform dispersion of MWNTs but also a strong interfacial adhesion with the matrix [67].

Blends of PLLA with methyl silicone resins (Me-MQ) were prepared after solvent evaporation at various Me-MQ contents. PLLA was high in strength and modulus, but its applications were limited partly due to its inherent brittleness (where M is the monofunctional siloxane unit $R_3SiO_{1/2}$ and Q is the tetrafunctional siloxane unit $SiO_{4/2}$). It was difficult to keep both toughness and transparency for modified PLLA without compromising its TS and biocompatibility. Biocompatible Me-MQ resin was able to toughen PLLA at low loadings. The elongation at break of neat PLLA was 7.8%. The elongation at break of PLLA blends increased to 67.9% and 217.7% at the contents of 1 and 2 wt% Me-MQ, respectively, without decrease of their TS [68].

3.2.2.2 Modification of polycarbonate properties with silicones

Polycarbonates (PCs) based on bisphenol A have a superior heat resistance, transparency, dimensional stability and self-extinction, exceptional impact resistance, and ductility even below RT. Their modulus, dielectric strength, and TS are comparable to other amorphous thermoplastics. The main limitations of PCs are their poor scratch, ultraviolet (UV), solvent resistance, and their sharply decreased impact strength at low temperature. Properties of PCs were modified with two kinds of silicone oils: PDMS and polydimethyldiphenylsiloxane (PDMDPS). A transition from brittle to ductile failure was observed with increasing PDMDPS content. At temperature of −30 °C, the impact strength increased from 8 to 52 kg_fcm/cm for neat PC and PC/PDMDPS blends, respectively [69].

Block copolymers of PCs with PDMS (PC/PDMS) are an interesting group of TPE, characterized by a great stiffness at low temperature. PC-*b*-PDMS copolymers were usually prepared in the reaction of phosgene with bisphenol A in the presence of PDMS containing terminal bisphenol groups [70]. Block PC-*b*-PDMS copolymers can be used as membranes for selective separation of gases [71]. Multiwalled CNTs (MWCNT)/PC NCs were prepared in reaction between α,ω-3-aminopropyl-PDMS and MWNTs functionalized with COOH groups. The siloxane-modified CNTs were dispersed well in the PC matrix, and the TS, flexural strength, flexural modulus, and flame retardancy of their composites were better than these MWNT-COOH/PC. Siloxane-modified MWNTs can improve the electrical properties of NCs at low loading in PC [72].

3.2.2.3 Modification of polyamide and polyimide properties with silicones and CFS

Inorganic–organic hybrids were prepared by melt mixing of POSS and polyamide 6 (PA 6) or by the in situ polymerization of ϵ-caprolactam in the presence of POSS. A very fine cubic dispersion of nanometric dimensions was obtained by the latter method. The nearly nanosized aromatic polyamide particles with amino groups, known as poly(aminoamide), were prepared by Yoshioka [73]. They were modified with the SCA, that is, glycidoxypropyl(triethoxysilane (GPTES), and further complexed with ZnO particles. The finally ZnO-based materials had specific optical properties: they showed excellent optical absorption properties in UV region and high transmittance in the visible region [74].

The incorporation of flexible PDMS segments into PIs gave new copolymers with low water absorption, atomic oxygen resistance, excellent adhesion, and good processability. Due to their unique properties, imide–siloxane copolymers have been especially attractive for applications in microelectronics and as structural adhesives. A series of amorphous poly(imide siloxane)s (PISs) with different PDMS contents and segmental lengths were prepared via condensation reaction [75]. Different morphologies of PIS films, including unilamellar vesicle, multilamellar vesicle, and sea-island, were also dependent on the content and the segmental length of PDMS, as well as the coexistence of large-scale and nano-scale phase separations of approximately 20 nm.

PI/polydiphenylsiloxane (DPS) composite films with high thermal stability near pure PI were prepared by a sol–gel method. Polysiloxanes well dispersed in PI matrix without macroscopic separation for the composite films were obtained with low content of DPS, while the large domain of polysiloxane was formed in films with high DPS content. The incorporation of DPS into PI improved the elongation at break but the composite films still remained with higher modulus and TS [76].

Aromatic PIs are very useful as gas separation membranes, owing to their gas selectivity and excellent thermal and mechanical properties. Recently, novel triamine-based hyperbranched PIs were applied for high-performance gas-separation materials. Modification of PIs with inorganic compounds was carried out in order to improve the gas transport properties. Hyperbranched PIs were prepared by polycondensation of a triamine 1,3,5-tris-(4-aminophenoxy)benzene and 4,4′-(hexafluoroisopropylidene) diphthalic dianhydride (6FDA) and were further modified by sol–gel reaction with (tetramethoxy)silane, methyl(trimethoxy)silane, or aminopropyl(trimethoxy)-silane (APTMS) [77].

New copolymers such as polysiloxane-imides (PSIs) were prepared from α,ω-(bisaminopropyl)dimethylsiloxane oligomers and aromatic dianhydrides: 1,2,4,5-benzenetetracarboxylic dianhydride and 6FDA [78]. Membranes made of these new PSIs were used to remove polar organics from water by pervaporation.

NCs of bismaleimide with surface-modified SiO_2 nanoparticles by amino-functionalized SCA showed better wear resistance and lower frictional coefficient than that with unmodified fillers [79].

3.2.2.4 Modification of polyurethanes and polyureas with silicones and CFPS

Multiblock thermoplastic polyurethane (TPU) copolymers contain hard segments (HS) and soft segments (SS) in their chemical structures. They were prepared by a polyaddition reaction of diisocyanates to hydroxyl-terminated oligomers (polyols) and short diols as chain extenders. The thermodynamic incompatibility between alternating segments usually leads to a two-phase structure with soft domains containing mostly polyol moieties and hard domains made up of diisocyanate-chain extender sequences.

Incompatibility between the two block components leads to a microphase separation and the hard domains cause a reinforcement to the system. The type of diisocyanate affects the strength of a final material. The most often applied isocyanates are 4,4′-methylenediphenyl diisocyanate, tolylene diisocyanate, bis(4-isocyanatocyclohexyl) methane, and isophorone diisocyanate (IPDI). TPUs are widely used for high-performance applications as membranes, medical implants, adhesives, and coatings, especially when a high TS and tear or good wear and abrasion resistance are required. Another special features of TPUs are their low-temperature elasticity, an electrical insulation, a good chemical resistance, and a smoothness for the touch [80]. As SS components in polyurethanes and polyurethaneureas, PDMS-diols or hydroxyalkenyl or aminopropyl or ethyl-piperazine-terminated PDMS were often applied. However, the TS and elongation at break of these materials compared to those based on polyether or PES SS are worse. Thus, various co-SS components, mainly based on poly(alkenyl oxide), were also applied. Poly(propylene oxide) (PPO) was useful as a second SS substrate because its solubility parameter (23.5 $J^{1/2}/cm^{3/2}$) is in between that of PDMS (15.6 $J^{1/2}/cm^{3/2}$) and urea (45.6 $J^{1/2}/cm^{3/2}$). The intersegmental mixing between urea segments and PPO affected the nature of interphase between the soft matrix and the hard urea microdomains [81].

Waterborne polyurethanes (WPUs) help reduce the emission of volatile organic compounds. However, WPUs are not enough resistant to water and show surface hydrophilicity and quite lower strength. These disadvantages were eliminated by using a different composition of SS. The anionic WPUs were prepared from hydroxyl-terminated PDMS, poly(tetramethylene oxide) as SS, IPDI as HS, and 2,2-bis (hydroxymethyl)propionic acid as the ionic center. The TS of WPU-PDMS films decreased with an increasing content of PDMS and a water repellency reached 80%, which was equal to the capability of silicone rubber (SR). These waterborne siloxane-containing PUs can be used for leather finishing, textile and plastic coatings, sealants, and a glass-fiber sizing [82]. A POSS-modified ionomeric polyurethane's aqueous dispersion showed an improved resistance to fire and to atomic oxygen as well as to the possibility to build nanostructured surfaces. Many of the POSS hybrids do yield elastomeric PU with improvements in physical and thermal properties at relatively low POSS contents. The waterborne PU hybrid dispersions were prepared by incorporating amino- and hydroxyl-terminated POSS macromers into the PU ionomeric backbones, using a homogeneous solution polymerization, followed by a solvent exchange with water. The absence of crystalline domains evidenced the incorporation of POSS macromer in the PU HS [83].

The gas permeability of PU membranes increased with the decrease in the HS content and the increase in the molecular weight of SS. In addition, a correlation was established between the gas permeability and the chemical nature of the polyols and chain extenders. The gas permeability of the PDMS–poly(propylene oxide) (PPO)–urethane/urea bi-SS membranes to CO_2, O_2, and N_2 increased with the increase in the PDMS content.

The $P(CO_2)/P(N_2)$ permeability ratio was higher for the bi-SS membranes in comparison with PDMS/PU membranes [84].

Some properties of TPUs (high tear, TS, good wear, and abrasion resistance) strongly decreased when the service temperature of the material exceeded 70–80 °C, mostly due to a creep phenomenon. The working temperature of commercial TPU was improved by a well-controlled chemical curing. A commercial PES-type TPU at first was reacted with a diisocyanate to form allophanate branching units, which at the second step was grafted with APTMS. The obtained TPU was cross-linked by a hydrolytic condensation of ≡Si-OMe groups, resulting in increased thermomechanical properties [85].

Polyurethane–poly(ε-caprolactone) nanofibrous membranes (NFMs) were modified with PDMS via an electrospinning technique. The morphologies and properties of these composite nanofibers were characterized by FTIR spectroscopy, thermogravimetric analysis (TGA), field-emission SEM, and mechanical testing. NFMs prepared at 70 °C had an optimized fibrous structure with a diameter of 514 ± 2 nm, a pore size of 0.55–0.65 mm, and a porosity of 77.7%. The NFMs modified with 5 wt% PDMS showed good waterproof properties (water contact angle 141 ± 1° at hydrostatic pressure 573.6 kPa), high breathability (air permeability rate 56.6 L/m^2 s and water vapor transmission rate 59.0 kg/m^2 day), a high mechanical strength (breakage stress 511.7 MPa), and excellent thermal stability. They should find promising practical applications [86].

3.2.3 Modifications of other polymers and polymeric materials by CFS and CFPS

Properties of many other polymers were modified with CFPS. For example, CFPS were used for the synthesis of conductive siloxane–pyrrole block copolymers. An incorporation of flexible PDMS chains into the rigid polymer chain improved mechanical properties and solubility in different nonpolar solvents. (*N*-Pyrrole)polysiloxane precursor, prepared in the reaction of *N*-glicydylpyrrole with aminopropyl groups of PDMS, was copolymerized with pyrrole (PY) by electrochemical methods, giving the block copolymer PY-PDMS. Classical conductive polymers are insoluble and infusible. A common way to facilitate their processing is syntheses of block or graft copolymers containing conductive segments and conventional segments, increasing the solubility of copolymer, for example, block copolymers of styrene and acetylene, styrene and thiophene, and graft copolymers of PY on polystyrene and on poly(methyl methacrylate). New multifunctional polysiloxanes containing mercapto-, thiocarbamide, and (hydroxyalkyl)ether linkages were applied for the modification of silica fillers used for the production of rubber tires. Copolymers containing siloxane segments play a role of connectors with organic polymers (compatibilizers). Siloxane moieties of copolymers migrate to the surface, while organic segments act as anchors of siloxane blocks in the polymeric matrix [87].

Bis-(3-triethoxysilyl)propyltetrasulfide) was used for the modification of natural and synthetic rubber [88]. Other CFS were applied for the preparation of silica in elastomers (in situ) and for the modification of calcium carbonate, which was used as the filler for elastomers [89].

Alkyd resin coatings modified with carbinol-terminated silicone oil and small amount of TiO_2 were very effective in saving wood-based panel products and bamboo composites from photodegradation, avoiding delamination in natural and artificial weathering tests. Decrease in gloss retention of the coating during exterior exposure depended strongly on the variation of relative humidity and time of wetness and much less on intensity of the solar radiation. The silicone-modified alkyd coatings exhibited excellent color retention, superior durability, no splitting, cracking, warping, chocking on exposure to weather, and UV light. Excellent resistance against photodegradation, weather damage, graying, termite attack, and rainwater with economical and outstanding coverage was observed [90].

Polyurethane fabrics were coated with hybrid silica films which were prepared from TEOS and 3-glycidyloxypropyl(trimethoxy)silane (GLYMO) by sol–gel method combined with dip coating. GLYMO-based treatments improved the thermal stability and saved textile feel of PU fabrics. These hydrophobic coatings were characterized by infrared, SEM, differential scanning calorimetry, TGA, energy-dispersive X-ray spectroscopy, and water contact angle measurement. SEM images showed that polyurethane surface and pores were completely modified with thin hybrid silica films, which were preserved for the most samples, except for the single treatment of GLYMO, due to the polar effect of the epoxy group [91].

3.3 Composites and nanocomposites of thermoplastic polymers

Composite materials consisting of a polymer matrix and inorganic fillers are widely used in order to improve the mechanical, thermal, barrier, and other properties of the neat polymers. Silane-modified fillers and nanofillers are very often applied for fabrication of composites and NCs.

3.3.1 Modification of fillers properties by CFS

The CFS very often play a role of *adhesion promotors* [92, 93] and they are also called as *silane coupling agents* (SCA) [94, 95]. These multifunctional silanes are directly introduced to polymer compositions (usually in a quantity 0.2–2 wt%) or in the form of water and water-alcoholic dispersions. The silyl functional groups of SCA (e.g., alkoxy group) undergo coupling reactions with hydroxyl groups of the fillers:

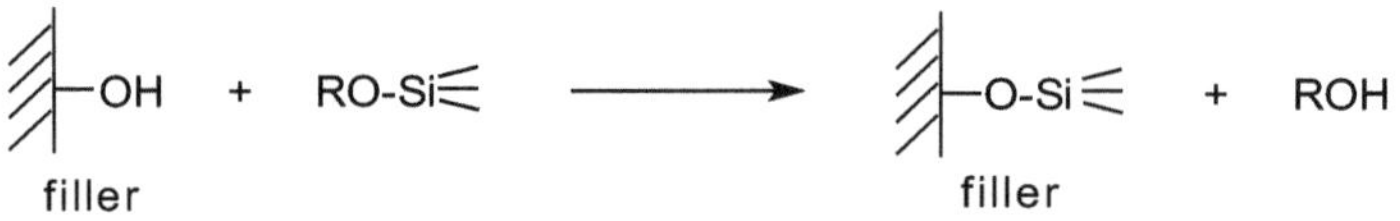

or hydrolysis reactions, leading to replacement of alkoxy groups by silanol groups Si-OH:

$$\equiv Si\text{-}OR + H_2O \longrightarrow \equiv Si\text{-}OH + ROH$$

The silanol groups Si-OH can form chemical bonds with inorganic materials (e.g., fillers, glass, or metals), while carbofunctional groups form chemical bonds with polymers (see Scheme 3.5) [96].

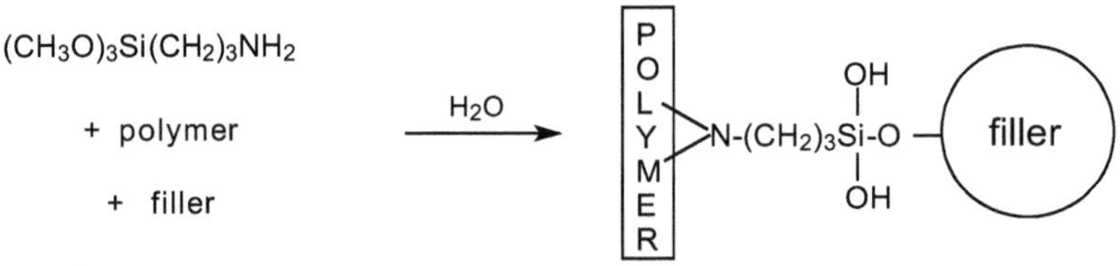

Scheme 3.5: An adhesion model in a system: polymer–carbofunctional silane–filler.

Moreover, the silanol functional groups Si-OH undergo:

(a) homocondensation reactions with a formation of thermodynamically more stable siloxane linkages Si–O–Si or
(b) Si-OH groups may react with OH groups of the filler surface.

Finally, on the surface of the fillers, very thin layers of cross-linked polysiloxanes are formed. An application of the fillers in plastics results from the necessity of an improvement of mechanical, thermal, electrical, and magnetic properties of the polymers. The most often used fillers are silica, aluminum oxide and hydroxide, metal oxides (e.g., ZnO, MgO, CaO, TiO_2, Fe_2O_3, and Pb_3O_4), calcium carbonate, calcium sulfate (gypsum), barium sulfate, and silicates of Ca, Al (kaolin), and Mg (talc and sepiolite). An addition of nanofillers profitably affects an increase in the abrasion resistance of transparent polymer coatings and enables preparation of homogeneously distributed nanoparticles in polymer matrices. The addition of 20–40 wt% of the silica and alumina caused an improvement in the abrasion resistance of composite materials. The unmodified silica and alumina have a large number of hydroxyl groups on their surfaces and thus they are hydrophilic materials with a high tendency to agglomeration. In the presence of nonpolar liquids, for example, acrylates, an aggregation of nanoparticles of these fillers takes place, which leads to "thickening" of their coatings based on acrylates. The surface modification of fillers with the CFS (*silanization*) substantially improves their dispersibility in polymers. The use of vinyl or methacryloxypropyl-substituted (trialkoxy) silanes for the modification of nanoparticles of SiO_2 and Al_2O_3 enabled the preparation of transparent abrasion-resistant coatings on their surfaces. Acrylic laquers containing nanoparticles of fillers are utilized as top, decorative layers of furniture and aluminum foils, and as varnishes for a floor and coatings.

Nanospheres of a mesoporous silica were functionalized with APTMS and a sol of gold [97]. By this way NCs with a very high specific surface area (~1,000 m^2/g) were obtained. (Trialkoxy)propylsilanes containing different carbofunctional groups such as amine, aldehyde, acrylate, isocyanate, thiol, ether, triacetoxydiamine, and (trialkoxy) silanes containing aldehyde and nitrile groups attached to a longer alkylene "connector" were grafted onto the surface of cobalt ferrite ($CoFe_2O_4$). They were used for the preparation of water dispersions of nanoparticles exhibiting supermagnetic properties [98].

Silicon nitride nanopowder (Si_3N_4) is the next important ceramic filler for various applications because of its high strength, low coefficient of thermal expansion, high thermal-shock resistance, and good wear resistance. It is a very useful filler for the improvement of mechanical properties and the wear resistance of polymer composites. To limit agglomeration of nanoparticles and improve their dispersivity in polymer matrix, the surface of this filler was modified with a novel macromolecular coupling agent copolymer BA-MMA-VTES [containing (triethoxy)silyl functional groups], which was bound covalently on the surface of nanosized Si_3N_4 particles and was used for the preparation of an organic coating layer [99].

Well-defined microporous carbon nanosphere MCNS materials were prepared by the Friedel–Crafts cross-linking reaction of the oligomeric POSS molecules (Ph-POSS) to form polymeric xPh-POSS, followed by its carbonization at 900 °C, and subsequent removal of monodispersed silica domains surrounded by molecular-scale carbon/silica interfaces (Figure 3.1). The obtained carbon nanospheres with a spherical morphology and average diameter of 360 nm exhibited very large surface area (1,176–2,894 m^2/g) and fast and selective sorption, demonstrating excellent adsorption and supercapacitance properties. These well-defined nanoporous carbons may find applications in fabrication of various polymer composites [100].

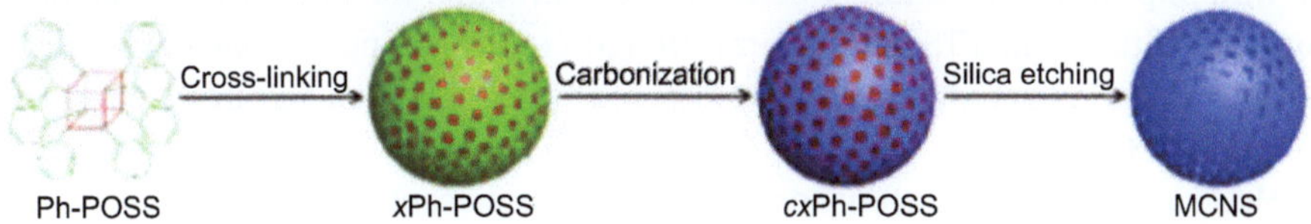

Figure 3.1: Schematic illustration of preparation of MCNS (microporous carbon nanosphere) materials. Reprinted (adapted) with permission from [100]. Copyright 2014, American Chemical Society.

Properties of hydroxylated CNTs (MWCNT-OH) were modified with $CH_2{=}CHSi(OMe)_3$ (VTMS), (3-mercaptopropyl)trimethoxysilane (MPTS), and [3-(2-aminoethylamino) propyl]trimethoxysilane (APTS). APTS was the most promising silane agent which strongly affected the properties of acrylonitrile–butadiene rubber (NBR) composites with silane-modified MWCNT-OH. The NBR vulcanizates showed increased cross-link density, TS, torque, and dynamic moduli [101].

3.3.2 Applications of silane-modified nanofillers in thermoplastic composites

In order to achieve satisfactory properties of polymer composites and NCs, good homogeneous dispersion of nanoparticles in the polymer matrices is required. An agglomeration of fillers can be avoided by grafting polymer chains onto the particles through covalent bonds, leading to organic–inorganic hybrid materials. Three grafting methods were applied: *grafting from*, *grafting to*, and *grafting through*. Living radical polymerization (LRP) or controlled radical polymerizations were used for grafting the polymers onto the inorganic particle surfaces. RAFT is the more applicable grafting technique than other LRP methods: nitroxide-mediated polymerization and atom transfer radical polymerization (ATRP). LRP enables to control the molecular architecture of polymers, and its procedures are simple and universal on monomers in comparison with ionic polymerizations [102].

3.3.2.1 Polyolefin composites modified with silanes and fillers

An isotactic PP is a commodity polymer with a wide range of applications. Various PO elastomeric impact modifiers were commonly used for improvement of its ductility and toughness. PP-based NC blends were compounded from finely dispersed reinforcing inorganic fillers and nanofillers: silica, $CaCO_3$, talc, $Mg(OH)_2$, $Al(OH)_3$, $BaSO_4$, and others. Due to the nonpolar character of the PP, functionalization of the polymer matrix played an important role. Moreover, modification of properties of nanofillers was carried out in order to achieve a good dispersion of the rigid nanoparticles in PP matrix and satisfactory mechanical properties of PP NCs. PP was functionalized by peroxide-initiated grafting of VTES, providing a PP-*g*-VTES derivative with improved compatibility between the matrix and the filler. The mechanical properties (first of all, a TS and flexural strength) of the composites were greatly enhanced, while impact strength was preserved when the silane-treated nanosilica was used [103].

Most surface modifiers act not only as effective dispersants for fillers. Surface treatment of mineral fillers with reactive silanes (or titanates) facilitates their incorporation into thermoplastic melts such as PP. The silane additives to functionalized polymers act as effective coupling agents in thermoplastic-based composites. Aluminum hydroxide is also an effective flame retardant for thermoplastics. In PP/$Al(OH)_3$ composites, three surface modifiers were used: a dicarboxylic acid anhydride, a silane-based system, and a maleated PB. At high filler content (>59% w/w), the PP/$Al(OH)_3$ composites modified with SCA showed the best properties [104]. Silica nanoparticles were also modified by grafting the foaming agent *p*-vinylphenylsulfonylhydrazide and blended with melted PP. The side sulfonylhydrazide groups were decomposed with volatile gas products and blowed up the agglomerated silica nanoparticles, while the remaining skeleton of the grafted polymer caused improvement of the filler–matrix interaction [105].

A great number of NCs were obtained by blending polymers with layered silicates (nanoclays), most often with montmorillonite (MMT). The layered silicate particles are usually hydrophilic and their interactions with nonpolar polymers are rather weak. There are two main goals of surface modification on MMT: (1) to expand the interlayer space, allowing large polymer molecules to migrate into the clay sheets, and (2) to improve the miscibility of MMT with the polymer through a good dispersion of layered structure within the polymer matrix. Layered silicates contain sheet-like platelets which are about 1 nm in thickness and 100–1,000 nm in length and width, thus they have high large surface area and aspect ratios. If the high degree of exfoliation or intercalation can be achieved, the high aspect ratio of the clay platelets allows significant reinforcement at relatively low contents of nanoclays. Three types of composites can be obtained by the dispersion of clays in polymer matrices: (1) conventional composites that contain clay tactoids dispersed as a segregated phase; (2) intercalated polymer–clay NCs that are formed by the infiltration of one or more molecular layers of polymer into the clay host galleries; (3) exfoliated polymer–clay NCs with low clay contents and monolithic structure, and a separation between clay layers that depend on the polymer content of the composite. Exfoliation is especially necessary for improvement specific properties that are affected by the degree of dispersion and the resulting interfacial area between polymer and clay nanolayers.

Silane grafting is one of the efficient methods to functionalization of POs, especially for the preparation of silane-grafted moisture-cured PE. PE-*g*-MAPTMS (methacryloxypropyl(trimethoxy)silane)/OMMT (organically modified MMT) NC was prepared by blending OMMT with (MAPTMS-grafted PE (PE-*g*-MAPTMS), which was obtained by melt grafting reaction [106]. The maleic-anhydride-grafted PE and PP are most often used for improvement of the interfacial bonding between the clay and polymers. Maleated PE (PE-*g*-MA)/aminosilane compatibilizer was very effective for the preparation of PE-clay NCs [107].

VTMS (1.6 wt%) was melt grafted onto HDPE with DCP (dicumyl peroxide, 0.08 wt%) and antioxidant (0.2 wt%) in the twin-screw extruder. Novel HDPE/SiO_2 hybrid composite materials were prepared from blends of HDPE and synthesized HDPE-*g*-VTMS by the HCl-catalyzed sol–gel process using 1–3 wt% of TEOS. The structure and thermal and mechanical properties of hybrid materials were studied by FTIR, X-ray photoelectron spectroscopy, SEM, thermogravimetry, differential thermal analysis, and tensile tests, respectively. Silica phase in the HDPE-*g*-VTMS hybrids showed a network nanostructure. The HDPE-*g*-VTMS functionalized with $(MeO)_3Si$ groups underwent hydrolytic polycondensation with TEOS and formed covalent bonds between organic and inorganic phases. The thermal stability and mechanical properties of HDPE-*g*-VTMS hybrids were improved by incorporation of silica networks. The formation of the HDPE-*g*-VTMS hybrid was beneficial for enhanced mechanical strength and the thermal stability in comparison with the neat HDPE [108].

3.3.2.2 Polyacrylate-based composites modified with silica or/and CFS

Silica is very often applied as a filler for polyacrylate composites. Polyacrylate-silica particles were usually prepared by polymerization carried out in emulsion or microemulsion. Emulsifier-free latexes were also developed. TEOS was used as a silica precursor for polyacrylate-silica hybrids and poly(vinyl alcohol) (PVA) served as a colloid stabilizer [109].

One of the methods for preparation of inorganic–organic composites is surfactant-free emulsion polymerization. Through conventional emulsion polymerization, PMMA–SiO_2 composite particles were prepared. The morphology of the composite particles (normal core–shell, multicore–shell, or raspberry-like) was dependent on the monomer–silica ratio, silica particle size, emulsifier content, and monomer feed method [110].

Sol–gel process was often applied for preparation of polymer-silica organic–inorganic hybrids, in which particles of nanosilica were dispersed in a polymer matrix. In polymer–silica hybrid films, silica formed hard segments inside a soft polymer matrix. SCAs facilitate miscibility of the polymer and silica. Highly transparent polyacrylic–silica hybrid thin films were obtained from MMA, trimethylol-propane-trimethacrylate, water-based monodispersed colloidal silica and coupling agent 3-(trimethoxysilyl)propyl methacrylate. A mixture of ethylene glycol and ethyl ether prevented the phase separation of the prepared hybrid materials. Silica domains in the hybrid film were in the range 20–40 nm, depending on the mole ratio of 3-(trimethoxysilyl)propyl methacrylate to colloidal silica. These hybrid films showed good thermal stability, hardness (in comparison with the PMMA/silica hybrid films), and a tunable refractive index, having potential applications as protective films for optical devices [111].

Covalent bonding of PMMA chains onto silica surface was achieved by grafting methods. PMMA–silica composites were prepared via sol–gel reaction. The base acrylic polymers were synthesized by copolymerization of MMA with 3-(trimethoxysilyl)propyl methacrylate or VTMS and obtained copolymers were then reacted with TEOS [112]. PMMA-grafted silica microparticles were obtained by radical photopolymerization of MMA, initiated from *N,N*-diethyldithiocarbamate groups previously anchored on the silica surface (using *grafting from* method) [113]. The same technique was also applied for grafting poly(octadecyl acrylate) (PODA) onto silica. Two different ATRP initiators were immobilized onto silica, and then ATRP was carried out from these initiators grafted onto silicas, giving comb-shaped polymer-grafted silica. The PODA–SiO_2 composites exhibited unique separation properties of long alkyl chains. Owing to a high selectivity toward polycyclic aromatic hydrocarbons (PAHs) in the ordered (crystalline) state, it was used as a stationary phase in high-performance liquid chromatography of PAHs [114].

SCAs alter the properties of material surfaces, which are modified by means of an organic functional group of specific silanes. Many kinds of hydrophobic SCA, including fluoro(trialkoxy)silanes, were applied for surface modification of silica-based and other materials. 3-Methacryloxypropyl trimethoxysilane was most often used as SCA for the surface treatment of fillers, especially in resin dental composites. Triethyleneglycol dimethacrylate and 2,2-bis [4-(2-hydroxy-3-methacryloxypropoxy)phenyl]propane were multifunctional monomers used as matrix resin in resin composites [115].

The silanization of nanosilica had an effect on the solvent (ethanol/water solution) sorption and solubility of dental composites [116]. Valuable reinforcing fillers for polymers are porous mesoporous silicas, for example, PMMA, namely, mesoporous silica was modified with Me_3SiCl/pyridine and the polymerization of MMA was carried out inside the silica pores. Some polymeric chains grew into the silica and the interaction of polymer with filler was increased. During melt blending and bulk polymerization, the silica pores were filled with polymer chains. The mechanical properties of this polymer composite were increased [117].

3.3.3 Modification of thermoplastic polymers with POSS reagents and fillers

POSS derivatives were used as versatile building blocks for the preparation of organic–inorganic hybrid polymers with tailored properties. The POSS molecules can be used as monomers, initiators, chain transfer agents (CTA) for a living polymerization, as well as modern fillers for polymer composites. Different POSS-containing homopolymers or copolymers were prepared from silsesquioxane monomers [118]. The POSS-containing hybrid homopolymers and triblock copolymers with the novel architecture were synthesized by a living and controlled-living polymerization of a POSS-based methacrylate monomer [119]. The star-shaped hybrid polymers were prepared by ATRP of MMA grafted from the octafunctional POSS initiators: octakis(2-bromo-2-methylpropionoxypropyldimethylsiloxy)octasilsesquioxane or the (octaethyl)-benzyl chloride analogue, derived from octakis(hydridodimethylsiloxy)octasilsesquioxane ($Q_8M^H{}_8$). Thermoplastic PMMA–POSS hybrid polymers with complete control of dispersion were obtained [120].

Tadpole-shaped organic–inorganic hybrid PNIPAAm was synthesized by RAFT polymerization of *N*-isopropylacrylamide (NIPAM) with the RAFT agent prepared from (aminopropyl)isobutyl polyhedral oligomeric silsesquioxane (POSS). Similarly, the organic–inorganic hybrid homopolymers and block copolymers were obtained by the RAFT polymerization of styrene [121].

The living/controlled polymerization technique was combined with *click chemistry*, based on the copper-catalyzed Huisgen 1,3-dipolar cycloaddition between alkynes and azides. The copper-catalyzed click reaction of alkyne-substituted POSS molecules and mono-, di-, and pentafunctional azido-terminal polymers made by ATRP gave monochelic (tadpole-shaped), ditelechelic (dumbbell-shaped) linear hybrid polymers, and pentatelechelic, star-shaped hybrids. Alternatively, the inverse procedure was also applied, that is, the reaction of azido-functional POSS molecule was "clicked" to alkyne-terminated polymers [122].

Dumbbell-shaped trifluoropropyl-substituted POSS derivatives substituted with aliphatic groups (ethyl, propyl, and hexyl) reduced their symmetries at low temperature. Due to their high symmetry and crystallinity, single POSS compounds formed optical transparent films (without cross-linking reagents), depending on the kind of aliphatic chains. These hybrid materials exhibited thermoplastic properties and low refractive index. They can be applied for coatings on different thermally unstable materials [123]. The star-shaped isobutyl-substituted caged silsesquioxane derivatives were prepared by the hydrosilylation of mono-vinyl-, allyl-, and octenylisobutyl-T8-silsesquioxane, respectively. They formed transparent or opaque films with high refractive indexes (~1.45) [124].

The synthesis of ladder poly(phenylsilsesquioxanes) (L-Ph-POSS) containing imide-building blocks as parts of the main parallel chains was performed for the first time by R. Chavez et al. The ladder structure of L-Ph-POSS was confirmed by SAXS measurements. L-Ph-POSS containing imide-building blocks exhibited high thermostability of up to 460 °C. Melting points of L-Ph-POSS were much lower than their decomposition temperatures, which might be very important for thermoplastic processing of these materials [125].

POSS molecules with functional groups such as octavinyl-POSS and octamethacrylate-POSS were used as additives to ethylene–propylene copolymer (EPM) and hydrogenated butadiene rubber (HNBR). Nanosilica Aerosil 380 and nanometric MMT were applied as reinforcing fillers. The POSS-modified EPM and HNBR compositions were prepared using melt mixing and they were chemically cross-linked with peroxides. The rheometric measurements indicated that octavinyl-POSS and octamethacrylate-POSS affected the curing behavior of rubbers due to increase of cross-link density. The POSS moieties participated via functional groups in curing reactions and cured rubber network formations. Significant enhancement of TS and modulus at 100% elongation (SE100) resulting from higher density of vulcanizates and nanometric reinforcement generated by grafted POSS particles was observed. The modified HNBR and EPM/HNBR vulcanizates containing methacrylate-POSS showed higher UV stability. The reinforcing effect of octamethacrylate-POSS on properties of EPM/HNBR blends was observed [126].

The surface properties of the most POSS molecules are hydrophobic. Modification of POSS molecules with the polar group gave oligomers compatible with many polymers. For instance, oligosilsesquioxane monomers functionalized with PEG (POSS-PEG) have hydrophilic properties and were soluble in water, thus they can be easily dispersed in polar monomers. The POSS oligomers containing hydroxyl groups served as initiators in the ring-opening polymerization of lactide, which was carried out in the presence of a stannous catalyst [1].

Photo-curable hybrid systems based on methacrylate-modified camelina oil (CO) were synthesized through a copolymerization reaction with hydrophilic dimethacrylated PEG macromonomers (with MW 300, 550, and 750 g/mol). The chain length of methacrylate-functionalized PEG directly influenced some key properties of the oil-based networks. Nanostructured POSS compounds bearing one or eight methacrylated groups were used as reagents and reinforcing fillers of oil-based networks and gave innovative organic–inorganic NCs. The number of methacrylic groups from the POSS cubes and the PEG chain length significantly affected the mechanical properties, for example, compression strength. Thus, lower PEG chain length was required to achieve higher compression strength for the final product. POSS monomers having more methacrylic groups led to a higher thermostability of the polymer networks [127].

Different kinds of hybrid polymer–POSS nanocomposites were described in the literature. POs, polyurethanes, polyamides, and epoxy resins were mainly used in these systems [128]. Many POSS-based polymers were recommended for biological and medical applications. Their studies have revealed desirable characteristics ranging from nontoxicity to biostability [129].

Star-shaped poly-L-lactide-PEG-polyhedral oligomeric silsesquioxane (SPPS) hybrid membranes were prepared by a photo-cross-linking method using inositol as a core. SPPS showed tunable mechanical properties (5.8–130 MPa in tensile modulus, 30–144% in elongation, beyond 90% recovery), good biomineralization activity, osteoblast biocompatibility, and biodegradation. These membrane hybrids can find future applications for guiding bone regeneration applications. Multiarmed biodegradable polymers have a great potential for biomedical applications, owing to their special structure and properties [130].

A series of hydrolytically degradable POSS–PEG hybrid hydrogels were prepared by covalently grafting hydrophobic POSS molecules onto hydrophilic star PEG-maleimide resin (M_n = 20 kDa) to form an amphipathic precursor which was further cross-linked with hydrolytically degradable PEG-diester-dithiol by Michael-type addition in aqueous phase. The gelation process was carried out without metal catalyst, UV irradiation, and organic solvents, which would be harmful to the cell. All the POSS–PEG hybrid hydrogels had a porous structure and high hydrophilic ability, and the grafted hydrophobic POSS as enhanced agents could tune the mechanical properties, swelling ratio, and degradation rate of the hydrogels [131].

POSS compounds are nontoxic and tasteless, and have good stability and biocompatibility. Thus, they can be used as “nano-building” units for the preparation of new organic–inorganic hybrid materials. POSS nanoparticles can be incorporated into hydrogels via simple physical blending, ionic interaction, and covalent bonding to prepare POSS hybrid hydrogels exhibiting the rigidity of inorganic materials and the processability of organic materials. These excellent POSS hybrid hydrogels can find applications as catalytic carriers, in drug delivery systems, in cell culture, and for bone regeneration [132].

Other examples of applications of POSS-modified thermoplastic polymers have been described in Section 2.5.1.

3.3.4 Composites based on epoxy resins modified with silanes, silicones, POSS, silica, and silicates

Properties of composites based on epoxy resins were very often effectively modified by the addition of reactive silanes, silicon containing monomers, polysiloxanes and other silicon polymers, silsesquioxanes, silica and silica-based fillers, MMT, other silicates, and numerous inorganic fillers. A substantial improvement of their mechanical properties, thermal and flame resistance, as well as very good corrosion resistance and antimicrobial protection was described in our extensive review [2]. Some newer examples are presented in this section.

PDMS with different molecular weights were used to modify diglycidyl ether of bisphenol A (DGEBA). A PDMS-*block*-DGEBA copolymer was prepared via condensation reactions between DGEBA, hydroxy-terminated PDMS, and SCA in the presence of a catalyst. The block copolymer played a role of a compatibilizer and PDMS with low molecular weight was well dispersed in the DGEBA matrix. The PDMS-modified DGEBA systems had higher impact strength and lower weight loss rate than pristine DGEBA [133].

NCs of epoxy-functionalized POSS derivatives such as the octaglycidyl and octaethylcyclohexenyl epoxide cubes, cured with diaminodiphenyl methane (DDM) showed increased thermal stability by 70 °C and T_g by about 60 °C, as compared to properties of similarly cured DGEBA epoxy resin, while their tensile properties were comparable to commodity epoxy resin systems [134].

The blends of silicone resin (SiR) containing Si–H, Si–CH=CH_2, and Si–OH groups with diglycidyl ether of hydrogenated bisphenol A (DGEHBA) were cured through platinum-catalyzed hydrosilylation and aluminum acetylacetonate-catalyzed polymerization. The ratio of SiR to DGEHBA was an important factor in the curing process. The Si–OH groups of SiR assisted polymerization of DGEHBA, reacted with the epoxy resin, and avoided phase separation. The cured hybrid resins were optically clear materials with a transmittance of about 87% at 400 nm for a thickness of 3 mm. UV resistance and thermal stability of these hybrids were highly dependent on their composition. The adhesive strength of the SiRs was significantly improved by a small fraction of DGEHBA with a marginal influence on the UV resistance. At the higher epoxy content, a marked negative effect on thermal stability was observed. The addition of thermal stabilizers was recommended for the high epoxy content in order to apply the hybrids for high-brightness light-emitting diode packaging [135].

Another type of epoxy resin, the tetraglycidyl-4,4′-diaminodiphenylmethane (TGDDM), was modified with phenyl silicone resins having amino groups (APSR) and was cured with (4,4′-diamino)diphenyl sulfone. The microstructure of the TGDDM/APSR resins was highly dependent on the content of amino groups of APSR and the content of the modifier. The toughness of the TGDDM resin was improved without decrease in the TS, the flexural strength, and the modulus. The thermal stability and water resistance were also improved, while the glass transition temperature decreased [136].

A silicone–epoxy hybrid resin was cured with APTES and APTES/APTMS mixture. The obtained T_g values of the cured resins were similar for both cure systems, although the use of APTMS as a curing agent in the mixture with APTES caused significant shortening of the curing time. The solvent and chemical resistance and the hardness and adhesion properties of the cured silicone–epoxy resin samples were excellent. However, the best properties were observed for the system cured with APTES hardener [137].

Cycloaliphatic epoxy oligosiloxane resin, synthesized by a nonhydrolytic sol–gel reaction, was cured with oxetane hardener and thermocationic polymerization initiator, giving epoxy–siloxane hybrimer, which showed excellent thermal resistance against yellowing. It was encapsulated on the blue LED without any cracks and distortion, exhibiting good light-emitting performance even after 360 h of thermal aging tests carried out at 120 °C and high humidity (85% at 85 °C). The epoxy–siloxane hybrimer may be useful as a high-performance LED encapsulant [138].

The steel surface pretreatment by the silane coatings containing functionalized graphene oxide (GO) nanofillers resulted in the increase of the adhesion strength and the decrease of the cathodic delamination and corrosion rate of top epoxy coating, while the addition of nonfunctionalized GO nanosheets did not have a significant effect. The GO nanofillers modified with different carbosilanes and TEOS significantly improved the corrosion resistance and adhesion properties of the top epoxy coating due to better compatibility with silane matrix, excellent barrier properties, and the formation of covalent bonds with the top epoxy coating [139, 140].

The anticorrosive epoxide NCs were obtained by modification of silicon nitride with methacryloxypropyl(trimethoxy)silane $CH_2{=}C(CH_3)COO(CH_2)_3Si(OCH_3)_3$ (KH-570). The dispersibility of silicon nitride in epoxy resin was enhanced after silane treatment. With the increase in immersion time, the corrosion resistance as well as adhesion strength of epoxy resin coating and unmodified silicon nitride coating decreased significantly. For the coating modified with silicon nitride, the corrosion resistance and adhesion strength still maintained 5.7×10^{10} Ωcm^2 and 7.6 MPa after 2,400 and 1,200 h immersion, respectively. The obtained NCs were applied to protect carbon steel from corrosion. The excellent corrosion resistance performance was observed, owing to the chemical interactions between functional groups from KH-570 and silicon nitride powders, which easily formed Si–O–Si bonds. The modified silicon nitride coating formed a strong barrier to corrosive electrolytes mainly due to the hydrophobicity of modified silicon nitride powder [141].

The silicone encapsulation materials with self-adhesion were prepared by hydrosilylation from epoxyphenylvinyl silicone resins (EPVS), which exhibited excellent adhesion properties as compared to commercial vinylsiloxane used for LED encapsulation, owing to epoxy groups, phenyl groups, and vinyl groups. The encapsulation materials based on EPVS resins showed excellent optical transparency, high thermal resistance, and superior mechanical properties. After thermal and UV light aging, the encapsulation materials also maintained high performance. Due to the excellent performance, the EPVS resins can be utilized in LED encapsulating formula [142].

The epoxy resin was also modified with a novel amine-functionalized hyperbranched polysiloxane (HBPSi) and cured with aromatic amine into cross-linked network without phase separation. At a 5 wt% HBPSi content, the TS, tensile modulus, and flexural modulus were increased by 17.6%, 13.7%, and 17.5%, respectively, compared with those of the neat epoxy resin. The elongation at break and impact strength were 63.3% and 49.1% higher than those of the neat epoxy resin, respectively. HBPSi also significantly improved thermal stability and increased the char yield of the material [143].

The epoxy-functionalized polysiloxane (PSE) was used to improve the cryogenic mechanical properties of bisphenol-F epoxy resin. The PSE had a better compatibility with epoxy resin. The cryogenic TS, failure strain, fracture toughness, and impact strength of the epoxy resin were improved significantly by adding the suitable amounts of PSE. Compared to the neat epoxy resin, the maximum TS (196.9 MPa, an improvement of 11.2%), failure strain (3.0%, the improvement of 33.8%), fracture toughness (3.1 $MPam^{1/2}$, the improvement of 30.7%), and impact strength (40.6 kJ/m^2, the improvement of 14.8%) at cryogenic temperature (90 K) were obtained by the incorporation 10 wt% PSE into the neat epoxy resin. The TS, Young's modulus, and fracture toughness of epoxy resin with the same PSE content at 90 K were higher than that at RT [144].

Copolymerization of a vinyl-containing bifunctional epoxy resin with vinyl-terminated polysiloxane (PDMS) via hydrosilylation reaction gave a silicone–epoxy block hybrid network with restricted micromorphology and outstanding macroscopic properties. The phase separation scale of PDMS/epoxy system was strictly restricted by a tailored molecular structure and the modified PDMS–epoxy hybrid polymer was transparent. The TS, elongation at break, and shear strength of the hybrid material were increased by 391%, 78%, and 292%, respectively. The initial thermal decomposition temperature of the hybrid material increased and gave quite high char residue at 800 °C. Thus, a transparent, high-performance PDMS–epoxy hybrid material can be applied for heat-resistant adhesives and transparent surface coatings [145].

Silica functionalized with ionic liquids (ILs) was prepared by modification of silica with 3-chloropropyl(trimethoxy)silane and 1-methylimidazole, which was carried out with silica dispersion in xylene at 80 °C for 12 h under reflux in nitrogen atmosphere, followed by drying under vacuum at 60 °C for 24 h (see Fig. 3.2). The IL supported on the silica (IL-silica) was used as a curing agent for an epoxy composite which showed improved dispersion and interfacial interaction. The IL-silica were highly dispersed in the epoxy resin, better than the unmodified silica. Thus, improved interfacial adhesion and enhanced overall mechanical and thermal properties of the epoxy composites were achieved [146].

Composites of epoxy resins, modified with octaaminophenyl POSS (OapPOSS) and polyphosphazene, cured with (4,4′-diamine)diphenylmethane (DDM) at 100–130 °C, showed significantly reduced peak of heat release rate and UL-94 V-0 rating as a flammability test. Polyphosphazene promoted char formation during combustion and released nonflammable gases such as CO_2, NH_3, and N_2 which diluted oxygen concentration and cooled the pyrolysis zone. Moreover, numerous phosphorus-containing species were formed during degradation and acted as free radical scavengers. Char residues were protected from thermal degradation by silica liberated from OapPOSS. These flame-retardant epoxy-based composites could find potential applications in the field of electrics and electronics [147].

The DOPO-POSS [POSS containing 9,10-dihydro-9-oxa-10-phosphaphenanthrene-10-oxide (DOPO)] and OPS/DOPO (octaphenyl polyhedral oligomeric silsesquioxane/DOPO) were used as flame retardants for TGDDM epoxy resins. These blends were cured with 4,4′-diaminodiphenylsulfone (DDS). The DOPO-POSS or OPS/DOPO exhibited the limited oxygen index (LOI) values of TGDDM/DDS resins greater than 34% and the UL-94 test reached V-0 rating. At the same content of Si and P elements in the epoxy resins, the OPS/DOPO showed better flame retardancy [148].

Epoxy–silica hybrid materials were prepared via in situ sol–gel process from bis (3-glycidyloxy)phenylphosphine oxide, TEOS, and diaminodiphenylmethane, in the presence of the catalyst *p*-toluenesulfonic acid. The glass transition temperatures of the hybrid epoxy composites increased with the silica content. The phosphorus–nanosilica synergistic effect on their flame-retardant properties gave the very high LOI increase up to a value of 44.5 [149].

A great number of other composites based on epoxy resins and containing POSS and phosphorous compounds showed also excellent flame-retardant properties [150].

Well-dispersed rubber nanodomains of silyl-diglycidyl ether-terminated PDMS in DGEBA epoxy were obtained by using a prepolymerization step. Rubber nanodomains of about 100 nm size were formed at the end of the curing reaction. Siloxane groups covalently attached to DGEBA led to flexibility of the molecular network and the T_g of the epoxy matrix significantly decreased. The dispersed siloxane phase at the nanoscale induced hydrophobic behavior in the NC and increased the mechanical properties (i.e., modulus, hardness, and toughness). The final NC also showed higher thermal stability and formed a char [151].

Another epoxy-containing silicone rubber (SR) network was prepared by hydrosilylation reaction between the epoxy resin prepolymer containing methacryloxy group, vinyl-terminated silicone oil, and PMHS oil. The epoxy-silane-grafted prepolymer was prepared from DGEBA and methacryloxypropyl(trimethoxy)silane at 130 °C in the presence of titanium tetraisopropoxide. The compatibility between

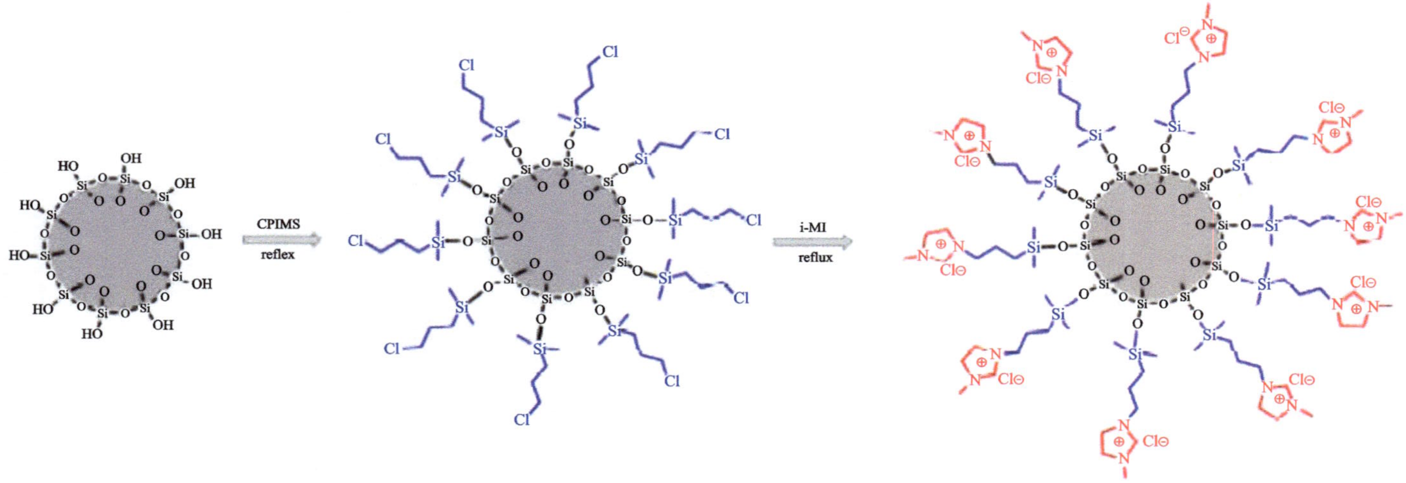

Figure 3.2: Scheme of supported ionic liquid silica (IL-silica). Reprinted (adapted) with permission from [146]. Copyright 2017, MDPI, Switzerland.

SR and epoxy resin was enhanced as a result of the good dispersion of vinyl-containing epoxy prepolymers in SR matrix. The adhesion and TSs of SRs modified with epoxy-silane prepolymer were greatly enhanced when compared with those of unmodified substrates. The formation of the epoxy-containing SR network altered the thermal degradation process of SR, giving a higher residue at 800 °C under nitrogen atmosphere [152].

The self-healing compositions of DGEBA epoxy matrix contained microcapsules filled with as a new healing agent, amine-functionalized PDMS (PDMS-a), and microcapsules filled with triethylenetetramine. Fractographic analysis showed the rupture of microcapsules and release of the healing agent, inducing a change in the mirror-like fracture plane morphology of the neat epoxy and healing of cracks. PDMS-a reduced the stiffness of the epoxy matrix, resulting in greater energy release rate values [153].

New amino-functionalized silica composites and nanoparticles were prepared by cohydrolysis of microgel based on NIPAM-*co*-vinyltrimethoxysilane (NIPAM-VTMS) with TEOS and APTES. The contents of amino group were increased on the surfaces of silica-containing composites. The amino-functionalized calcined silica nanoparticles and NIPAM-VTMS-polysiloxane showed better dispersion capability than silica NP particles existed in the epoxy hybrid coatings. The amino-functionalized hybrid silica epoxy coatings had good adhesion, and mechanical and anticorrosive properties as compared to hybrid materials without amino-functionalization. The amino-functionalized NIPAM-VTMS-polysiloxane were blended with epoxy matrix and showed smart self-healing performance at weight content at least 6 wt% after exposure to salt spray humidity of seawater. Their self-healing properties probably resulted from the salt sensitivity of amino-functionalized NIPAM-VTMS-polysiloxane composite and improvement of the adhesion strength due to the formation of Fe–O–Si and Fe–NH–Si covalent bonds at the interface of epoxy coating [154].

References

[1] J.J. Chruściel and E. Leśniak, Chapter 9: Modification of thermoplastics with reactive silanes and siloxanes, In: Thermoplastic Elastomers, ISBN 979-953-307-353-7, InTech-Open Science, Rijeka, Croatia, 155–192, on-line: 28. 03.2012.

[2] J.J. Chruściel and E. Leśniak, Modification of epoxy resins with functional silanes, polysiloxanes, silsesquioxanes, silica and silicates, *Progress in Polymer Science*, 2015, 41, 67–121.

[3] C.J. Brinker and G.W. Scherer, *Sol-Gel Science: The Physics and Chemistry of Sol-Gel Processing*, Academic Press, 1990.

[4] H. Schmidt, Organically modified silicates by sol-gel processes, *Materials Research Society Symposia Proceedings*, 1984, 32, 327.

[5] H. Schmidt, Inorganic-organic composites by sol-gel techniques, *Journal of Sol-Gel Science and Technology*, 1994, 1, 217–231. ISSN 0928-0707.

[6] Y. Abe and T. Gunji, Oligo- and polysiloxanes, *Progress in Polymer Science*, 2004, 29, 149–182. ISSN 0079-6700.
[7] N.L.D. Filho, H.A. De Aquino, G. Pires and L. Caetano, Relationship between the dielectric and mechanical properties and the ratio of epoxy resin to hardener of the hybrid thermosetting polymers, *The Brazilian Society of Chemistry*, 2006, 17, 533–541.
[8] T. Nedelčev, I. Krupa, K. Csomorová, I. Janigová and J. Rychlý, Synthesis and characterization of the new silane-based antioxidant containing 2,6-di-*tert*-butylphenolic stabilizing moiety, *Polymers for Advanced Technologies*, 2007, 18, 157–164.
[9] B. Arkles, Tailoring surfaces with silanes, *Chemtech*, 1977, 7, 766–780.
[10] Y. Nomura, A. Sato, S. Sato, H. Mori and T. Endo, Synthesis of novel moisture-curable polyurethanes end-capped with trialkoxysilane and their application to one-component adhesives, *Journal of Polymer Science. Part A, Polymer Chemistry*, 2007, 45, 2689–2704.
[11] C.D. Han, T. Van De Weghe, P. Shete and J.R. Haw, Effects of coupling agents on the rheological properties, processability, and mechanical properties of filled poly-propylene, *Polymer Engineering & Science*, 1981, 21, 196–204.
[12] K. Okuno, R.T. Woodhams and P.P. Mica Reinforced, *Polymer Engineering & Science*, 1975, 15, 308–315.
[13] C. Scott, H. Ishida and F.H.J. Maurer, Infrared analysis and Izod impact testing multicomponent polymer composites: Polyethylene/EPDM/filler systems, *Journal of Materials Science*, 1987, 22, 3963–3973.
[14] J.P. Trotignon, J. Verdu, R. Boissard and A. Vallois, *Polymer Composites*, Walter de Gruyter, Berlin, 1986, 191.
[15] Z. Demjen and B. Pukanszky, Effect of surface coverage of silane treated $CaCO_3$, on the tensile properties of polypropylene composites, *Polymer Composites*, 1997, 18, 741–746.
[16] Z. Demjen, B. Pukanszky and J. Nagy, Evaluation of interfacial interaction in polypropylene/surface treated $CaCO_3$ composites, *Composites, Part A*, 1998, 29, 323–329.
[17] Z. Demjen, B. Pukanszky and J. Nagy Jr., Possible coupling reactions of functional silanes and polypropylene, *Polymer*, 1999, 40, 1763–1773.
[18] B. Pukanszky, Interfaces and interphases in multicomponent materials: Past, present, future, *European Polymer Journal*, 2005, 41, 645–662.
[19] M.I.H. Mondal, M.K. Islam and F. Ahmed, Textile performance of functionalized cotton fiber with 3-Glycidoxypropyltriethoxysilane, *Journal of Textile Engineering*, 2018, 8, 337. DOI: 10.4172/2165-8064.1000337.
[20] D. Munteanu, *Chapter 5: Moisture cross-linkable silane-modified polyolefins*, In: *Reactive Modifiers for Polymers*, S. Al-Malaika (Ed.), Blackie Academic & Professional, London-New York, 1997, 196–265.
[21] J. Morshedian and P.M. Hoseinpour, Polyethylene cross-linking by two-step silane method: A review, *Iranian Polymer Journal*, 2009, 18, 103–128. http://journal.ippi.ac.ir.
[22] B. Thomas and M. Bowrey, *Wire Journal*, 1977, 10, 88.
[23] P. Swarbrick, W.J. Green and C. Maillefer, Manufacture of extruded product, US Pat. 4,117,195 (1978).
[24] D.C. Eagles, Water crosslinkable copolymer, *Wire Technology International*, 1990, 18(15).
[25] A. Smedberg, T. Hjeftberg and B. Gustafsson, Characterisation and crosslinking properties of a poly-(ethylene-*co*-divinylsiloxane), *Polymer*, 2004, 45, 4845–4855.
[26] H. Malz and C. Tzoganakis, Hydrosilylation of terminal double bonds in polypropylene through reactive processing, *Polymer Engineering & Science*, 1998, 38, 1976–1984.
[27] J. Long, C. Tzoganakis and P. Chen, Surface characteristics of hydrosilylated polypropylene, *Journal of Applied Polymer Science*, 2003, 88, 3117–3131.

[28] J. Long, C. Tzoganakis and P. Chen, Surface characteristics of hydrosilylated poly-propylenes: Effect of co-catalyst and reaction temperature, *Polymer Engineering & Science*, 2004, 44, 56–71.

[29] S.-H. Zhu, N.T. McManus, C. Tzoganakis and A. Penlidis, Effect of a Poly(dimethyl-siloxane) modified polyolefin additive on the processing and surface properties of LLDPE, *Polymer Engineering & Science*, 2007, 47, 1309–1316.

[30] S.-H. Zhu and C. Tzoganakis, Surface properties of hydrosilylated polyolefins annealed in supercritical carbon dioxide, *Polymers for Advanced Technologies*, 2008, 19, 258–269.

[31] W. Zhang, S. Deodhar and D. Yao, Processing properties of polypropylene with a minor addition of silicone oil, *Polymer Engineering & Science*, 2010, 50, 1340–1349.

[32] A. Tsuchida, C. Bolin, F.G. Sernetz, H. Frey and R. Mülhaupt, Ethene and propene copolymers containing silsesquioxane side groups, *Macromolecules*, 1997, 30, 2818–2824.

[33] J. Kerres and H. Strathmann, Synthesis and properties of a new AB-cross-linked copolymer membrane system, *Journal of Applied Polymer Science*, 1993, 50, 1405–1421.

[34] W. Kujawski, J. Kermes and R. Roszak, Application of AB-crosslinked polymers composed of styrene/isoprene–siloxane copolymers to pervaporative removal of volatile organic compounds from water, *Journal of Membrane Science*, 2003, 218, 211–218.

[35] X. Guo, R. Farwaha and G.L. Rempel, Catalytic hydrosilylation of diene-based polymers. 1. Hydro-silylation of polybutadiene, *Macromolecules*, 1990, 23, 5047–5054.

[36] A. Iraqi, S. Seth, C.A. Vincent, D.J. Cole-Hamilton, M.D. Watkinson, I.M. Graham and D. Jeffrey, Catalytic hydrosilation of polybutadienes as a route to functional polymers, *Journal of Materials Chemistry*, 1992, 2, 1057–1064.

[37] B.P.S. Chauhan and B. Balagam, Silyl functionalization of polyolefins, *Macromolecules*, 2006, 39, 2010–2012. ISSN 0024-9297.

[38] B.P.S. Chauhan, B. Balagam, M. Raghunath, A. Sarkar and E. Cinar, Pt-Nanocluster catalysts and their practical utility in generation of functionalized organosilicon polymers, *Polymer Preprints*, 2008, 49(2), 897–898.

[39] N.T. McManus, S.-H. Zhu, C. Tzoganakis and A. Penlidis, Grafting of ethylene–ethyl acrylate–maleic anhydride terpolymer with amino-terminated polydimethylsiloxane during reactive processing, *Journal of Applied Polymer Science*, 2006, 101, 4230–4237.

[40] K. Zhang, L. Gao and Y. Chen, Organic-inorganic hybrid materials by self-gelation of block copolymer assembly and nanoobjects with controlled shapes thereof, *Macromolecules*, 2007, 40(16), 5916–5922.

[41] C.G. Gamys, E. Beyou and E. Bourgeat-Lami, Micellar behavior of well-defined polystyrene-based block copolymers with triethoxysilyl reactive groups and their hydrolysis-condensation, *Journal of Polymer Science. Part A, Polymer Chemistry*, 2010, 48(4), 784–793.

[42] K. Ni, E. Bourgeat-Lami, N. Sheibat-Othman, G. Shan and G. Fevotte, Preparation of hybrid nanocapsules, *Macromolecular Symposia*, 2008, 271(1), 120–128.

[43] L.-Y. Shi, Z. Shen and X.-H. Fan, Order-order transition in a rod-coil diblock copolymer induced by supercritical CO_2, *Macromolecules*, 2011, 44(8), 2900–2907.

[44] T. Wan, C. Wu, X.-L. Ma, L. Wang, J. Yao and K. Lu, Microstructure and properties of silane monomer-modified styrene-acrylate nanocoatings, *Polymer Bulletin*, 2009, 62(6), 801–811.

[45] A.H. Fawcett, A.B. Foster, M. Hania, M. Hohn, G.O. McCaffery, J.L. Mazebedi and E. Mullen, *Polymer Preprints*, 2001, 42(1), 213.

[46] M.G. Assadi, M. Makham and Z. Tajrezaiy, Synthesis and characterization of some organosilicon derivatives of poly 2-hydroxyethyl methacrylate with cubane as a cross-linking agent, *Journal of Organometallic Chemistry*, 2005, 690, 4755–4760.

[47] R. Rodriguez, M.J. Barandiaran and J.M. Asua, Polymerization strategies to overcome limiting monomer concentration in silicone-acrylic miniemulsion polymerization, *Polymer*, 2008, 49, 691–696.

[48] D. Nagao, M. Hashimoto, K. Hayasaka and M. Konno, Synthesis of anisotropic polymer particles with soap-free emulsion polymerization in the presence of reactive silane coupling agent, *Macromolecular Rapid Communications*, 2008, 29, 1484–1488.
[49] K. Daimatsu, Y. Anno, H. Sugimoto, E. Nakanishi, K. Inomata, T. Ikeda and K. Yokoi, Preparation, morphology, and physical properties of transparent PSt hybrid materials containing silicone macromonomer, *Journal of Applied Polymer Science*, 2008, 108, 362–369.
[50] N. Oz and A. Akar, Redox initiation system of ceric salt and α,ω-dihydroxy-poly (dimethylsiloxane)s for vinyl polymerization, *Journal of Applied Polymer Science*, 2006, 102, 2112–2116.
[51] J.-T. Zhang, C.-J. Pan, T. Keller, R. Bhat, M. Gottschaldt, U.S. Schubert and K.D. Jandt, Monodisperse, temperature sensitive microgels crosslinked by Si-O-Si bonds, *Macromolecular Materials and Engineering*, 2009, 294, 396–404.
[52] Y.-P. Wang, K. Yuan, Q.-L. Li, L.-P. Wang, S.-J. Gu and X.-W. Pei, Preparation and characterization of poly(N-isopropylacrylamide) films on a modified glass surface via surface initiated redox polymerization, *Materials Letters*, 2005, 59, 1736–1740.
[53] D. Pavlovic, J.G. Linhardt, J.F. Kunzler and D.A. Shipp, Synthesis of amphiphilic multiblock and triblock copolymers of PDMS and poly(N,N-dimethacrylamide), *Journal of Polymer Science. Part A, Polymer Chemistry*, 2008, 46, 7033–7048.
[54] Z. Luo, T. He, H. Yu, L. Dai and A. Novel, ABC triblock copolymer with very low surface energy: Poly(dimethylsiloxane)-*block*-poly(methyl methacrylate)-*block*-poly(2,2,3,3,4,4,4-heptafluorobutyl-methacrylate), *Macromolecular Reaction Engineering*, 2008, 2, 398–406.
[55] W. Darras, O. Fichet, F. Perrot, S. Boileau and D. Tyssie, Polysiloxane-poly(fluorinated acrylate) interpenetrating polymer networks: Synthesis and characterization, *Polymer*, 2007, 48, 687–695.
[56] E. Martinelli, M.K. Sarvothaman, G. Galli, M.E. Pettitt, M.E. Callow, J.A. Callow, S.L. Conlan, A.S. Clare, A.B. Sugiharto, C. Davies and D. Williams, Poly(dimethyl siloxane) (PDMS) network blends of amphiphilic acrylic copolymers with poly(ethylene glycol)-fluoroalkyl side chains for fouling-release coatings. II. Laboratory assays and field immersion trials, Biofouling, 2012, 28(6), 571–582.
[57] M. Ohnishi, T. Uno, M. Kubo and T. Itoh, Synthesis and radical polymerization of dissymmetric fumarates with alkoxyethyl and bulky siloxy groups, *Journal of Polymer Science: Part A; Polymer Chemistry*, 2009, 47, 420–432.
[58] E. Pouget, E.H. Garcia and F. Ganachaud, Direct synthesis of PVA-g-PDMS in microsuspension, *Macromolecular Rapid Communications*, 2008, 29, 425–430.
[59] J. Tonnar, E. Pouget, P. Lacroix-Desmazes and B. Boutevin, Synthesis of poly(vinyl acetate)-*b*-PDMS-b-poly(vinyl acetate) triblock copolymers by iodine transfer polymerization, *European Polymer Journal*, 2008, 44, 318–328.
[60] Y. Chen and H. Kim, Poly(vinylidene fluoride) grafted with 3-trimethoxysilylpropyl-methacrylate for silyl functional membranes, *Reactive & Functional Polymers*, 2008, 68, 1499–1506.
[61] J. Guiot, B. Ameduri, B. Boutevin and T. Lannuzel, Original crosslinking of poly(vinylidene fluoride) via trialkoxysilane containing cure-site monomers, *Journal of Polymer Science. Part A, Polymer Chemistry*, 2006, 44, 3896–3910.
[62] J. Kozakiewicz, I. Ofat and J. Trzaskowska, Silicone-containing aqueous polymer dispersions with hybrid particle structure, *Advances in Colloid and Interface Science*, 2015, 223, 1–39.
[63] Y. Poojari and S.J. Clarson, Lipase-catalyzed synthesis and properties of silicone aromatic polyesters and silicone aromatic polyamides, *Macromolecules*, 2010, 43, 4616–4622.

[64] L. Guo, Z. Zhang, Y. Zhu, J. Li and Z. Xie, Synthesis of polysiloxane-polyester copolymer by lipase-catalyzed polycondensation, *Journal of Applied Polymer Science*, 2008, 108, 1901–1907.

[65] V.V. Antic, M.V. Pergal, M.N. Govedarica, M.P. Antic and J. Djonlagic, Copolymers based on poly(butylene terephthalate) and polycaprolactone-*b*-PDMS-*b*-polycaprolactone, *Polymer International*, 2010, 59, 796–807.

[66] C.A. Schoener, C.B. Weyand, R. Murthy and M.A. Grunlan, Shape memory polymers with silicon-containing segments, *Journal of Materials Chemistry*, 2010, 20, 1787–1793.

[67] G.-X. Chen and H. Shimizu, Multiwalled carbon nanotubes grafted with POSS and its dispersion in poly(L-lactide) matrix, *Polymer*, 2008, 49(4), 943–951.

[68] X. Shi, Z. Chen and Y. Yang, Toughening of poly(L-lactide) with methyl MQ silicone resin, *European Polyme Journal*, 2014, 50, 243–248.

[69] S.-R. Kim and D.-J. Kim, Effect of silicone oil on the morphology and properties of polycarbonate, *Journal of Applied Polymer Science*, 2008, 109, 3439–3446.

[70] H.A.M. Van Aert, L. Nelissen, P.J. Lemstra and D.J. Brunelle, Poly(Bisphenol A carbonate)-PDMS multiblock copolymers, *Polymer*, 2001, 42, 1781–1788.

[71] D.G. LeGrand, *US Pat.* 3 679 774 (1972).

[72] M. Wang, B. Li, J. Wang and P. Bai, Preparation and properties of polysiloxane grafting multi-walled carbon nanotubes/polycarbonate nanocomposites, *Polymers for Advanced Technologies*, 2011, 22, 1738–1746. DOI: 10.1002/pat.1665.

[73] Y. Yoshioka, Preparation of poly(amino-amide) particles complexed with ZnO particles using silane coupling agents, *Journal of Polymer Science. Part A, Polymer Chemistry*, 2009, 47(19), 4908–4918.

[74] J. Alongi, O. Monticelli, S. Russo, J. Galy and J.-F. Gerard, Morphology of inorganic-organic systems consisting of POSS and polyamide 6, *e-Polymers*, 2009, No. 028.

[75] C.-K. Ku and Y.-D. Lee, Microphase separation in amorphous poly(imide siloxane) segmented copolymers, *Polymer*, 2007, 48, 3565–3573.

[76] C. Lu, Z. Wang, F. Liu, J. Yan and L. Gao, Microstructure and properties of new polyimide/siloxane composite films, *Journal of Applied Polymer Science*, 2006, 100, 124–132.

[77] T. Suzuki, Y. Yamada and K. Itahashi, 6FDA-TAPOB hyperbranched polyimide-silica hybrids for gas separation membranes, *Journal of Applied Polymer Science*, 2008, 109, 813–819.

[78] M. Krea, D. Roizard, N. Moulai-Mostefa and D. Sacco, New copolyimide membranes with high siloxane content designed to remove polar organics from water by pervaporation, *Journal of Membrane Science*, 2004, 241, 55–64.

[79] H. Yan, P. Li, R. Ning, X. Ma and Z. Zhang, Tribological properties of bismaleimide composites with surface-modified SiO_2 nanoparticles, *Journal of Applied Polymer Science*, 2008, 110, 1375–1381.

[80] B. Arkles, Look what you can make out of silicones (biomedical applications of silicones), *Chemtech*, 1983, 13, 542–555.

[81] J.P. Sheth, E. Yilgor, B. Erenturk, H. Ozhalici, I. Yilgor and G.L. Wilkes, Structure-property behavior of PDMS based segmented polyurea copolymers modified with poly(propylene oxide), *Polymer*, 2005, 46(19), 8185–8193.

[82] R.-S. Chen, C.-J. Chang and Y.-H. Chang, Study on siloxane-modified polyurethane dispersions from various PDMS, *Journal of Polymer Science. Part A, Polymer Chemistry*, 2005, 43, 3482–3490.

[83] A.K. Nanda, D.A. Wicks, S.A. Madbouly and J.U. Otaigbe, Nanostructured polyurethane/POSS hybrid aqueous dispersions prepared by homogeneous solution polymerization, *Macromolecules*, 2006, 30, 7037–7043.

[84] D.P. Queiroz and M.N. De Pinho, Structural characteristics and gas permeation of PDMS/Poly (Propylene Oxide) Urethane/urea bi-soft segment membranes, *Polymer*, 2005, 46, 2346–2353.
[85] S. Dassin, M. Dumon, F. Mechin and J.-P. Pascault, Thermoplastic polyurethanes (TPUs) with grafted organosilane moieties: A new way of improving thermo-mechanical behavior, *Polymer Engineering and Science*, 2002, 42, 1724–1739.
[86] X. Gu, N. Li, H. Gu, X. Xia and J. Xiong, Polydimethylsiloxane-modified polyurethane–poly(ε-capro-lactone) nanofibrous membranes for waterproof, breathable applications, *Journal of Applied Polymer Science*, 2018, 135(23), 46360.
[87] E. Kalaycioglu, L. Toppare, Y. Yagci, V. Harabagiu, M. Pintela, R. Ardelean and B.C. Simionescu, Synthesis of conducting H-type polysiloxane-polypyrrole block copolymers, *Synthetic Metals*, 1998, 97, 7–12.
[88] N. Sombatsompop, E. Wimolmala and T. Markpin, Fly-ash particles and precipitated silica as fillers in rubbers. II. Effects of silica content and Si69-Treatment in natural rubber/styrene–butadiene rubber vulcanizates, *Journal of Applied Polymer Science*, 2007, 104(5), 3396–3405.
[89] M. Zaborski, Nanotechnology and polymeric nano-composites, *Przemysl Chemiczny*, 2003, 82(8–9), 544–547.
[90] S.C. Sahoo, A. Sil, K. Thanigai and C.N. Pandey, Use of silicone based coating for protection of wood materials and bamboo composites from weathering and UV degradation, *Journal of the Indian Academy of Wood Science*, 2011, 8, 143–147.
[91] M.S. Skoc, J. Macan and E. Pezelj, Modification of polyurethane-coated fabrics by sol–gel thin films, *Journal of Applied Polymer Science*, 2014, 131, 39914.
[92] Z.W. Kornetka, *Polimery*, 1987, 32, 104–109.
[93] J. Chruściel, Proadhezyjne silany i tytaniany w technologii tworzyw sztucznych i gumy, *Rynek Chemistry*, 1997, 11–12, 31–34. In polish.
[94] E.P. Pleuddeman, *Silane Coupling Agents*, Plenum Press, New York, USA, 1991.
[95] M. Xanthos, *Functional Fillers for Plastics*, J. Wiley & Sons, 2010.
[96] K.H. Haas and H. Wolter, Synthesis, properties and applications of inorganic–organic copolymers (ORMOCER®s), *Current Opinion in Solid State & Materials Science*, 1999, 4, 571–580.
[97] P. Botella, A. Corma and M.T. Navarro, Single gold nanoparticles encapsulated in monodispersed regular spheres of mesostructured silica produced by pseudo-morphic transformation, *Chemistry of Materials*, 2007, 19, 1979–1983.
[98] R. De Palma, S. Peeters, M.J. Van Bael, H. Van Den Rul, K. Bonroy, W. Laureyn, J. Mullens, G. Borghs and G. Maes, Silane ligand exchange to make hydrophobic superparamagnetic nanoparticles water-dispersible, *Chemistry of Materials*, 2007, 19, 1821–1831.
[99] R. Xia, M. Li, Y. Zhang, Q. Zhu and J. Qian, Synthesis of tercopolymer BA–MMA–VTES and surface modification of nano-size Si_3N_4 with this macromolecular coupling agent, *Journal of Applied Polymer Science*, 2008, 107, 1100–1107.
[100] Z. Li, Y. Liang, R. Fu, D. Wu and K. Matyjaszewski, Synthesis of well-defined microporous carbons by molecular-scale templating with POSS sacrificial moieties, The *Journal of the American Chemical Society*, 2014, 136(13), 4805–4808.
[101] B. Szadkowski, A. Marzec and M. Zaborski, Effect of *in situ* silanization of multiwalled carbon nanotubes on the properties of NBR/MWCNTOH composites, *Polymer-Plastics Technology and Materials*, 2019, 58(12), 1327–1341.
[102] K. Chrissopoulou and S.H. Anastasiadis, Polyolefin/layered silicate nanocomposites with functional compatibilizers, *European Polymer Journal*, 2011, 47, 600–613.
[103] M. Bailly and M. Kontopoulou, Preparation and characterization of thermoplastic olefin/ nanosilica composites using a silane-grafted PP matrix, *Polymer*, 2009, 50, 2472–2480.

[104] C.M. Liauw, G.C. Lees, S.J. Hurst, R.N. Rothon and D.C. Dobson, The effect of filler Surface modification on the mechanical properties of aluminium hydroxide filled polypropylene, *Plastics, Rubber, Composites Processing & Applications*, 1995, 24, 249–260.
[105] L.F. Cai, X.B. Huang, M.Z. Rong, W.H. Ruan and M.Q. Zhang, Effect of grafted polymeric foaming agent on the structure and properties of nanosilica/PP composites, *Polymer*, 2006, 47, 7043–7050.
[106] C. Lu, Z. Wang, F. Liu, J. Yan and L. Gao, Microstructure and properties of new polyimide/siloxane composite films, *Journal of Applied Polymer Science*, 2006, 100, 124–132.
[107] S. Sanchez-Valdes, J. Mendez-Nonell, F.J. Medellin-Rodriguez, E. Ramirez-Vargas, J.G. Martinez-Colunga, H. Soto-Valdez, L. Munoz-Jimenez and G. Neira-Velazquez, Effect of PEgMA/Amine silane compatibilizer on clay dispersion of PE-clay nanocomposites, *Polymer Bulletin*, 2009, 63, 921–933.
[108] W. Ma, F. Gong, C. Liu, G. Tao, J. Xu and B. Jiang, SiO_2 reinforced HDPE hybrid materials obtained by the sol–gel method, *Journal of Applied Polymer Science*, 2014, 131, 39891.
[109] Y.-D. Shen, Y.-N. Zhao and X.-R. Li, Polyacrylate/silica hybrids prepared by emulsifier-free emulsion polymerization and the sol-gel process, *Polymer Bulletin*, 2009, 63, 687–698.
[110] X. Cheng, M. Chen, S. Zhou and L. Wu, Preparation of SiO_2/PMMA composite particles via conventional emulsion polymerization, *Journal of Polymer Science. Part A, Polymer Chemistry*, 2006, 44(12), 3807–3816.
[111] -Y.-Y. Yu and W.-C. Chen, Transparent organic-inorganic hybrid thin films prepared from acrylic polymer and aqueous monodispersed colloidal silica, *Materials Chemistry and Physics*, 2003, 82, 388–395.
[112] W. Brostow, T. Datashvili and K. Hackenberg, synthesis and characterization of poly(methyl acrylate) + SiO_2 hybrids. *e-Polymers*, No. 054 (2008).
[113] D. Derouet and C.N.H. Thuc, Synthesis and characterization of poly(methyl methacrylate)-grafted silica microparticles, *Journal Applied Polymer Science*, 2008, 109, 2113–2127.
[114] A.K. Malik, M.M. Rahman, M. Czaun, M. Takafuji and H. Ihara, Facile synthesis of high-density poly(octadecyl acrylate)-grafted silica for reversed-phase high-performance liquid chromatography by surface-initiated atom transfer radical polymerization, *Journal of Chromatography A*, 2008, 1187(1–2), 119–127.
[115] T. Nihei, Dental applications for silane coupling agents, *Journal of Oral Science*, 2016, 58(2), 151–155.
[116] M.M. Karabela and I.D. Sideridou, Effect of the structure of silane coupling agent on sorption characteristics of solvents by dental resin-nanocomposites, *Dental Materials*, 2008, 24, 1631–1639.
[117] L.D. Pérez, L.F. Giraldo, W. Brostow and B.L. López, Poly(methyl acrylate) plus mesoporous silica nanohybrids: Mechanical and thermophysical properties, *e-Polymers*, No. 029 (2007).
[118] J.D. Lichtenhan, Y.A. Otonari and M.J. Carr, Linear hybrid polymer building blocks: Methacrylate-functionalized polyhedral oligomeric silsesquioxane monomers and polymers, *Macromolecules*, 1995, 28, 8435–8450.
[119] J. Pyun, K. Matyjaszewski, J. Wu, G.M. Kim, S.B. Chun and P.T. Mather, ABA triblock copolymers containing polyhedral oligomeric silsesquioxane pendant groups: Synthesis and unique properties, *Polymer*, 2003, 44, 2739–2750.
[120] R.O.R. Costa, W.L. Vasconcelos, R. Tamali and R.M. Laine, Organic/inorganic nanocomposite star polymers via ATRP of methyl methacrylate using octa-functional silsesquioxane cores, *Macromolecules*, 2001, 34, 5398–5407.
[121] W. Zhang, X. Zhung, X. Li, Y. lin, J. Bai and Y. Chen, Preparation and characterization of organic-inorganic hybrid polymers containing polyhedral oligomeric silsesquioxane via RAFT polymerization, *Reactive & Functional Polymers*, 2009, 69, 124–129.

[122] W. Zhang and A.H.E. Müller, Synthesis of tadpole-shaped POSS-containing hybrid polymers via “click chemistry”, *Polymer*, 2010, 51, 2133–2139.
[123] H. Araki and K. Naka, Synthesis of dumbbell-shaped trifluoropropyl-substituted POSS derivatives linked by simple aliphatic chains and their optical transparent thermoplastic films, *Macromolecules*, 2011, 45, 6039–6045.
[124] Y. Yasumoto, T. Yamanaka, S. Sakurai, H. Imoto and K. Naka, Design of low-crystalline and low-density isobutyl-substituted caged silsesquioxane derivatives by star-shaped architectures linked with short aliphatic chains, *Polymer Journal*, 2016, 48, 281–287. DOI: 10.1038/pj.2015.114.
[125] R. Chavez, E. Ionescu, C. Fasel and R. Riedel, Imide-containing ladder polyphenylsilsesquioxanes with high thermal stability and thermoplastic properties, *Journal of Applied Polymer Science*, 2014, 131, 40085.
[126] M. Lipińska and M. Zaborski, Influence of POSS addition on properties of ethylene-propylene EPM and hydrogenated butadiene-acrylonitrile HNBR composites containing layered silicate and nanometric silica, *International Research Journal of Advanced Engineering and Science*, 2017, 2(4), 220–230.
[127] B. Balanuca, A. Lungu, A.-M. Hanganu, L.R. Stan, E. Vasile and H. Iovu, Hybrid nanocomposites based on POSS and networks of methacrylated camelina oil and various PEG derivatives, *The European Journal of Lipid Science and Technology*, 2014, 116, 458–469.
[128] S. Kalia and K. Pielichowski, Polymer/POSS Nanocomposites and Hybrid Materials, Preparation, Properties, Applications, Springer Nature Switzerland AG 201.
[129] J. Wu and P.T. Mather, POSS polymers: Physical properties and biomaterials applications, *Polymer Reviews*, 2009, 49(1), 25–63.
[130] M. Xie, J. Ge, B. Lei, Q. Zhang, X. Chen and P.X. Ma, Star-shaped, biodegradable, and elastomeric PLLA-PEG-POSS hybrid membrane with biomineralization activity for guiding bone tissue regeneration, *Macromolecular Bioscience*, 2015, 15, 1656–1662.
[131] J. Shen, H. Li, C. Lu, G. Yang, F. Wang, J. Nie, X. Hu, N. Dong and J. Shi, Hydrolytically degradable POSS-PEG hybrid hydrogels prepared in aqueous phase with tunable mechanical properties, swelling ratio and degradation rate, *Reactive & Functional Polymers*, 2018, 123, 91–96.
[132] S. Liu, R. Guo, C. Li, C. Lu, G. Yang, F. Wang, J. Nie, C. Ma and M. Gao, POSS hybrid hydrogels: A brief review of synthesis, properties and applications, *European Polyme Journal*, 2021, 143, 110180.
[133] F. Zhao, Q. Sun, D.P. Fang and K.D. Yao, Preparation and properties of polydimethylsiloxane-modified epoxy resins, *Journal of Applied Polymer Science*, 2000, 76, 1683–1690.
[134] R.M. Laine, J. Choi and I. Lee, Organic-inorganic nanocomposites with completely defined interfacial interactions, *Advanced Materials*, 2001, 13(11), 800–803.
[135] Y. Zhang, X. Yang, X. Zhao and W. Huang, Synthesis and properties of optically clear silicone resin/epoxy resin hybrids, *Polymer International*, 2012, 61, 294–300.
[136] Y. Zhang, C. Shang, X. Yang, X. Zhao and W. Huang, Morphology and properties of TGDDM/DDS epoxy systems toughened by amino-bearing phenyl silicone resins, *Journal of Material Science*, 2012, 47, 4415–4427.
[137] P. Karamanolevski, A. Buzarovska and G. Bogoeva-Gaceva, The effect of curing agents on basic properties of silicone-epoxy hybrid resin, *Silicon*, 2017. DOI: 10.1007/s12633-016-9490-0.
[138] H.Y. Kim, J.Y. Bae, Y.H. Kim, Y.B. Kim and B.-S. Bae, Low temperature curable epoxy siloxane hybrid materials for LED encapsulant, *Journal of Applied Polymer Science*, 2014, 131, 39968.
[139] N. Parhizkar, T. Shahrabi and B. Ramezanzadeh, A new approach for enhancement of the corrosion protection properties and interfacial adhesion bonds between the epoxy coating and steel substrate through surface treatment by covalently modified amino functionalized graphene oxide film, *Corrosion Science*, 2017, 123, 55–75.

[140] N. Parhizkar, B. Ramezanzadeh and T. Shahrabi, Corrosion protection and adhesion properties of the epoxy coating applied on the steel substrate pre-treated by a sol-gel based silane coating filled with amino and isocyanate silane functionalized graphene oxide nanosheets, *Applied Surface Science*, 2018, 439, 45–59.

[141] Y. Zhang, M. Zhao, J. Zhang, Q. Shao, J. Li, H. Li, B. Lin, M. Yu, S. Chen and Z. Guo, Excellent corrosion protection performance of epoxy composite coatings filled with silane functionalized silicon nitride, *Journal of Polymer Research*, 2018, 25, 130.

[142] Y. Liu, X. Luan, Y. Fenga, X. Tan, Y. Han and X. Sun, Self-adhesive epoxy modified silicone materials for light emitting diode encapsulation, *Advances in Polymer Technology*, 2017, 28, 1473–1479.

[143] H. Liu, X. Gao, B. Deng and G. Huang, Simultaneously reinforcing and toughening epoxy network with a novel hyperbranched polysiloxane modifier, *Journal of Applied Polymer Science*, 2018, 135, 46340.

[144] S. Li, H. Wang, M. Liu, C. Peng and Z. Wu, Epoxy-functionalized polysiloxane reinforced epoxy resin for cryogenic application, *Journal of Applied Polymer Science*, 2019, 136, 46930.

[145] Y. Wang, H. Wub, Z. Chen, S. Zhoua, Y. Chen, M. Liang and H. Zou, Silicone-epoxy block hybrid network to achieve high-performance and transparent polydimethylsiloxane materials, *Reactive & Functional Polymers*, 2020, 150, 104537.

[146] C. Zhang, X. Mi, J. Tian, J. Zhang and T. Xu, Supported ionic liquid silica as curing agent for epoxy composites with improved mechanical and thermal properties, *Polymers*, 2017, 9, 478. DOI: 10.3390/polym9100478.

[147] Y. Shi, B. Yu, Y. Zheng, J. Guo, B. Chen, Z. Pan and Y. Hu, A combination of POSS and polyphosphazene for reducing fire hazards of epoxy resin, *Advances in Polymer Technology*, 2018, 29, 1242–1254.

[148] W. Zhang, X. Li and R. Yang, Study on flame retardancy of TGDDM epoxy resins loaded with DOPO-POSS compound and OPS/DOPO mixture, *Polymer Degradation and Stability*, 2014, 99, 118–126.

[149] G.H. Hsiue, Y.L. Liu and H.H. Liao, Flame-retardant epoxy resins: An approach from organic–inorganic hybrid nanocomposites, *Journal of Polymer Science, Part A: Polymer Chemistry*, 2001, 39, 986–996.

[150] Z. Li and R. Yang, Study of the synergistic effect of polyhedral oligomeric octadiphenylsulfonyl-silsesquioxane and 9,10-dihydro-9-oxa-10-hosphaphenanthrene-10-oxide on flame-retarded epoxy resins, *Polymer Degradation and Stability*, 2014, 109, 233–239.

[151] J.A.A. Casarrubias, A. Reyes-Mayer, R. Guardian-Tapia, P. Castillo-Ocampo and A. Romo-Uribe, Rubber nanodomains reinforced epoxy resin, *MRS Advances*, 2016. DOI: 10.1557/adv.2016.297.

[152] Y. Wang, B. Zhang, S. Zhou, Y. Chen, M. Liang and H. Zou, Preparation of high-performance epoxy-containing silicone rubber via hydrosilylation reaction, *Journal of Applied Polymer Science*, 2020, 137, 48397.

[153] W.R.K. Weihermann, M.M. Meier and S.H. Pezzin, Microencapsulated amino-functional polydimethylsiloxane as autonomous external self-healing agent for epoxy systems, *Journal of Applied Polymer Science*, 2019, 136, 47627.

[154] A.M. Atta, A.O. Ezzat, A.M. El-Saeed, A.M. Tawfeek and N.I. Sabeela, Self-healing of chemically bonded hybrid silica/epoxy for steel coating, *Progress in Organic Coatings*, 2020, 141, 105549.

Chapter 4
Polysilanes

Polysilanes (or polysilylenes) (PS) have only silicon atoms in the backbone, and organic groups, hydrogen atoms, and other substituents bound to silicon (see Fig. 4.1). Most often, they are prepared by dechlorination of dichlorosilane monomers with metallic sodium. Insoluble and infusible poly(diphenylsilane) $(Ph_2Si)_n$ was prepared by Kipping [1, 2], and poly(dimethylsilylene) ($(Me_2Si)_n$) by Burkhard [3].

Figure 4.1: Chemical structures of inorganic and organic polysilanes [4].

A mixture of pyrophoric and explosive inorganic oligosilanes ($H(SiH_2)_nH$, PHS, $n = 1–15$) was prepared by decomposition of magnesia silicide (Mg_2Si) with phosphoric acid at elevated temperatures. They were next trapped at −70 °C using an apparatus with special valves and storage vessels [5–7]. Alternatively, a high molecular weight (MW) inorganic PHS was prepared by hydrogenation (with $LiAlH_4$) of a mixture of perhalogenated PS such as cyclic Si_nCl_{2n} and linear Si_nCl_{2n+2}, which were obtained by redistribution of $SiCl_4$ in the presence of H_2 in a plasma process [8].

A radical ring-opening polymerization (ROP) of octachlorocyclotetrasilane (Si_4Cl_8) gave an insoluble, a moisture-sensitive low MW perchloropolysilane (~3,500 g/mol), which was used as a precursor for PHS. A further functionalization with isopropanol increased its solubility and allowed a MW determination by size-exclusion chromatographic method [9].

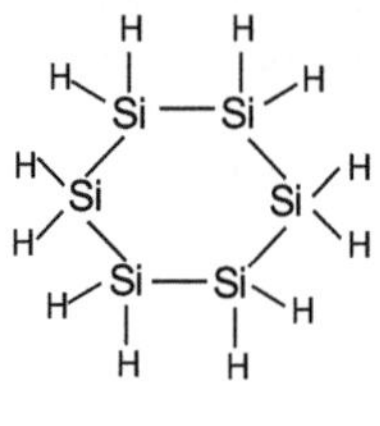

Figure 4.2: Chemical structure of cyclohexasilane (Si_6H_{12}).

A liquid Si–H functional cyclohexasilane (Si_6H_{12}, Figure 4.2) was obtained by P. Boudjouk et al. via the reduction of the (tetradecachloro)cyclohexasilane dianion complex (**1**) (Figure 4.3) with $LiAlH_4$, which was prepared in the reaction of

https://doi.org/10.1515/9783110643671-004

pentaethyldiethylenetriamine (PEDETA) with $HSiCl_3$ and was isolated in high yield and purity [10].

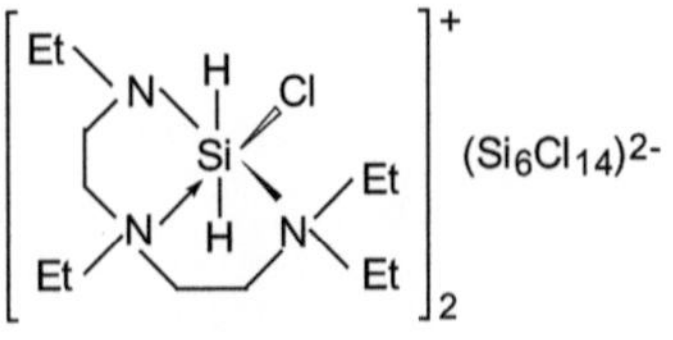

Figure 4.3: A structure of (tetradecachloro)cyclohexasilane dianion complex $[Et_2(CH_2CH_2)N(Et)(CH_2CH_2)NEt_2SiH_2Cl]^+{}_2$ $(Si_6Cl_{14})^{2-}$ [10].

Cyclohexasilane (Si_6H_{12})-based inks served as liquid precursor to silicon-containing electronic materials such as n-type and p-type amorphous silicon (a-Si:H) films, which can be useful for rectifying diodes and field-effect transistors. Solar cells, transistors, and so on could be prepared by printing Si_6H_{12}-based inks, followed by heat and/or UV treatment or laser light radiation [11]. By the electrospinning method, a-Si nanowires (for lithium ion battery anodes) were obtained from Si_6H_{12}-polymer inks [12].

In the reaction of calcium disilicide ($CaSi_2$) with dilute hydrochloric acid, a yellow solid of siloxene with the chemical formula Si_2H_2O or $[Si_6H_3(OH)_3]_x$ was prepared by C.S. Bradley in 1900 [13]. However, its chemical structure, including the intrapolymer Si–Si bond order, was not confirmed so far [14].

An increased interest in PS caused a very important discovery of Yajima, who pyrolyzed insoluble poly(dimethylsilane) $-(Me_2Si)_n-$ through Kumada's rearrangement into meltable and soluble polycarbosilane, which at high temperature (~1,100 °C), under vacuum, formed thermally stable and extraordinary strong β-SiC fibers as ceramic residue [15]. Most often, PS are prepared by the Würtz–Fittig reductive coupling reaction of chlorosilanes with sodium or lithium (Figure 4.4) in boiling solvents (toluene, benzene, or tetrahydrofuran (THF)) [1, 2, 16]. Soluble homo- and copolymers, mainly methyl-containing PS, were described in the late 1970s [17–19]. Although by this method it was difficult to control syntheses of PS and polymers with different structures, MW, and polydispersities of MW (PDI), it is still the most often used for the synthesis of PS. The Wűrtz–Fittig reaction is very sensitive on reaction temperature, on solvents and additives, on the kind of the alkali metals used (Na, Li, K, Na/K alloy), and on their dispersity [20–25]:

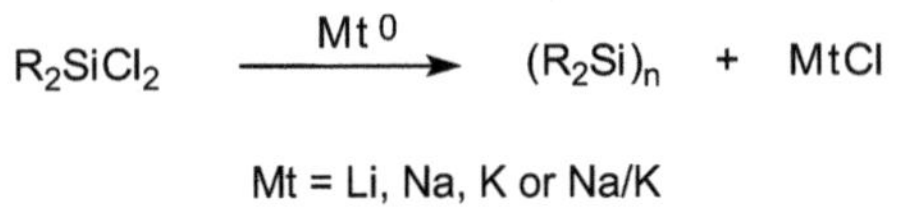

Figure 4.4: The Würtz-type reductive coupling of chlorosilanes.

With the Würtz–Fittig reductive coupling method, it was difficult to obtain PS having a low dispersion of their MW. Usually polymodal MW distributions were observed (Fig. 4.5) [26].

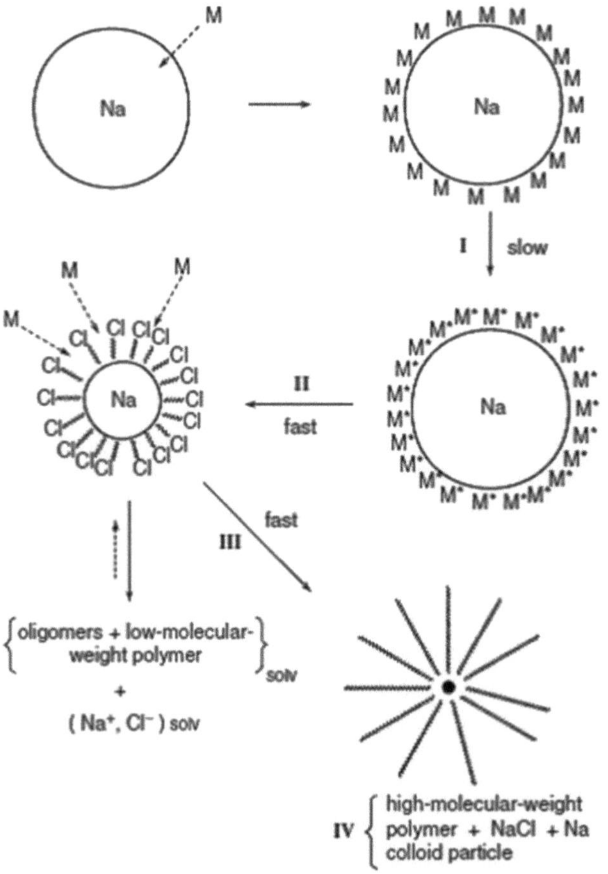

Figure 4.5: Schematic of the progress of the alkali-metal-mediated reductive coupling of dihalosilane. Reprinted with permission from [26]. Copyright 1996 by American Chemical Society.

West et al. prepared a soluble topological PS analogue of polystyrene with weight average MW (M_w) of 5×10^4 g/mol, by reacting equimolar amounts of methylphenyl (dichloro)silane ($MePhSiCl_2$) and dimethyl(dichloro)silane (Me_2SiCl_2, DMDCS) with sodium. Cross-linking by UV irradiation and pyrolysis of "polysilastyrene" gave only ~30 wt% ceramic yield [18, 27–31].

Many factors determine the Würtz-type reductive coupling of (dichloro)organosilanes by alkali metals. In aromatic solvents, boiling at higher temperature, under reflux, the bimodal MW distributions of the PS were observed, probably due to a defect-diffusion-controlled polymerization at the alkali metal surface. Frequently PS syntheses were carried out in THF at ambient temperature, providing higher yields and generally much narrower MW distributions of PS. The higher yields were assigned to the ability of the solvent to stabilize the anionic chain carriers. The narrower polydispersity of MWs that resulted from syntheses in THF was attributed to the suppression of defect diffusion rates at lower temperatures [21, 26].

Through a homogeneous coupling reaction of (dichloro)diphenylsilane with (dichloro)methylsilane, poly(phenylmethylsilane) copolymers with various contents of methylhydrosilyl groups (Scheme 4.1.), having very narrow MW distribution and a low content of linear and cyclic oligomers were obtained. The reaction of chlorosilanes with a sodium–potassium alloy complex with a crown ether (18-crown-6) was carried out in THF solution at −75 °C. This PS was characterized by ^1H NMR, ^{13}C NMR, Fourier-transform infrared (FT-IR), and UV/visible (vis) spectroscopies and gel permeation chromatography (GPC) [32].

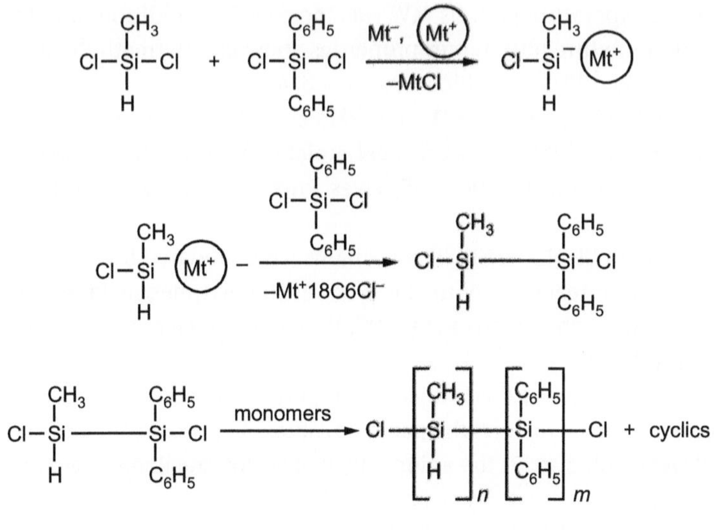

Scheme 4.1: Synthesis of a poly(phenylmethylsilane) copolymer. Reprinted with permission from [32]. Copyright 2001 by J. Wiley and Sons.

Polyhydrosilanes with various contents of methylhydrosilyl groups were synthesized by homogeneous coupling of methyldichlorosilane with diphenyldichlorosilane. Spectral analysis techniques combined with atomic force and polarized light microscopy were used to elucidate the mechanism of the silicon particle formation within the polymeric matrix. Poly[diphenylsilane-*co*-methyl(H)silane], prepared by Würtz coupling of $MeHSiCl_2$ and Ph_2SiCl_2 with sodium, reduced $HAuCl_4$ to Au^0 in solution at 4 °C. The formed small Au nanoparticles (5–15 nm diameter) with length-to-width aspect ratios up to 10^3 served as seeds for the heterogeneous nucleation and anisotropic growth that takes place at 25 °C and gave crystalline needle-like polymer–gold nanostructures [33].

Functionalized disilenes having Si=Si unit have been very useful precursors for synthesis of cyclic, polycyclic, and cluster-like structures [34].

Functionalization of PS distinctly affected their solubility, glass temperature (T_g), cure rate, and yield of the ceramic residue. Vinyl-substituted PS were synthesized from methylvinyl(dichloro)silane, dimethyl(dichloro)silane, and trimethyl(chloro)silane with sodium in inert solvents. The application of methyl(dichloro)silane ($MeHSiCl_2$, MDCS) in coupling with sodium in a mixture of toluene and THF led to PS functionalized with Si-vinyl and Si–H groups [35]. Many other polymethylsilane -$[Me(H)Si]_n$- (PMS) precursors of Si–C were prepared by the condensation of MDCS, which has an ideal ratio of Si:C = 1:1 [36–58]. The Si–H groups take place in cross-linking reactions with curing agents (e.g., boron and metallocene compounds) and improve the ceramic yields [56]. However, it was found that PMS is very sensitive to oxygen and it can be flammable in air even at ambient temperature [43]. Its MW was increased by addition of polyborazine to the reaction mixture [60]. Better properties showed poly(methylhydro-*co*-dimethyl)silanes (PMS-*co*-PDMS), which were good matrix resins for ceramic composites [61, 62]. Radiation curing of PMS and PMS-*co*-PDMS improved ceramic yields [62]. PMS –$(MeSiH)_n$–, which has an ideal stoichiometry for conversion to SiC, prepared by dechlorination of MDCS [65] gives lower ceramic yields during pyrolysis.

New PS containing both Si–H and Si–SR side chains were prepared by dehydrocoupling and hydrosilylation reactions of the Si–H bonds in poly(phenylsilane) catalyzed by $B(C_6F_5)_3$. They were characterized by GPC, IR, UV–vis, elemental analysis, and ^1H, ^{13}C, and ^{29}Si-NMR [63].

As new methods of syntheses of PS (1) the dehydropolycondensation reaction of trihydro- and dihydrosilanes [64, 65], (2) the polymerization of cyclic precursors disilenes (*masked disilenes*) [66], and (3) the anionic ROP of cyclotetrasilanes were applied [67].

Dehydropolycondensation (dehydrocoupling) of trihydro- and dihydro-substituted silanes was discovered by J.F. Harrod et al. [65, 68]. This reaction gave linear and cyclic PS with relatively low MW [64, 65, 68–77]:

$$n\ PhSiH_3 \xrightarrow{Cp_2MY_2} H(PhSiH)_xH + (PhSiH)_y + (x+y)H_2 \tag{4.1}$$

$$\text{where: } M = Ti, Zr, Hf;\ Y = F, OPh, NMe_2.$$

In dehydropolycondensation of $RSiH_3$ in the presence of early transition metal catalysts, side degradation products and cyclics were also formed [75].

Dehydrogenative coupling of primary monoalkylsilanes ($RSiH_3$) carried out toward dimethyltitanocene (Cp_2TiMe_2) or other catalysts led to low MW PS (with a degree of polymerization: DP = 10–20), mostly with linear structures [39, 65, 76]. The dehydrocoupling reaction of $RSiH_3$ in the presence of η^2-alkynyl titanocene or -zirconocene is an important salt-free, high-yield synthesis method for high-purity PS [64, 77]. Zirconocene and hafnocene hydrides were also used as suitable catalysts for the dehydrocoupling polymerization from hydridosilanes [64, 70, 73].

Most often phenylsilane ($PhSiH_3$) was used as a monomer in dehydrocoupling reaction. Other *m*- and *p*-phenyl-substituted primary silanes were used as well for this purpose: *m*-Me, *p*-Me, *p*-MeO, *p*-F, *p*-Me_2N, *m*-CF_3, *p*-CF_3, *p*-MeS, 3,5-Me_2, and 3,5-$(CF_3)_2$ [78–83]. Polyphenylsilanes with varying MWs and MW distributions were prepared, depending on the used catalyst. Corey et al. obtained PS with M_n of 2,300, M_w of 3,850, and a PDI of 1.7 [84]. Tilley et al. polymerized $PhSiH_3$ into polyphenylsilanes with M_n of 2,310 and M_w of 12,030 (PDI 5.2) [78, 85]. The complex (1-MeInd)Ni(PPh_3)Cl reacted with cationic initiators $AgBF_4$, $AlCl_3$, methylaluminoxane, or $LiAlH_4$ to give intermediates which catalyzed the dehydropolymerization of $PhSiH_3$ [81]. The dehydrogenative coupling of phenylsilane to poly(phenylsilane) was also catalyzed by two different types of zirconocene catalysts. The obtained poly (phenylsilanes) were characterized by ^{1}H-NMR, GPC, FT-IR, and UV/vis. They were shown to be good photoinitiators for acrylate and methacrylate monomers [82]. Gaseous methylsilane ($MeSiH_3$) is less useful due to preparative difficulties and the insolubility of PMS [82]. Other primary alkyl or alkylarylsilanes were successfully decoupled (*n*-hexyl, *n*-octyl, *n*-dodecyl, Si_6H_{11}, etc.) [69, 86–89].

The branched structure of oligosilanes obtained by dehydrocoupling reaction of $PhSiH_3$, catalyzed by $Cp_2Ti(OAr)_2$ [Ar=(2,6-(iPr)$_2C_6H_3$)] was confirmed by advanced NMR techniques including liquid chromatography (LC) combined with NMR (LC-NMR), and especially with ^{1}J, ^{2}J, ^{29}Si INADEQUATE experiments. The presence of the branched units =SiPh–, not containing Si–H bond, was described [90].

Other suitable methods used for the synthesis of PS include the anionic polymerization of masked disilenes [66, 91–93], ROP of strained cyclosilanes [67], and reduction reaction of chlorosilanes with Mg in the presence of LiCl and a Lewis acid under mild condition [94].

H. Sakurai et al. developed an anionic living polymerization method of the so-called *masked disilenes* leading to PS in the presence of catalytic amounts of anionic initiators such as organolithium compounds (*n*-BuLi, sec-BuLi, MeLi, or PhLi) or alkali metal alkoxides. Termination was achieved in the reaction with an alcohol (Scheme 4.2) [66, 91–93].

This method allowed for better control of polymer structure of PS, their MW and MW distributions, than Würtz-type coupling reaction. MWs of PS with highly ordered structures ranged from 3,400 to 140,000 g/mol, depending on the reaction conditions (for different substituents: Ph, Me; Bu_2N, Me; *n*-Bu, Me; *n*-Hex, *n*-Hex, etc.). Some PS were insoluble in organic solvents [69, 95–97].

Polysilylenes were also synthesized by electrochemical methods [95–100]. Rates of coupling reaction were accelerated by ultrasounds [101–104]. Newer PS were prepared [105–115] and their yields at room temperature ranged from 50% to 83% [116].

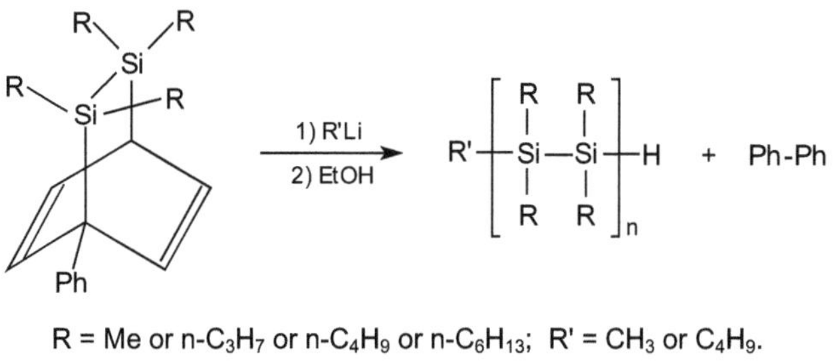

Scheme 4.2: Anionic polymerization of masked disilenes. Adapted with permission from [91]. Copyright 1990 by American Chemical Society.

A new precious cyclic monomer 1,2,3,4-tetramethyl-1,2,3,4-tetraphenylsilane ($(MePhSi)_4$) was prepared in ~90% yield by two-step synthesis – by a partial dearylation of octaphenyl cyclooctasilane ($(Ph_2Si)_4$) with trifluoromethanesulfonic acid, followed by substitution of four "triflate" groups (CF_3SO_3) with methyl groups in reaction with Grignard reagent MeMgBr (in methylene chloride or in benzene solution), under oxygen and moisture-free conditions. *All-trans* isomer was a predominant component of a mixture of stereoisomers [117–119].

During attempted ROP of $(MePhSi)_4$ in the presence of $Pd(Ph_3P)_4$ only a dimer $(MePhSi)_8$ was formed, while toward other palladium ($PdCl_2$, $PdCl_2(Ph_3P)_2$, $Pd(dba)_2$, dba=PhC=CHCOCH=CHPH), or Pt catalysts (e.g. $H_2PtCl_6{\cdot}H_2O$, platinum vinyl complexes) $(MePhSi)_4$ underwent mainly dimerization reaction, accompanied by small amounts of higher oligomers with maximal values of MWs ranging from 3,000 to 36,700 g/mol [120].

Similar to Würtz coupling, from trihaloorganosilanes ($RSiX_3$), polysilynes with the formula $[RSi]_n$ were prepared:

$$3\,Na + RSiCl_3 \longrightarrow [RSi]_n + 3\,NaCl \tag{4.2}$$

The methyl and hexyl derivatives were described by P.A. Bianconi [121]. A dark polysilyne yellow powder formed a clear colloidal nonviscous suspension with some solvents (ether, THF, toluene, etc.) [122]. It can be deposited as a film or coating on various substrates. Poly(methylsilyne) was thermally decomposed to silicon carbide.

Amorphous $[RSi]_n$ alkyl polysilynes were also prepared by reductive condensation from mixtures of $RSiX_3$ with (dihalo)organosilanes (R_2SiX_2) with a liquid Na/K alloy – hydrocarbon emulsions under ultrasonic irradiation [123]. Sonication avoided a surface passivation; it allowed more precise control of reaction rate, stoichiometry, MW, and decreased formation of incompletely substituted and insoluble polysilynes. This way the first soluble polysilynes with the composition $[RSi]_n$ (with alkyl substituents containing at least three carbon atoms) were prepared. All spectroscopic data confirmed sp^3 hybridization of alkylsilyne (RSi) fragments, irregularly connected through Si–Si single bonds into random networks.

Poly(methylsilyne) and poly(cyclohexylsilyne-*co*-methylsilyne) branched copolymers were prepared by the electrochemical reduction of methyl(trichloro)silane or its mixtures with cyclohexyl(trichloro)silane [124]. Poly(methylsilyne) was recommended as a potential soluble precursor for SiC. Nonionic water-soluble polysilynes were synthesized by L.W. Jenneskens et al. [125].

Polysilynes with *n*-propyl, *n*-butyl, isobutyl, amyl, cyclohexyl, and phenyl substituents at silicon were examined as resist materials for deep UV lithography. Photolysis of polysilyne films on air using 193, 215, or 254 nm light resulted in the formation of an organosiloxane [126, 127].

A new preceramic polymer, poly(silyne-*co*-hydridocarbyne), was electrochemically synthesized from a monomer containing both silicon and carbon in its structure. This copolymer was soluble in $CHCl_3$, CH_2Cl_2, and THF. It was easily pyrolyzed to silicon carbide by heating at 1,000 and 750 °C under an inert atmosphere, or at 750 °C to SiO_2 under air atmosphere [128]. Poly(silyne-*co*-hydridocarbyne) was characterized with FT-IR, Raman and UV/vis spectroscopy, ^{1}H-NMR, and GPC, while the obtained ceramic materials were characterized with optical microscopy, SEM, X-ray, and Raman spectroscopies.

It was found that polysilynes are better for the anode material than PS. They absorb Li ions by means of a redox reaction [129].

New tetrafunctional monomers for PS synthesis, having a star structure $Si(SiMe_2X)_4$ (X=Cl, Br), were prepared by halogenation of tetrakis(dimethylsilyl)silane ($Si(SiMe_2H)_4$) with carbon tetrachloride, *n*-BuBr or $CHBr_3$. These monomers were used for synthesis of PS with different nonlinear topologies: branched, spiro, and bicyclic architectures [130–132].

PS are unstable in air and moisture atmosphere and easily undergo degradation under UV light irradiation. Their properties significantly depend on MW, as well as on the nature of the organic groups attached to silicon and the conformation (branching) of the polymer chain. Due to insolubility, nonmeltability, and intractability of PS, their processability is quite difficult [133].

PS have a number of unique properties and interesting applications [16, 24, 27, 134–137, 141–143]. Chains of PS exhibit internal conductivity, thanks to conjugation of silicon atoms. Their doping with iodine and other additives provided conducting materials [24, 27, 136]. PS show unique photochemical and optoelectronic properties as a result of the extensive delocalization of σ electrons along silicon chains (σ conjugation) [24, 27]. PS absorb UV light >300 nm, and the absorption maxima depend on the nature of the substituents and chain conformation. Some of the photochemical properties depend on the regularity of their chemical structure and configuration of the substituents. Therefore, there was a need and interest to find new methods of synthesis of PS other than polycondensation dichlorosilanes with an alkali metal, resulting in polymers having the statistical structure with a multiple structural defects.

PS are useful as photoresists in microlithography, as well as they have found applications as photoconducting polymers, as third-order nonlinear optical materials, and as valuable precursors for the synthesis of silicon carbide [16, 137, 144]. PS are also applied for fabrication of semiconductors, especially bilayer resists and antireflection layers with high etching properties [138].

References

[1] F.S. Kipping and J.E. Sands, XCIII.—Organic derivatives of silicon. Part XXV. Saturated and unsaturated silicohydrocarbons, Si_4Ph_8, *Journal of the Chemical Society*, 1921, 119, 830.

[2] F.S. Kipping, CCCVIII. - Organic derivatives of silicon. Part XXX. Complex silicohydrocarbons $[SiPh_2]n$, *Journal of the Chemical Society*, 1924, 125, 2291.

[3] C.A. Burkhard, Polydimethylsilanes. *Journal of the American Chemical Society*, 1949, 71, 963.

[4] A. Feigl, A. Bockholt, J. Weis and B. Rieger, Modern Synthetic and Application Aspects of Polysilanes: An Underestimated Class of Materials?, *Advances in Polymer Science*, 2011, 235, 1.

[5] F. Fehér, D. Schinkitz and J. Schaaf, Ein Verfahren zur Darstellung höherer Silane, *Zeitschrift für anorganische und allgemeine Chemie*, 1971, 383, 303.

[6] F. Fehér, D. Schinkitz and G. Wronka, Beiträge zur Chemie des Siliciums und Germaniums. XVII. Über die Aufbewahrung von Silanen in Behältern aus Stahl, *Zeitschrift für anorganische und allgemeine Chemie*, 1971, 384, 226.

[7] F. Fehér, D. Schinkitz and H. Strack, Beiträge zur Chemie des Siliciums und Germaniums. Darstellung von Rohsilan in einem 1-l-Reaktor, *Zeitschrift für anorganische und allgemeine Chemie*, 1971, 358, 202.

[8] G.A. Auner, C. Bauch, G. Lippold and R. Deltschew, (a) *Pat. DE* 102006034061, A1, 2008, (b) *Can. Pat.* 115: 115263, 2008.

[9] J.R. Koe, D.R. Powell, J.J. Buffy, S. Hayase and R. West, Perchloropolysilane: X-Ray Structure, Solid-State 29Si NMR Spectroscopy, and Reactions of $[SiCl_2]n$, *Angewandte Chemie*, 1998, 37, 1441.

[10] S.-B. Choi, B.-K. Kim, P. Boudjouk and D.G. Grier, Amine-promoted disproportionation and redistribution of trichlorosilane: formation of tetradecachlorocyclohexasilane dianion, *Journal of the American Chemical Society*, 2001, 123, 8117.

[11] S. Han, X. Dai, P. Loy, J. Lovaasen, J. Huether, J.M. Hoey, A. Wagner, J. Sandstrom, D. Bunzow, O.F. Swenson, I.S. Akhatov and D.L. Schulz, Printed silicon as diode and FET materials - Preliminary results, *Journal of Non-Crystalline Solids*, 2008, 354, 2623.

[12] D.L. Schulz, J. Hoey, J. Smith, A. Elangovan, X. Wu, I. Akhatov, S. Payne, J. Moore, P. Boudjouk, L. Pederson, J. Xiao and J.-G. Zhang, Si_6H_{12}/Polymer Inks for Electrospinning a-Si Nanowire Lithium Ion Battery Anodes, *Electrochemical and Solid-State Letters*, 2010, 13, A143.

[13] C.S. Bradley, New Silicon Compounds Obtained by the Use of the Electric Furnace, *Chemistry News*, 1900, 82, 149.

[14] (a) https://gaz.wiki/wiki/pt/Polysilicon_hydride; (b). J.W. Mellor, *A Comprehensive Treatise on Inorganic and Theoretical Chemistry, Vol. VI*, Longmans, Green and Co, 1947, 226.

[15] S. Yajima, J. Hayashi and M. Omori, Continuous silicon carbide fiber of high tensile strength, *Chemistry Letters*, 1975, 4, 931.

[16] D.R. Miller and J. Michl, Polysilane high polymers, *Chemical Reviews*, 1989, 89, 1359.

[17] J.P. Wesson and T.C. Williams, Organosilane polymers: I. Poly(dimethylsilylene), *Journal of Polymer Science: Polymer Chemistry Edition*, 1979, 17, 2833.

[18] R. West, L.D. David, P.I. Djurovich, K.L. Stearley, K.S.V. Srinivasan and H. Yu, Phenylmethylpolysilanes: formable silane copolymers with potential semiconducting properties, *Journal of the American Chemical Society*, 1981, 103, 7352.

[19] R.E. Trujillo, Preparation of Long-Chain Poly(Methylphenylsilane), *Journal of Organometallic Chemistry*, 1980, 198, C27.

[20] T. Kawabe, M. Naito and M. Fujiki, Multiblock Polysilane Copolymers: One-Pot Wurtz Synthesis, Fluoride Anion-Induced Block-Selective Scission Experiments, and Spectroscopic Characterization, *Macromolecules*, 2008, 41, 1952.

[21] R.G. Jones and S.J. Holder, High-yield controlled syntheses of polysilanes by the Wurtz-type reductive coupling reaction, *Polymer International*, 2006, 55, 711.

[22] U. Herzog and R. West, Heterosubstituted Polysilanes, *Macromolecules*, 1999, 32, 2210.

[23] B. Lacave-Goffin, L. Hevesi and J. Devaux, Graphite–potassium, a new reagent for the synthesis of polysilanes, *Journal of the Chemical Society, Chemical Communications*, 1995, 769.

[24] R.G. Jones, R.E. Benfield, P.J. Evans and A.C. Swain, *Journal of the Chemical Society, Chemical Communications*, 1995, 1465.

[25] P. Kochs, Chapter 4: *Poly(organosilanes), Poly(carbosilanes), Poly(organosilazanes), in Silicon in Polymer Synthesis*, H.R. Kricheldorf, Ed., Springer-Verlag, Berlin-Heidelberg, 1996, 223–51.

[26] R.G. Jones, U. Budnik, S.J. Holder and W.K.C. Wong, Reappraisal of the Origins of the Polymodal Molecular Mass Distributions in the Formation of Poly(methylphenylsilylene) by the Wurtz Reductive-Coupling Reaction, *Macromolecules*, 1996, 29, 8036.

[27] R. West, The polysilane high polymers, *Journal of Organometallic Chemistry*, 1986, 300, 327.

[28] K.S. Mazdiyasni, R. West and L.D. David, Characterization of Organosilicon-Infiltrated Porous Reaction-Sintered Si_3N_4, *Journal of the American Ceramic Society*, 1978, 61, 504.

[29] R. West, L.D. David, P.I. Djurovich, H. Yu and R. Sinclair, Polysilastyrene: Phenylmethylsilane-dimethylsilane Copolymers as Precursors to Silicon Carbide, American *Ceramic Society Bulletin* 1983, 62, 899.

[30] R. West, in *Ultrastructure Processing of Ceramics, Glasses and Composites*, L.L. Hench and D.R. Ulrich, Eds., Wiley, New York, 1984, 235.

[31] R. West, Y.H. Zhang, P.I. Djurovich and H. Stüger, *Science of ceramic chemical* processing, L.L. Hench and D.R. Ulrich, Eds., Wiley, New York, 1986, Vol. 36, 337.

[32] G. Sacarescu, L. Sacarescu, R. Ardeleanu, P. Kurcok and Z. Jedliński, Si-H Functional Polysilanes via a Homogeneous Reductive Coupling Reaction, *Macromolecular Rapid Communications*, 2001, 22, 405.

[33] L. Sacarescu, M. Simionescu, G. Sacarescu, V. Harabagiu, Polyhydrosilane Mediated Synthesis of One-Dimensional Gold Nanostructures, J. Inorg. Organomet. Polym., 2013, 23, 621–628.

[34] A. Rammo and D. Scheschkewitz, Functional Disilenes in Synthesis, Chemistry – European Journal, 2018, 24, 6866–6885.

[35] C.L. Schilling Jr., Polymeric Routes to Silicon Carbide, *British Polymer Journal*, 1986, 8, 355.

[36] M. Birot, J. Pillot and J. Dunogues, Comprehensive Chemistry of Polycarbosilanes, Polysilazanes, and Polycarbosilazanes as Precursors of Ceramics, *Chemical Reviews*, 1995, 95, 1443.

[37] C.L. Schilling Jr., J.P. Wesson and T.C. Williams, Polycarbosilane Precursors for Silicon-Carbide, *American Ceramic Society Bulletin*, 1983, 62, 912.

[38] D. Seyferth, T.G. Wood, H.J. Tracy and J.L. Robisoe, Near-Stoichiometric Silicon Carbide from an Economical Polysilane Precursor, *Journal of the American Ceramic Society*, 1991, 74, 1300.

[39] Z. Zhang, F. Babonneau, R.M. Laine, Y. Mu, J.F. Harrod and J.A. Rahn, Poly(methylsilane) - A High Ceramic Yield Precursor to Silicon Carbide, *Journal of the American Ceramic Society*, 1991, 74, 670.

[40] T. Kobayashi, T. Sakakura, T. Hayashi, M. Yumura and M. Tanaka, Neodymium-Catalyzed Dehydrogenative Condensation of Methylsilane and Ceramization of the Resulting Polymer, *Chemistry Letters*, 1992, 21, 1157.

[41] F.I. Hurwitz, T.A. Kacik, X. Bu, J. Masnovi, P.J. Heimann and K. Beyene, Pyrolytic conversion of methyl- and vinylsilane polymers to Si-C ceramics, *Journal of Materials Science*, 1995, 30, 3130–6.

[42] Y. Mu, R.M. Laine and J.F. Harrod, Some further observations on polymethylsilane as a precursor for silicon carbide, *Applied Organometallic Chemistry*, 1994, 8, 95.

[43] B. Boury, N. Bryson and G. Soula, Borate-Catalyzed Thermolysis of Polymethylsilane, *Chemistry of Materials*, 1998, 10, 297.

[44] B. Boury, N. Bryson and G. Soula, Stoichiometric silicon carbide from borate-catalyzed polymethylsilane–polyvinylsilane formulations, *Applied Organometallic Chemistry*, 1999, 13, 419.

[45] P. Czubarow, T. Sugimoto and D. Seyferth, Sonochemical Synthesis of a Poly(methylsilane), a Precursor for Near-Stoichiometric SiC, *Macromolecules*, 1998, 31, 229.

[46] M.F. Gozzi, M.D.C. Calves and I.V.P. Yoshida, Near-stoichiometric silicon carbide from a poly(methylsilylene)/tetra-allylsilane mixture, *Journal of Materials Science*, 1999, 34, 155.

[47] M. Scarlet, I.S. Butler and J.F. Harrod, Nitrogenation of Silicon Carbide Layers Deposited on Silicon Single-Crystal Wafers via Pyrolysis of Poly(methylsilane), *Chemistry of Materials*, 1995, 6, 1214.

[48] T. Iseki, M. Narisawa, K. Okamura, K. Oka and T. Dohmaru, Reflux heat-treated polymethylsilane as a precursor to silicon carbide, *Journal of Materials Science Letters*, 1999, 18, 185.

[49] K.W. Chew, A. Sellinger and R.M. Laine, Processing Aluminum Nitride-Silicon Carbide Composites via Polymer Infiltration and Pyrolysis of Polymethylsilane, a Precursor to Stoichiometric Silicon Carbide, *Journal of the American Ceramic Society*, 1999, 82, 857.

[50] Z. Zhang, C.S. Scotto and R.M. Laine, Processing stoichiometric silicon carbide fibers from polymethylsilane. Part 1 Precursor fiber processing, *Journal of Materials Chemistry*, 1998, 8, 2715.

[51] P. Czubarow and D. Seyferth, Application of poly(methylsilane) and Nicalon® polycarbosilane precursors as binders for metal/ceramic powders in preparation of functionally graded materials, *Journal of Materials Science*, 1997, 32, 2121.

[52] M.F. Gozzi and I.V.P. Yoshida, Thermal and Photochemical Conversion of Poly(methylsilane) to Poly(carbosilane), *Macromolecules*, 1995, 28, 7232.

[53] F.C. Schilling, T.W. Weidman and A.M. Joshi, Solid State Characterisation of *Polysilanes* Containing the Si-H Bond, *Journal of the American Chemical Society*, 1991, 113, 6711.

[54] G.D. Soraru, F. Babonneau and J.D. Mackenzie, Structural evolutions from polycarbosilane to SiC ceramic, *Journal of Materials Science*, 1990, 25, 3886.

[55] R.J.P. Corriu, D. Leclercq, P.H. Mutin, J. Planeix and A. Vioux, Mechanism of pyrolysis of polycarbosilanes: poly(silylethylene) and poly(dimethylsilylethylene), *Organometallics*, 1993, 12, 454.

[56] R.M. Laine, A. Sellinger, K.W. Chew, Highly processable hyperbranched polymer precursors to controlled chemical and phase purity fully dense SiC, *US Patent* 6,133,396 (2000); and referencies cited therein.

[57] Q. Liu, H.J. Wo, R. Lewis, G.E. Maciel and L.V. Interrante, An Efficient Process of Cross-Linking Poly(methylsilane) for SiC Ceramics, *Chemistry of Materials*, 1999, 11, 2038.

[58] A.T. Hemida, M. Birot, J.P. Pillot, J. Dunogues and P. Pailler, Synthesis and characterization of new precursors to nearly stoichiometric SiC ceramics: Part II A homopolymer ro, *Journal of Materials Science*, 1997, 32, 3457.

[59] T. Iseki, M. Narisawa, Y. Katase, K. Oka, T. Dohmaru and K. Okamura, An Efficient Process of Cross-Linking Poly(methylsilane) for SiC Ceramics, *Chemistry of Materials*, 2001, 13, 4163.

[60] J.G. Kho, M.D. Soo and D.P. Kim, Polymethylsilane post-treated with a polyborazine promotor as a precursor to SiC with high ceramic yield, *Journal of Materials Science Letters*, 2000, 19, 303.

[61] M. Narisawa, K. Okamura, T. Iseki, K. Oka and T. Dohmaru, Polysilane-Based Precursors for SiC/SiC Composites, *Ceramic Transactions*, 2002, 144, 141.

[62] M. Narisawa, T. Yoshida, T. Iseki, Y. Katase, K. Okamura, K. Oka and T. Dohmaru, γ-Ray Curing of Poly(methylsilane) and Poly(methylsilane-dimethylsilane) for Improved Ceramic Yields, *Chemistry of Materials*, 2000, 12, 2686.

[63] P.T.K. Lee, M.K. Skjel, and L. Rosenberg, Borane-Catalyzed Si–H Activation Routes to Polysilanes Containing Thiolato Side Chains, Organometallics, 2013, 32, 1575–1578.

[64] Q. Wang and J.Y. Corey, Dehydrocoupling reactions of hydrosilanes with group 4 metallocenes Cp_2MY_2 (M = Ti, Zr, Hf; Y = F, OPh, NMe_2), *Canadian Journal of Chemistry*, 2000, 78, 1434, and references cited therein.

[65] J.F. Harrod, Chapter 7: *Polymerization of Group 14 Hydrides by Dehydrogenative Coupling, in: Inorganic and Organometallic Polymers*, edited by M. Zeldin, K.J. Wynne and H.R. Allcock, *ACS Symposium Series*, 1988, Vol. 360, p. 89–100, and references cited therein.

[66] K. Sakamoto, K. Obata, H. Hirata, M. Nakajima and H. Sakurai, Chemistry of organosilicon compounds. 257. Novel anionic polymerization of masked disilenes to polysilylene high polymers and block copolymers, *Journal of the American Chemical Society*, 1989, 111, 7641.

[67] M. Cypryk, Y. Gupta and K. Matyjaszewski, Anionic ring-opening polymerization of 1,2,3,4-tetramethyl-1,2,3,4-tetraphenylcyclotetrasilane, *Journal of the American Chemical Society*, 1991, 113, 61046.

[68] J.F. Harrod and S.S. Yun, Silyltitanocene complexes as catalysts for the hydrogenation, isomerization, and hydrosilylation of olefins, *Organometallics*, 1987, 6, 1381.

[69] C. Aitken, J.F. Harrod and E. Samuel, Polymerization of primary silanes to linear polysilanes catalyzed by titanocene derivatives, *Journal of Organometallic Chemistry*, 1986, 279, 11.

[70] J.Y. Corey, X.-H. Zhu, T.C. Bedard and L.D. Lange, Catalytic dehydrogenative coupling of secondary silanes with Cp_2MCl_2-BuLi, *Organometallics*, 1991, 10, 924.

[71] J.Y. Corey and X.-H. Zhu, Reactions of hydrosilanes and olefins in the presence of Cp_2MCl_2/BuLi, *Organometallics*, 1992, 11, 672.

[72] F. Lunzer and C. Marschner, Chemoselectivity in the Dehydrocoupling Synthesis of Higher Molecular Weight Polysilanes, *Materials*, 2010, 3, 1125.

[73] H.G. Woo and T.D. Tilley, Dehydrogenative polymerization of silanes to polysilanes by zirconocene and hafnocene catalysts. A new polymerization mechanism, *Journal of the American Chemical Society*, 1989, 111, 8043.

[74] M.-H. Kim, J. Lee, S.-Y. Mo, J.-H. Kim and H.-G. Woo, Dehydrogenative Polymerization of Hydrosilanes to Silicon Polymers, *Journal of the Chosun Natural Science*, 2010, 3(2), 61.

[75] T.D. Tilley, The coordination polymerization of silanes to polysilanes by a "sigma.-bond metathesis" mechanism. Implications for linear chain growth, *Accounts of Chemical Research*, 1993, 26, 22, and references cited therein.

[76] E. Samuel and J.F. Harrod, Synthesis and characterization of a novel bis(cyclopentadienyl)-titanium hydride complex, *Journal of the American Chemical Society*, 1984, 106, 1859.

[77] L.S. Chang and J.Y. Corey, Dehydrogenative coupling of diarylsilanes, *Organometallics*, 1989, 8, 1885.

[78] H.G. Woo, J.F. Walzer and T.D. Tilley, sigma.-Bond metathesis mechanism for dehydropolymerization of silanes to polysilanes by d0 metal catalysts, *Journal of the American Chemical Society*, 1992, 114, 7047.

[79] J.P. Banovetz, H. Suzuki and R.M. Waymouth, Dehydrogenative coupling of substituted phenylsilanes: synthesis of poly[((trifluoromethyl)phenyl)silanes], *Organometallics*, 1993, 12, 4700.

[80] H. Hashimoto, S. Obara and M. Kira, Substituent effects on catalytic synthesis and properties of poly(phenylsilane), *Chemistry Letters*, 2000, 188.

[81] F.-G. Fontaine, T. Kadkhodazadeh and D. Zargarian, Nickel indenyl complexes as precatalysts for dehydropolymerization of phenylsilane, *Chem. Commun.*, 1998, 1253–1254.

[82] B.A. Goldslager and S.J. Clarson, Dehydrogenative Coupling of Poly(phenylsilane) by Metallocene Catalysis, *Journal of Inorganic and Organometallic Polymers*, 1999, 9(2), 123.

[83] Y. Obora and M. Tanaka, Dehydrocoupling polymerization of arylsilanes with chloro(aryloxy) bis(cyclopentadienyl)zirconium complex catalysts, Journal of Organometallic Chemistry, 2000, 595, 1.

[84] B.J. Grimmond and J.Y. Corey, Amino-functionalized zirconocene $(C_5H_4CH(Me)NMe_2)_2ZrCl_2$, acatalyst for the dehydropolymerization of PhSiH3 Probing of peripheral substituent effects on catalyst dehydropolymerization activity, *Inorganica Chimica Acta*, 2002, 330, 89.

[85] T. Imori and T.D. Tilley, The influence of catalyst structure on the dehydropolymerization of phenylsilane, *Polyhedron*, 1994, 13, 2231.

[86] Y. Mu, J.F. Harrod and F. John, Synthesis of Poly(methylsilylene) by Catalytic Dehydrocoupling with Cp_2MM_2 (M = Ti, Zr) Catalysts, Inorganic and Organometallic Oligomers and Polymers, Proceedings of the 33rd IUPAC Symposium on Macromolecules, 1991, 33, 23.

[87] L. Rosenberg and D.N. Kobus, Dehydrogenative coupling of primary alkyl silanes using Wilkinson's catalyst, *Journal of Organometallic Chemistry*, 2003, 685, 107.

[88] R. Shankar, A. Saxena and A.S. Brar, Synthesis and reactivity of novel oligosilanes bearing functional carbosilane side chains, *Journal of Organometallic Chemistry*, 2001, 628, 262.

[89] E. Hengge, P. Gspaltl and E. Pinter, Si-H-containing cyclosilanes and their behaviour in the dehydrogenative polymerization reaction, *Journal of Organometallic Chemistry*, 1996, 521, 145.

[90] J. Pinkas, I. Císařová, J. Karban, J. Schraml and J. Sýkora, Identification of branched oligosilanes in the phenylsilane dehydrocoupling reaction, *Journal of Organometallic Chemistry*, 2012, 710, 20.

[91] K. Sakamoto, M. Yoshida and H. Sakurai, Highly ordered high-molecular weight alternating polysilylene copolymer prepared by anionic polymerization of masked disilene, *Macromolecules*, 1990, 23, 4494.

[92] H. Sakurai, K. Sakamoto, Y. Funda and M. Yoshida, *Chapter 2: Anionic Polymerization of Masked Disilenes to Polysilanes*, in *Inorganic and Organometallic Polymers II, Advanced Materials and Intermediates*, P. Wisian-Neilson, H.R. Allcock and K.J. Wynne (Eds.), ACS Symp. Series, vol. 572, 1994, pp. 8–31.

[93] K. Sakamoto, M. Yoshida and H. Sakurai, Anionic polymerization of masked disilenes. Regioselective synthesis of monomers, structures of polymers and mechanism of initiation for polysilylenes of the type $(R^1R^2SiSiMe_2)_n$, *Polymer*, 1994, 35(23), 4990–7.

[94] S. Kashimura, Y. Tane, M. Ishifune, Y. Murai, S. Hashimoto, T. Nakai, R. Hirose and H. Murase, Practical method for the synthesis of polysilanes using Mg andLewis acid system, *Tetrahedron Letters*, 2008, 49, 269.

[95] E. Hengge and G. Litscher, Eine neue elektrochemische Methode zur Bildung von SiSi-Bindungen, *Angewandte Chemie*, 1976, 88, 414.

[96] E. Hengge and G. Litscher, Zur elektrochemischen Bildung von Di-, Oligo-und Polysilanen, *Monatshefte für Chemie*, 1978, 109, 1217.

[97] E. Hengge and W. Kalchauer, Neue Wege zu Polysilanen, *Monatshefte für Chemie*, 1990, 121, 793.

[98] M. Ishifune, S. Kashimura and Y. Kogai, Electroreductive synthesis of oligosilanes and polysilanes with ordered sequences, *Journal of Organometallic Chemistry*, 2000, 611, 26.

[99] S. Kashimura, M. Ishifune, H. Bu, M. Takebayashi and S. Kitajima, Electroreductive Synthesis of Some Functionalized Polysilanes and Related Polymers, *Tetrahedron Letters*, 1997, 38, 4607.

[100] M. Elangovan, A. Muthukumaran and M. Anbukulandainathan, Novel electrochemical synthesis and characterisationof poly(methyl vinylsilane) and its co-polymers, *European Polymer Journal*, 2005, 41, 2450.

[101] K. Matyjaszewski, D. Greszta, J.S. Hrkach and H.M. Kim, Sonochemical Synthesis of Polysilylenes by Reductive Coupling of Disubstituted Dichlorosilanes with Alkali Metals, *Macromolecules*, 1995, 28, 59.

[102] H.K. Kim and K. Matyjaszewski, Sonochemical Synthesis of. Polysilanes, *Polymer Preprints*, 1988, 29, 168.

[103] H.K. Kim and K. Matyjaszewski, Preparation of polysilanes in the presence of ultrasound, *Journal of the American Chemical Society*, 1988, 110, 3321.

[104] P. Boudjouk, B.H. Han and K.R. Anderson, Sonochemical and electrochemical synthesis of tetramesityldisilene, *Journal of the American Chemical Society*, 1982, 104, 4992.

[105] Z.J. Peng, W.J. Si, S.W. Lin, H.Z. Miao, H.J. Fu, M.N. Xie, H. Ma, H.M. Zeng and Z.S. Su, Synthesis and characterization of branched copolysilanes of high molecular weight with high steric hindrance, *European Polymer Journal*, 2002, 38, 1635.

[106] O.V. Mukbaniani, T.N. Tatrishvili and M.G. Karchkhadze, Synthesis and investigation of the properties of poly(phenyl-α-naphthylsilylene)–dimethylsilylene copolymers, *Journal of Applied Polymer Science*, 2002, 85, 1047.

[107] D. Vyprachticky and V. Cimrova, New Synthesis, Electroluminescence, and Photophysical Properties of Poly[(formylphenyl)methylsilanediyl] and Its Derivatives, *Macromolecules*, 2002, 35, 3463.

[108] O. Minge, N.W. Mitzel and H. Schmidbaur, Synthetic Pathways to Hydrogen-Rich Polysilylated Arenes from Trialkoxysilanes and Other Precursors, *Organometallics*, 2002, 21, 680.

[109] J. Chojnowski, M. Cypryk and J. Kurjata, Organic polysilanes interrupted by heteroatoms, *Progress in Polymer Science*, 2003, 28, 691.

[110] Y. Hatanaka, Unusual behavior of silicon oligomers and polymers havingfunctional groups, *Journal of Organometallic Chemistry*, 2003, 685, 207.

[111] A. Watanabe, Optical properties of polysilanes with various silicon skeletons, *Journal of Organometallic Chemistry*, 2003, 685, 122.

[112] M.S. Cho, B.H. Kim, J.I. Kong, A.Y. Sung and H.G. Woo, Synthesis, catalytic Si/Si dehydrocoupling, and thermolysis of polyvinylsilanes $[CH_2CH(SiH_2X)]_n$ (X/H, Ph), *Journal of Organometallic Chemistry*, 2003, 685, 99.

[113] D. Bratton, S.J. Holder, R.G. Jones and W.K.C. Wong, *Journal of Organometallic Chemistry*, 2003, 685, 60.

[114] X.M. Liu, T. Lin, J. Huang, X.T. Hao, K.S. Ong and C. He, Hyperbranched blue to red light-emitting polymers with tetraarylsilyl cores: Synthesis, optical and electroluminescence properties, and ab initio modeling studies, *Macromolecules*, 2005, 38, 4157.

[115] T. Iwamoto, Novel Stable Silicon-Based π-Electron Systems: Synthesis, Unique Structure, and Properties of Cyclic, Spiroconjugated, and Cumulative Silicon–Silicon Doubly-Bonded Compounds, *Bulletin of the Chemical Society of Japan*, 2005, 78, 393.

[116] S.J. Holder, M. Achilleos and R.G. Jones, Room-Temperature, High-Yield Route to Poly(*n*-alkylmethylsilane)s and Poly(di-*n*-hexylsilane), *Macromolecules*, 2005, 38, 1633.
[117] M. Cypryk, J. Chruściel, E. Fossum and K. Matyjaszewski, Ring-opening polymerization of strained cyclotetrasilanes as a new route towards well defined polysilylenes, *Macromolecular Chemistry, Macromolecular Symposia*, 1993, 73, 167.
[118] J. Chruściel, M. Cypryk, E. Fossum and K. Matyjaszewski, Fluorine-19 NMR studies of the reaction of octap8enylcyclotetrasilane with triflic acid, *Organometallics*, 1992, 11, 3257.
[119] E. Fossum, J. Chruściel and K. Matyjaszewski. *Chapter 4: Stereostructure of Polysilanes by Ring-Opening Polymerization, in: Inorganic and Organometallic Polymers II, Advanced Materials and Intermediates*, P. Wisian-Neilson, H.R. Allcock and K.J. Wynne (Eds.), ACS Symp. Series, vol. 572, 1994, pp. 32–42.
[120] J. Chruściel and K. Matyjaszewski, Ring-Opening of 1,2,3,4-Tetramethyl-1,2,3,4-tetraphenylcyclo-tetrasilane in the Presence of Transition Metal Catalysts, *Journal of Polymer Science: Part A: Polymer Chemistry*, 1996, 34, 2243.
[121] P.A. Bianconi and T.W. Weidman, Poly(n-hexylsilyne): synthesis and properties of the first alkyl silicon $[RSi]_n$ network polymer, *Journal of the American Chemical Society*, 1988, 110, 2342.
[122] *U.S. Patent* 6,989,428 B1, 2006.
[123] T.W. Weidman, P.A. Bianconi and E.W. Kwock, Sonochemical Na/K alloy emulsion synthesis of polysilynes, *Ultrasonics*, 1990, 28, 310.
[124] L.A. Vermeulen and K. Huang, Electrosynthesis of new polysilynes: poly(methylsilyne) and poly(cyclohexylsilyne- co-methylsilyne), *Polymer*, 2000, 41, 441.
[125] T.J. Cleij, S.K.Y. Tsang and L.W. Jenneskens, Facile Sonochemical Synthesis of Graphite Intercalation Compounds, *Macromolecules*, 1999, 32, 3286.
[126] R.R. Kunz, M.W. Horn, R.B. Goodman, P.A. Bianconi, D.A. Smith and C.A. Freed, Polysilyne thin films as resists for deep ultraviolet lithography, *Journal of Vacuum Science and Technology B, Nanotechnology and Microelectronics: Materials, Processing, Measurement, and Phenomena*, 1990, 8, 1820.
[127] R.R. Kunz, M.W. Horn, G.M. Wallraff, P.A. Bianconi, R.D. Miller, R.W. Goodman, D.A. Smith, J.R. Eshelman and E.J. Ginsberg, Surface-Imaged Silicon Polymers for 193-nm Excimer Laser Lithography, *Japanese Journal of Applied Physics*, 31(Part 1), 1992, 12B, 4327.
[128] Y. Nur and L. Toppare, Synthesis of Poly(silyne-co-hydridocarbyne) for Silicon Carbide Production, *Journal of Macromolecular Science, Part A, Pure and Applied Chemistry*, 2013, 50, 923.
[129] F.R. Villatoro, Recent advances in lithium-ion rechargeable batteries, http://francisthemulenews.wordpress.com/
[130] K. Matyjaszewski, J. Chruściel, H.K. Kim and J. Maxka, Polysilanes with Nonlinear Topologies: Branched, Spiro, and Bicyclic Architectures, *Polymer Preprints*, 1993, 34(1), 67, and references cited therein.
[131] J. Maxka, J. Chruściel, M. Sasaki and K. Matyjaszewski, Polysilanes with Various Architectures, *Macromolecular Chemistry, Macromolecular Symposia*, 1994, 77, 79.
[132] K. Matyjaszewski, J. Chruściel, J. Maxka and M. Sasaki, Branched Polysilanes from Tetrafunctional Monomers, *Journal of Inorganic and Organometallic Polymers*, 1995, 5, 261.
[133] G. Mera, M. Gallei, S. Bernard and E. Ionescu, Ceramic Nanocomposites from Tailor-Made Preceramic Polymers, *Nanomaterials*, 2015, 5, 468.
[134] K. Matyjaszewski, M. Cypryk, H. Frey, J. Hrkach, K. Kim, M. Moeller, K. Ruehl and M. White, Synthesis and Characterization of Polysilanes, *Journal of Macromolecular Science, Part A – Chemistry*, 1991, A28, 1151.

[135] R. West, Organopolysilanes, in *Comprehensive Organometallic Chemistry*, G.A. Wilkinson, F.G.A. Stone and E.W. Abel, Eds., Pergamon Press, Oxford, 1983, Vol. 9, 365.
[136] R. West and J. Maxka. Chapter 2: Polysilanes High polymers: An Overview, in *Inorganic and Organometallic Polymers*, edited by M. Zeldin, K.J. Wynne and H.R. Allcock, ACS *Symposium Series*, 1988, Vol. 360, p. 41.
[137] K.J. Wynne, Chapter 1: *An introduction to inorganic and organometallic polymers, Inorganic and Organometallic Polymers*, edited by M. Zeldin, K.J. Wynne and H.R. Allcock, *ACS Symposium Series*, 1988, Vol. 360, pp. 1–4.
[138] S. Hayase, Polysilanes for semiconductor fabrication, Polysilanes for semiconductor fabrication, Progress in Polymer Science, 2003, 28, 359–381.

Chapter 5
Polycarbosilanes

Polycarbosilanes (PCS), which contain silicon and carbon atoms in their main chains, and organic substituents and/or hydrogen atoms bound to Si and C, are one of the most important preceramic polymers. Many publications and patents have been devoted to PCS. Different kinds of PCS were prepared by Yajima et al. via thermal rearrangement of polysilanes at temperature above 400 °C under pressure and argon atmosphere (see Scheme 5.1) [1–3]. In Yajima's process, poly(dimethylsilane) $-(Me_2Si)_n-$ was transferred on heating under reduced pressure into soluble and fiber-forming PCS [2], which is a silicon analogue of polyisopropylene $-[CH_2SiMe(H)]_n-$ [4, 5]:

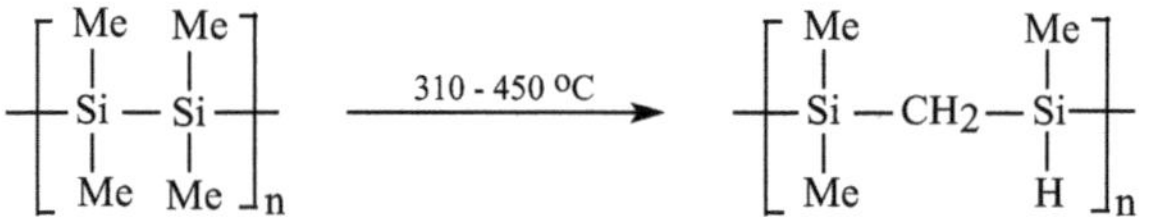

Scheme 5.1: Rearrangement of poly(dimethylsilane) into polycarbosilane.

This process proceeds similarly to Kumada's rearrangement of hexamethyldisilane [6]:

$$Me_3SiSiMe_3 \xrightarrow{600\ ^\circ C} Me_3SiCH_2SiMe_2H \tag{5.1}$$

Since $-(Me_2Si)_n-$ was insoluble in most lower boiling organic solvents and was infusible, it was considered as a nonattractive polymer before Yajima's discovery. Yajima's PCS with an idealized approximate structure $-[CH_2MeSi(H)]_n$- formed fibers of diameter 8–20 μm (by melt spinning) and underwent cross-linking and cyclization processes under drastic reaction conditions. It was proposed that it was transformed into cyclolinear and branched structures (Figure 5.1) [2, 3, 7, 8].

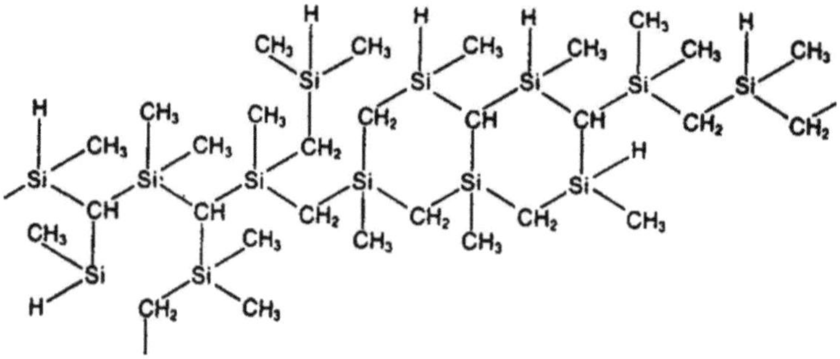

Figure 5.1: A chemical structure of PCS based on chemical composition and proposed by Yajima [2]. Reprinted (adapted) from [2] with permission. Copyright 1978, by Springer Nature.

https://doi.org/10.1515/9783110643671-005

The structure from Figure 5.1 does not contain Si–Si bonds but formation of a certain number was confirmed later [9, 10]. The structure of meltable and processable PCS prepared by pyrolysis of poly(dimethylsilane) was not linear but also cyclolinear and partially branched and was dependent on reaction conditions. This conclusion was based on the chemical composition, and Fourier-transform infrared (FT-IR) and ^{29}Si-MAS NMR spectra [2, 7, 11–15]. At 200 °C in air, the Si–H groups were successible to oxidation, followed by condensation processes. A cross-linked layer on a surface protected fibers from deformations at higher temperatures [10]. At high temperature (1,000–1,100 °C), these PCS fibers underwent cross-linking reactions through Si–H groups, followed by pyrolysis into fibers with β-SiC structure [2, 5–7, 9, 16], which are one of the strongest inorganic fibers, known under the commercial names of Nicalon™ and High Nicalon™. Cross-linking of PCS was achieved by thermal curing in air atmosphere or e-beam curing [2, 9, 16–18]. Poly(borodiphenylsiloxane) was used as a catalyst of Yajima's process [19], leading to PCS containing some boron atoms in their structures [8].

By thermal decomposition of polydimethylsilane $-(Me_2Si)_n-$ at 450 °C under high pressure (10 MPa) for 6 h, PCS was synthesized, which was soluble in xylene and was the precursor of SiC fiber. The composition, structure, and properties of PCS were characterized by infrared (IR), gel permeation chromatography (GPC), NMR, the measurements of softening point, elemental analysis, thermogravimetric analysis (TGA), X-ray diffraction (XRD), and oxidative reaction activity. PCS with a Si–C backbone had M_n ~1,587. IR and ^{1}H- and ^{29}Si-NMR spectra showed the presence of SiC_4 and SiC_3H structure units containing $Si–CH_3$, $Si–CH_2–Si$, and Si–H groups. An empirical formula of $SiC_{1.87}H_{7.13}O_{0.03}$ was based on the elemental analysis data [14].

Thermooxidative stability of multiwalled carbon nanotubes (MWCNTs) coated with SiC (derived from polydimethylsilane as PCS precursor) was improved, as a result of heating to ~1300°C under an inert atmosphere [20]. During pyrolysis of $-(Me_2Si)_n-$, which was carried out in the presence of zeolite ZSM-5 in nitrogen atmosphere, a liquid PCS with molecular weight (MW) of 350–530 g/mol was obtained and some Si–Si bonds remained unchanged [14]. By thermal decomposition of $-(Me_2Si)_n-$ at high temperature and under high pressure, PCS with a Si–C backbone and a number average MW M_n of ~1,600 g/mol was synthesized. It had approximate elemental composition of $SiC_{1.94}H_{5.01}O_{0.028}$, as compared to PCS with similar softening point prepared under normal pressure, but higher Si–H bond content and reaction activity, higher MW, and higher ceramic yield, and lower ratio of SiC_3H and SiC_4 units. A TGA of the PCS at 1,200 °C in a N_2 flow showed that the ceramic yield reached 78.9% [14]. At 900–1,200 °C at normal pressure, a nanocrystalline β-SiC, containing small amount of α-cristobalite, was formed with 85% yield from $-(Me_2Si)_n-$ [21].

Polyphenylcarbosilane (PPCS) was synthesized by thermal rearrangement of poly(methylphenylsilane) at 350–430 °C (Figure 5.2).

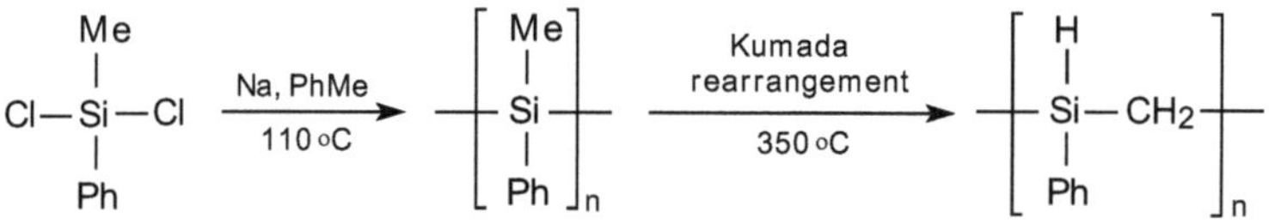

Figure 5.2: Preparation of polyphenylcarbosilane by Würtz coupling followed by thermal rearrangement [22].

It was characterized by FT-IR, ^{29}Si-, ^{13}C-, and ^{1}H-NMR spectroscopy, TG, XRD, and GPC analyses. The average M_w of the PPCS was 2,500 g/mol, and it was easily soluble in organic solvents. TGA data indicated that PPCS was thermally stable of up to 200 °C. Its pyrolysis at 1,200 °C led to β-SiC ceramics. The ceramic yield based on TGA data was ~60% [22]. PCS containing allyl and hydrosilyl functional groups (SMP-10) is commercially available [23].

Schilling et al. prepared branched PCS of different molecular compositions by Wűrtz coupling of methylvinyldichlorosilane (MVDCS, $MeViSiCl_2$) and Me_3SiCl or $MeViSiCl_2$ and Me_2SiCl_2 or $MeHSiCl_2$ and Me_3SiCl with potassium in tetrahydrofuran (THF) [15, 24, 25].

Dichlorodisilyl-substituted methanes, prepared from dichloromethane, magnesium, and chlorosilanes in THF, were also used as comonomers for the syntheses of PCS. Dunogues et al. prepared PCS by the reductive coupling of (dichlorodisilyl)-methane with alkali metal [26, 27]. Similarly, the dehalogenation of diorgano(dichloro)silanes with dihalomethanes and alkali metals afforded PCS [28, 29].

The copolymerization of MVDCS and styrene was carried out with various monomer feed ratios with sodium in toluene at 110 °C. Mainly insoluble polymers were obtained in 84–95% yields by the copolymerization of MVDCS and styrene with monomer to styrene ratios of 1:0.25 and 1:0.5. The copolymers prepared with monomer ratios of 1:1, 1:3, and 1:7 contained 3%, 26%, and 47% of soluble fractions, respectively. The copolymers (Fig. 5.3) were characterized with FT-IR, ^{1}H-NMR, GPC, TGA, pyrolysis/gas chromatography, X-ray diffraction (XRD), and elemental analysis (%Si). The soluble fractions were mainly composed of polystyrene segments [30].

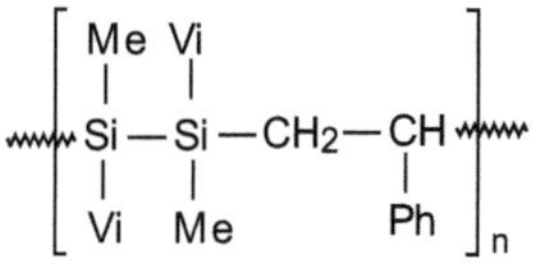

where: Vi = -CH=CH$_2$.

Figure 5.3: A segment of a chemical structure of $-(MeViSi)_m(CH_2CHC_6H_5)_n-$ [30].

Other methods of syntheses of PCS include Grignard coupling reactions of (chloromethyl)chlorosilanes [31], hydrosilylation of vinylhydridosilanes, and dehydrocoupling reaction of trimethylsilane (Me_3SiH) [13]. Interrante and Whitmarsh prepared a highly branched Si–H-functionalized PCS by the Grignard coupling of (chloromethyl) trichlorosilane ($ClCH_2SiCl_3$) with magnesium in diethylether, followed by reduction with $LiAlH_4$ [31]. In the initial step of the polymerization, a nearly quantitative formation of the Grignard compound Cl_3SiCH_2MgCl took place. The coupling of Grignard reagent proceeded almost exclusively through *head-to-tail* (Si–C) manner, and due to its trifunctional $SiCl_3$ "tail," a complicated, branched, PCS polymer was formed, which contained the following structural units: $-SiCl_3CH_2-$, $-SiCl_2CH_2-$, $=SiClCH_2-$, and $\equiv SiCH_2-$. The chlorofunctional PCS underwent side reactions with ether, which led to incorporation of small amounts of ethyl and ethoxy functionality. During the reduction step, the ethoxy groups were eliminated, giving the polymer with the approximate formula $[SiH_{1.85}Et_{0.15}CH_2]_n$, which was characterized by ^1H, ^{13}C, and ^{29}Si-NMR, IR, GPC, and elemental analysis. This polymer was a precursor to near-stoichiometric silicon carbide [31].

A poly(methylsilane carbosilane) copolymer (PMSCS) was synthesized by Würtz-type copolycondensation of methyldichlorosilane ($MeHSiCl_2$) with (chloromethyl)dichloromethylsilane ($ClCH_2MeSiCl_2$) and terminated with vinylmagnesium chloride. Through pyrolysis of the PMSCS at 1,000 °C under argon, silicon carbide with nearly stoichiometric C:Si ratio and low oxygen content was prepared in 64% of ceramic yield. The properties of the PMSCS and the C:Si ratio of its derived ceramic were tuned by changing the ratios of both monomers and the end-block agent. PMSCS can be used as a SiC ceramic precursor for the fabrication of SiC matrix, coating, and adhesives [32]. A new PMSCS was also prepared by copolycondensation of $MeHSiCl_2$, $ClCH_2MeSiCl_2$, and (dichloromethyl)methylsilane ($Cl_2CHMeSiH_2$), with $ClCH_2MeSiH_2$ or Me_3SiCl as a terminating reagent. An incorporation of $Cl_2CHMeSiH_2$ and capping with $ClCH_2MeSiH_2$ markedly increased yield of the ceramic residue. Pyrolysis at 1,000 °C of the $H_2MeSiCH_2$-capped PMSCS with $CHMeSiH_2$ structure unit in main chain gave SiC ceramic with 1.21 of C/Si atomic ratio and 78 wt% ceramic yield. The CH_xMeSiH_2-containing PMSCS, prepared by inexpensive method, is a new cost-effective SiC ceramic precursor having excellent pyrolytic properties, good storage stability in air, and beneficial processability [33]. Poly(hydro)carbosilanes were also obtained by the reduction of halogenated poly(methylphenylsilmethylene) and poly(diphenylsilmethylene) with $LiAlH_4$ [34, 35].

Ring-opening polymerization (ROP) of 1,3-disilacyclobutanes (DSCB) catalyzed by Pt-containing complexes is other important method of PCS syntheses [36–38]. The DSCB and its derivatives were polymerized in the presence of $H_2PtCl_6.6H_2O$ at 80 °C. The obtained PCS was soluble in cyclohexane. The catalyst was removed by filtration and volatile compounds – by evacuation at 100 °C/0.5 mmHg, giving a brown rubber. Poly[(dimethylsilylene)methylene $-(Me_2SiCH_2)_n-$ ($M_n \approx 250{,}000$) with linear structure was converted into a poly[(methylsilylene)methylene] ($M_n \approx 2{,}300$). Reaction of trimethylchlorosilane with poly(dimethylsilylene)-*co*-methylene] $-(Me_2SiCH_2)_n-$ in the presence of a catalytic amount of $AlCl_3$ gave quantitatively a novel chlorinated PCS, $-(MeSiClCH_2)_n-$ [39, 40].

Following Kriner's procedure [40], the ROP of 1,1,3,3-tetramethyl-1,3-disilacyclobutane with Pt catalyst gave poly(dimethylsilmethylene) $-[(CH_3)_2SiCH_2]_n-$, with high number average MW (M_n of 250,000 g/mol) and polydispersity (PDI) M_w/M_n = 1.45. This PCS was sensitive to thermooxidative degradation at 200 °C in the presence of air but was thermally resistant almost up to 450 °C in inert atmosphere, exhibiting only 3% weight loss, while it decomposed rapidly between 450 and 550 °C, giving a low ceramic yield. A high MW linear atactic PCS, poly(methylhydrosilylene-*co*-methylene) $[(CH_3)Si(H)CH_2]_n-$, which is also called as polysilapropylene, was prepared by the ROP of 1,3-dichloro-1,3-dimethyl-1,3-disilacyclobutane, followed by reduction with $LiAlH_4$. The structures of monomeric precursors and PCS were studied by spectroscopic methods (IR, MS, ^{1}H-, ^{13}C-, and ^{29}Si-NMR), and GPC. The pyrolysis of the poly(silapropylene) at 400 °C led to 66% ceramic yield [41].

Poly(dichlorosilaethylene) was prepared by ROP of 1,1,3,3-tetrachloro-1,3-disilacyclobutane (I) catalyzed by chloroplatinic acid or Karstedt's catalyst (benzene solution of a platinum–$(ViMe_2Si)_2O$ complex). Direct reduction of this product in benzene led to the corresponding linear poly(silaethylene) $-(SiH_2CH_2)_n$, viscous liquid, which was soluble in hydrocarbons. The GPC indicated a monomodal MW distribution (M_n = 12,300, M_w = 33,000 g/mol, vs. polystyrene standards). Its pyrolysis in N_2 at 1,000 °C gave remarkably high ceramic yields of SiC (87 wt%). The powder XRD study of the ceramic product, obtained after pyrolysis to 1,000 °C in N_2, heated at that temperature for 1 h, confirmed the formation of β-SiC. The average crystallite size of SiC was 2.5 nm, as calculated by the line broadening of the XRD pattern [37]. The IR spectrum of the ceramic residue obtained after pyrolysis to 1,200 °C showed only one absorption band at 823 cm^{-1}, consistent with the formation of SiC [8]. These studies revealed that high MW linear PCS (LPCS), when substituted with H on Si, can undergo relatively facile thermally induced cross-linking prior to, or accompanying, chain scission. Moreover, along with the results based in earlier studies of poly(silapropylene) [41], it was concluded that an initially cross-linked structure is a requirement for high ceramic yields.

Poly(1,1-dimethylsilabutane) with M_n = 2,400 g/mol and narrow PDI of MW (M_w/M_n = 1.10) was prepared in 99% yield by a living anionic polymerization of 1,1-dimethylsilacyclobutane with 5 mol% of butyllithium, in THF–hexane (1: 1) at −48 °C. Similarly, 1,1-diethylsilacyclobutane and 1-methyl-1-phenylsilacyclobutanewere polymerized. Addition of styrene to the living poly(1,1-dimethylsilabutane) gave a poly (1,1-dimethylsilabutane-*b*-styrene) block copolymer. Similarly, the living polymerization of 1,1-diethylsilacyclobutane in THF–hexane at −48 °C was carried out. However, the polymerization of 1-methyl-1-phenyl-silacyclobutane in THF at −78 °C did not show a living nature [42].

The polymerization of disilacyclobutanes gave high elastomeric, thermostable poly(silamethylenes) and poly(silatrimethylenes). Thermal polymerization (T-ROP) of 1,3-dimethyl-1,3-diphenyl-1,3-disilacyclobutane (DDDC) was initiated at 170 °C and 1,1,3,3-tetraphenyl-1,3-disilacyclobutane (TPDC) at ~300 °C [43–45]. Sometimes, thermal initiation was combined with other impacts, for example, laser irradiation, to support heterogeneous catalysts (e.g., nanoparticles of Pt, Cu, and Ag) [45]. In the case of thermal ROP of cyclocarbosilanes, MWs of PCS were controlled by the addition of small amounts of trimethyl(chloro)silane to the reaction mixture [43]. The T-ROP of DDDC at 230 °C in the absence of Me_3SiCl gave PCS with M_w = 2,640,000 (M_w/M_n = 4.1) and a yield of 75%. At Me_3SiCl:monomer ratio of 0.08, 0.37, and 3.9, the yields of PCS with different MWs 648,000 (M_w/M_n = 4.1), 269,000 (M_w/M_n = 2.1), and 90,000 (M_w/M_n = 2.0) were 79%, 81%, and 55%, respectively. These PCS can be used as precursors for fabrication of SiC ceramics and photoresists.

Poly(phenylsilylenemethylene)s were prepared by ROP of phenyl-substituted 1,3-disilacyclobutanes with or without catalyst. Copper compounds were used as catalysts for ROP of TPDC and DDDC, regardless of the valence of copper and the counter anion. However, the catalytic activities of the copper compounds in polymerization of DDDC were quite low in comparison with platinum or rhodium catalysts. Copolymerization of these two cyclic dimers without catalyst gave copolymers with $[MePhSiCH_2]/[Ph_2SiCH_2]$ ratios very close to the starting compositions. It was suggested from the dimer reactivity ratio that DDDC copolymerized with TPDC giving almost random copolymers, while the dimer sequential length distributions in the copolymers were slightly different from an ideal distribution of two dimeric units [43].

1,3-Disilacyclobutane [cyclic$(H_2SiCH_2)_2$] was polymerized in the presence of H_2PtCl_6, providing preceramic polymer, which is a silicon analogue of linear polyethylene [46].

The ROP of cyclic carbosilane monomers is the most effective method of PCS synthesis. It concerns the ROP of strained cyclocarbosilanes, ring-opening metathesis polymerization of silylcycloolefins, and vinyl-type addition polymerization (AP) of silyl-norbornenes [47].

PCS-containing reactive functional groups such as Si–Cl, Si–OR, and Si–H were obtained by the ROP of appropriate cyclobutane monomers. The ethoxy-substituted PCS $-[Si(OEt)_2CH_2]_n-$ underwent hydrolytical condensation with water and gave a gel with inorganic Si–O–Si and organic Si–CH_2–Si network structure bridges as units, which was further pyrolyzed into silicon oxycarbide ceramic material SiO_xC_{4-x} [48].

Asymmetrically substituted poly(silylenemethylene)s (PSM) of the structures $[SiHRCH_2]_n$ and $[SiMeRCH_2]_n$ were synthesized by the ROP of DSCB containing *n*-alkyl substituents (R=C_2H_5, *n*-C_3H_7, *n*-C_4H_9, *n*-C_5H_{11}, and *n*-C_6H_{13}, or a phenyl group) bound to silicon. All of these PSMs had high MW (M_w = 11,200 g/mol), except for a silicon analogue of polystyrene, $(SiHPhCH_2)_n$ (M_w = 70,900–108,900 g/mol), and quite high PDIs of 2.5–4.6. Their configurations were atactic. The aryl-substituted PSM showed enhanced thermal stability and higher glass transition temperature (T_g = −37.6 °C) than for the alkyl-substituted PSMs (T_gs = −91.4 ÷ −61.2 °C). The alkyl-substituted $[SiMeRCH_2]_n$ PSM reached a maximum T_g at −61 °C (for the R = *n*-Pr-substituted polymer). PSM had lower T_gs than their all-carbon analogues [49].

Another series of new linear PSM was prepared by ROP of a substituted disilacyclobutane of the structure $[SiRR'CH_2]_n$ with a wide range of different substituents R and R′. Symmetrically disubstituted PCS (with R = R′ = F, alkyl, and alkoxy) formed crystalline solid phases, and various amorphous atactic polymers had different R and R′ groups. The linear analogue of linear polyethylene $[SiH_2CH_2]_n$ served as the high-yield precursor to SiC [50].

R.J.P. Corriu [51, 52] prepared PCS directly from alkenylsilanes – by platinum-catalyzed hydrosilylation, for example, from vinyl(hydro)dichlorosilane $CH_2{=}CHSiHCl_2$, followed by reduction of Si–Cl groups with $LiAlH_4$. A low C:Si ratios in the polymer seemed to be a profitable factor, while availability of starting monomer might be a serious problem for future practical applications.

Polymerization of vinylsilane $CH_2{=}CHSiH_3$ with dimethyltitanocene catalyst at room temperature for 30 days gave 82 wt% of polymers, consisting of a low MW PCS (M_n = 540, M_n = 1,800) as a major product (in 74% yield) and polysilane as a minor product (in 26% yield). The hydrosilylation of alkenylsilanes was the main reaction [53]. Pyrolysis of PCS obtained from alkenylsilanes at 1,400 °C led to relatively high char yields. Formation of crystalline β-SiC was optimum for a copolymer of an alkylsilane and an alkenylsilane with a silane/carbosilane backbone ratio of 85/15 and a C/Si ratio of 1.3/1 [53, 54].

Polymerization of methylsilane with a titanocene catalyst gave a polysilane, while poly(vinylsilane) (PVS) polymers had mainly PCS backbone, accompanied with some polysilane structure. The pyrolysis path at high temperature (650–1,400 °C) and char yield was dependent primarily on the backbone structure and slightly on polymer MW. PVS formed a carbon-rich Si–C ceramic with homogeneously dispersed C on a sufficiently fine level providing resistance to oxidation on heating in air to 1,400 °C. Copolymerization of methyl- and vinylsilane gave stoichiometric SiC. Polymers of methylsilane were sensitive to oxidation and sometimes were pyrophoric [55].

Allyl-functionalized hyperbranched PCS was synthesized by UV-activated polyhydrosilylation of methyl(diallyl)silane $MeHSi(CH_2CH{=}CH_2)_2$, AB_2 monomer, with bis(acetylacetonato)-platinum(II) as the catalyst. The polymerization process was monitored by real-time FT-IR spectroscopy and the obtained hyperbranched PCs were characterized using ^1H, ^{13}C ^{29}Si-NMR, and size exclusion chromatography/multi-angle laser light scattering (SEC/MALLS). The hydrosilylation activated by UV irradiation was much faster than thermal process [56].

Transparent liquid oligo(vinylsilane)s with structures $[-CH_2CH(SiH_3)-]_x[-CH_2CH_2SiH_2-]_y$ and $[-CH(CH_3)SiH_2-]$, respectively, and MW (M_n) of 500–1,500 g/mol were prepared in high yields with a radical initiator catalyst and by anionic polymerizations of vinylsilane ($CH_2{=}CHSiH_3$). A slightly soluble in a solvent white solid polymer with the structure $[-CH_2CH(SiH_3)-]_n$ was obtained by the coordinated anionic polymerization with Ziegler–Natta catalyst. These polymers had the reactive Si–H bonds and were stable under air. By their pyrolysis at high temperature under inert gas, silicon carbide was obtained [57].

Unsaturated hybrid PCS with silicon atoms adjacent to π-conjugated segments (phenylene, ethenylene, or diethylene) [58–62] were synthesized by coupling reactions [60, 63], thermal cyclopolymerization [64, 65], ROP reactions including anionic [66–69], thermolytic, and catalytic coordination techniques [9]. Novel unsaturated polycarbosilanes were synthesized by the catalytic acyclic diene metathesis (ADMET) reaction of unsaturated oligosilanes in the presence of the Grubbs catalyst – a ruthenium carbene complex ($RuCl_2(PCy_3)_2(=CHPh)$) [70].

The carbosilane polymers containing C=C double bonds in their chains and reactive Si–H bonds gave high char yields upon pyrolysis. Anionic ROP (AROP) of 1-silacyclo-3-pentene was catalyzed by methyllithium in HMPA and gave low MW poly(1-sila-*cis*-pent-3-ene). Its pyrolysis gave high char yields: up to 450 °C – ~90%, and above this temperature ~60% [71]. Similarly, poly(1-phenylsila-*cis*-pent-3-ene) was synthesized by ROP of 1-phenylsila-*cis*-pent-3-ene with *n*-BuLi (Figure 5.4). Its pyrolysis gave only ~22% char yield in nitrogen and ~38% in air [72].

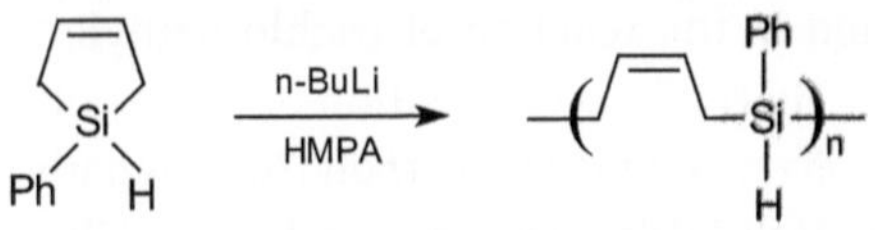

Figure 5.4: ROP of 1-phenylsila-*cis*-pent-3-ene with *n*-BuLi [71].

The obtained unsaturated PCS were analyzed by FT-IR, ^{13}C and ^{29}Si-NMR, UV, GPC, TGA, energy dispersive spectra (EDS), end group determination, and elemental analysis [71, 72].

Low MW PCS (M_n = 6,000–11,000 g/mol) of regular structure containing pendant vinyl groups were prepared by the ROP of 1-vinyl-1-silacyclobutane with *n*-butyllithium, in THF or hexamethyl phosphoramide solution, with 60–88% yield. Similarly, poly(1-phenyl-1-vinyl-1-silabutane) and poly(1,1-divinyl-1-silabutane) were synthesized. The thermal stability of these 1-vinyl-1-silabutane polymers was analyzed by TGA in a nitrogen atmosphere. Poly(1-methyl-1-vinyl-1-silabutane) and poly(1-phenyl-1-vinyl-1-silabutane) were thermally stable up to 400 °C. At 400–450 °C, they almost completely decomposed, giving <12% char yield. Poly(1,1-divinyl-1-silabutane) was thermally stable to almost 450 °C. Thermal degradation of poly(1,1-divinyl-1-silabutane) at 450–500 °C in an inert atmosphere gave a higher char yield (45%), but no further weight loss occurred up to 750 °C, probably due to cross-linking processes [73].

PCS containing ethylenic and acetylinic functionality were prepared from poly (dichlorosilmethylene)s [74]. New poly(dialkenylsilylenemethylene) copolymers of the structure $-[R_2SiCH_2]_n$ [R = $CH{=}CH_2$, $CH_2CH{=}CH_2$, $CH_2C(CH_3){=}CH_2$, $(CH_2)_2CH{=}CH_2$, $(CH_2)_3CH{=}CH_2$] were synthesized by the reaction of poly(dichlorosilylene-*co*-methylene) with appropriate alkenylmagnesium chlorides or bromides in diethyl ether–THF solution. These copolymers were characterized by elemental analysis, FT-IR, ^1H-,

^{13}C-, and ^{29}Si-NMR. Their MWs (M_n = 780–1,850 g/mol, M_w = 30,800–254,300 g/mol, M_w/M_n = 31.7–270.5) were determined by HPLC analysis. Pyrolysis of these copolymers, carried out in argon atmosphere, afforded ceramic residue (SiC), with char yields ranging from 45.9 to 68.5 wt% at 1,000 °C and 22.1–55.1 wt% at 1,700 °C [74]. Habel et al. also elaborated the synthesis of poly(alkynylsilylenemethylene)s $-[R_2SiCH_2]_n$ (where R = $-C{\equiv}CH$, $-C{\equiv}C(CH_2)_nCH_3$, or $-C{\equiv}C-C_6H_5$) [75].

Poly(siladiacetylene)s were synthesized by the reaction of diorganodichlorosilanes ($RR'SiCl_2$) (where R = Me, Vi, or aryl group) or 1,2-dichlorodisilane ($ClMe_2SiSiMe_2Cl$) with dilithioacetylene or diacetylene Grignard reagents. Their MWs determined by GPC (M_w) ranged from 1,600 to 4,060 g/mol and PDIs (M_w/M_n): 1.24–1.51. These organosilicon polymers were insulators, exhibiting poor conductivity values, ranging from 10^{-12} to 10^{-15} S/cm. After doping with the acceptor $FeCl_3$, their conductivity increased up to the range 10^{-3}–10^{-5} S/cm, while doping with the stronger acceptors (SbF_5 or HNO_3) led to higher conductivity values [60].

The silylene–acetylene polymers $-[R_2SiC{\equiv}C]_n$ were prepared by quenching dilithioacetylene, LiC≡CLi (which was obtained in the reaction of trichloroethylene $Cl_2C{=}CHCl$ with n-butyllithium) and R_2SiCl_2. Thermal decomposition of these polymers gave high yield of a char, composed of silicon carbide and carbon [76]. Similarly, silylene–diacetylene polymers of relatively high MWs were prepared in excellent yields by "one-pot" method elaborated by Maghsoodi and Barton, by dechlorination of hexachlorobutadiene to dilithiobutadiyne with four equivalents of n-BuLi in THF, followed by quenching the resulting solution with different dichlorosilanes or 1,3-dichlorotetramethyldisilane [77].

Thermally stable PSM having siloxane cross-linking moieties with quite low T_g values, ranging from 15 to 20 °C were prepared by Ogawa et al. [78].

Protodephenylation of phenylated polysilanes as well as PSM with triflic acid gave new functionalized compounds [79].

A series of linear and hyperbranched PCS was prepared by hydrosilylation reaction of monomers containing two Si–H groups [1,4-bis(dimethylsilyl)benzene, diethylsilane (Et_2SiH_2), and dihydrosilsesquioxane, A_2] with monomers containing di-, tri-, and tetravinyl groups (B_n, n = 2–4) in the presence of Karstedt's catalyst. The corresponding linear polycarbosilanes (LPCS) and hyperbranched PCS (HBPC) with a number-average MW (M_n) 2,200–51,500 g/mol were obtained in 34–94% yield. The values of refractive index (n_D) of the synthesized LPCS and HBPC ranged from 1.460 to 1.711. High refractive index hyperbranched PCS were prepared by polyaddition of A_2 with B_n (n = 3 and 4) and the values of n_D increased with increase M_n, PDI (M_w/M_n), and glass transition temperature (T_g), probably due to increased density and the presence of microgels [80].

Cross-linked PCS films with superior mechanical properties, insensitive to moisture, were prepared by spin-coating and thermal curing of hexylene-bridged disilacyclobutane rings (Scheme 5.2) [81]. The obtained films showed good thermal stability and hydrophobicity, a low dielectric constant (k = 2.5), and were free of mesopores and supermicropores.

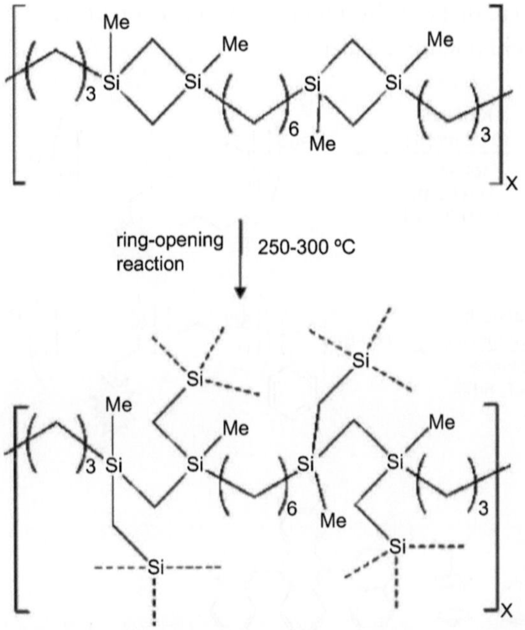

Scheme 5.2: Preparation of cross-linked PCS films via ROP of disilacyclobutane units. Reprinted from [81] with permission. Copyright 2012, by American Chemical Society.

Due to their unique molecular structure, the elastic stiffness and fracture resistance of these films substantially exceeded those of traditional porous organosilicate glasses. The films were remarkably hydrolytically stable and should find numerous demanding applications, providing protection against moisture and harsh environments [81].

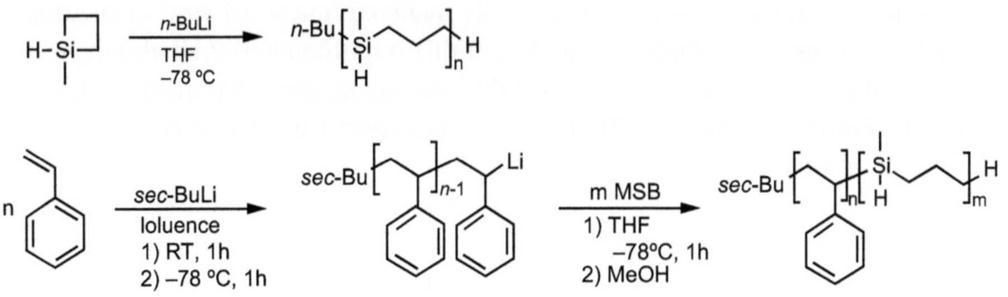

Figure 5.5: Anionic ROP of 1-methylsilacyclobutane (MSB) (top) and block copolymerization by sequential monomer addition of styrene and MSB (bottom). Reprinted from [82] with permission. Copyright 2015, by Royal Chemical Society.

Polyhydrocarbosilane $[Si(H)MeCH_2CH_2CH_2]_n$ (PHCS) was synthesized by AROP of 1-methyl-1-silacyclobutane $[Si(H)MeCH_2CH_2CH_2]$ (MSB), while block copolymers of polystyrene and PHCS (PS-*b*-PHCS) were prepared by anionic polymerization of styrene with *sec*-BuLi, followed by AROP of MSB (Figure 5.5) [82].

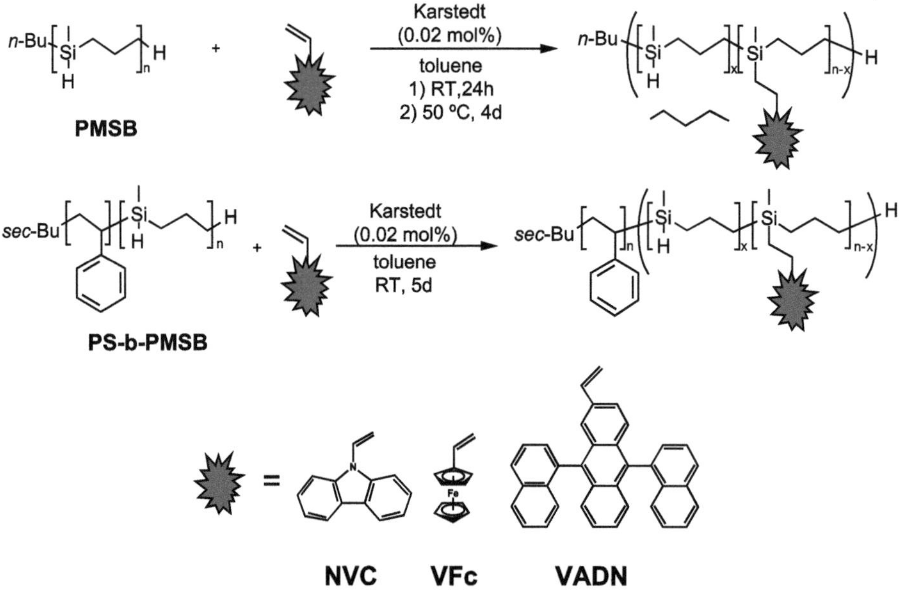

Figure 5.6: Hydrosilylation reaction of PMSB with functional vinyl compounds using Karstedt's catalyst (top) and hydrosilylation reaction of diblock copolymers consisting of polystyrene and PMSB (PS-*b*-PMSB). Reprinted from [82] with permission. Copyright 2015, by Royal Chemical Society.

The PS-*b*-PHCS were further modified by hydrosilylation reaction with 26–43 mol% (with respect to the Si–H content) of sterically hindered functional vinyl compounds: *N*-vinyl-carbazole, vinylferrocene, and 9,10-di(1-naphthalenyl)-2-vinylanthracene (Figure 5.6) [82]. These novel functional diblock copolymers contained 28–63 mol % modified units, as it was confirmed by ^{1}H NMR spectroscopy and showed microphase separation [82].

Cyclolinear PCSs (CLPCS) containing 1,3-disilacyclobutane (DSCB) rings, that is, cyclo-{RSi(CH_2)$_2$SiR}–, bridged by linear chains of various linking groups were prepared by two different synthesis methods: Grignard coupling reactions and ADMET polymerization. Strained rings served as cross-linking sites. The ring-chain CLPCS had the general formula [–{RSi(CH_2)$_2$SiR}–Y–]$_n$, where R and Y can be a variety of different groups, such as R = $-CH_3$, $-C_6H_5$, and $-C_6H_4(CH_3)$; Y = $-(CH_2)_x-$, $-C_6H_4$-(CH=CH)-C_6H_4–, and $-C_6H_4-$. Their physical properties were dependent on the kind of the silicon substituents, R, and the bridging group, Y. They showed a relatively low glass transition temperature (−75 ÷ −55 °C), ability to the formation of thin films and coatings via spin coating, and the tendency to undergo thermal cross-linking at temperature 200–300 °C. Their curing was faster in contact with Cu surface at temperature of 160 °C. They could find applications as protective, insulating coatings, photoluminescent films, and in microelectronics processing (as low-k materials, $k < 2.2$) [83].

In conclusion, it seems quite obvious that PCS exhibit a special range of promising physicochemical properties. Depending on their macromolecular structure, the PCS can be glassy, elastomeric, or thermosetting materials having T_g values in the range from −100 to +350 °C. They can be applied as precursors for polymer-derived ceramics (see Chapter 8), photoresist components, thermostable adhesives, liquid crystals, electrosensitive materials, highly permeable membranes for separation of light and hydrocarbon gases, and so on.

The major limitation in wide practical applications of PCS is concerned with quite complex synthetic methods and their costs. It seems necessary to elaborate technologically suitable methods of preparation of monosilacyclobutane, disilacyclobutane, silaferroceno-, and sila-arenophane monomers, and to find the efficient catalysts for vinyl type addition polymerization (AP) of norbornenes with bulky substituents [47].

References

[1] S. Yajima, M. Omori, J. Hayashi, K. Okamura, T. Matsuzawa and C.F. Liaw, Simple synthesis of the continuous SiC fiber with high tensile strength, *Chemistry Letters*, 1976, 5(4), 551.

[2] S. Yajima, Y. Hasegawa, J. Hayashi and M. Imura, Synthesis of continuous silicon carbide fibre with high tensile strength and high Young's modulus, *Journal of Materials Science*, 1978, 13(12), 2569.

[3] S. Yajima, *American Ceramic Society Bulletin*, Special Heat-Resisting Materials from Organometallic Polymers, 1983, 62(8), 893.

[4] S. Yajima, J. Hayashi, M. Omori and K. Okamura, Development of a silicon carbide fibre with high tensile strength, *Nature*, 1976, 261(8), 683.

[5] Y. Hasegawa, M.J. Imura and S. Yajima, Synthesis of continuous silicon carbide fibre, Part 2 Conversion of polycarbosilane fibre into silicon carbide fibres, *Journal of Materials Science*, 1980, 15(3), 720.

[6] K. Shiina and M. Kumada, Thermal Rearrangement of Hexamethyldisilane to Trimethyl (dimethylsilylmethyl)silane, *Journal of Organic Chemistry*, 1958, 23(1), 139.

[7] Y. Hasegawa, T. Kobori and K. Fukuda, *U.S. Pat.* 4,590,253, 1986.

[8] W. Habel, C. Nover and P. Sartori, *Chemische Zeitung*, 1990, 114, 305.

[9] Y. Hasegawa and K. Okamura, Synthesis of continuous silicon carbide fibre, *Journal of Materials Science*, 1983, 18(12), 3633.

[10] J. Chojnowski, *Polimery*, Krzemowe polimery preceramiczne, 1987, 32(5), 167. In Polish.

[11] D. Seyferth, Chapter 3: *Polycarbosilanes – An overview, in ACS Symposium Series*, 1988, 360, pp. 21–42.

[12] M. Birot, J.-P. Pilot and J. Dunogues, Comprehensive chemistry of polycarbosilanes, polysilazanes, and polycarbosilazanes as precursors of ceramics, *Chemical Reviews*, 1995, 95(5), 1443.

[13] Y. Kim, D.G. Shin, H.R. Kim, H.S. Park and D.H. Riu, Preparation of Polycarbosilane using a Catalytic Process and Its Practical Uses, *Key Engineering Materials*, 2005, 287, 108.

[14] X. Cheng, Z. Xie, Y. Song, J. Xiao and Y. Wang, Structure and properties of polycarbosilane synthesized from polydimethylsilane under high pressure, *Journal of Applied Polymer Science*, 2006, 99(3), 1188.

[15] C.L. Schilling Jr., J.P. Wesson and T.C. Williams, Polycarbosilane precursors for silicon carbide, *Journal of Polymer Science, Polymer Symposia*, 1983, 70(1), 121.

[16] R.M. Laine and F. Babonneau, Preceramic polymer routes to silicon carbide, *Chemistry of Materials*, 1993, 5(3), 260.

[17] S. Yajima, Y. Hasegawa, K. Okamura and T. Matsuzawa, Development of high tensile strength silicon carbide fibre using an organosilicon polymer precursor, *Nature*, 1978, 273(6), 525.

[18] G. Mera, M. Gallei, S. Bernard and E. Ionescu, Ceramic Nanocomposites from Tailor-Made Preceramic Polymers, *Nanomaterials*, 2015, 5(2), 468.

[19] S. Yajima, J. Hayashi and K. Okamura, Pyrolysis of a polyborodiphenylsiloxane, *Nature*, 1977, 266(6), 521.

[20] R.K. Gupta, R. Mishra, K. Mukhopadhyay, R.K. Tiwari, A. Ranjan, and A. Saxena, A New Technique for Coating Silicon Carbide Onto Carbon Nanotubes Using a Polycarbosilane Precursor, Silicon, 2009, 1, 125–129. DOI 10.1007/s12633-009-9016-0

[21] M. Lodhe, N. Babu, A. Selvam and M. Balasubramanian, Synthesis and characterization of high ceramic yield polycarbosilane precursor for SiC, *Journal of Advanced Ceramics*, 2015, 4(4), 307–311. DOI: 10.1007/s40145-015-0165-x.

[22] Y.J. Lee, J.H. Lee, S.R. Kim, W.T. Kwon, H. Oh, J.-H.P. Klepeis, S.J. Teat and Y.H. Kim, Synthesis and characterization of novel preceramic polymer for SiC, *Journal of Materials Science*, 2010, 45, 1025–1031.

[23] S. Kaur, G. Mera, R. Riedel and E. Ionescu, Effect of boron incorporation on the phase composition and high-temperature behavior of polymer-derived silicon carbide, *Journal of the European Ceramic Society*, 2016, 36, 967–977.

[24] C.L. Schilling Jr., J.P. Wesson and T.C. Williams, Polycarbosiiane Precursors for Silicon Carbide, *American Ceramic Society Bulletin*, 1983, 62(8), 912.

[25] C.L. Schilling Jr., T.C. Williams and J.P. Wesson, *U.S. Pat. 4,414,403*, 1983.

[26] J.P. Pillot, E. Bacque, J. Dunogues and P. Orly, *German Patent DE3717450 A1*, 1987.

[27] E. Bacque, M. Birot, J.P. Pillot, P. Lapouyade, P. Gerval, C. Biran and J. Dunogues, Synthesis of disilylmethanes and polysilacarbosilanes, precursors of silicon carbide-based materials, *Journal of Organometallic Chemistry*, 1996, 521(1–2), 99.

[28] B. Van Aefferdan, W. Habel and P. Sartori, *Chemische Zeitung*, 1991, 115, 1336.

[29] C.K. Whitmarsh and L.V. Interrante, Synthesis and structure of a highly branched polycarbosilane derived from (chloromethyl)trichlorosilane, *Organometallics*, 1991, 10(5), 1336.

[30] G. Ambadas, S. Packirisamy, T.S. Radhakrishnan and K.N. Ninan, Synthesis, characterization, and thermal properties of poly(methylvinylsilylene-*co*-styrene), *Journal of Applied Polymer Science*, 2004, 91, 3774–3784.

[31] W. Habel, B. Harnack, C. Nover and P. Sartori, Synthese, charakterisierung und folgereaktionen von verzweigten poly(phenylcarbosilanen), *Journal of Organometallic Chemistry*, 1994, 467(1), 13.

[32] X. Wang, Y. Chen, Y. Li and C. Xu, Evaluation of poly(methylsilane-carbosilane) synthesized from methyldichlorosilane and chloromethyldichloromethylsilane as a precursor for SiC, *Journal of Applied Polymer Science*, 2018, 135, 46610.

[33] X. Wang, Y. Chen, Y. Li and C. Xu, A new SiC precursor with high ceramic yield: Synthesis and characterization of CH_xMeSiH_2-containing poly(methylsilanecarbosilane), *Journal of Applied Polymer Science*, 2019, 136, 47618.

[34] W. Habel, L. Mayer and P. Sartori, Synthese_und_charakterisierung_von_hydridopoly carbosilanen" Synthese und charakterisierung von hydridopoly(carbosilanen), *Journal of Organometallic Chemistry*, 1994, 474(1–2), 63.

[35] J. Smith and L. Troy, *US Patent* 4631179 A, 1986.

[36] H.J. Wu and L.V. Interrante, Preparation of poly (dichlorosilaethylene) and poly (silaethylene) via ring-opening polymerization, *Macromolecules*, 1992, 25(6), 1840.

[37] P. Kochs, Chapter 4: *Poly(organosilanes), poly(carbosilanes), poly(organosilazanes), in Silicon in Polymer Synthesis*, H.R. Kricheldorf (Ed.), Springer-Verlag, Berlin-Heidelberg, 1996, 223–287.

[38] T.L. Smith Jr., *US Patent* 4631179, 1986.

[39] E. Bacque, J. Pillot, M. Birot and J. Dunogues, New Polycarbosilane Models. 1. Poly [(Methylchlorosilylene)Methylene], a Novel, Functional Polycarbosilane, *Macromolecules*, 1988, 21(1), 30.

[40] W.A. Kriner, The Preparation of Cyclic Siliconmethylene Compounds, *Journal of Organic Chemistry*, 1964, 29(6), 1601.

[41] H.-J. Wu and L.V. Interrante, Preparation of a polymeric precursor to silicon carbide via ring-opening polymerization: synthesis of poly[(methylchlorosilylene)methylene] and poly(silapropylene), *Chemistry of Materials*, 1989, 1(5), 564.

[42] K. Matsumoto, H. Shimazu, M. Deguchi and H. Yamaoka, Anionic ring-opening polymerization of silacyclobutane derivatives, *Journal of Polymer Science. Part A, Polymer Chemistry*, 1997, 35(15), 3207.

[43] T. Ogawa, Polymerization and copolymerization behaviour of phenyl-substituted 1,3-disilacyclobutane, *Polymer*, 1998, 39(13), 2715.

[44] T. Ogawa and M. Murakami, Synthesis, Thermal and Mechanical Properties of Poly(methylphenylsilmethylene)s, *Chemistry of Materials*, 1996, 8(6), 1260.

[45] G. Song, M. Yamagushi, O. Nishimura and M. Suzuki, Investigation of metal nanoparticles produced by laser ablation and their catalytic activity, *Applied Surface Science*, 2006, 253(6), 3093.

[46] R.J.P. Corriu, Ceramics and Nanostructures from Molecular Precursors, *Angewandte Chemie, International Edition*, 2000, 39(8), 1376.

[47] E. Sh. Finkelshtein, N.V. Ushakov and M.L. Gringolts, *Chapter: Polycarbosilanes based on silicon-carbon cyclic monomers, advances in polymer science*, 2011, 235, in *Silicon Polymers*, A.M. Muzafarov (Ed.), Springer-Verlag, Berlin – Heidelberg, 2010, 111–159, Published online: 26 Feb. 2010. DOI: 10.1007/12_2009_39.

[48] L.V. Interrante, Q. Liu, I. Rushkin and Q. Shen, Poly(silylenemethylenes) – A novel class of organosilicon polymers, *Journal of Organometallic Chemistry*, 1996, 521(1–2), 1.

[49] Q.H. Shen and L.V. Interrante, Preparation and thermal properties of asymmetrically substituted poly(silylenemethylene)s, *Journal of Polymer Science. Part A, Polymer Chemistry*, 1997, 35(15), 3193.

[50] L.V. Interrante, I. Rushkin and Q. Shen, Linear and hyperbranched polycarbosilanes with Si-CH2-Si bridging groups: A synthetic platform for the construction of novel functional polymeric materials, *Applied Organometallic Chemistry*, 1998, 12(10–11), 695.

[51] B. Boury, L. Carpenter and R.J.P. Corriu, A New Way to SiC Ceramic Precursors by Catalytic Preparation of Preparation Polymers, *Angewandte Chemie, International Edition English*, 1990, 29(7), 785.

[52] B. Boury, R.J.P. Corriu, D. Leclercq, P.H. Mutin, J.-M. Planeix and A. Vioux, Poly(vinylsilane): A Precursor to Silicon Carbide. 1.1 Preparation and Characterization, *Organometallics*, 1991, 10 (5), 1457.

[53] J.F. Harrod, Chapter 7: *Polymerization of group 14 hydrides by dehydrogenative coupling, in Inorganic and Organometallic Polymers*, M. Zeldin, K.J. Wynne and H.R. Allcock (Eds.), *ACS Symposium Series*, 1988, Vol. 360, 89–100, and references cited therein.

[54] J. Masnovi, X.-Y. Bu, K. Beyene, P.J. Heimann, T.A. Kacik, A.H. Andrist and F.I. Hurwitz, Syntheses, structures and properties of polycarbosilanes formed directly by polymerization of alkenylsilanes, NASA-CR-191927; https://ntrs.nasa.gov./search.jsp?R=19930012000

[55] F.I. Hurwitz, T.A. Kacik, X.-Y. Bu, J. Masnovi, P.J. Heimann and K. Beyene, Pyrolytic conversion of methyl- and vinylsilane polymers to Si-C ceramics, *Journal of Materials Science*, 1995, 30(12), 3130–3136.

[56] G.-B. Zhang, J. Kong, X.-D. Fan, X.-G. Li, W. Tian and M.-R. Huang, UV-activated hydrosilylation: A facile approach for synthesis of hyperbranched polycarbosilanes, *Applied Organometallic Chemistry*, 2009, 23, 277–282.

[57] M. Itoh, K. Iwata, M. Kobayashi, R. Takeuchi and T. Kabeya, Preparations and Properties of Poly(vinylsilane)s, *Macromolecules*, 1998, 31(17), 5609.

[58] K. Nate, T. Inoue, H. Sugiyama and M. Ishikawa, Organosilicon deep UV positive resist consisting of poly(p-disilanylenephenylene), *Journal of Applied Polymer Science*, 1987, 34, 2445.

[59] T. Iwahara, S. Hayase and R. West, Synthesis and properties of ethynylene-disilanylene copolymers, *Macromolecules*, 1990, 23(5), 1298.

[60] R.J.P. Corriu, C. Guerin, B. Henner, A. Jean, F. Garner and A. Yassar, Organosilicon polymers: synthesis of poly[(silanylene)diethynylene]s with conducting properties, *Chemistry of Materials*, 1990, 2, 351.

[61] J. Ohshita, D. Kanaya, M. Ishikawa and T. Yamanaka, Polymeric organosilicon systems: XVIII. Synthesis and photochemical properties of poly[2,5-(disilanylene)thienylenes], *Journal of Organometallic Chemistry*, 1989, 369(2), C18.

[62] J. Ohshita and A. Kunai, Polymers with alternating organosilicon and π-conjugated units, *Acta Polymerica*, 1998, 49(8), 379.

[63] E. Bacque, J.P. Pillot, M. Birot, J. Dunogues and G. Bourgeois, Nouveaux polycarbosilanes modèles. 3-Synthèse et caractérisation de carbosilanes linéaires fonctionnels, *Journal of Organometallic Chemistry*, 1988, 346(2), 147.

[64] Y.H. Kim, Y.S. Gal, U.Y. Kim and S.K. Choi, Cyclopolymerization of dipropargylsilanes by transition-metal catalysts *Macromolecules*, 1988, 21(7), 1991.

[65] J. Shinar, S. Ijadimaghsoodi, Q.X. Ni, Y. Pang and T.J. Barton, Synthesis and study of a polysilole *Synthetic Metals*, 1989, 28(1–2), C593.

[66] X.H. Zhang, Q.S. Zhou, W.P. Weber, R.F. Horvath, T.H. Chan and G. Manuel, Anionic ring-opening polymerization of sila- and germacyclopent-3-enes, *Macromolecules*, 1988, 21(6), 1563.

[67] Y.T. Park, Q.S. Zhou and W.P. Weber, Anionic ring opening polymerization of 3,4-benzo-1,1-dimethyl-1-silacyclopentene, *Polymer Bulletin*, 1989, 22(4), 349.

[68] Y.T. Park, G. Manuel and W.P. Weber, Anionic Ring Opening Polymerization of 1,1,3-Trimethyl-1- silacyclopent-3-ene. Effect of Temperature on poly(1,1,3-Trimethyl-1-sila-cis-pent-3-ene) Microstructures, *Macromolecules*, 1990, 23(7), 1911.

[69] Q.S. Zhou and W.P. Weber, Anionic Polymerization of 1-methyl-1-silacyclopent-3-ene. Characterization of poly(1- Methyl-1-sila-cis-pent-3-ene) by 1H, 13C and 29Si NMR Spectroscopy. Mechanism of Polymerization, *Macromolecules*, 1990, 23(7), 1915.

[70] G. Mera, Contributions to The Synthesis of Silicon-Rich Oligocarbosilanes and Their Use as Precursors for Electrically Conductive Films. Ph.D. Thesis, Ruhr-Universität Bochum, Bochum, Germany, 2005.

[71] Q.S. Zhou, Y.T. Park, G. Manuel and W.P. Weber, Anionic ring opening polymerization of 1-silacyclopent-3-ene, *Polymer Bulletin*, 1990, 23(5), 491.

[72] X. Liao, Y.H. Ko, G. Manuel and W.P. Weber, Synthesis and microstructure of poly(1-phenyl-1-sila-cis-pent-3-ene), *Polymer Bulletin*, 1991, 25(1), 63.

[73] C.X. Liao and W.P. Weber, Synthesis and Characterization of Poly(1-methyl-1-vinyl-1-silabutane), Poly(1-phenyl-1-vinyl-1-silabutane) and Poly(1,1-divinyl-1-silabutane), *Macromolecules*, 1992, 25(6), 1639.

[74] W. Habel, A. Oelscchlager and P. Sartori, Synthese und charakterisierung von poly (dialkenylsilylen-co-methylenen) als grundmaterialien von SiC-Fasern, *Journal of Organometallic Chemistry*, 1995, 486(1–2), 267.

[75] W. Habel, A. Oelschlager and P. Sartori, Synthese und charakterisierung von poly (dialkinylsilylen-co-methylenen), *Journal of Organometallic Chemistry*, 1995, 494(1–2), 157.

[76] S. Ijadi-Maghsoodi, Y. Pang and T. Barton, Efficient, "one-pot" synthesis of silylene–acetylene and disilylene–acetylene preceramic polymers from trichloroethylene, *Journal of Polymer Science. Part A, Polymer Chemistry*, 1990, 28(5), 955.

[77] I.S. Maghsoodi and T.J. Barton, Synthesis and Study of Silylene-Diacetylene Polymers, *Macromolecules*, 1990, 23(20), 4485.

[78] T. Ogawa, S.D. Lee and M. Murakami, Synthesis of Siloxane-Crosslinked Polysilylenemethylenes by Chlorodephenylation, *Journal of Polymer Science, Polymer Chemistry Edition*, 2002, 40(3), 416.

[79] W. Uhlig, Synthesis, functionalization, and cross-linking reactions of organosilicon polymers using silyl triflate intermediates, *Progress in Polymer Science*, 2002, 27(2), 255.

[80] H. Kudo, Y. Fujiwara, M. Miyasaka and T. Nishikubo, Synthesis of polycarbosilanes by A2 + B_n ($n = 2, 3,$ and 4) type hydrosilylation reaction and evaluation of their refractive index properties, *Journal of Polymer Science. Part A, Polymer Chemistry*, 2010, 48(24), 5746.

[81] Y. Matsuda, J.S. Rathore, L.V. Interrante, R.H. Dauskardt and G. Dubois, Moisture-Insensitive Polycarbosilane Films with Superior Mechanical Properties, *ACS Applied Materials & Interfaces*, 2012, 4, 2659.

[82] V. Rittscher and M. Gallei, A convenient synthesis strategy for microphaseseparating functional copolymers: the cyclohydrocarbosilane tool box, *Polymer Chemistry*, 2015, 6(31), 5653.

[83] L.V. Interrante and J.S. Rathore, Cyclolinear polycarbosilanes - a recent addition to the class of organosilicon polymers, *Dalton Transactions*, 2010, 39, 9193. DOI: http://dx.doi.org.0000424c27e9.han.p.lodz.pl/10.1039/C0DT00142B.

Chapter 6
Polysilazanes

6.1 Polysilazanes and polycarbosilazanes

Polysilazanes (PSZ) contain silicon and nitrogen atoms in their chains and hydrogen atoms or organic substituents connected to silicon and nitrogen. They contain Si–N–Si linkages and are precursors to amorphous silicon nitride and silicon carbonitride ceramics. PSZ-containing organic substituents bound to silicon are also called as *polycarbosilazanes* [1, 2]. PSZ with various chemical structures (see Figures 6.1–6.3) such as perhydropolysilazanes (PHPSZ), polymethyl(hydro)silazanes (PMHSZ), poly[methyl (hydro)-*co*-dimethyl]silazanes (PMHDMS), and poly[methyl(hydro)-*co*-methyl(vinyl)]silazanes (PMVSZ) are commercially available [3]. Depending on the presence of different organic or inorganic functional groups in their chemical structures, PSZ form a wide family of products with interesting properties, which find many practical applications.

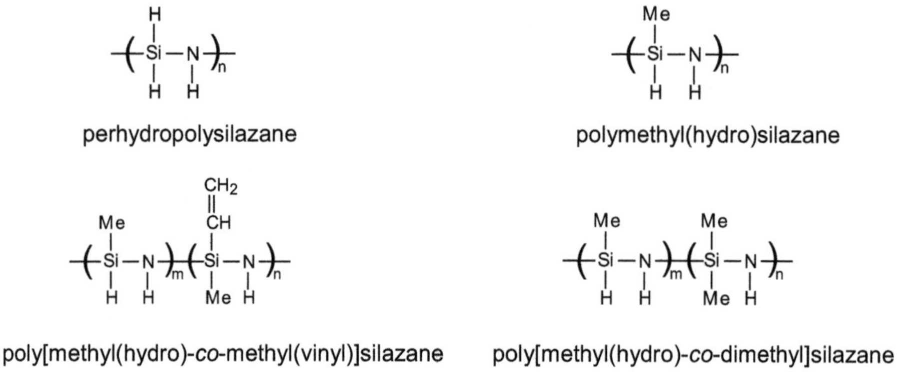

Figure 6.1: Different chemical structures of commercial polysilazanes.

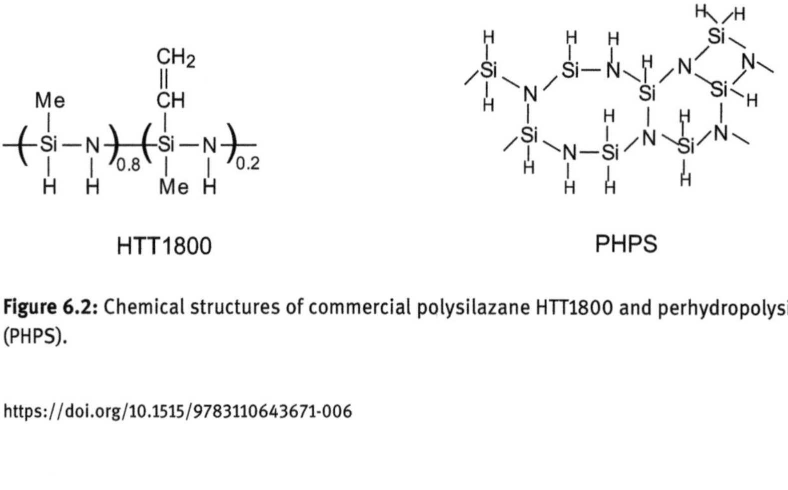

Figure 6.2: Chemical structures of commercial polysilazane HTT1800 and perhydropolysilazane (PHPS).

https://doi.org/10.1515/9783110643671-006

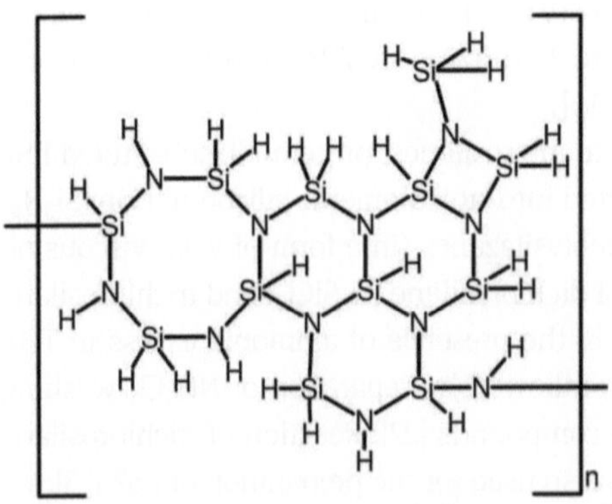

Figure 6.3: An idealized hypothetical molecular structure of perhydropolysilazane [4].

Carbon-free PSZ – silicon diimide ($Si(NH)_2$) – was prepared through ammonolysis of silicon tetrachloride in a liquid phase [5–7]:

$$SiCl_4 + 6\,NH_3 \longrightarrow Si(NH)_2 + 4\,NH_4Cl\ (\text{at } 0\ °C) \tag{6.1}$$

It polymerizes at 360 °C into polymeric silicon diimide ($[Si(NH)_2]_n$). At 1,000–1,250 °C in inert gas atmosphere, $[Si(NH)_2]_n$ decomposes into amorphous silicon nitride (Si_3N_4) and gaseous ammonia [8, 9]:

$$3\,Si(NH)_2 \longrightarrow Si_3N_4 + N_2 + 3\,H_2\ (\text{at } 1,250\ °C) \tag{6.2}$$

Recently, a mesoporous silicon diimide gel was obtained by ammonolysis of tris(dimethylamino)aminosilane [$(Me_2N)_3SiNH_2$] [10].

Another carbon-free precursor for silicon nitride is PHPSZ, which was prepared by ammonolysis of dichlorosilane (H_2SiCl_2) with gaseous ammonia in dichloromethane or diethyl ether. The obtained low molecular linear and cyclic oligomers cross-linked rapidly, liberating hydrogen and giving highly viscous or glassy PSZ [11], which under thermal treatment at 1,150 °C in nitrogen atmosphere formed a mixture of silicon nitride such as α-Si_3N_4 (in ~70% yield), β-Si_3N_4, and silicon. However, in ammonia atmosphere, the formation of elemental silicon from PHPSZ was decreased. Ammonolysis of $MeHSiCl_2$ in diethyl ether at 0 °C gave cyclic silazanes ($(MeSiHNH)_n$, $n = 3$ and 4) and a linear polymer [11]. The application of pyridine derivative ($H_2SiCl_2\ (NC_5H_5)_2$) slowed down the high reactivity of dichlorosilane in reaction with ammonia [12, 13]. The Si–H and N–H moieties easily underwent dehydrocondensation reaction with elimination of H_2 [5]. Pyridine catalyzed the dehydrocondensation reaction between ≡Si–H and =N–H functional groups and disproportionation of –NHSi≡ groups. An application of C_5H_5N had several advantages: it was responsible for stabilization of dichlorosilane and increase in molecular weights. Pyridine catalyzed the thermal cross-linking and caused incorporation of very small amounts of carbon into the ceramic [14–19]. The presence of ammonia or hydrazine avoided the formation of a free silicon during pyrolysis of PHPSZ [19, 20].

By the co-ammonolysis of dichloro- and trichlorosilanes, preceramic-substituted PSZ were obtained, which were thermally converted into stoichiometric silicon nitride (Si_3N_4) [21]. PHPSZ and N-alkyl-substituted hydridopolysilazanes (in a form of wax, viscous oil or powder) were also prepared in reactions of dichlorosilane (H_2SiCl_2) and trichlorosilane ($HSiCl_3$) with *n*-butylamine or *t*-butylamine in the presence of ammonia excess in THF (tetrahydrofuran) solution, at 0 °C to reflux, followed by separation of NH_4Cl, washing with the solvent and evaporation of volatile compounds [22]. Reaction of trichlorosilane with hexamethyldisilazane [$(Me_3Si)_2NH$] was also used for the preparation of PSZ [23].

Generally, PSZ and other Si–N-containing silicone polymers show higher thermostability and mechanical properties than polysiloxanes [5]. However, poly(perhydridosilazane)s have poor storage stability [24]. PHPSZ was very sensitive to moisture, and dense silica was formed through hydrolytic condensation reactions, for example, in the presence of ammonia solution [3]. For instance, PHPSZ $-(H_2SiNH)_n-$ rapidly cross-linked even at 0 °C, while $-(H_2SiNMe)_n-$ was stable for a long time in the absence of moisture and oxygen [25]. Moisture-curing of PHPSZ, polymethyl(hydro)silazane, poly[methyl(hydro)-*co*-dimethyl]-silazanes, poly[methyl(hydro)-*co*-methyl(vinyl)] silazanes, and trialkoxysilyl-substituted poly(methyl-*co*-dimethyl)silazane to preceramic materials, at temperatures ranging from room temperature to about 150 °C, required the presence of 80% relative humidity and a catalyst (ammonia or piperidine derivative). Hydrolysis of Si–H and Si–N bonds in $-(H_2SiNH)-$ units led to trisilanol structures which further cross-linked via condensation reactions into silica structure, while polyorganosilazanes formed layers composed of cross-linked silicone resins. PSZ films of about 3 μm thickness were coated on polycarbonate (PC) plates and cured at 50°C in air of about 80% relative humidity for 1 h. Blends obtained from PHPSZ showed higher curing rate (and improved surface microhardness up to ~450 N/mm^2) than those prepared from polyorganosilazanes (~250 N/mm^2). These materials seemed to be very promising for technical coating applications, exhibiting reduced gas permeability and improved corrosion resistance [26]. Pyrolysis of $-(H_2SiNH)_n-$ at 1,150 °C gave brown crystalline solid composed of a mixture of α- and β-Si_3N_4 [27].

The aminolysis of dichlorosilane with methylamine, carried out by Seyferth and Wiseman (see Scheme 6.1), after removal of the cyclotetramer (by distillation), gave a mixture of linear oligo-*N*-methylsilazanes with a low molecular weight ($M_n \approx 600$, $x \approx 10$).

Unlike in the case of PHPSZ, a low char yield (38%) was obtained, which was composed of silicon nitride and was free of carbon (based on IR spectra). It was concluded that the presence of N–Me bonds prevented thermal cross-linking [27] and confirmed that Si–H bonds alone were not able to induce the formation of a branched backbone under pyrolysis without condensation with N–H bonds.

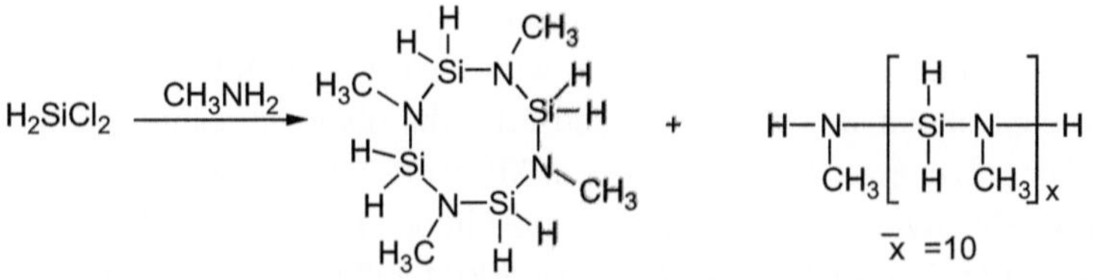

Scheme 6.1: The aminolysis of dichlorosilane with methylamine [27]. Reprinted (adapted) from [5]. Copyright 1995, by American Ceramic Society.

Polyhydridosilazanes [28, 29] and poly(methylvinylsilazanes) (PMVSZ) [30–33] were chemically modified by reactions with transition metal alkoxides ($M(OR)_4$, M = Ti, Zr, Hf). The reaction of polyhydridosilazane with $Ti(On\text{-}Bu)_4$ took place at the N–H groups upon formation of Si–Ti linkages. In commercial PMVSZ (HTT1800), both functional groups (N–H and Si–H) reacted with $Hf(On\text{-}Bu)_4$ and it was confirmed by Raman and ^{1}H-NMR (nuclear magnetic resonance) spectroscopy [29]. The chemical modifications of PHPSZ as well as vinyl- and hydro-substituted PSZ with amido complexes of Ti [28, 29, 34, 35], Zr, and Hf [36, 37] were also described. Ceramic nanocomposites with compositions MN/Si_3N_4, MN/SiCN, MCN/SiCN, and so on (where M = Ti, Zr, Hf) were prepared from modified PSZ precursors. They exhibited ultrahigh thermal stability, depending on the pyrolysis conditions.

Synthetic routes to PSZs have been described in excellent reviews [1, 2, 38]. Most often, they were prepared by ammonolysis or aminolysis of (halogeno)organosilanes with a general formula R_xSiCl_{4-x}. From dichlorosilanes (R_2SiCl_2), mixtures of cyclic oligomers and low-molecular-weight linear PSZ were synthesized, which were further cross-linked usually by the thermal method [27, 39–41].

In the case of silazanes containing both kinds of functional groups such as Si–H and N–H, cross-linking reactions were catalyzed by bases (e.g., potassium hydride, KH), followed by termination reactions, for instance, with methyl iodide (MeI) [39].

Oligomers of the structure $-(MeHSiNH)_x-$, prepared by ammonolysis of $MeHSiCl_2$, in reaction with KH, underwent dehydrocyclization, giving soluble, tractable polymers, containing combined octa- and tetra-membered silazane ring structures of the average composition $(MeSiNH)_{0.39}(MeHSiNMe)_{0.04}(MeSiN)_{0.57}$ [11]. Its chemical structure was proposed by Seyferth et al. (Scheme 6.2) [41].

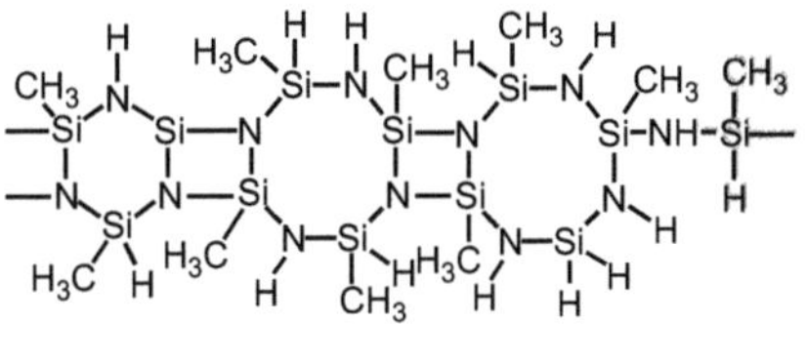

Scheme 6.2: The proposed structure of ladderlike polysilazanes obtained from the reaction of $(MeHSiNMe)_n$ with KH. Reprinted (adapted) from [5]. Copyright 1995, by American Ceramic Society.

Similar vinyl- or allyl-functionalized oligosilazane (OSZ) copolymers were synthesized by ammonolysis of mixtures of $MeHSiCl_2$ with $Me(CH_2{=}CH)SiCl_2$ or $Me(CH_2{=}CHCH_2)SiCl_2$ in different molar ratios (3:1, 4:1, and 6:1). The obtained soluble, liquid products were further polymerized in the presence of KH in dry THF, followed by termination reactions with MeI giving white powders, soluble in organic solvents (THF, hexane, benzene, and toluene), with MWs in the range 800–1,200 g/mol and molecular composition $[(MeSiHNH)(Me(CH_2{=}CH)SiNH)_{2.3}]_x$ [39, 40].

PSZ copolymers with the chemical structure $[(SiH_2NH)_3MeHSiNH]_n$ were prepared by ammonolysis of mixtures of three equivalents of dichlorosilane, H_2SiCl_2, with one equivalent of methyldichlorosilane, $MeSiHCl_2$. Alternatively, dichlorosilane reacted with ammonia and methylamine in a 3:1 ratio to give $[(SiH_2NH)_3SiH_2NMe]_n$. Their molecular structure was confirmed by ^1H-, ^{13}C-, ^{29}Si-NMR, and IR spectroscopy. These PSZ were further cross-linked by base-catalyzed dehydrocoupling reactions through Si–H and N–H functional groups. Insoluble cross-linked PSZ were characterized by elemental analysis and spectroscopic methods: solid-state ^1H, ^{13}C, and ^{29}Si CP-MAS NMR and IR [42].

W. Fink synthesized a mixture of α,ω-dichloro-substituted cyclodisilazane (CDSZ) monomer and homological oligomers (Scheme 6.3) by heating of hexamethylcyclotrisilazane $(Me_2SiNH)_3$ with dimethyldichlorosilane up to 240–340 °C [43].

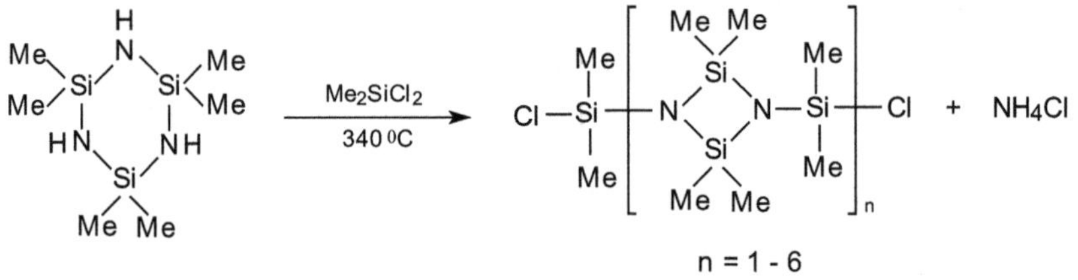

Scheme 6.3: Synthesis of methyl derivatives of α,ω-dichlorocyclodisilazanes [43].

L.W. Breed et al. and U. Wannagat with coworkers prepared monomeric dichlorocyclodisilazane derivative in 66% yield through thermal ring contraction of $(Me_2SiNH)_3$ with Me_2SiCl_2 to the CDSZ ring, during heating within 16 h, which proceeded with the elimination of ammonium chloride [44, 45].

A polycondensation of *N,N'*-bis-(trimethylsilyl)dimethyldiphenylcyclodisilazane (I) and *N,N'*-bis-(trimethylsilyl)tetraphenylcyclodisilazane(II) took place in the presence of catalytic amounts of metallic potassium and proceeded with untypical liberation of benzene molecule and formation of polymer chains containing CDSZ rings tethered by phenylene moieties (Scheme 6.4) [46].

The polycondensation of bis(diethylamino)dimethylsilane $Me_2Si(NEt)_2$ with *para*- and *meta*-phenylene diamines proceeded with elimination of diethylamine and gave cyclic oligomers with high yields. Their heating at temperature above 400 °C resulted in the formation of cross-linked PSZ. With *o*-phenylenediamine, bimolecular

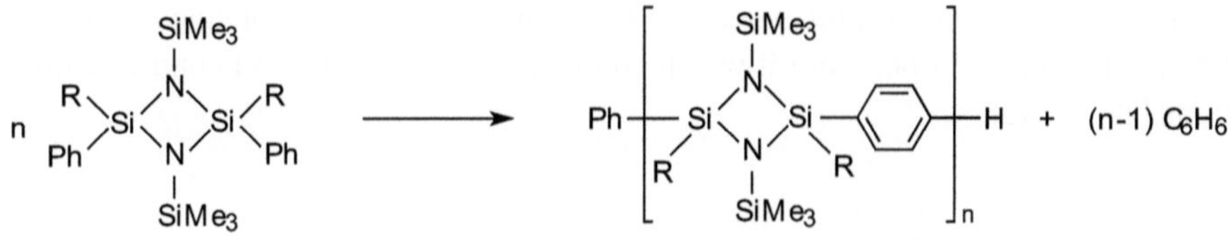

where: R = Me (I), Ph (II).

Scheme 6.4: Synthesis of poly[*N*,*N*′-bis-(trimethylsilyl)cyclodisilazane-*co*-phenylene] derivatives [46].

cyclization led to diazadimethylsilaindane, which underwent further rearrangements [47, 48]. Similar reactions of (*N*,*N*′-diphenyl)tetramethylcyclodisilazane with phenylenediamines, carried out in the melt at temperatures over 200 °C, gave oligomers composed of linear and cyclic disilazane units tethered by phenylene groups. The reactions of *meta*- and *para*-phenylenediamine with $Me_2Si(NEt)_2$ at temperature over 300 °C led to the cross-linked, infusible, and insoluble polymeric material. A progress of growth of the polymer molecules, which included CDSZ ring cleavage, transaminations, and recyclizations, was studied by elemental analysis and spectroscopic methods of the intermediate low molecular and oligomeric volatile and nonvolatile products of these reactions [48, 49].

Organic hydrosilanes (Ph_2SiH_2, $PhSiH_3$, Et_3SiH, *n*-$C_6H_{13}SiH_3$, etc.) reacted with N–H bonds of ammonia or amines, giving PSZ (Scheme 6.5) [25, 38, 50, 51]:

$$R_2SiH_2 \; + \; R'NH_2 \longrightarrow \text{-}(R_2Si\text{-}NR')_n\text{-} \; + \; 2H_2\uparrow$$

Scheme 6.5: Synthesis of PSZ by dehydrocoupling of organic hydrosilanes with amines.

PSZ of the structures MeNH–$[H_2SiNMe]_n$–H and MeNH–$[H_2SiNMe]_x[HSi(NHMe)NMe]_y$H were prepared by the transition-metal-catalyzed dehydrocoupling reactions between Si–H bonds in hydrosilanes and N–H bonds in low MW silazane oligomers, which were obtained by ammonolysis of appropriate chlorosilane monomers [25, 38, 50, 51]. PSZ prepared according to Scheme 6.6 increased their degree of polymerization up to ≈18–20 (from $x \approx 10$) and contained less than 10% cyclosilazane oligomers [25]:

$$MeNH(H_2SiNMe)_xH \longrightarrow \text{"PSZ"} \; + \; H_2\uparrow$$

Scheme 6.6: Polycondensation of low MW silazane oligomers.

In the presence of hundred ppm of $Ru_3(CO)_{12}$ in THF at temperatures of 60–90 °C, the composition of PSZ was dependent on the reaction time. After 65 h of heating, M_n changed from 1,200 Da to only 2,300 Da, while M_w reached 25,000 Da, and it was accompanied by distinct viscosity increase from 1–3 poise to 100 poise, due to cross-linking and gelation processes. The *N*-methyl-PSZ precursor of silicon nitride, -$[H_2SiNMe]_n$-, formed fibers and it was recommended for binder and coating applications [25, 38, 50].

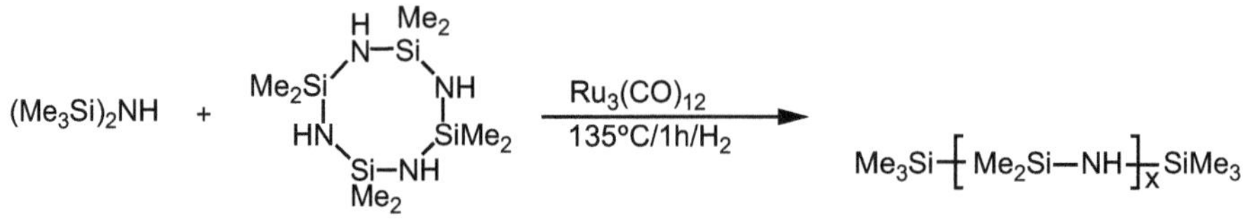

Scheme 6.7: The oligomerization of $(Me_2SiNH)_4$ in the presence of $(Me_3Si)_2NH$, catalyzed by $Ru_3(CO)_{12}$ [50]. Reprinted (adapted) from [5]. Copyright 1995, by American Ceramic Society.

The ring-opening oligomerization of cyclosilazanes (Scheme 6.7) was promoted by transition metal catalysts: different metal carbonyls, Pt or Pd supported on carbon, PtO_2, $AlCl_3$ but also with H_2SO_4 [50, 52]. Moreover, redistributions of $(Me_2SiNH)_n$ with alkali metals (e.g., Na, K), metal hydrides (NaH, KH), or bases ($NaNH_2$), as well as oligomerization of $(HMe_2Si)_nNH$ with NH_3 in the presence of $Ru_3(CO)_{12}$ were observed [50].

D. Seyferth et al. obtained only low-molecular-weight linear oligomers ($M_n \leq$ 3,000 g/mol) by the ring-opening polymerization (ROP) of permethylcyclodisilazane monomer, 1,1′,2,3,3′,4-hexamethylcyclodisilazane [cyclo-$(Me_2SiNMe)_2$], which was carried out with alkyllithium reagents (RLi or Me_3CONa/RLi). The *N*-isopropyl-substituted CDSZ monomer underwent only ring-opening reaction but oligomers were not formed [53].

Burns et al. obtained PSZ-containing silacyclobutane units in a main chain through ammonolysis and co-ammonolysis reactions. In the first method, silacyclobutasilazane was equilibrated with chlorosilane providing a mixture of silazanes terminated with chlorosilyl groups, which was treated with ammonia to give the desired substituted silacyclobutasilazane. Alternatively, a mixture of 1,1-dichloro-1-silacyclobutane with chlorosilane was directly reacted with ammonia, leading to silazane precursors with different molecular compositions (see Scheme 6.8) [54, 55].

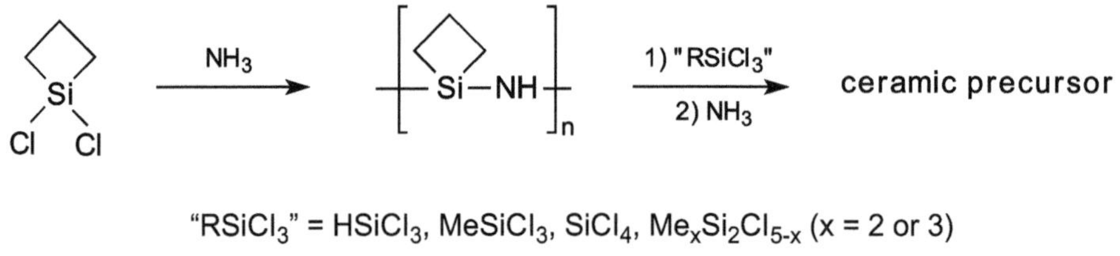

Scheme 6.8: Synthesis of silazane oligomers by ammonolysis of 1,1-dichloro-1-silacyclobutane, followed by ROP and further functionalization with chlorosilanes and ammonia.

On heating to 200–250 °C, these silazane oligomers cross-linked into resins through ROP of silacyclobutane ring. The cross-linking step was a combination of ROP of the silacyclobutyl units and N–H addition across Si–C bonds. Below 1,200 °C, Si–C–N–O ceramics of various compositions were obtained in 60–80% yields. When the pyrolysis was continued up to 1,800 °C, the ceramic residue (obtained with ~50% yield) was mostly composed of β-SiC as it was determined by X-ray diffraction (XRD).

A great achievement has been an elaboration of the ROP of CDSZ by Soum et al. [56]. The living anionic and cationic ROP of various cyclosilazane monomers was initiated with triflic acid, its ester derivatives, naphthalene-sodium (or MeLi, *t*-BuLi, or PhLi) giving either polymers or linear and cyclic oligomers, depending on the structure of both the nitrogen and the silicon substituents and on the size of the ring. It is also a very useful synthetic route to PSZ. From *N*-methyl-substituted CDSZ having methyl or vinyl substituents at silicon (Me_2, MeVi, Vi_2), high-molecular-weight PSZ were prepared, while with cyclotrisilazanes and cyclotetrasilazanes an equilibrated polymerization took place. The obtained white crystalline PSZ were soluble in organic solvents. CDSZ with *tert*-butyl, isopropyl, or ethyl groups on nitrogen only oligomerized under similar conditions.

The ring-opening oligomerization of hexamethylcyclotrisilazane $(Me_2SiNH)_3$ and octamethylcyclotetrasilazane $(Me_2SiNH)_4$ was initiated by methyl triflate in 1,2-dichloroethane and led to an equilibrium mixture of cyclosilazanes $(Me_2SiNH)_n$ ($3 \leq n \leq 13$) [57].

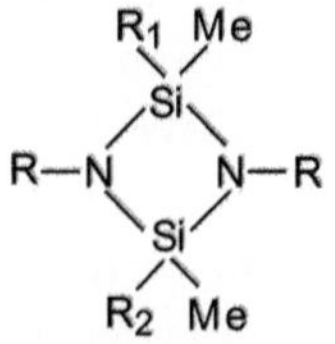

R = Me, Et, i-Pr, t-Bu, Ph, etc.

$R_1 = R_2$ = Me (D_{2NR}); $R_1 = R_2$ = Vi or Et ($D^{R1}{}_{2NR}$); R_1 = Me, R_2 = Vi, Allyl, Et, or Ph ($D^{R1}{}_{2NR}D_{NR}$)

Figure 6.4: Chemical structure of cyclodisilazanes.

The structure of substituents at nitrogen (Figures 6.4 and 6.5) determined the possibility of their polymerization. Highly strained CDSZ with methyl groups on N atoms underwent cationic or anionic ROP, while other N-alkyl- or N-aryl-substituted CDSZ only oligomerized or did not react at all. Their reactivity in ROP changed in the order [58]:

$$D_{2NMe} \gg D_{2NEt} > D_{2NiPr} \gg D_{2Nt\text{-}Bu} \sim D_{2NPh}.$$

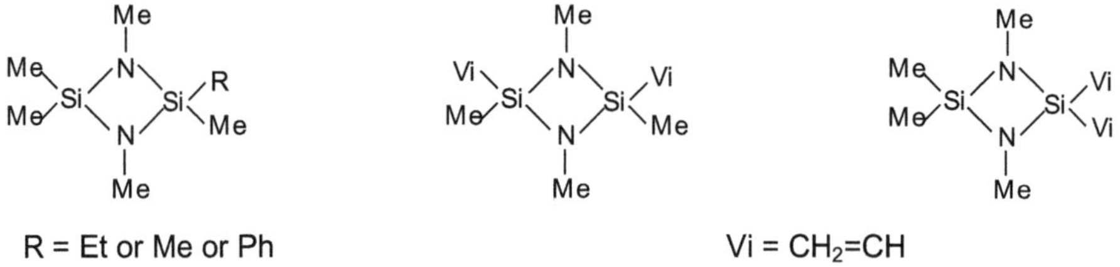

Figure 6.5: Chemical structures of cyclodisilazanes [58].

Methyl triflate was found to be the best initiator for cationic polymerization of CDSZ in polar solvents at room temperature, giving PSZ with MWs (M_n) of up to 30,000 g/mol, accompanied with thermodynamically stable six- and eight-membered ring oligomers. Long-chain PSZ (with MWs of ~100,000 g/mol and narrow polydispersities of MWs ~1.1), free of cyclic silazanes, were achieved by anionic polymerization of *N*-methylcyclodisilazanes in the presence of organosodium and organolithium reagents in polar and nonpolar solvents [58].

The kind of the substituent on nitrogen played an important role in the polymerization of CDSZ. The anionic ROP of several CDSZ having hindered substituents (phenyl or allyl) on silicon atoms initiated with benzyllithium in toluene–THF was not the living process. In the case of CDSZ with less sterically hindered groups, the polymerization proceeded through living mechanism, and the reactivity of CDSZ was much different. The presence of vinyl groups on monomers substantially increased the rate of polymerization. Thus, PSZ with MWs ranging from 900 to 15,400 g/mol (and polydispersity $M_w/M_n = 1.1–1.2$) were obtained [59].

PSZ with block and random structures were prepared by the anionic copolymerization of various CDSZ monomers. Their molecular composition was dependent on the substituents at silicon atoms in the monomers. The mechanical and thermal properties of various PSZ copolymers were also determined [60].

The solid-state structure and transition properties of three poly(*N*-methylcyclodisilazanes) were characterized by the Fourier-transform infrared (FTIR) and NMR spectroscopies, differential scanning calorimetry (DSC), XRD, and dielectric relaxation techniques. Semicrystalline PSZ showed two endothermic DSC transitions. The transition properties of the alternating copolymer -($SiMe_2$-NMe-SiMeVi)$_n$- were similar to those of the corresponding asymmetric homopolymer -(SiMeVi-NMe-SiMeVi)$_n$- (Vi=CH=CH_2), but its X-ray spectrum was close to that of the PSZ with the symmetric structure -($SiMe_2$-NMe-$SiMe_2$)$_n$- [61].

A hyperbranched organopolysilazane with a relatively high molecular weight of ~26,000 Da, highly soluble in all organic solvents, was prepared from the reaction of cyclohexyl(trichloro)silane ($CySiCl_3$) with lithium azide (Li_3N) in diglyme [62].

OSZ with cyclolinear structure (Figure 6.6) were synthesized by the dehydrocoupling reaction of diphenylsilane (Ph_2SiH_2) with 1,1,3,3,5,5-hexamethylcyclotrisilazane [$(Me_2SiNH)_3$, D_3^N], catalyzed by KH, carried out in THF [63]. Next polymer chains were terminated in the reaction with methyliodide. These OSZ were characterized by spectroscopic methods (IR, ^{1}H-, and ^{29}Si-NMR) and vapor pressure osmometry (M_n = 1,344 g/mol).

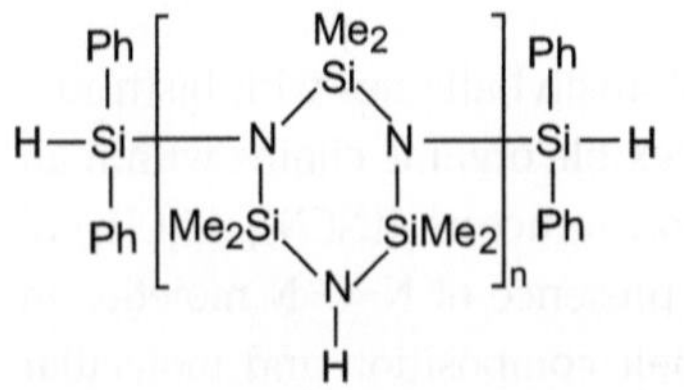

Figure 6.6: Proposed cyclolinear structure of oligosilazanes prepared by polycondensation of Ph_2SiH_2 with D_3^N [63].

An excellent review of synthetic methods of an extremely wide family of cyclo(carbo) silazanes and their different derivatives was presented by P. Szolcsányi et al. [64]. Their functionalization, rearrangements, and physicochemical properties were described, including hydrolytic stability. These compounds serve for further preparation of numerous composite ceramic materials of a general composition $Si_xC_yN_z$ and polymeric materials [64].

Polycarbosilazanes, containing alternating silicon, carbon, and nitrogen atoms in their backbone, were prepared by reductive coupling of a mixture of Me_2SiCl_2 and 1,3-dichloro-1,3-dimethyldisilazane with sodium [65]. This polymer was pyrolyzed into SiC–Si_3N_4 mixed nonoxide ceramics. Hydrosilylation between 1,1,3,3-tetramethyldisilazane and 1,3-divinyl-1,1,3,3-tetramethyldisilazane occurred rapidly at 90 °C in the presence of Karstedt's catalyst (Pt-divinyltetramethyldisiloxane) giving large macrocyclic products. Similarly, 1,1,3,3-tetramethyldisilazane reacted with trimethylsilylated or methylated derivative of 1,3-divinyl-1,1,3,3-tetramethyldisilazane [66].

6.2 Polysilylcarbodiimides

Polysilylcarbodiimides (PSCDI), containing Si-N=C=N structural units in a polymer chain, were first synthesized by Ebsworth, Wannagat, Birkofer, and coworkers [67–70]. Pump and Rochow prepared PSCDI in reactions of di, tri-, or tetrachlorosilanes with silver cyanamide [71], while Klebe and Murray synthesized PSCDI through transsilylation of bis(trimethylsilylcarbodiimide) with chlorosilanes [72]. PSCDI were also obtained by the polycondensation reaction of cyanamide with chlorosilanes [73]. The polycondensation reaction of chlorosilanes with bis(trimethylsilylcarbodiimide) catalyzed by pyridine is the most useful method for the synthesis of PSCDI [74–81]. From dichlorosilanes, cyclic or linear PSCDI were obtained.

New oligomeric cyclic silylcarbodiimides, useful as ceramic precursors for the pyrolytic preparation of SiCN, were prepared in the reaction of the appropriate dichlorosilanes with cyanamide [82].

Pyridine-catalyzed reactions of methyldichlorosilane with bis(trimethylsilyl)carbodiimide proceeded with formation of Me_3SiCl and involved disproportionation of Si–H bonds leading to a cross-linked anhydrous gel. The dried gels formed amorphous or crystalline xerogels. Their pyrolysis at 1,200 °C gave SiCN ceramics, containing amorphous Si_3N_4 and carbon [79].

By sol–gel polycondensation reactions of bis(trichlorosilyl)alkanes with bis(trimethylsilyl)carbodiimide, hybrid polymers consisting of flexible organic chains within an inorganic silsesquicarbodiimide network of the general structure $[(NCN)_{1.5}Si(CH_2)_xSi(NCN)_{1.5}]_n$ (where $x = 2$, 6, and 8) were prepared. The presence of N=C=N moieties in xerogel structures was confirmed by FTIR spectra. Their composition and molecular structures were analyzed by solid-state ^{13}C CP MAS- and ^{29}Si CP MAS-NMR spectroscopies, elemental analysis, and XRD. The morphology of the xerogels was analyzed by scanning electron microscopy and transmission electron microscopy (TEM). Moreover, the pore structure of these materials was studied by the gas adsorption (Brunauer–Emmett–Teller) method (surface areas around 100 m^2/g) – the surface area decreased with increasing length of the alkylene spacing group [80].

PSCDI show some similarities to polysiloxanes [75, 76, 83]. PSCDI are useful as insulator coatings, high-temperature-stable pigments, stabilizing agents for polyurethanes and polyvinylchloride [72], and sealing materials resistant to irradiation [84]. Moreover, PSCDI can be used for the synthesis of organic cyanamides, carbodiimides, and heterocycles [85]. PSCDI are usually sensitive to air and moisture [73]. Their chemical resistance against moisture can be significantly improved by incorporation of bulky aromatic substituents at silicon [77]. Homological series of poly(phenylsilylcarbodiimides) (Figure 6.7)

$$-[\mathrm{PhRSi-N=C=N}]_n- \quad (\mathrm{R = Ph, Me, H, Vi})$$

were prepared in the reactions of appropriate substituted phenyldichlorosilanes with bis(trimethylsilylcarbodiimide) in the presence of pyridine as the catalyst [77, 86].

The nanostructure of novel carbon-rich silicon carbonitride (SiCN) ceramics fabricated via thermolysis of poly(methylphenylsilylcarbodiimide), $-[Ph(CH_3)Si\text{-}N{=}C{=}N]_n-$, at temperatures of 1,300, 1,500, 1,700, and 2,000 °C was studied by micro-Raman spectroscopy, X-ray powder diffractometry, and small-angle X-ray scattering. This SiCN-based ceramics contained nanodomains composed of free carbon, SiC, and Si_3N_4. The SiCN ceramics were structurally and chemically stable up to 1,500 °C and did not show steady-state creep despite their amorphous structure but showed viscoelasticity at high temperatures [81].

R. Riedel et al. tried to apply poly(phenylvinylsilylcarbodiimide)-derived SiCN ceramics as an anode material for lithium-ion batteries. It was prepared by the thermal

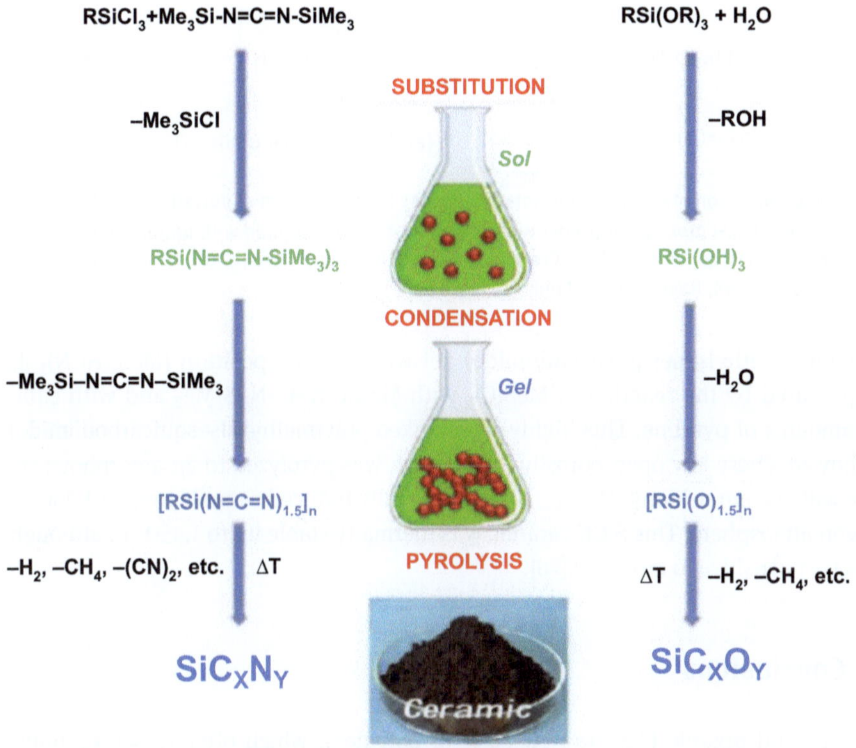

Figure 6.7: Synthesis polyphenylsilylcarbodiimides of different molecular composition [77]. Reprinted with permition from [1] under the terms and conditions of the Creative Commons Attribution license (CC BY 4.0 rule). Copyright 2015, MDPI, Basel, Switzerland.

treatment of poly(phenylvinylsilylcarbodiimide) under argon atmosphere at different temperatures: 1,100–2,000 °C [87].

The thermal decomposition of SiCN ceramics obtained by pyrolysis of poly(diphenylsilylcarbodiimide) precursor gave micro- and mesoporous materials with a high specific surface area (up to 568 m^2/g). High-resolution TEM studies showed that the pores were present only in the carbon phase and confirmed that the micro- and nanostructures were composed of SiC, which was homogeneously dispersed in a graphene-like carbon matrix. The presence of pores was associated with the carbon phase.

This novel porous ceramics was chemically resistant and could find many practical applications, for example, for selective gas separation and purification membranes, as catalyst support at high temperatures and for hydrogen storage [88].

Branched polysilsesquicarbodiimides, were prepared from trichlorosilanes and bis(trimethylsilylcarbodiimide). The sol–gel process of trichlorosilanes with bis(trimethylsilylcarbodiimide) is similar to the sol–gel process of trialkoxysilane with water (Figure 6.8) [1, 73].

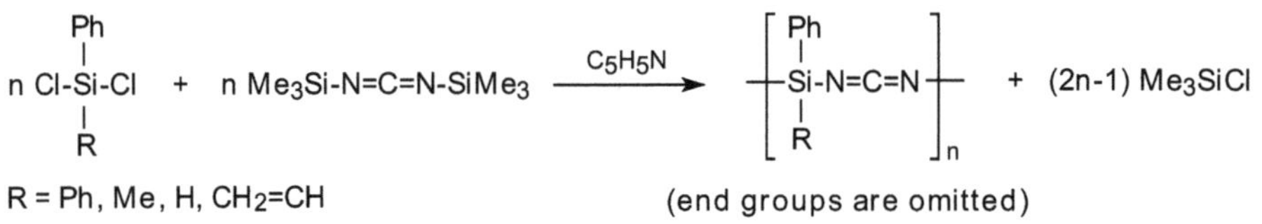

Figure 6.8: Comparison of the nonoxide sol–gel process for preparation of polysilylcarbodiimides and the oxide sol–gel process for preparation of polysiloxanes. Reprinted with permission from [1] under the terms and conditions of the Creative Commons Attribution license (CC BY 4.0 rule). Copyright 2015, MDPI, Basel, Switzerland.

A poly(methylsilsesquicarbodiimide) gel with the composition $[MeSi(NCN)_{1.5}]_n$ was prepared by the reaction of $MeSiCl_3$ with $Me_3Si–N=C=N–SiMe_3$ and with catalytic amounts of pyridine. This highly cross-linked poly(methylsilsesquicarbodiimide) gel showed a very low open porosity (<1 m^2/g). It was pyrolyzed to an amorphous ceramic with a composition $SiC_{1.1}N_{1.6}$, by the thermally induced ceramization at 1,200 °C in argon atmosphere. This SiCN ceramic was thermally stable up to 1,450 °C, although it was very sensitive to moisture [78].

6.3 Conclusions

Inorganic and organic PSZ are new class of polymers, which offer versatile, multifunctional properties and variety of possible applications. They are excellent precursors for modern high-tech ceramics. As coatings, they provide very good protection for metal and organic substrates against chemicals and mechanical action [26]. PSZ are very useful for the preparation of fibers, coatings, and 3-D-continuous fiber-reinforced ceramic-matrix composites that cannot be prepared by traditional ceramic-processing methods [1, 2, 89].

PHPSZ undergo hydrolysis and condensation reactions leading to the formation of dense SiO_2. Chemical bonding of the reactive Si–H and N–H functional groups to polar groups on substrates, as well as a dense coating structure, are responsible for the good protection of various metals against corrosion and the protection of paints against graffiti. Organopolysilazanes react in a similar way as PHPSZ [3]. Alternatively, PHPSZ may be thermally cured [26]. The pencil hardness of coatings obtained by thermal curing of PHPSZ was 7H and hardness values from indentation measurements was 2,000 N/mm^2. Such values are much higher than for organic polymers, for example, PC or poly(methyl methacrylate) ($H_v \gg 100–200$ N/mm^2). Owing to reactive Si–H and N–H bonds, PHPSZ showed very good adhesion to most polar substrates such as glass, metals, or plastics and provided good protection against corrosion, scratching, chemical degradation, and thermal oxidation [3].

PHPSZ modified with hydrophilic promoters were used for coating of painted automotive surfaces and windows and exhibited surface energies of 10–20 mN/m toward water. Superhydrophilic thin coatings (0.5–2 μm), obtained by incorporation of TiO_2 nanoparticles, provided good surface protection and kept clean car windows and outdoor housing tiles. At present, PHPSZ is also often used in the electronics industry as a precursor to silica in the manufacture of dense or porous surface coatings on silicon wafers. Such coatings were also very effective in protection of steel, anodized aluminum, copper, bronze, and other alloys against corrosion. Coatings based on PHPSZ showed exceptional barrier properties to oxygen permeation (0.2 cm^3m^2/day bar) and to water permeation (1 cm^3m^2/day bar) and could be used in the packaging industry [3].

The functionalization of PSZ with organic groups extends possibilities of their practical applications [26]. PMVSZ can be thermally cured to solid materials through heating to 180–200 °C or at lower temperatures in the presence of free radical initiators (dialkyl peroxides, diperoxyester, alkyl peroxyesters, peroxyketals, and peroxycarbonates) or by cross-linking induced with UV, electron beam, laser, or microwave radiation [3].

PMVSZ are usually used as ceramic precursors to ceramic matrix composites (CMC), which are prepared by pyrolysis at temperatures of 1,000 °C or greater under controlled atmospheres, where PSZ is transformed either into silicon nitride (Si_3N_4) and silicon carbide (SiC) ceramics. Blends of organopolysilazanes with conventional organic polymers form the fire-resistant and tough polymer matrix composites and CMCs, which are useful in high-temperature applications. They include components for the aerospace, defense, and underbonnet automotive markets, and electromechanical microelectronic devices. The PMVSZ and poly(dimethylsilazanes) can be blended with metal oxide powders (SiO_2, Al_2O_3, ZrO_2), nonoxide ceramic powders (SiC, B_4C), and/or metal powders (Al, Zn, Cu) for preparations of high-temperature-resistant coatings (above 1,000 °C), in air or corrosive atmospheres without delaminating from powder fillers. They can be also used as heavy-duty coatings resistant against high-temperature oxidation and chemicals – in high-performance motor sport vehicles, gas turbine engines, aerospace components, and metal and glass goods manufacturing. The PSZ coatings cured at ambient temperatures also showed high thermal properties and mechanical durability [3].

PSZ clear coats (~5 μm thick) are very often applied for efficient protection of vehicles and trains against corrosion and graffiti (from aggressive markers and spray paints). Polymer composites, plastics, and metal parts can also be protected, owing to the exceptional adhesion of these coatings to many surfaces. The low-viscosity PSZ coatings can be sprayed or wiped under ambient conditions. They become dry-to-touch after 1 h and get full performance after a few days. Due to the low surface energy (γH_2O = 105 mN/m), such coatings are impermeable to paint and marker ink and form nonadherent droplets. Their dense structure avoids deposition of any chemical on the surface. Moreover, removal of graffiti may be facilitated with the use of simple cleaning solvents. Overcoating of the polyester/polyacrylic varnishes with a PSZ or PSZ/polyurethane hybrid clearcoats prevents yellowing of the rims from very aggressive phenolic dust particles, liberated by car brakes. The rims remain dust-free even after 1 year of intensive use [3].

PSZ clearcoats may be directly applied on metal surfaces, since they show very good adhesion to metals (aluminum, copper, or brass) without the need of a primer coating. The Si–H and N–H groups form very strong bonds with metals. PSZ clearcoats can be also blended with fillers, pigments, and additives and are suitable for colored coatings, which show low surface energy and good protection against weathering and corrosion resistance. Formulations of PSZ with dyes and pigments give colored coatings exhibiting stable properties [3].

References

[1] G. Mera, M. Gallei, S. Bernard and E. Ionescu, Ceramic nanocomposites from tailor-made preceramic polymers, *Nanomaterials*, 2015, 5(2), 468.

[2] E. Kroke, Y.L. Li, C. Konetschny, E. Lecomte, C. Fasel and R. Riedel, Silazane derived ceramics and related materials, *Materials Science and Engineering, R: Reports*, 2000, 26(4–6), 97.

[3] C. Vu, F. Osterod, S. Brand and K. Ryan, Silicon and nitrogen, *European Coatings Journal* of Vincentz Network, 2008, 2, 1.

[4] https://en.wikipedia.org/wiki/Polysilazane

[5] M. Birot, J.-P. Pilot and J. Dunogues, Comprehensive chemistry of polycarbosilanes, polysilazanes, and polycarbosilazanes as precursors of ceramics, *Chemical Reviews*, 1995, 95 (5), 1443.

[6] D.V. Tsu and G. Lucovsky, *Silicon nitride and silicon diimide grown by remote plasma enhanced chemical vapor deposition, Journal of Vacuum Science* and Technology A, 1986, 4, 480; https://doi.org/10.1116/1.573875

[7] U. Wannagat, A. Schervan, M. Jansen, H.-P. Baldus and A. Eiling, Silicon diimide, a process for its preparation and silicon nitride obtained therefrom, *US Pat.* 5 258 169, 1993.

[8] O. Glemser and P. Naumann, Über den thermischen abbau von siliciumdiimid $Si(NH)_2$, *Zeitschrift für Anorganische und Allgemeine Chemie*, 1959, 298(3–4), 134.

[9] K.C.S. Richetto, C.R.M. Silva and S.A. Baldacin, Qualitative analysis of silicon nitride synthesized by ammonolysis of silicon tetrachloride, *Materials Science Forum*, 2003, 416(1), 688.

[10] S.P. Kitney, S.M. Sajedin, V. Rocher, F. Cheng and S.M. Kelly, Silicon diimide gel as an efficient stationary phase in thin layer chromatography for acid-sensitive organic compounds, *Chemical Communications*, 2017, 53(80), 11080.

[11] D. Seyferth, G.H. Wiseman and C. Prudhomme, A liquid silazane precursor to silicon-nitride, *Journal of American Ceramic Society*, 1983, 66(1), C13.

[12] M. Arai, S. Sakurada, T. Soda and H. Tomizawa, Preceramic polysilazane to silicon-nitride, *American Ceramic Society, Abstracts of Papers*, 1987, 193, 41.

[13] M. Arai, T. Funayama, I. Nishii and T. Isoda, *Japanese Patent* 62 125,015, 1987, *Chemical Abstracts*, 1987, 107, 160180.

[14] O. Funayama, M. Arai and T. Isoda, *European Patent Appl.* 303,498, 1989, *Chemical Abstracts*, 1989, 111, 134998.

[15] M. Arai, S. Sakurada, T. Isoda and T. Tomizawa, Preceramic polysilazane to silicon-nitride, *Polymer Preprints*, 1987, 28, 407.

[16] K.K. Toa Nenryo Kogyo, *Japanese Patent* 84 207,812, 1984, *Chemical Abstracts*, 1985, 102, 206067.

[17] K.B. Schwartz, D.J. Rowcliffe, Y.D. Blum and R.M. Laine, *Better ceramics through chemistry II*, in Materials Research Society, Vol. 73, C.I. Brinker, D.E. Clark and D.R. Ulrich (Eds.), Pittsburgh, 1986, 407.

[18] T. Isoda, H. Kaya, H. Nishii, O. Funayama, T. Suzuki and Y. Tashiro, Perhydropolysilazane precursors to silicon nitride ceramics, Journal of Inorganic and Organometallic Polymers and Materials, 1992, 2(1), 151.

[19] O. Funayama, T. Isoda, H. Kaya, T. Suzuki and Y. Tashiro, Characterization of Preceramic Perhydropolysilazane, *Polymer Preprints*, 1991, 32, 542.

[20] I. Nishii, H. Aoki, O. Funayama, S. Suzuki, Y. Tashiro, M. Arai and T. Isoda, *Japanese Patent* 01 045,817, 1989, *Chemical Abstracts*, 1989, 111, 101868.

[21] C.R. Blanchard and S.T. Schwab, X-ray-diffraction analysis of the pyrolytic conversion of perhydro-polysilazane into silicon-nitride, *Journal of American Ceramic Society*, 1994, 77(7), 1729.

[22] S.T. Schwab, Polysilazane precursors for silicon nitride and resultant products, *US Pat.* 5,294,425, 1994.

[23] R. Baney and G. Chandra, *Encyclopedia of Polymer Science and Engineering*, F.G.A. Stone and W.A.G. Graham (Eds.), Wiley, New York, 1988, Vol. 13, 312.

[24] B.J. Aylett, *Special Ceramics*, E.P. Popper (Ed.), Academic Press, New York, 1964, 105.

[25] R.M. Laine, Transition metal catalysed synthesis of oligo-and polysilazanes, *Platinum Metals Reviews*, 1988, 32(2), 64.

[26] F. Bauer, U. Decker, A. Dierdorf, H. Ernst, R. Heller, H. Liebe and R. Mehnert, Preparation of moisture curable polysilazane coatings Part I. Elucidation of low temperature curing kinetics by FT-IR spectroscopy, *Progress in Organic Coatings*, 2005, 53(3), 183.

[27] D. Seyferth and G.H. Wiseman, Polysilazane routes to silicon nitride, *Polymer Preprints*, 1984, 25(1), 10.

[28] Y. Iwamoto, K. Kikuta and S. Hirano, Crystallization and microstructure development of Si_2N_4-Ti(C,N)-Y_2O_3 ceramics derived from chemically modified perhydropolysilazane, *Journal of Ceramic Society of Japan*, 2000, 108(12), 1072.

[29] Y. Iwamoto, K. Kikuta and S. Hirano, Synthesis of polytitanosilazanes and conversion into Si_3N_4–TiN ceramics, *Journal of Ceramic Society of Japan*, 2000, 108(4), 350.

[30] E. Ionescu and R. Riedel, Chapter: *Polymer processing of ceramics, in Ceramics and Composites* Processing Methods, N. Bansal and A. Boccaccini (Eds.), Wiley, Hoboken, NJ, USA, 2012, 235–270.

[31] A. Saha, S.R. Shah and R. Raj, Oxidation behavior of SiCN-ZrO_2 fiber prepared from alkoxide-modified silazane, *Journal of American Ceramic Society*, 2004, 87(8), 1556.

[32] B. Papendorf, K. Nonnenmacher, E. Ionescu, H.-J. Kleebe and R. Riedel, Strong influence of polymer architecture on the microstructural evolution of hafnium-alkoxide-modified silazanes upon ceramization, *Small*, 2011, 7(7), 970.

[33] E. Ionescu, B. Papendorf, H.-J. Kleebe, H. Breitzke, K. Nonnenmacher, G. Buntkowsky and R. Riedel, Phase separation of a hafnium alkoxide-modified polysilazane upon polymer-to-ceramic transformation – A case study, *Journal of European Ceramic Society*, 2012, 32(9), 1873.

[34] Y. Iwamoto, K. Kikuta and S. Hirano, Si_3N_4-TiN-Y_2O_3 ceramics derived from chemically modified perhydropolysilazane, *Journal of Materials Research*, 1999, 14(11), 4294.

[35] M.C. Bechelany, V. Proust, C. Gervais, R. Ghisleni, S. Bernard and P. Miele, *In situ* controlled growth of titanium nitride in amorphous silicon nitride: A general route toward bulk nitride nanocomposites with very high hardness, *Advanced Materials*, 2014, 26(38), 6548.

[36] J. Yuan, X. Luan, R. Riedel and E. Ionescu, Preparation and hydrothermal corrosion behavior of Cf/SiCN and Cf/SiHfBCN ceramic matrix composites, *Journal of European Ceramic Society*, 2015, 35(12), 3329.

[37] J. Yuan, S. Hapis, H. Breitzke, Y.P. Xu, C. Fasel, H.-J. Kleebe, G. Buntkowsky, R. Riedel and E. Ionescu, Single-source-precursor synthesis of hafnium-containing ultrahigh-temperature ceramic nanocomposites (UHTC-NCs), *Inorganic Chemistry*, 2014, 53(19), 10443.

[38] R.M. Laine, Y.D. Blum, D. Tse and R. Glaser, Chapter 10: *Synthetic routes to organosilazanes and polysilazanes, polysilazane precursors to silicon nitride, in Inorganic and Organometallic Polymers*, Vol. 360, M. Zeldin, K. Wynne and H.R. Allcock (Eds.), ACS Symposium Series, 1988, 124–142.

[39] D. Seyferth and G.H. Wiseman, High-yield synthesis of Si_3N_4/SiC ceramic materials by pyrolysis of a novel polyorganosilazane, *Journal of American Ceramic Society*, 1984, 67(7), C132.

[40] D. Seyferth, G.H. Wiseman, J.M. Schwark, Y.-F. Yu and C.A. Poutasse, Chapter 11: Oligosilicon polymers as precursors for silicon-containing ceramics, in Inorganic and Organometallic Polymers, Vol. 360, M. Zeldin, K. Wynne and H.R. Allcock (Eds.), ACS Symposium Series, 1988, 143–155.

[41] N.R. Dando, A.J. Perrotta, C. Strohmann, R.M. Stewart and D. Seyferth, Methylhydridopolysilazane and its pyrolytic conversion to Si_3N_4/SiC Ceramics, *Chemistry of Materials*, 1993, 5, 1624.

[42] M. Hörz, A. Zern, F. Berger, J. Haug, K. Müller, F. Aldinger and M. Weinmann, Novel polysilazanes as precursors for silicon nitride/silicon carbide composites without "free" carbon, *Journal of European Ceramic Society*, 2005, 25(2–3), 99.

[43] W. Fink, Beiträge zur Chemie der Si-N-Bindung XV [1]. α,ω-Dimethylchlorsilyl-substituierte Cyclodisilazane, *Helvetica Chimica Acta*, 1968, 51(5), 1011.

[44] L.W. Breed, H.L. Budde and R.L. Elliot, Functionally substituted cyclodisilazane derivatives, *Journal of Organometallic Chemistry*, 1966, 6(6), 676.

[45] U. Wannagat, E. Bogush and P. Geymayer, Substituierte N,N'-Bis(silyl)-cyclodisilazane, *Monatshefte für Chemie*, 1971, 102(6), 1825.

[46] G.V. Kotrelev, A.A. Zhdanov, N.A. Tebeneva, T.V. Strelkova and V.V. Kazakova, Polycondensation of methylphenylsilanes and cyclodisilazanes in the presence of metallic potassium, *Polymer Science U.S.S.R.*, 1988, 30(9), 1999.

[47] J. Kulpiński and Z. Lasocki, Polycondensation of bis(diethylamino)dimethylsilane with phenylenediamines, *Acta Polymerica*, 1991, 42(3), 86.

[48] Z. Lasocki, B. Dejak, J. Kulpinski, E. Leśniak, S. Piechucki and M. Witekowa, Chapter 13: *Silicon-nitrogen-containing rings and polymers*, in *Inorganic and Organometallic Polymers, ACS Symposium Series*, Vol. 360, 1988, pp. 166–178; DOI: 10.1021/bk-1988-0360.ch013

[49] J. Kulpiński and Z. Lasocki, Silazane oligomers and polymers with phenylene spacing groups, *Journal of Polymer Science, Part A: Polymer Chemistry*, 1992, 30(3), 461.

[50] Y.D. Blum and R.M. Laine, Catalytic methods for the synthesis of oligosilazanes, *Organometallics*, 1986, 5(10), 2081.

[51] Y.D. Blum, K.B. Schwartz and R.M. Laine, Preceramic polymer pyrolysis - Part 1 Pyrolytic properties of polysilazanes, *Journal of Materials Science*, 1989, 24(5), 1707.

[52] M.T. Zoeckler and R.M. Laine, Homogeneous catalytic formation of carbon-nitrogen bonds. 2. Catalytic activation of the silicon-nitrogen bond, *Journal of Organic Chemistry*, 1983, 48, 2539.

[53] D. Seyferth, M.J. Schwark and M.R. Stewart, Stoichiometric and catalytic ring opening of hexaalkyl-cyclodisilazanes by organoalkali reagents, *Organometallics*, 1989, 8(8), 1980.

[54] G.T. Burns, C.K. Saha, G.A. Zank and H.A. Freemann, Polysilacyclobutasilazanes: preceramic polymers for the preparation of sintered silicon carbide monoliths, *Journal of Materials Science*, 1992, 27(8), 2131.

[55] C.K. Saha and H.A. Freemann, Characterization of pyrolysed silacyclobutasilazane, *Journal of Materials Science*, 1992, 27(17), 4651.

[56] E. Duguet, M. Schappacher and A. Soum, High molar mass polysilazane: A new polymer, *Macromolecules*, 1992, 25(19), 4835.

[57] E. Duguet and A. Soum, Cationic ring-opening oligomerization of hexamethylcyclotrisilazane and octamethylcyclotetrasilazane, *Macromolecular Chemistry and Physics*, 1995, 196(2), 645.

[58] M. Bouquey, C. Brochon, S. Bruzaud, A.-F. Mingotaud, M. Schappacher and A. Soum, Ring-opening oligomerization of nitrogen-containing cyclic organosilicon monomers, *Journal of Organometallic Chemistry*, 1996, 521(1–2), 21.

[59] C. Cazalis, A.-F. Mingotaud and A. Soum, Anionic ring-opening polymerization of cyclodisilazanes, Influence of the silicon substituent on the kinetics of polymerization, *Macromolecular Chemistry and Physics*, 1997, 198(11), 3441.

[60] M. Bouquey and A. Soum, Anionic ring-opening copolymerization of cyclodisilazanes, *Macromolecular Chemistry and Physics*, 2001, 202(9), 1232.

[61] H. Tang, R.E. Prud'homme, A.-F. Mingotaud, M. Schappacher and A. Soum, Solid-state characterization and properties of poly(*N*-methylcyclodisilazanes), *Macromolecules*, 1997, 30 (5), 1400.

[62] J.E. Bender IV, W. Chin and M.M. Banaszak Holl, Synthesis and characterization of a soluble, highly branched organo-silicon-nitride polymer, *Chemistry of Materials*, 1999, 11(1), 154.

[63] C.-H. Xu, N. Zhou and Z.-M. Xie, Synthesis and characterization of silazane oligomer containing 1,1,3,3,5,5-hexamethylcyclotrisilazane, *Chinese Journal of Polymer Science*, 2000, 18(5), 449.

[64] M. Kavala, A. Hawkins and P. Szolcsányi, Syntheses and properties of cyclosilazanes and cyclocarbosilazanes, *Journal of Organometallic Chemistry*, 2013, 732, 58–91.

[65] E. Bacque, J.P. Pillot, M. Birot, J. Dunogues, G. Bougeois and M. Petraud, Synthesis and chemical properties of 1,3-dichloro-1,3-dihydridodisilazanes, *Journal of Organometallic Chemistry*, 1994, 481, 167.

[66] T. Rasul and D.Y. Son, Hydrosilylation reactions of tetramethyldisilazanes and their derivatives, *Journal of Organometallic Chemistry*, 2002, 655, 115.

[67] E.A. Ebsworth and M.J. Mays, Zur Reaktion zwischen halogensilanen und silbercyanamid, *Angewandte Chemie International Edition*, 1962, 74(3), 117.

[68] E.A. Ebsworth and M.J. Mays, Preparation and properties of disilylcyanamide, *Journal of Chemical Society*, 1961, 0, 4879.

[69] J. Pump and U. Wannagat, Bis-(trimethylsilyl)carbodiimid, *Angewandte Chemie International Edition*, 1962, 1(3), 117.

[70] L. Birkofer, L. Ritter and P. Richter, Uber siliciumorganische verbindungen 13: *N, N'*-bis-trimethylsilylcarbodiimid, *Tetrahedron Letters*, 1962, 3(5), 195.

[71] J. Pump and E.G. Rochow, Silycarbodiimide 4: Sila-polycarbodiimide, *Zeitschrift für Anorganische und Allgemeine Chemie*, 1964, 330(1–2), 101.

[72] J.F. Klebe and J.G. Murray, Organosiliconcarbodiimide polymers and process for their preparation. US Patent 3,352,799 (1967).

[73] R. Riedel, E. Kroke, A. Greiner, A.O. Gabriel, L. Ruwisch, J. Nicolich and P. Kroll, Inorganic solid-state chemistry with main group element carbodiimides, *Chemistry of Materials*, 1998, 10(10), 2964.

[74] S. Widgeon, G. Mera, Y. Gao, E. Stoyanov, S. Sen, A. Navrotsky and R. Riedel, Nanostructure and energetics of carbon-rich SiCN ceramics derived from polysilylcarbodiimides: Role of the nanodomain interfaces, *Chemistry of Materials*, 2012, 24(6), 1181.
[75] P. Colombo, G. Mera, R. Riedel and G.D. Sorarù, Polymer-derived ceramics: 40 years of research and innovation in advanced ceramics, *Journal of American Ceramic Society*, 2010, 93 (7), 1805.
[76] R. Riedel, G. Mera, R. Hauser and A. Klonczynski, Silicon-based polymer-derived ceramics: Synthesis properties and applications – A review, *Journal of Ceramic Society of Japan*, 2006, 114, 425.
[77] G. Mera, R. Riedel, F. Poli and K. Muller, Carbon-rich SiCN ceramics derived from phenyl-containing poly(silylcarbodiimides), *Journal of European Ceramic Society*, 2009, 29(13), 2873.
[78] A.O. Gabriel, R. Riedel, S. Storck and W.F. Maier, Synthesis and thermally induced ceramization of a non-oxidic poly(methylsilsesquicarbodiimide) gel, *Applied Organometallic Chemistry*, 1997, 11(10–11), 833.
[79] D.S. Kim, E. Kroke, R. Riedel, A.O. Gabriel and S.C. Shim, An anhydrous sol–gel system derived from methyldichlorosilane, *Applied Organometallic Chemistry*, 1999, 13(7), 495.
[80] S. Nahar-Borchart, E. Kroke, R. Riedel, B. Boury and R.J.P. Corriu, Synthesis and characterization of alkylene-bridged silsesquicarbodiimide hybrid xerogels, *Journal of Organometallic Chemistry*, 2003, 686(1–2), 127.
[81] G. Mera, A. Tamayo, H. Nguyen, S. Sen and R. Riedel, Nanodomain structure of carbon-rich silicon carbonitride polymer-derived ceramics, *Journal of American Ceramic Society*, 2010, 93 (4), 1169.
[82] A. Kienzle, A. Obermayer, A. Simon, R. Riedel and F. Aldinger, *Chemische Berichte*, 1993, 126 (12), 2569.
[83] H.D. Schadler, L. Jager and I. Senf, Pseudoelement compounds 5: Pseudochalcogens – An attempt of an empirical and theoretical characterization, *Zeitschrift für Anorganische und Allgemeine Chemie*, 1993, 619, 1115.
[84] G.A. Razuvaev, A.S. Gordetsov, A.P. Kozina, T.N. Brevnova, V.V. Semenov, S.E. Skobeleva, N.A. Boxer and Y.I. Dergunov, Synthesis of monomeric and oligomeric carbodiimides with polysilane and siloxane fragments, *Journal of Organometallic Chemistry*, 1987, 327(3), 303.
[85] V.I. Gorbatenko, M.N. Gertsyuk and L.I. Samarai, Synthesis of 2,4-bis(trimethylsilyloxy)-6-chloro-1,3,5-triazine, *Zhournal Organitscheskoi Khimii*, 1977, 13, 899.
[86] R.M. Morcos, G. Mera, A. Navrotsky, T. Varga, R. Riedel, F. Poli and K. Muller, Enthalpy of formation of carbon-rich polymer-derived amorphous SiCN ceramics, *Journal of American Ceramic Society*, 2008, 91(10), 3349.
[87] M. Graczyk-Zajac, G. Mera, J. Kaspar and R. Riedel, Electrochemical studies of carbon-rich polymer-derived SiCN ceramics as anode materials for lithium-ion batteries, *Journal of European Ceramic Society*, 2010, 30(15), 3235.
[88] R.M. Prasad, G. Mera, K. Morita, M. Muller, H.J. Kleebe, A. Gurlo, C. Fasel and R. Riedel, Thermal decomposition of carbon-rich polymer-derived silicon carbonitrides leading to ceramics with high specific surface area and tunable micro- and mesoporosity, *Journal of European Ceramic Society*, 2012, 32(2), 477.
[89] A. Lukacs III, Polysilazane precursors to advanced ceramics, *American Ceramic Society Bulletin*, 2007, 86(1), 9301.

Chapter 7
Other silicon-containing polymers

Many kinds of different silicon-containing polymers and especially copolymers containing polysiloxane segments have been described in recent decades. They include a variety of siloxane copolymers containing only siloxane units, for example, polydimethylsiloxane (PDMS)–polydiphenylsiloxane and polymethylphenylsiloxane copolymers, siloxane–alkylene oxide block and graft copolymers, different polysiloxane–polyester copolymers (e.g., with polycarbonates, polyphthalates, polyalkyl esters), polyimides (PI), polyamides, polyureas, and polyurethanes (PUs), polysiloxane–polyalkylene copolymers, polysiloxane–polyalkyne copolymers, polysiloxane–silyl–arylene copolymers, liquid-crystalline (LC) siloxane networks, boron-containing copolymers, inorganic silicates and borates, organically modified silicates, and others [1, 2]. Different kinds of organosiloxane copolymers exhibit a game of useful properties: hydrophobicity, thermal stability, UV, atomic and oxygen plasma resistance, biocompatibility, and satisfactory mechanical properties [1].

Some newer examples of copolymers containing siloxane moieties are presented in this chapter.

7.1 Poly(methylhydrosiloxane) copolymers

A very important group of silicones are multifunctional polysiloxanes, containing reactive hydrosilane groups (Si–H). Poly(methylhydrosiloxanes) (PMHS) are inorganic–organic hybrid polymers with inorganic backbone, built alternatively of silicon and oxygen atoms. Hydrogen atoms and methyl groups, and also other organic groups as well, are attached to silicon atoms in PMHS. The PMHS are mostly colorless oils that find many practical applications [1–11]. Most of them are random (more frequently) or block (seldom) copolymers having Me_2SiO and MeHSiO units in macromolecules [3]. The Si–H functionalities serve as substrates for further functionalizations of polysiloxanes and other silicon-based polymers.

7.1.1 Methods of syntheses of poly(methylhydrosiloxanes)

Many examples of synthetic description of PMHS of linear and cyclic structure and their practical applications are known from the literature [3–6]. The PMHS and copolymers containing H-siloxane units are prepared by the following methods:

(1) hydrolytic polycondensation of methyldichlorosilane ($(CH_3)HSiCl_2$, MDS) or its copolycondensation with other chlorosilanes [4, 5, 12–17];

https://doi.org/10.1515/9783110643671-007

(2) cationic polymerization and copolymerization of methylhydrocyclosiloxanes $[(CH_3)HSiO]_n$ [17–20] or their equilibration (repolymerization) with linear siloxanes [21–27]; and
(3) heterocondensation of siloxanolates of alkali metals [28–30] or siloxanediols [6, 31–33] with hydrochlorosilanes.

PMHS of random structures, having only $(CH_3)HSiO$ mers, are easily prepared by a hydrolytic polycondensation of methyldichlorosilane ($(CH_3)HSiCl_2$), while cohydrolysis of $MeHSiCl_2$ with Me_2SiCl_2 leads to copolysiloxanes containing mers $(CH_3)HSiO$ (D^H) and $(CH_3)_2SiO$ (D) as well (Scheme 7.1). In order to terminate polysiloxanes with unreactive trimethylsiloxane groups ($(CH_3)_3SiO_{0.5}$) and to limit their molecular weights (MWs), small amounts of the monofunctional monomer trimethylchlorosilane $(CH_3)_3SiCl$ must be introduced into the reaction medium. Due to relatively high reactivity of the Si–H bond, hydrolysis reactions of (hydro)chlorosilanes are usually carried out at temperatures below 0 °C.

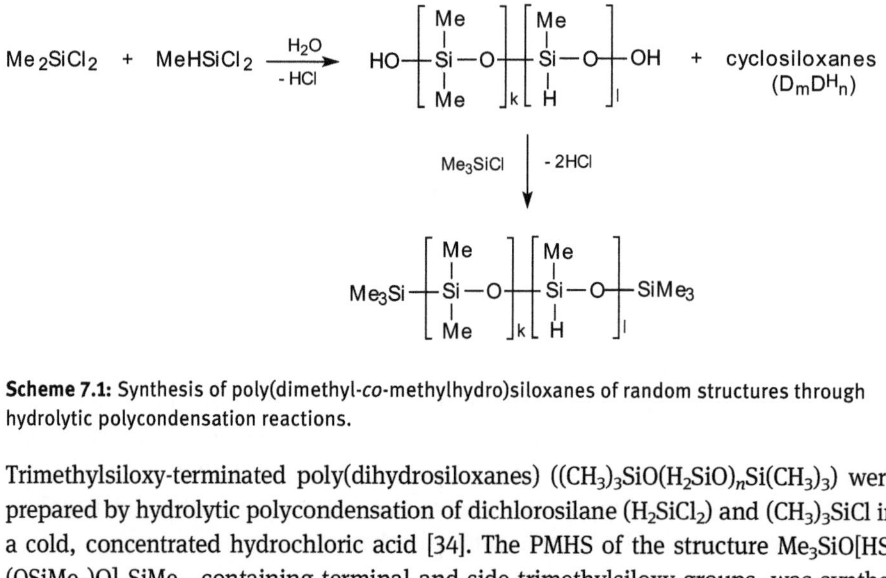

Scheme 7.1: Synthesis of poly(dimethyl-*co*-methylhydro)siloxanes of random structures through hydrolytic polycondensation reactions.

Trimethylsiloxy-terminated poly(dihydrosiloxanes) ($(CH_3)_3SiO(H_2SiO)_nSi(CH_3)_3$) were prepared by hydrolytic polycondensation of dichlorosilane (H_2SiCl_2) and $(CH_3)_3SiCl$ in a cold, concentrated hydrochloric acid [34]. The PMHS of the structure $Me_3SiO[HSi(OSiMe_3)O]_nSiMe_3$, containing terminal and side trimethylsiloxy groups, was synthesized by cohydrolysis of $HSiCl_3$ and Me_3SiCl with 80% H_2SO_4. As a side product, 1% of octahydrosilsesquioxane $(HSiO_{1.5})_8$ was isolated [35].

PMHS with terminal Si–H groups can be prepared by a catalytic equilibration of octamethylcyclotetrasiloxane (D_4) with 1,1′,3,3′-tetramethyldisiloxane ($M^H{}_2$, M^H=$HMe_2SiO_{0.5}$) with 50% H_2SO_4 or acid activated clays, by reduction of appropriate dichlorosiloxanes with metal hydrides, for example, $LiAlH_4$ [32, 36], LiH, or NaH in the presence of Et_3B [37] and by the heterocondensation of appropriate siloxanediols (in the presence of an amine as an acceptor of HCl) [32] or alkali metal siloxanediolate with dimethylchlorosilane ($H(CH_3)_2SiCl$, Scheme 7.2) [28–30, 33].

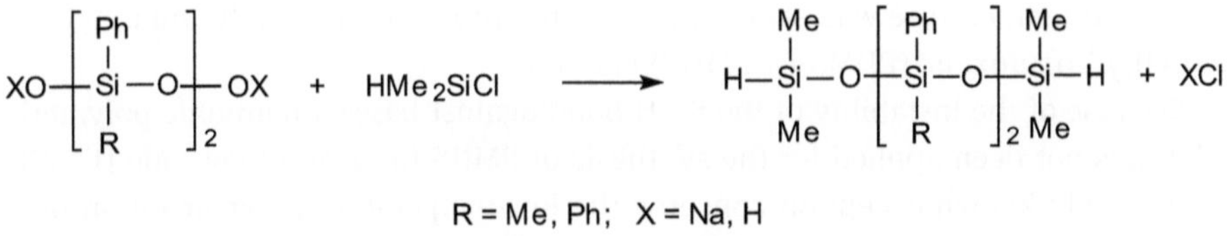

Scheme 7.2: Synthesis of Si–H-terminated poly(dimethylsiloxanes) through heterocondensation reactions.

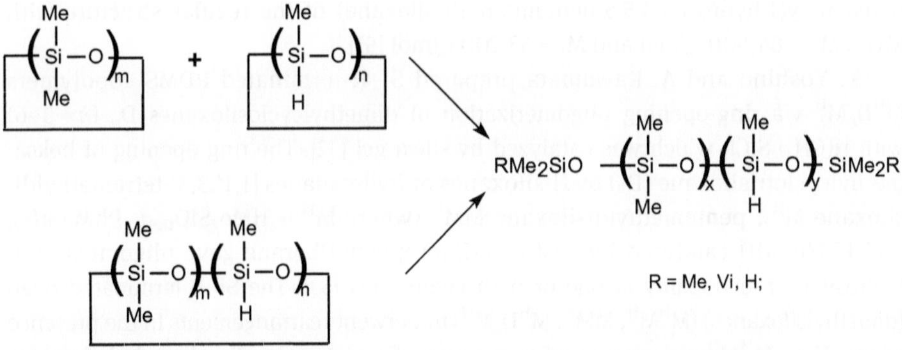

Scheme 7.3: Synthesis of poly(methylhydrosiloxanes) through cationic polymerizations of cyclosiloxanes: $D_mD^H{}_n$ or $D^H{}_n$ with D_m.

Very important synthetic route to PMHS with a random structure of the siloxane chain is the cationic polymerization of cyclic monomers $D^H{}_n$ ($n = 4$–6) or $D_mD^H{}_n$, and copolymerization of $D^H{}_n$ with D_n ($n = 3$–5) as well (Scheme 7.3) [18, 19], most often in the presence of hexamethyldisiloxane ($Me_3SiOSiMe_3$) as a chain terminating agent [38–40]. The cationic polymerization provides a very good control of a size of macromolecules and proceeds in the presence of protonic acids (e.g., H_2SO_4, CF_3SO_3H, $C_6H_5SO_3H$, CF_3COOH, or CCl_3COOH), cationic exchange resins [18, 19, 23, 24, 41], dimethylsilyl sulfate [42], acidic salts (e.g., $(NH_4)_2SO_4$), Lewis acids, and commercial clays activated with protonic acids [22, 23, 43].

The rates of the cationic polymerization and the microstructure of the polysiloxane chain of PMHS strongly depend on the monomer structures and reaction conditions [13, 18, 19]. At low temperature (−78 °C), in the presence of CF_3SO_3H or H_2SO_4, the polymerization of $D^H{}_4$ with M_2 was stereospecific, leading to the isotactic PMHS, with small addition of the syndiotactic polymer. However, within a longer reaction time and at higher temperature, both stereopolymers underwent rearrangement to the random PMHS. The stereochemistry of the siloxane chain was thermodynamically controlled and at temperature of 25 °C, only atactic PMHS was formed [44].

Equilibration polymerization of PDMS with linear hydrosiloxane polymer and hexamethyldisiloxane ($Me_3SiOSiMe_3$) and phosphonitrile catalyst ($[Cl_3P(NPCl_2)_2PCl_3]^+PCl_6^-$)

at elevated temperature was also applied for the preparation of poly(dimethyl-*co*-methylhydro)siloxane (PDMS-*co*-PMHS) [45–49].

Because of the instability of the Si–H bond against bases, an anionic polymerization has not been applied for the synthesis of PMHS for a very long time [17, 43, 50]. The only known exception concerns the low-temperature polymerization of a new monomer pentamethylcyclotrisiloxane (D_2D^H) in the presence of $Ph_2Si(OLi)_2$ in tetrahydrofuran (THF) at −79 °C. The starting cyclic monomer D_2D^H was prepared with 51% yield – by heterocondensation of 1tetramethyldisiloxane-1,3-diol with $MeHSiCl_2$ in the presence of amine. In this way, W.P. Weber and J.K. Paulasaari prepared poly(1-hydro-1,3,3,5,5-pentamethyltrisiloxane) of the regular structure with MWs: M_w = 84,600 g/mol and M_n = 53,200 g/mol [51].

K. Yoshino and A. Kawamata prepared Si–H-terminated PDMS copolymers $M^HD_nM^H$ via ring-opening oligomerization of dimethylcyclosiloxanes D_n (n = 3–6) with $H(CH_3)_2SiCl$, which was catalyzed by silica gel [52]. The ring opening of heksamethylcyclotrisiloxane (D_3) by H-siloxanes or hydrosilanes [1,1′,3,3′-tetramethyldisiloxane $M^H{}_2$, pentamethyldisiloxane MM^H (where M^H = $HMe_2SiO_{0.5}$), $PhMeSiH_2$, and $PhMe_2SiH$] catalyzed by tris(pentafluorophenyl)borane gave oligomers with hydrosilane functionality at one or both chain ends [53]. The Si–H-terminated oligo (dimethylsiloxanes) (M^HM^H, MM^H, $M^HD_2M^H$) underwent rearrangements in the presence of $B(C_6H_5)_3$. M^HM^H was converted to a series of α,ω-dihydrooligodimethylsiloxanes $M^HD_nM^H$, Me_2SiH_2, and small amounts of D_3 and D_4. M^HDM^H, $M^HD_3M^H$, $M^HD_5M^H$, D_3, and Me_2SiH_2 were formed from $M^HD_2M^H$. The dismutation of MM^H proceeded with formation of MD_nM (as major products), MD_nM^H (as minor products), and Me_3SiH [54].

Chruściel [6, 55] elaborated a sequential condensation method of syntheses of three homological series of new poly(dimethyl-*b*-methylhydro)siloxanes (PDMS-*b*-PMHS), containing single and multiple MeHSiO units with a structure described by a general formula:

$$RMe_2SiO\left[(Me_2SiO)_m(MeHSiO)_k\right]_n(Me_2SiO)_mSiMe_2R$$

where:

(a) R = OH; (b) R = Me; m = 2, 6, 10, 14, ~50; k = 1–4; n = 5, 10, 15, 20.

This "*one-pot*" method was based on a nonstoichiometric polycondensation of the appropriate poly(dimethylsiloxane-α,ω-diols) $HO(Me_2SiO)_mH$ with $Cl(MeHSiO)_{k-1}SiHMeCl$, followed by termination reactions with Me_3SiCl, when PDMS-*b*-PMHS chains were terminated with $(CH_3)_3SiO_{0.5}$ groups. MWs (M_w) of PDMS-*b*-PMHS ranged from 4,900 to 122,000 and their polydispersities were as follows: 1.9–4.8 for m = 2 and 8–20 for m = 6–50. New model linear and ring H-siloxane oligomers were isolated: cyclic $(D_2D^H)_2$, $(D_2D^H{}_2D_2D^H{}_2)$, $D_2D^H{}_4$, $(D_2D^H{}_2D_2D^H{}_2D_2D^H{}_2)$] and linear oligomers [$MD_2D^HD_2M$, $M(D_2D^H)_2D_2M$, $MD_2D^H{}_2D_2M$, $MD_6D^HD_6M$, $M(D_2D^H{}_2)_2D_2M$. Their characterization by

Fourier-transform infrared (FT-IR), ^{1}H-NMR (nuclear magnetic resonance), and ^{29}Si-NMR helped to determine the regular microstructure of PDMS-*b*-PMHS block copolymers. All products were also characterized by the elemental analysis. The microstructure of random PDMS-*co*-PMHS prepared by the hydrolytic polycondensation of appropriate chlorosilanes was also studied by ^{29}Si-NMR [6, 55]. It changed with a decrease in the content of Si–H groups and an increase in their MWs. In determinations of tacticities of PDMS-*co*-PMHS, ^{29}Si-NMR data from spectra of the above-listed model oligomeric PMHS [6] and literature data published by R.K. Harris [22] were very helpful. The chemical shifts of Si atoms of the terminal $Me_3SiO_{0.5}$ groups (M) changed in tetrads, depending on substituents at silicon atoms of the siloxane chain, which interact through six Si–O bonds. The chemical shifts of Si atoms of Me_2SiO (D) units were sensitive on substituent effects, which transfer through four bonds in two directions of the siloxane chain. Thus, the δ ^{29}Si values for units D changed in pentades, while δ ^{29}Si-NMR values of Si atoms of central MeHSiO units ($\underline{D^H}$) underwent changes in triads ($D\underline{D^H}D$, $D\underline{D^H}D^H$, $D^H\underline{D^H}D^H$), and magnetic interactions of substituents were transfered through two bonds in both directions. A big complexity of NMR spectra of random copolysiloxanes may also be affected by microtacticity of the siloxane chain [6].

The microstructure of PDMS-*co*-PMHS, obtained by the cationic copolymerization of D_4 and $D^H{}_4$ with hexamethyldisiloxane and by the equilibration of PMHS with D_4 in the presence of clays containing 1% of HCl, was also studied by G. Sauvet et al., using ^{29}Si-NMR techniques such as INVGATE and INEPTRD [56, 57]. Chemical shifts of mers D and D^H in appropriate heptades have been characterized. The microstructure of siloxane chain ends was not analyzed. Moreover, it was shown that in hydrosilylation reactions of PDMS-*co*-PMHS with glicydyl ether, catalyzed by H_2PtCl_6, reactivity of the Si–H groups in isolated (single) mers D^H (i.e., in triads $D\underline{D^H}D$) was significantly lower than in mers adjacent to at least one unit D^H (i.e., in triads $D\underline{D^H}D^H$ and $D^H\underline{D^H}D^H$), presumably due to steric reasons [58].

Liquid-branched PMHS of new random structures (PMHS-Q), containing quadruple branching units $SiO_{4/2}$ (Q), MeHSiO (D^H) and Me_2SiO (D) chain building units (or only mers MeHSiO), and terminal groups $Me_3SiO_{0.5}$ (M) were prepared by the hydrolytic polycondensation of appropriate organic chlorosilanes and tetraethoxysilane, in diethyl ether medium at temperature below 0 °C. Chemical structures of a first series of PMHS-Q is described by the following general formula:

$$(SiO_{4/2})_y[CH_3(H)SiO]_n\left[(CH_3)_3SiO_{0.5}\right]_p$$
$$(\text{where } y = 1-3,\ n = 48-50,\ p = 2y+2),$$

while branched PDMS-*co*-PMHS had the general formula:

$$(SiO_{4/2})_y\left[(CH_3)_2SiO\right]_m[CH_3(H)SiO]_n\left[(CH_3)_3SiO_{0.5}\right]_p$$
$$(\text{where } y = 1-3,\ m = n = 49-52,\ p = 2y+2)$$

Volatile low-MW siloxanes were removed by a vacuum distillation at 150–155 °C. Yields of PMHS-Q ranged from 55 to 69 wt%, after removal of low MW linear and cyclic siloxane oligomers. Their dynamic viscosities, measured in the Brookfield HBDV + IIcP cone-plate viscometer, were low (10.7–13.1 cP). MWs of PMHS-Q (M_n = 2,440–6,310 g/mol, M_w = 5,750–10,350 g/mol) and polydispersities of MWs (M_w/M_n = 2.0–2.8) were determined by size-exclusion chromatography (SEC). All polymers were characterized by FT-IR, ^{1}H- and ^{29}Si-NMR, and an elemental analysis. Based on ^{29}Si-NMR data, the microstructure of siloxane chains was analyzed and compared with literature data [59]. The PMHS-Q have been applied as cross-linking agents in silicone elastomers and rubber compositions [60].

7.1.2 Chemical modifications of poly(methylhydrosiloxanes) and their copolymers

Many kinds of polysiloxane graft copolymers were prepared from PMHS. For instance, polyaddition of fluoroalkenes to the Si–H bonds of PMHS gave high viscous oils, waxes, and solids with very good surface properties and low surface energy, dependent on chemical structure of both substrates [61, 62]. Polyaddition of PMHS to allyl derivatives [(phenyl)allyl- and (glicydyl)allyl-ethers, allyl methacrylate, allyl chloride, and allylamine], 1-octene, and other olefins containing different functional groups, catalyzed by platinum complexes and $Ru(CO)_{12}$, is a useful method for the chemical modification of PMHS properties [63, 64]. Very effective catalysts of addition of hydrosilanes and hydrosiloxanes to vinylsilanes, dienes, and alkynes are Pt and Rh compounds supported on polyamides having a pyridine moiety [65]. New polymeric materials were prepared by addition of fullerene molecules to PMHS, followed by cross-linking [66, 67]. Biodegradable multiblock copolymers were prepared by hydrosilylation of poly(lactic acid) or poly(α-hydroxyacids) having terminal allyl groups, with α,ω-dihydro(polydimethylsiloxanes) ($M^H{}_2D_n$) having number average MWs in the range M_n = 1,700–9,000 g/mol [68].

1,3,5,7-Tetramethylcyclotetrasiloxane [$(HMeSiO)_4$, $D^H{}_4$] served as a core in the synthesis of styrenic and (meth)acrylic star polymers by atom transfer radical polymerization (ATRP) [69]. Liquid silicone resins (QM^H) containing tetrafunctional branching units $SiO_{4/2}$ (Q) and reactive groups M^H ($M^H = HMe_2SiO_{0.5}$) were applied as reinforcing agents for silicone rubbers [70]. Functionalized polysiloxanes having glicydyl, polyether (PE), (hydroxyl)polyether, aminopropyl, and $-CH_2CH_2Me_2SiCl$ pendant groups were prepared by hydrosilylation of appropriate alkene derivatives with heptamethyltrisiloxane (M_2D^H, D^H = MeHSiO) and silsesquioxane ($Q_8M^H{}_8$) in the presence of Rh(I) complexes having siloxyl ligands [71, 72].

7.1.2.1 Polymer electrolytes based on polymethylsiloxane copolymers

A large group of siloxane-based polymer electrolytes are graft, comblike copolymers of linear PMHS and methylhydrocyclosiloxanes with oligoethers or polyethers. A cross-linked polymer was prepared by the reaction of linear PMHS with PEG (poly (ethylene glycol)) monomethyl ether (average MW 350 g/mol) and PEG (average MW 300 g/mol), which were catalyzed with $ZnCl_2$, zinc dioctoate, $SnCl_2$, Et_3N, and Na_2CO_3 or K_2CO_3. It formed ionic conductive complexes with lithium triflate ($LiSO_3CF_3$) or other alkali metal salts ($MtSO_3CF_3$, Mt = Na, K, Rb, Cs). An ionic conductivity was dependent on salt concentration (1–25 wt%) and temperature [73]. Similarly, comblike polymethylsiloxanes (PMS) (and model cyclic tetramer) having oligo(oxyethylene) side chains were obtained from PMHS with MW 2,600 g/mol (or from $D^H{}_4$) and methoxy-terminated PEG (MPEG) of different MWs (250, 350, 550, 750, and 1,000 g/mol) in THF solution with zinc octanoate.

Side-chain crystallization was observed in all polymers except PMS with five oxygen atoms in a side chain. Copolymers containing 5,12, 16, and 22 oxygen atoms in the side chain were insoluble in water [74]. By hydrosilylation reaction of PMHS with the corresponding monoallyl glycols derivatives of $HO(CH_2CH_2O)_nH$ (n = 0, 8 or 15) in which the hydroxyl group was protected with tetrahydropyran were prepared next comb-like polysiloxanes with pendant oligo(ethylene glycol)s chains. Crystallization was slower than for polymers containing side chains with end methoxy groups [75].

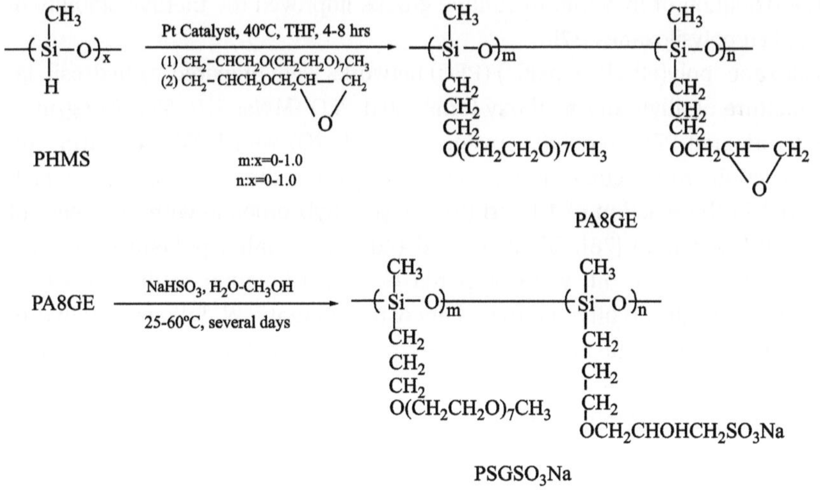

Scheme 7.4: Synthesis of conductive comblike polysiloxane materials. Reprinted with permission from [76]. Copyright 1993, American Chemical Society.

Solvent-free polymeric alkali-metal ion-conductive materials based on comblike polysiloxane-containing oligo(oxyethylene) side chains and pendant sulfonate groups were prepared by the sequential hydrosilylation of an oligo(ethyleneglycol) terminated with allyl and methoxy groups, and allyl glycidyl ether with poly(hydromethylsiloxane), followed by sulfonation of the oxirane ring (Scheme 7.4) [76]. Their conductivities (σ) reached values of only 10^{-6} S/cm at 25 °C and 10^{-5} S/cm at 70 °C, owing to the tightness of the alkali-metal sulfonate ion pair. The σ values increased in the order Li < Na < K < Cs, with a maximum at a ratio of ethylene oxide (EO) units to cation: 13. Addition of linear and macrocyclic polyethers strongly enhanced σ, by stretching the interionic $SO_3^-M^+$ ion pair bond and by decreasing the glass transition temperature (T_g) of a modified polymer, and the most effective polymer was tetra(ethylene glycol) (TEG). At a doping level TEG: Na = 1 the value of σ raised to 2×10^{-5} S/cm at 25 °C and to 10^{-4} S/cm at 70 °C [76].

Polysiloxanes containing thioether and two kinds of heterocyclic polymerizable functions (epoxy or oxetane) were prepared by grafting allyl or vinyl heterocyclic compounds onto poly(dimethyl-*co*-methyl mercaptopropyl)siloxane chains, having ~5 mol% of MeHSiO mers, through a thiol-ene radical addition reaction, which were carried out with 2,2'-azobisisobutyronitrile (AIBN) in acetonitrile at 80 °C. Some mercaptopropyl side groups remained unchanged. All the products were characterized by ^{1}H-NMR, size-exclusion chromatography (SEC), and viscosimetry. These new polymers were thermally cross-linked with an isophorone diamine in the presence of cationic photoinitiators at different temperatures (30–90 °C), providing elastomers with a low T_g and a good thermostability. Thermal and mechanical properties were examined by dynamical mechanical analysis (DMA) and thermogravimetry (TG). By oxidation of thioether functions with *m*-chloroperbenzoic acid, the corresponding sulfones were obtained. The oxidation of thioether to sulfone groups improved the thermal stability of the cross-linked polysiloxanes [77].

Polysiloxane–poly(ethylene oxide) (PEO) networks were prepared by hydrosilylation of a mixture of allyl- and methoxy-terminated PEO (MWs: 350, 550, 750 g/mol) and allyl-terminated PEO-*co*-poly(propylene oxide) (PPO) with PDMS-*co*-PMHS, followed by reaction with a cross-linking agent, containing three NCO groups, which was prepared in the reaction of 1,1,1-tri-(hydroxymethyl) propane with an excess of toluene 2,4-diisocyanate [78]. Disubstituted ("double-comb") polysiloxane–PEO electrolytes, in a form of oil, were prepared by hydrosilylation of dichlorosilane with oligo(ethylene glycol) allyl methyl ethers with different MW, followed by hydrolytic polycondensation in the presence of Et_3N and subsequent workup. The obtained network polymer electrolytes with internal plasticizing ether chains were recommended as excellent polymer electrolytes for fabrication of lithium batteries [79].

A variety of disubstituted (*double-comb*) polysiloxane polymers containing linear, branched, and cyclic oligoethylene-oxide units, $-(OCH_2CH_2)_n-$, in the side chains (Scheme 7.5) and as a part of the siloxane backbone, were prepared. Some copolymers were synthesized from mixtures of linear oligomers of EO. They were doped with lithium bis(trifluoromethylsulfonyl)-imide ($LiN(SO_2CF_3)_2$) and conductivities of the polymer–salt complexes were determined as a function of temperature and doping level. The maximum conductivity of these polymers at 25 °C was 2.99×10^{-4}, for a copolymer containing equimolar amounts of side chains with $n = 5$ and 6 [80].

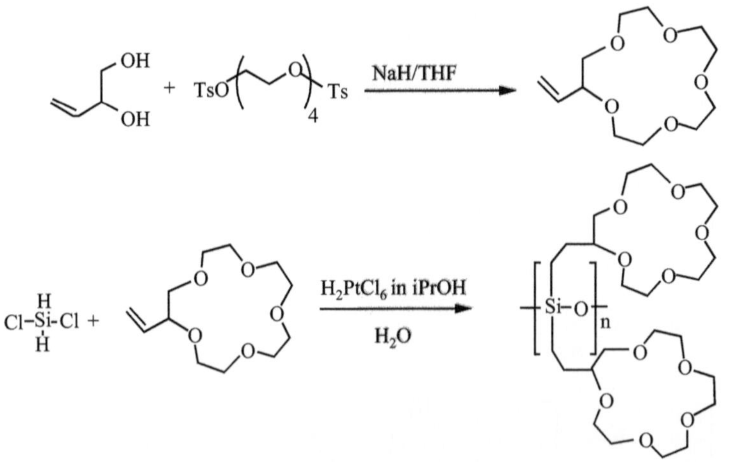

Scheme 7.5: Synthesis of poly[[bis-2-(15-crown-5)ethyl]siloxane]. Reprinted with permission from [80]. Copyright 2002, Springer Nature.

Other liquid polysiloxanes containing crown ether functionalities were synthesized by hydrosilylation reaction of allyl derivatives of the aza- or diaza-crown ethers with PDMS-*co*-PMHS with the degree of polymerization (DP) = 50–1,000 and M_n = 3,560–73,230 g/mol, in the presence of Pt catalyst [81].

Comb-like conductive polysiloxanes containing oligo(oxyethylene) and quaternary ammonium groups as side chains were also prepared by hydrosilylation of PMHS with PEG allyl methyl ether and (*N,N*-dimethyl)allylamine, followed by quaternization reaction with methyl iodide (see Scheme 7.6). A maximum conductivity (1.57×10^{-5} S/cm at 25 °C) showed a polysiloxane with 30% content of quaternary ammonium side groups. The T_g of these electrolytes was also dependent on the content of quaternary ammonium groups attached to the polysiloxane backbone [82].

Polysiloxanes with pendant poly(oxyethylene) (PEO) side chains were prepared by the dehydrocoupling reaction of PMHS or tetramethylcyclotetrasiloxane ($D^H{}_4$) with 2-(2-(2-(2-(2-methoxyethoxy)ethoxy)ethoxy)ethoxy)ethanol and PEG methyl ether and metal catalysts: $Rh(Ph_3P)_3Cl$, $Pd_2(dba)_3$, and tin(II) 2-ethylhexanoate. The polysiloxanes were characterized by FT-IR, ^{1}H-, ^{29}Si-NMR, and GPC (gel permeation chromatography). Significant differences in the structures of the prepared siloxane polymers were dependent on the kind of the catalyst [83]. Novel network polymer electrolytes based on the cross-linked cyclosiloxane precursor were prepared by hydrosilylation reactions of tri(ethylene glycol) methyl allyl ether and diallyl PEO cross-linking agent with PDMS-*co*-PMHS or pentamethylcyclopentasiloxane ($D^H{}_5$), in the presence of Karstedt catalyst. Their ionic conductivities were measured after doping with $LiN(SO_2CF_3)_2$. The highest conductivity ($\sigma_{37} = 1.14 \text{x} 10^{-4}$ S/cm) was observed for the electrolyte with a doping ratio EO:Li^+ of 28:1 and a T_g reached the lowest value −122.2 °C [84]. Branched polysiloxanes with oligoether chains were also obtained by

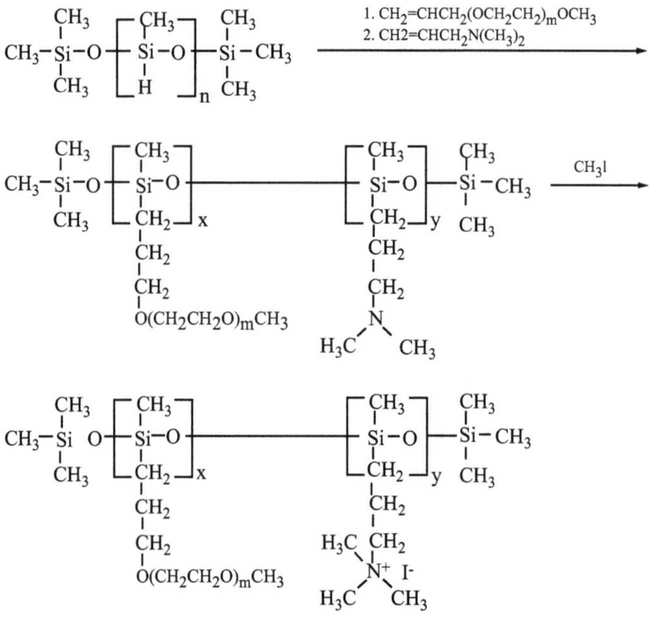

Scheme 7.6: Synthesis of polysiloxanes with oligo(oxyethylene) side chains and quaternary ammonium groups. Reprinted with permission from [82]. Copyright 2002, Springer Nature.

hydrosilylation and dehydrocondensation reactions using short-chain PMHS as a substrate. The conductivity of the two short-chain siloxane derivative complexes with LiN $(SO_2CF_3)_2$ was higher than for longer chain derivatives of PMHS with higher MW. The siloxane–PEO copolymers exibited higher conductivity than the trimethylene-grafted polysiloxane derivative, presumably due to the lower viscosity of this polymer [85]. Ether derivatives of $D^H{}_5$ containing oligo(ethylene glycol) substituents and a short-chain siloxane derivative oligo(ω-methoxytri(oxyethylene)) methylsiloxane (MD_6N3M) (obtained from $MD^H{}_6M$) were prepared by platinum-catalyzed hydrosilylation and by $B(C_6F_5)_3$-catalyzed dehydrocondensation reactions. The oxygen-linked cyclic siloxane-EO derivative, doped with $LiN(SO_2CF_3)_2$, exhibited higher conductivity than trimethylene-grafted siloxane–salt complex. The substituted linear oligomeric siloxane MD_6N3M showed a lower T_g than the pentasiloxanes and much higher conductivity at the same Li^+ concentration. The conductivity–temperature relationship indicated a free volume mechanism of ion transport [86].

Cross-linked solid polymer electrolytes (CPEs) containing oligoethylene oxide chains (EO), $CH_3(OCH_2CH_2)_3(CH_2)_3$-, within the network structures were prepared by hydrosilylation reactions of an α,ω-diallyl-terminated oligo(ethylene glycol) with partially substituted oligoethyleneoxy precursors, which were obtained from stereoisomers of pentamethylcyclopentasiloxane ($D^H{}_5$). The conductivities of the network siloxane/lithium salts complexes with various EO/Li ratios were determined by impedance experiments. The CPEs with 36.0% cross-linking density exhibited higher conductivity than those with 43.8% cross-linking density. The optimum conductivities (σ) 9.24×10^{-5} S/cm at 25 °C and 2.11×10^{-4} S/cm at 37 °C were observed for the latter cyclosiloxane-EO derivative, doped with lithium bis(oxalato)borate. These complexes showed higher conductivity than $LiN(SO_2CF_3)_2$-doped siloxanes at the same salt content [87]. Si–H-functional trisiloxane (MD^HM) and tetrasiloxane ($MD^H{}_2M$) were grafted with oligo(ethylene oxide) chains, -$(CH_2CH_2O)_n$- ($n = 2$–7), via hydrosilation and dehydrocoupling reactions. The low viscosity products were characterized by ^1H, ^{13}C, and ^{29}Si NMR. They were doped with lithium bis(oxalato)borate or lithium bis(trifluorosulfonyl) imide. Very low T_gs before and after doping were observed from differential scanning calorimetry (DSC) analysis. The obtained electrolytes showed ambient temperature conductivities ranging from 2×10^{-4} to 6×10^{-4} S/cm, which were the highest observed for liquid "polymer" electrolytes. Temperature-dependent conductivity studies revealed that the EO chains contributed to an ion transport [88]. New graft polysiloxane copolymers with mixed substituents were also prepared by hydrosilylation of 4-allyloxymethyl-[1,3]dioxolan-2-one and tri (ethylene glycol) allyl methyl ether with PMHS. The effect of the incorporation of carbonate groups on ionic transport, viscosity, and thermal properties was studied. The all mixed substituted polysiloxane polymers with different carbonate content, doped with $LiN(SO_2CF_3)_2$, showed higher conductivity than the polysiloxanes having pure carbonate or pure oligomeric ethyleneoxy substituents. The maximum conductivity (1.62×10^{-4} S/cm) was observed for the polymer containing 8.5% carbonate groups at a doping level of EO: $LiN(SO_2CF_3)_2 = 15$. The polymers containing larger contents of carbonate groups showed lower conductivity, probably due to their increased viscosity and higher T_g [89].

The hydrosilylation of PMHS with tetraethylene glycol allyl methyl ether and varying molar ratios of allyltrimethoxysilane (ATMS) (5–40 mol%) gave a series of comb-like polysiloxanes. The ATMS side groups enabled cross-linking of these solvent-free graft polysiloxane electrolyte membranes, after doping with lithium trifluoromethylsulfonate (triflate) or lithium bis(trifluoromethylsulfonyl)imide. These polymer membranes containing 10 wt% lithium triflate had a maximum conductivity of 4.6×10^{-5} S/cm at 30 °C for 10 mol% ATMS substitution. With the constant ATMS substitution (10 mol%), the salt content varied, giving a maximum conductivity of 1.4×10^{-4} S/cm at 30 °C for 12.5 wt% lithium triflate [90].

A similar new hybrid polymer electrolyte was prepared by cross-linking PEG diglycidyl ether (DE) and 3-glycidoxypropyltrimethoxysilane (GPTMS) with α,ω-diaminopoly(propyleneoxide) and was characterized by FT-IR, ^{13}C solid-state NMR, and thermal analysis (DSC and thermal gravimetric analysis (TGA)). These hybrid films were mixed with the liquid electrolyte (1 M $LiClO_4$ in propylene carbonate), giving plasticized polymer electrolytes. The effects of hybrid composition, liquid electrolyte content, and temperature on the ionic conductivity of hybrid electrolytes were studied. TGA results showed excellent thermal stability of these hybrid networks. The sample with the weight ratio of GPTMS/PEG-DE of 1:2, containing a

75 wt% liquid electrolyte exhibited the ionic conductivity of 5.3×10^{-3} S/cm at 95 °C and 1.4×10^{-3} S/cm at 15 °C. This homogenous film showed a good mechanical strength and good chemical stability. These polymer electrolyte hybrids containing 45 wt% liquid electrolyte showed the maximum conductivity for the sample with the weight ratio of GPTMS/PEG-DE of 1:10 [91].

Random and block PEG-*co*-PDMS brushes were prepared from oligo(ethylene glycol) methyl ether methacrylate (M_n = 475 g/mol) and (monomethacryloxy)propyl-terminated PDMS (6–9 cSt), using ATRP method, by Synytska et al. Their effect on the wettability and adhesion properties in wet and under water conditions was studied. All PEG-*co*-PDMS brushes were nonadhesive in water. They showed biocompatibility and can be used in cosmetics [92].

Well-defined amphiphilic diblock copolymers EG_x–FA_z and pentablock copolymers FA_z–EG_x–$(PDMS)_y$–EG_x–FA_z containing polysiloxane (PDMS), polyethylene glycol (EG), and (perfluorohexyl)ethyl polyacrylate (FA) blocks were prepared by ATRP of FA monomer initiated from a difunctional bromo-terminated macroinitiator. These diblock copolymers resisted protein adhesion better than did pentablock copolymers. Blends of the pentablock copolymer with PDMS resulted in increased protein adsorption, while the PDMS-matrix coatings showed high removal percentages of sporelings of the green fouling alga *Ulva linza*. The block copolymers or their blends with various contents of a PDMS matrix were used for preparation of antibiofouling coatings [93].

7.1.2.2 Liquid-crystalline siloxane materials derived from poly(methylhydrosiloxanes) and other silicon polymers

One of the main features of low MW liquid crystals and liquid-crystalline (LC) polymers is the presence of anisometric molecular structures (mesogenic groups) which are able to form LC phases (mesophases), under suitable conditions of temperature, pressure, and concentration. Most of the mesogenic groups consist of rigid rod-, board- (or lath-), and disk-shaped molecular structures, which play a role of specific "building blocks." A large number of various mesophases were described in the literature: nematic, smectic, and columnar LC phases. They are usually formed by melting of crystalline organic solids (or cooling of an isotropic melt) and are called as thermotropic liquid crystals. The temperature at which the transition between the mesophase and the isotropic phase occurs is called the clearing (T_{cl}) or isotropization temperature (T_{iso}) [94].

Although silicones (mostly PDMS) have outstanding flexibility of the siloxane chain [4, 10 (pp. 208–222), 95], some derivatives of polysiloxanes and other organosilicon polymers exhibit LC properties. The presence of longer alkyl substituents, phenyl groups in siloxanes, phthalocyaninato complexes, or ring structures of hydrogen bonds [96] led to mesomorphic properties. Some LC polysiloxanes with no mesogenic groups as well as LC elastomers are also known. Polydialkylsiloxanes with linear alkyl substituents, except PDMS, formed mesophases [97, 98], which were classified as plastic or conformational disordered (condis) crystals [99–101]. Cross-linking of linear polydialkylsiloxanes with peroxides or polydialkylsiloxane-α,ω-diols with poly(ethyl silicate) in the presence of a metalloorganic catalyst gave mesomorphic elastomers [102].

Most of LC polysiloxanes are the side-chain-type structures bearing calamitic or discotic side groups tethered to the backbone through a flexible spacer groups. The main-chain LC polymers contain flexible siloxane segments combined with rigid rod-type mesogens. One of the most often used methods of the preparation of side-chain LC polysiloxanes (SCLCP) and main-chain LC polysiloxanes is the hydrosilylation of unsaturated mesogens [103].

The highly flexible Si–O bonds in SCLCP are materials exhibiting much lower T_g than LC polyacrylates and polymethacrylates (PMA). The hydrosilylation reaction of mesogenic (or promesogenic) alkenes with PMHS, catalyzed by platinum complexes at low temperatures (25–45 °C), forms materials with mesomorphic properties and chiral smectic C or cholesteric mesophases. Mostly low MW $Me_3Si[OSi(Me)H]_nOSiMe_3$ (M_n = 700–800 g/mol) having quite low polydispersity (polydispersity index, PDI ~ 2) were applied for this purpose. The mesogens had a different flexible spacer between cholesteric group [103, 104]. SCLCP may find potential optical applications as switches, as filters, or in data storage systems. The copolymers bearing dichroic dyes, azobenzene [105] or anthraquinone [106] derivatives, and copolymers containing cholesterol and menthol or a mesogen and nonmesogenic chiral systems were also used for their preparation [107]. The length of the flexible spacer affected the properties of LC materials. The hydrosilylation reaction was also used for the synthesis of highly branched star, comb-like dendritic systems [108, 109], and polymers with increased density of side-chain mesogens [110].

The hydrogen bonded SCLCPs based on polysiloxane backbone and interactions of 3-carboxypropyl side groups of polysiloxanes and imidazole ring in azobenzene [111, 112] or stilbazole derivatives [113]. The LC polysiloxanes with PDI > 2 were prepared by hydrolysis and condensation reactions of (3-cyanopropyl)methyldichlorosilane (and dimethyldichlorosilane for copolymers) [114].

Monodisperse polyhedral LC silsesquioxanes were prepared by the hydrosilylation reaction of mesogenic alkenes with hydrosilsesquioxanes. The LC phase behavior of these materials is dependent on the structure of the mesogenic side chains and affects their T_gs, the type of LC phases, and phase transition temperatures over a wide temperature range [115].

Two series of LC materials with different structures of siloxane cores (cyclooctasiloxane ring or octasilsesquioxane cage) containing methoxyphenylbenzoate mesogenic groups were prepared by hydrosilylation of alkene mesogenic side-chain precursors (4-(ω-alkenyloxy)-4′-methoxyphenylbenzoates) with octamethyloctahydrocyclooctasiloxane ($D^H{}_8$) and octakis(dimethylsiloxy)octasilsesquioxane. The spacers in mesogen molecules contained 3, 4, 5, 8, or 11 methylene (CH_2) groups. The chemical structures of all compounds were analyzed by means of FT-IR and ^{1}H- and ^{29}Si-NMR spectroscopy. Most of the obtained siloxane oligomers showed LC phase. The effects of the spacer and the type of siloxane core, and the thermal and mesogenic properties of the LC siloxane oligomers were studied by TGA, DSC, polarized optical microscopy

(POM), and wide-angle X-ray diffraction (XRD) methods. The phase transition temperatures were strongly dependent on the siloxane core and length of the flexible spacer. The melting point changed a little by increasing the length of the spacer chain in cyclosiloxane series, so that a clearing point increased slightly and led to the stabilization of the mesomorphic properties. The effect of changing the siloxane frameworks from cyclic cores to cube cores gave comparable stabilities of mesophases. The LC silsesquioxanes exhibited a good thermal stability in air, and they may find various practical applications [116].

The side-chain discotic polysiloxane bearing 2,3,6,7-tetrakis(hexyloxy)-10-methoxytriphenylene-11-undecanoate moieties was prepared by hydrosilylation reaction of appropriate undecene-functionalized triphenylene derivative with PMHS in toluene using Karstedt's catalyst at 60–80 °C. The phase behavior and thermooptical properties of the obtained LC polysiloxane were examined by polarizing optical microscopy, thermooptical analysis, DSC, and X-ray scatterring studies. A columnar planar alignment of LC in the layers was determined. This LC polymer can be used in optoelectronic devices [117].

New LC polyethers (PEs) containing siloxane segments were prepared by polycondensation reactions between α,ω-bis(chloromethyl)polydimethylsiloxanes (with different MWs) and 2,6-bis(4-hydroxybenzylidene)cyclohexanone. Their properties and mesophase behavior were studied by TGA, DSC, and polarizing light microscopy. Some of the obtained compounds exhibited thermotropic LC properties, depending on the length of the siloxane spacer. A decrease in the transition temperatures was observed with an increase in the spacer length [118].

New hydrogen-bonding supramolecular polymers were prepared from 4,4′-bipyridine and different silicon-containing diacids. Their thermotropic behavior was studied with polarizing optical microscopy and DSC. The mesophase transitions of these supramolecular liquid crystals were dependent on the structure of the starting siloxane/silane diacid [119].

Two series of carbosilane LC dendrimers (from first to fifth generations) containing 8, 16, 32, 64, and 128 terminal chiral mesogenic groups, respectively, were prepared by Shibaev, Muzafarov, and coworkers. The abovementioned LC dendrimers had the same T_g around –5 °C. With increasing generation number, the isotropization temperature increased and the enthalpy of this phase transition decreased. The strongest effect of spherical molecular architecture on the phase behavior of the LC dendrimers was observed for high generations. Two levels of a structural organization were observed for the LC dendrimer of the fifth generation, which formed different supramolecular nanostructures of columnar type in addition to smectic-like arrangement of mesogenic groups. The LC dendrimers of the lower generations (G-1 to G-3) formed a ferroelectric SmC* phase up to about 180 °C, while the LC dendrimers of the higher generations (G-4 and G-5) exhibited a rectangular columnar mesophase. Electrical measurements on the ferroelectric LC dendrimers showed that an increase in the generation number led to a decrease in the value of the spontaneous polarization and an increase in switching time [120, 121].

A different molecular packing was observed in the LC carbosilane dendrimers having 128 periphery polar cyanbiphenyl groups (CB) and polar (butoxyphenyl)benzoate groups inside of molecularly thin surface films. Similar dendrimers with 128 (butoxyphenyl)benzoate end groups formed two-dimensional organized layers due to the microphase separation of the flexible cores and mesogenic terminal groups. Complete disruption of the regular lamellar ordering was observed for dendrimer with shorter and more polar CB groups with a weaker trend toward layered (smectic) ordering, owing to microphase separation of core and shell structures [122].

Five generations of carbosilane dendrimers containing 8, 16, 32, 64, and 128 mesogenic photochromic groups were prepared and characterized by Shibaev et al. [123, 124]. The molecular structures of three generations of the LC dendrimers are presented in Figure 7.1. The dendrimers of lower generations usually formed the layered (smectic) structure with the antiparallel packing of mesogenic groups and H-aggregates. LC dendrimers of higher generations formed columnar phases owing to the packing of disklike dendritic molecules as cylindrical columns. These azobenzene-containing carbosilane LC dendrimers showed photoactive properties under the action of irradiation [123, 124].

Photooptical studies and kinetics of dichroism growth revealed a distinct difference between the behaviors of LC dendrimers of different generations. For the dendrimers of lower generations, the efficiency of orientation strongly decreased by increasing the generation number. These properties were probably caused by a partial destruction of the smectic structure with increasing size of the matrix, while the dendrimers of the fourth and fifth generations showed high photoinduced dichroism: D reached 0.6 for the dendrimer of the fifth generation. Presumably the branched centrosymmetric structure of dendrimers controlled their photooptical properties, which were much different from those of the comb-shaped polymers [124].

Silazane oligomers with promesogenic groups on both ends were prepared by the hydrosilylation reaction of well-defined telechelic carbosilazane oligomers, end-fitted with vinylsilane groups, with promesogenic compounds containing one reactive hydrosilyl Si–H group. The effects of the structure of the resulting mesogen end-capped oligomers and their phase behavior were studied by DSC and polarizing optical microscopy (POM) [125].

LC materials containing siloxane segments also form main-chain or side-chain three-dimensional networks through cross-linking processes: elastomers or resins, depending on temperature and cross-link density. Most of LC polysiloxane elastomers contain mesogens attached to the siloxane backbone [10 (pp. 208–222), 126–130]. Such networks may be swollen by chiral compounds with formation of cholesteric elastomers. Alternatively, Si–H-terminated PDMS was used as a cross-linking agent for LC linear polymer (Scheme 7.7) [131, 132].

LC main-chain polymers having mesogenic side chains exhibiting ferroelectric behavior were cross-linked with siloxanes as well [133–135]. The mesogens were usually oriented by stretching or compressing the LC elastomer. After reducing the

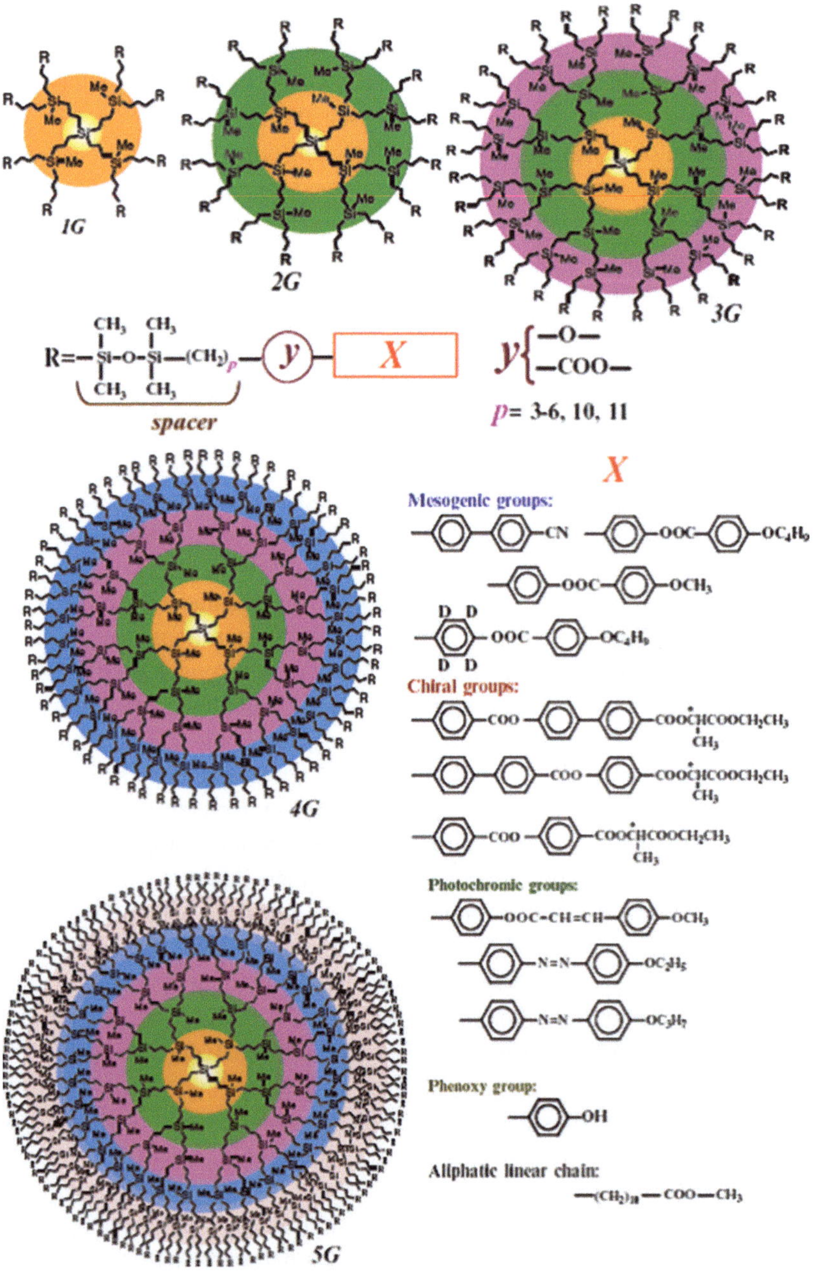

Figure 7.1: The general structures of five generations of LC carbosilane dendrimers with different end (periphery) mesogenic and nonmesogenic groups; y is the linking group; X denotes the mesogenic (or nonmesogenic) terminal group; *m* is the number of terminal groups (*m* = 8, 16, 32, 64, 128). Reprinted with permission from [124]. Copyright 2009, Springer Nature.

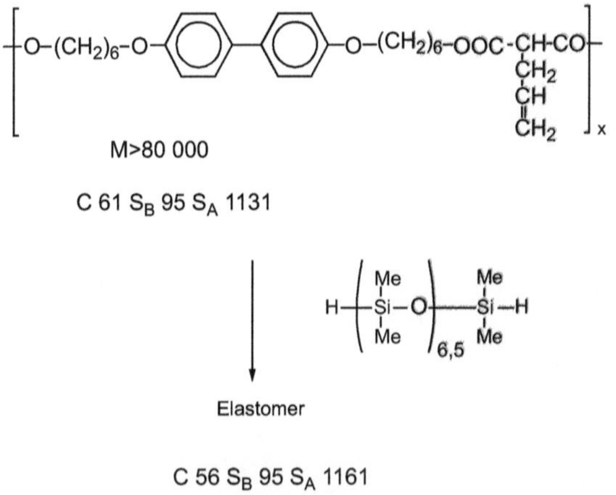

Scheme 7.7: Cross-linking of allyl-functionalized LC linear polymer through hydrosilylation reaction with α,ω-dihydro(oligodimethylsiloxane) and some physical properties of LCs. Reprinted with permission from [10]. Copyright 1996, Springer Nature.

stress, these materials relaxed again into the nonoriented state. However, LC elastomers that preserved the oriented state were obtained by a two-step reaction using difunctional mesogens terminated with alkene groups of different reactivity in the hydrosilylation reaction (Scheme 7.8) [136].

Highly cross-linked cholesteric LC silicone resins were obtained by thermal or photochemical radical polymerization of methacrylic groups attached to mesogens tethered with cyclosiloxanes. The prepared LC materials with tailored properties exhibited S_A-, S_C-’ S_C*-, and N-phases. These unique liquid crystals should find special, temperature-independent, technical applications, for example, as filters or polarizers [137].

A special type of LC siloxanes with complex structures were synthesized by polycondensation of dihydroxy-substituted silicon phthalocyanines (Si-PhtC). Their solubility was improved by incorporation of alkoxy substitutents into Si-PhtC structures [138–141].

7.1.3 Comprehensive applications of poly(methylhydrosiloxanes)

PMHS are widely used as hydrophobic, antiadhesive, and antifoaming materials [3–5,7], cross-linking agents [4, 5, 7–9, 21, 142–150], components of dental composites [151], and ingredients of cosmetics, which are concidered as physiologically inert [152, 153]. Papers and parchments coated with silicone compositions cross-linked with PMHS are most important antiadhesive materials for production of *silicone-release coatings* and for the manufacture of packages for sticky food and industrial products [154].

In summary, most important applications of PMHS include:

1. cross-linking of silicone elastomers at room temperature [3–11,155,156];
2. oxidation or hydrolysis of side Si–H groups and consecutive cross-linking of preceramic polymers [157–160] and hydrophobic finishes on textiles [4, 5, 7];

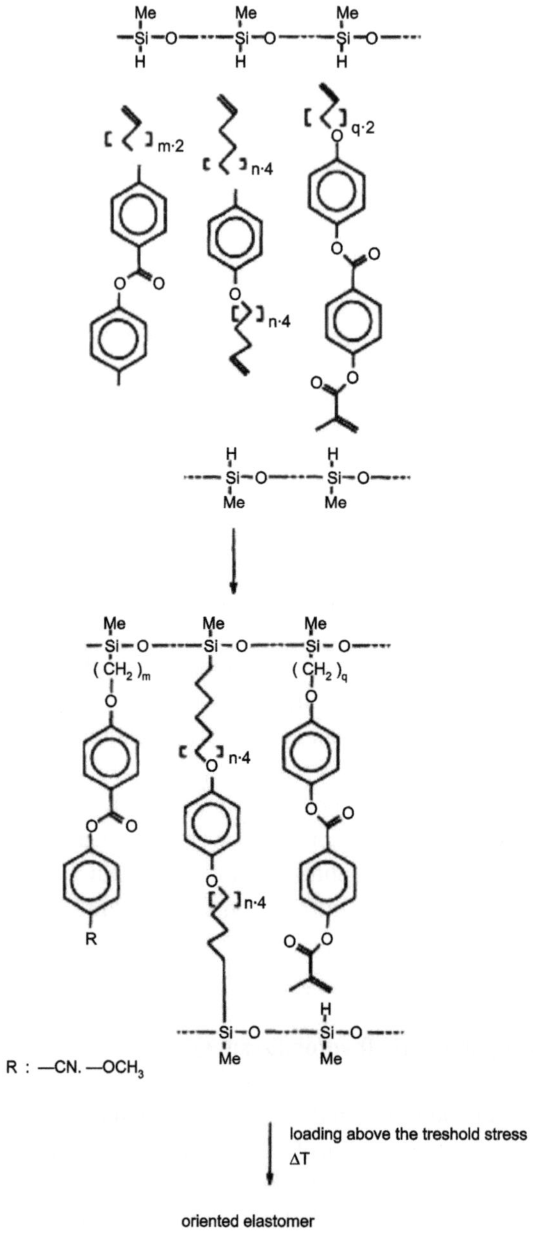

Scheme 7.8: Preparation of LC silicone elastomer via hydrosilylation cross-linking of PMHS with telechelic methacryloxy-functionalized LC linear oligomer. Reprinted with permission from [10]. Copyright 1996, Springer Nature.

3. synthesis of comb polysiloxanes with different pendant groups, used as LC organosilicon polymers [7, 95, 161–173], surfactants, ingredients of cosmetics, and so on [4, 5];
4. synthesis of highly branched organosilicon polymers (dendrimers) [169–171,174–177] and star polymers [178, 179];
5. grafting of macroinitiators [180], medicines [181], and stabilizers [182] on PMHS and synthesis of interpenetrating polymer networks (IPN) with good antiadhesive [61, 62, 183] and conductive properties (after doping with salts) [184];
6. preparation of solid polyelectrolytes [185] and biomaterials [186–190];
7. synthesis of carbofunctional polysiloxanes from α,ω-dihydro(polydimethylsiloxanes) $M^H{}_2D_n$ [$M^H = H(CH_3)_2SiO_{0.5}$, $D = (CH_3)_2SiO$] [191];
8. surface modification of inorganic fillers and supports [192–196];
9. modification of polymer properties [197–200];
10. reduction of organic compounds, for example, alkenes [201], aldehydes and ketones [202, 203], nitrocompounds [203], or organic phosphorous esters [204–206].

7.2 Other copolymers containing siloxane segments

The synthesis of organic–inorganic hybrid polymers gives the possibility to combine the advantages of organic polymer (elasticity, formability) and inorganic material (hardness, strength, high scratch and chemical resistance, as well as good thermal stability). An incorporation of siloxane, silsesquioxane, and silicate segments into a wide family of copolymer structures has been often utilized by many chemists and researchers who work in the field of materials science and gave a huge number of new materials with valuable physicochemical properties for numerous potential applications.

7.2.1 Copolymers containing only siloxane backbone

Most of the linear polysiloxanes, available on the market, are composed of dimethylsiloxane chain building units. Polysiloxane copolymers containing different kinds of mers have been synthesized (1) by hydrolytic polycondensation of appropriate chlorosilane monomers, (2) by condensation reactions of bifunctional silanes or oligosiloxanes containing different kinds of functional groups (e.g., Si–Cl + Si–OH, Si–OH + Si–OR, Si–H + Si–OH, Si–H + Si–CH = CH_2, and Si–OH + C–OH), or (3) by copolymerization of cyclosiloxanes bearing different substituents at silicon. All these methods have been reviewed in numerous monographs and publications [1, 2, 4, 5, 10, 207–209]. In this section, only some newer examples of “pure-siloxane copolymers” are described.

Silicone copolymers of the type $[(SiR_2O)_m(SiR'_2O)_n]$ were prepared by the copolymerization of mixtures of cyclosiloxanes. For example, the random copolymer -$[(Me_2SiO)_m(Ph_2SiO)_n]$- was obtained by heating a mixture of $(Me_2SiO)_4$ and $(Ph_2SiO)_4$ in the presence of KOH as the catalyst [4, 209, 210] (Scheme 7.9). The MWs of the copolymers ranged from 100,000 to 200,000 g/mol. These copolymers were characterized by ^{29}Si-NMR: δ $SiPh_2O$ signals were present at around −47 ppm and the δ $SiMe_2O$ signals – at −20 ppm.

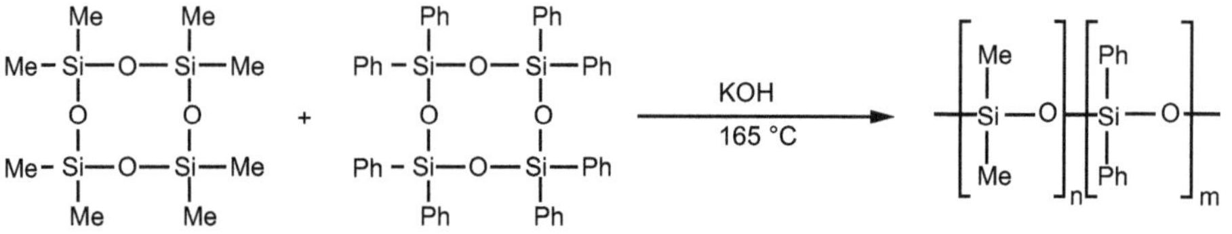

Scheme 7.9: Copolymerization of a mixture of $[Me_2SiO]_4$ and $[Ph_2SiO]_4$.
Reprinted with permission from [209]. Copyright 2005, Springer Nature.

Copolymerization of a mixture of $(Me_2SiO)_4$ and $(SiMeViO)_4$ (Vi = CH = CH_2) afforded poly(dimethylsiloxane-*co*-methylvinylsiloxane) with a random distribution of the Me_2SiO and MeViSiO units [211]. Other copolymers having a regular sequence of mers were prepared by the anionic polymerization of two methylvinylcyclotrisiloxanes: $[(OSiMeVi)(OSiMe_2)_2]$ and $[(SiVi_2O)(SiMe_2O)_2]$ [212, 213] (Scheme 7.10).

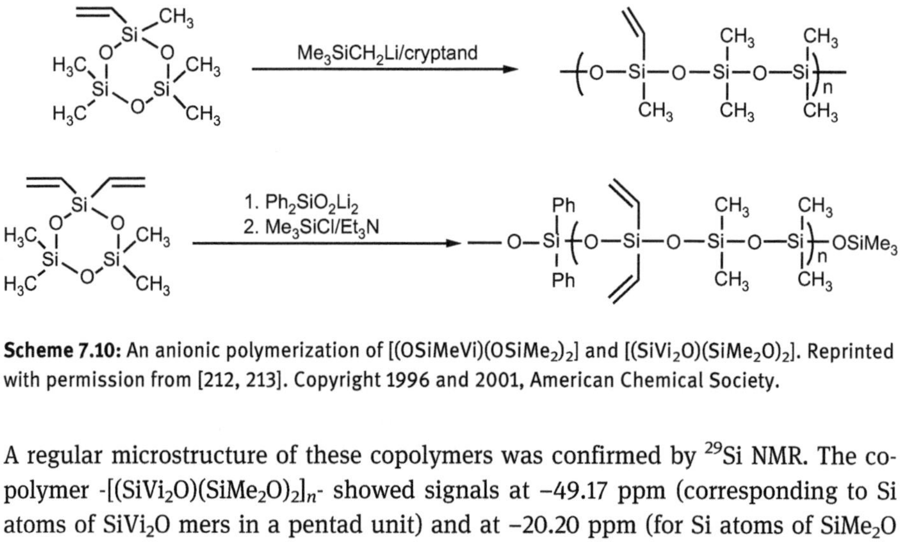

Scheme 7.10: An anionic polymerization of $[(OSiMeVi)(OSiMe_2)_2]$ and $[(SiVi_2O)(SiMe_2O)_2]$. Reprinted with permission from [212, 213]. Copyright 1996 and 2001, American Chemical Society.

A regular microstructure of these copolymers was confirmed by ^{29}Si NMR. The copolymer -$[(SiVi_2O)(SiMe_2O)_2]_n$- showed signals at −49.17 ppm (corresponding to Si atoms of $SiVi_2O$ mers in a pentad unit) and at −20.20 ppm (for Si atoms of $SiMe_2O$ mers in a pentad unit). However, by cationic polymerization of $[(SiVi_2O)(SiMe_2O)_2]$ with triflic acid (CF_3SO_3H), a random copolymer was prepared [213].

Similar polysiloxane diblock copolymers containing only polysiloxane chains, with MW distributions (M_w/M_n) around 1.4, were synthesized by the sequential anionic copolymerization of hexamethylcyclotrisiloxane with 1,3,5,7-tetramethyl-1,3,5,7-tetravinylcyclotetrasiloxane or 1,3,5-trimethyl-1,3,5-trivinylcyclotrisiloxane. The vinyl groups of the poly(dimethylsiloxane)-*b*-poly(methylvinylsiloxane) copolymers were selectively functionalized by hydrosilylation reactions (with trimethoxysilane, triethoxysilane or dimethylchlorosilane) or epoxidation reactions with 3-chloroperbenzoic acid [214].

A series of amphiphilic polysiloxanes containing multicationic groups were prepared by copolymerization of (octamethyl)cyclotetrasiloxane with *N,N′*-(γ-dimethylaminopropyl)-γ-aminopropyl(dimethoxy)silane (Scheme 7.11) followed by quaternization with benzyl chloride. They showed good surface activity, self-emulsifying property, and better wettability of polyester fabrics in comparison with the amphiphilic polysiloxanes, in which these properties were higher than it was found with commercial aminofunctional polysiloxane softener AEAP-PDMS. The novel amphiphilic polysiloxanes containing multicationic groups may find many practical applications in improving the wettability of composites and biomaterials [215].

The vinyl-terminated poly[methylhydrosiloxane-*co*-vinylmethylsiloxane] $M^{Vi}(D^H)_x D^{Vi})_y M^{Vi}$ (PMHVS) was prepared by cationic ring-opening polymerization (CROP) of tetramethyl(tetravinyl)cyclotetrasiloxane ($D^{MeVi}{}_4$), tetramethylcyclotetrasiloxane ($D^H{}_4$) with the end-capping reagent tetramethyldivinyldisiloxane ($M^{Vi}{}_2$) according to Scheme 7.12. [216]. The thermal decomposition of PMHVS into SiOC ceramic at various heating rates was studied by TG combined with mass spectrometry [216, 217].

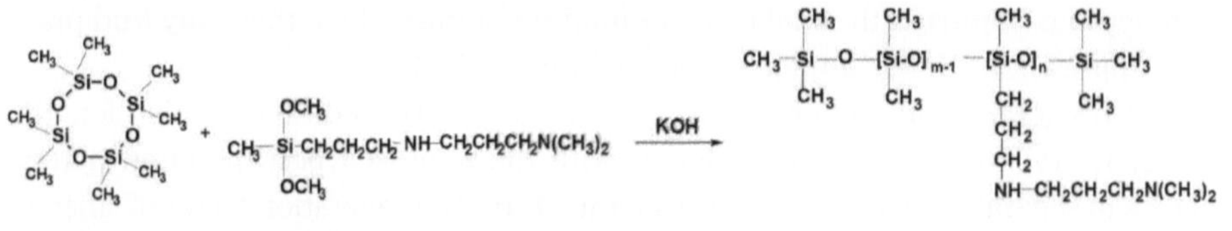

Scheme 7.11: Copolymerization of (octamethyl)cyclotetrasiloxane and *N,N′*-(γ-dimethylaminopropyl)-γ-aminopropyl(dimethoxy)silane. Adapted with permission from [215]. Copyright 2010, Springer Nature.

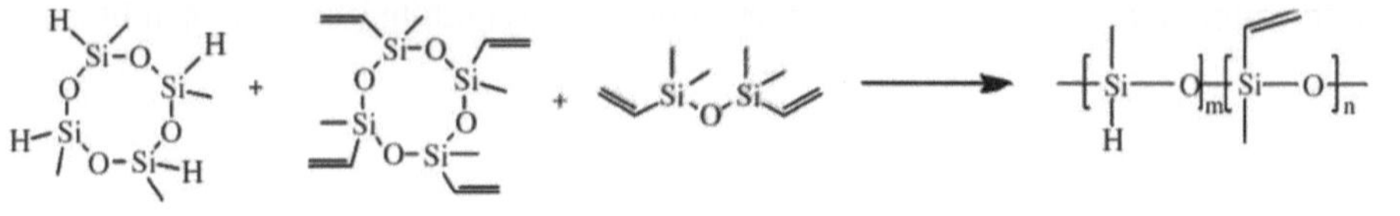

Scheme 7.12: Synthesis of PMHVS by CROP. Reprinted with permission from [216]. Copyright 2014, Springer Nature.

The siloxane derivatives can be used for fabrication of medical implants. For example, the gold electrodes of acoustic wave sensors were coated with poly(mercaptopropylmethylsiloxane) or (octaphenyl)cyclotetrasiloxane. The flow-through adsorption of different proteins to the two siloxane surfaces and a gold electrode was detected by the acoustic network analysis. The adsorption of all proteins to the three surfaces was found to be irreversible [218].

Siloxane–polyhedral oligomeric silsesquioxane (POSS) copolymers were prepared by the dehydrocondensation of octahydrooctasilsesquioxane $(HSiO_{3/2})_8$ ($T^H{}_8$) with diphenylsilanediol, tetraphenyldisiloxane-1,3-diol, or oligodimethylsiloxane-α, ω-diols ($HO(Me_2SiO)_nH$, $n = 6, 13, 55$) in a molar ratio of 1:2 or 1:4 in the presence of diethylhydroxylamine, followed by trimethylsilylation with Me_3SiCl and Et_3N (Scheme 7.13) [219].

Weight average MWs of these siloxane copolymers ranged from 9,000 to 45,000 g/mol. By spin-coating of the solutions of condensation products, coating films were prepared. Their hardness was found to increase up to 6H with increasing curing temperature. The subsequent pyrolysis of the concentrated coating solutions at 650 °C gave silica gels with a specific surface area of 449 m^2/g [219].

Copolycondensation of diphenylsilanediol and 3-methacryloxypropyl(trimethoxy) silane (MATS) with different amounts of trisilanolphenyl-POSS, carried out at 85–100 °C, was catalyzed by $Ba(OH)_2$ and gave a series of thermally stable silicone resins (PSSQ, Scheme 7.14), which were further cured with phenyltris(hydrodimethylsiloxy)silane ($PhSi(OSiMe_2H)_3$). These resins showed enhanced T_g and lower coefficient of thermal expansion with the increasing POSS content. Incorporation of POSS unit into methacrylate (MA)-modified polysiloxane led to excellent transparency and improved thermal discoloration resistance. Thus, they may find practical applications, for example, in optoelectronics [220].

PMHS grafted with pendant alkyl groups $-(CH_2)_nCH_3$ of different chain lengths ($n = 8, 12, 16$) were used as membranes with the improved pervaporation performance in separation of ethyl acetate/water mixture. The separation factor of silicone membrane was dependent on alkyl grafting ratio and alkyl chain length because it caused the enhanced preferential sorption of ethyl acetate. The membrane grafted with shorter alkyl groups was preferred for pervaporation when grafting ratio was above 6.9%, while the best pervaporation performance in separation of 1% ethyl acetate/water mixture at 40 °C was achieved by 9% octyl-grafted silicone membranes.

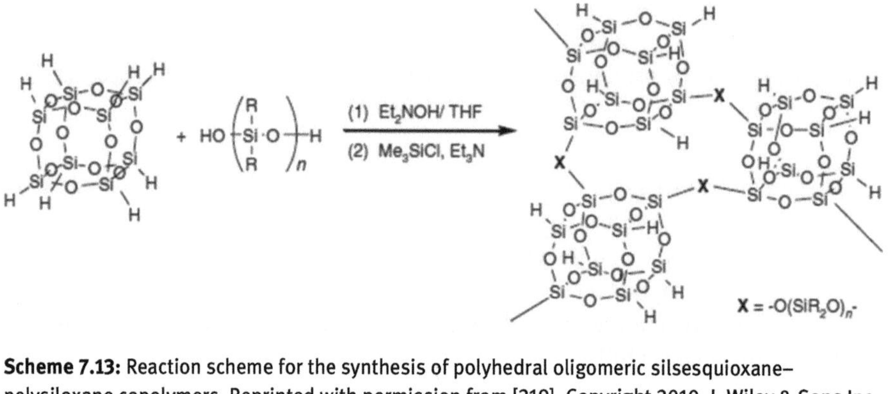

Scheme 7.13: Reaction scheme for the synthesis of polyhedral oligomeric silsesquioxane–polysiloxane copolymers. Reprinted with permission from [219]. Copyright 2010, J. Wiley & Sons Inc.

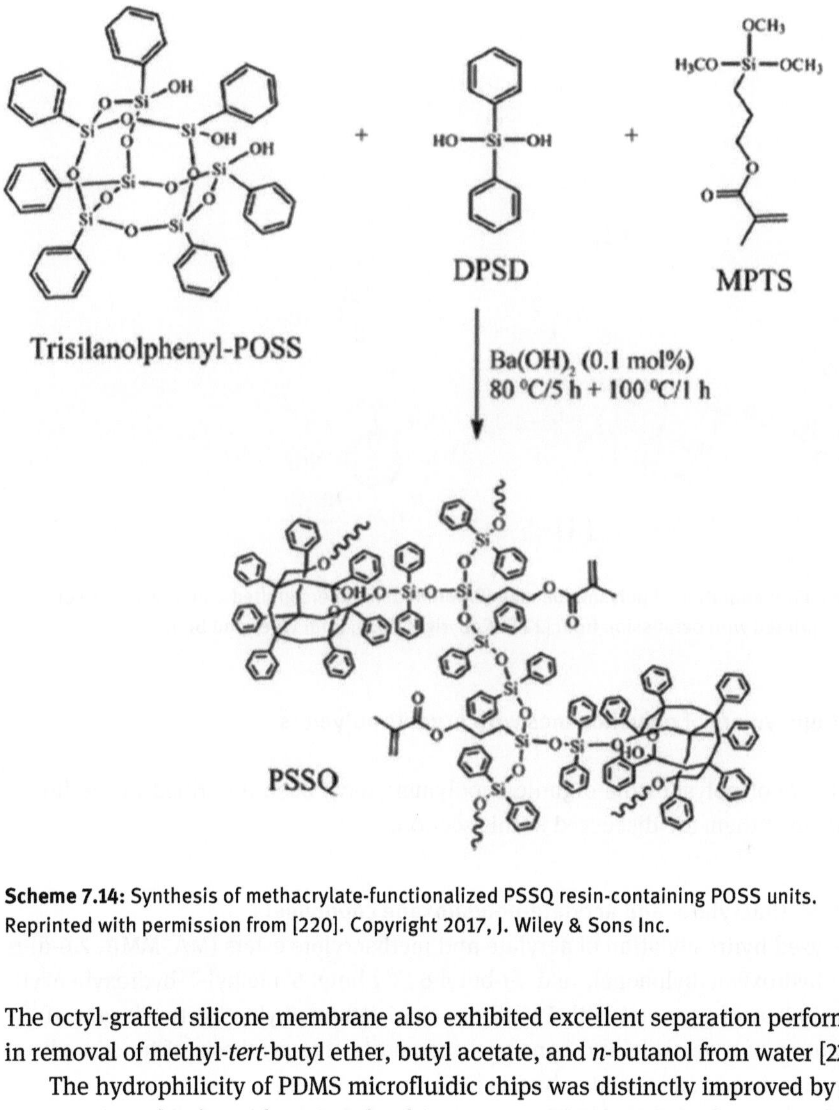

Scheme 7.14: Synthesis of methacrylate-functionalized PSSQ resin-containing POSS units. Reprinted with permission from [220]. Copyright 2017, J. Wiley & Sons Inc.

The octyl-grafted silicone membrane also exhibited excellent separation performance in removal of methyl-*tert*-butyl ether, butyl acetate, and *n*-butanol from water [221].

The hydrophilicity of PDMS microfluidic chips was distinctly improved by grafting a series of polyamidoamine dendrimers onto PDMS activated on a surface by oxygen plasma and modified with γ-glycidoxypropyl(trimethoxy)silane (γ-GPS) (Scheme 7.15). A dense and uniform coating was generated, which showed the contact angle (CA) of 31.9° in comparison with 108.2° for the unmodified PDMS. The modified PDMS microfluidic chips were used for the separation of arginine and phenylalanine and showed good stability and long service life. Their reproducibility was also extraordinary [222].

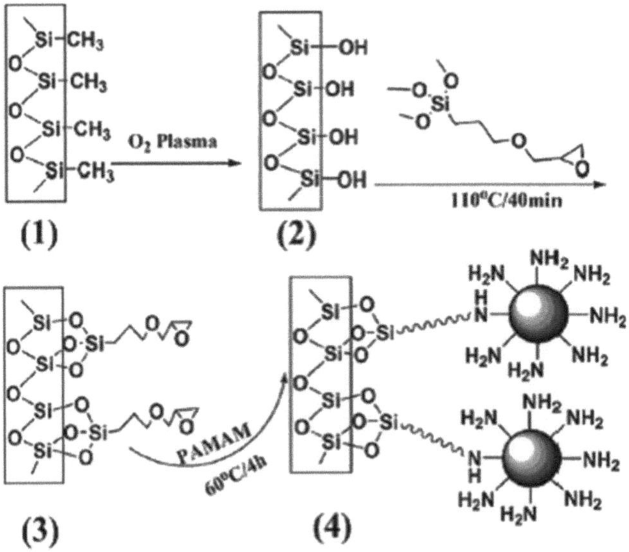

Scheme 7.15: Preparation of polyamidoamine (PAMAM) dendrimers grafted onto the surface of PDMS. Reprinted with permission from [222]. Copyright 2016, John Wiley and Sons.

7.2.2 Copolymers of polysiloxanes with organic polymers

Many kinds of polysiloxane-organic copolymers have been described in the literature. Some of them are discussed in this section.

7.2.2.1 Methacrylate- and acrylate-polysiloxane copolymers

Pt-catalyzed hydrosilylation of acrylate and methacrylate esters (MA, MMA, 2,6-di-*t*-butyl-4-hydroxymethylphenol, and 2-*t*-butyl-6-(3′-*t*-butyl-5′-methyl-2′-hydroxybenzyl)-4-methyl-phenoyl acrylate) with PMHS or 1,3,5,7-tetramethylcyclotetrasiloxane (D^H_4) has features of a versatile and beneficial method for the synthesis of functionalized siloxane macromonomers and polymers [223].

The full and graft IPNs based on PDMS and PMMA (poly(methyl methacrylate)) was obtained by an in situ sequential method. Most samples were opaque; however, some of them were transparent. Their T_g, stress–strain behavior, hardness, surface properties, and the permeability to oxygen and nitrogen were studied [224]. A PMMA-*g*-PDMS graft copolymer was prepared via a miniemulsion copolymerization of highly hydrophobic methacryl-functionalized PDMS macromonomer with methyl methacrylate (MMA), using AIBN as an initiator, SDS as a surfactant, and a PMMA-*b*-PEO block copolymer as a cosurfactant. Addition of methanol was used in order to reduce the interfacial tension. Latex particles were obtained in a high yield, and the PDMS macromonomer was completely incorporated into copolymer. The resulting latex was characterized by ^{1}H-NMR and by quasielastic light scattering [225].

Random and block copolymers of MMA and 3-(heptamethyl cyclotetrasiloxanyl) propyl methacrylate (HCPM) were prepared by radical copolymerization. Two distinct T_gs, which were determined from DSC studies, indicated the presence of heterogeneous phases in HCPM-MMA block copolymers, due to PMMA block and poly-HCPM block, while single T_gs in the homogeneous phases in HCPM-MMA random copolymers decreased with increasing HCPM content. Coefficients of the oxygen and nitrogen gas permeability of HCPM-MMA random copolymer films measured at 23 °C steeply increased with increasing HCPM contents, although those of HCPM-MMA block copolymers only slightly increased (<50%). The gas pearmeability of both kinds of these copolymers was compared with tris(trimethylsiloxy)silylmethacrylate (MTTS) copolymers. The gas permeability coefficients of MTTS-MMA random copolymers prevailed over those of HCPM-MMA random copolymers despite the same number of four Si atoms, probably because of its free volume effect. The diffusion and solubility coefficients were dependent on the HCPM content, and the effect of cross-linking on their gas permeability was observed [226].

Amphiphilic comb-like diblock copolymers of PDMS and oligo(ethylene glycol) methyl ether methacrylate were prepared by the ATRP method. Suitable macroinitiators for ATRP of oligo(ethylene glycol) methyl ether methacrylate were obtained by the anionic ring-opening polymerization (ROP) of hexamethylcyclotrisiloxane, followed by reaction with 3-(chlorodimethylsilyl)propyl 2-bromo-2-H-methyl-propanoate. The copolymers with moderate polydispersity of MWs ($M_w/M_n > 1.3$) were synthesized at room temperature using *N,N,N′,N″,N‴*-pentamethyldiethylenetriamine complex with Cu(I)Br in *n*-propanol with Cu(II)Br. The optimum reaction conditions in *n*-propanol/water mixture or in toluene at 90 °C gave block copolymers of the desired MWs and narrow polydispersity of MWs ($M_w/M_n < 1.1$). These block copolymers were characterized by ^{1}H- and ^{13}C-NMR spectroscopy and SEC [227].

Branched PMMA/polysiloxane copolymers were used for preparation of composites and their coatings. A copolymer such as PMMA-VTES (vinyl(triethoxy)silane) containing MMA and VTES was prepared by free radical polymerization and then condensed with oligomers obtained by a hydrolytic condensation of methyl(triethoxy)silane to form PMMA/polysiloxane composites. They were cured in an oven (at 75 °C) giving appropriate coatings. The polymers were characterized by gel permeation chromatography and FT-IR spectroscopy. The surface property, water contact angle (CA), hardness, thermostability, and optical property of the coatings were studied by scanning electron microscopy (SEM), pencil hardness testing, TGA, and UV–visible spectroscopy. The thermal stability, defined as decomposition temperatures corresponding to a 5% weight loss (T_5), decreased from 365 to 314 °C, and the pencil hardness of the coatings was reduced from 6H to 2H, as a result of increased content of MMA units. Moreover, it caused an increase in the flexibility, the water CA (from 94.8° to 102.2°), and an improvement of adhesion properties. The transparent

PMMA/polysiloxane coatings showed a smooth surface, excellent scratch resistance, a strong adhesion property, and high thermal stability [228].

Well-defined triblock copolymers of poly(methyl methacrylate-*b*-dimethylsiloxane-*b*-methyl methacrylate) (MMA-*b*-DMS-*b*-MMA) were prepared via activators generated by electron transfer ATRP, using the bromo-terminated macroinitiator, which was prepared by direct addition reaction of difunctional PDMS having MMA end groups with hydrobromic acid in acetic acid. The water CAs decreased gradually with the increasing PMMA block content. The self-assembly behaviors of the triblock polymer were studied by transmission electron microscopy (TEM), SEM, and dynamic light scattering measurement. The copolymers self-assembled into various complex morphologies in different solvents and their morphologies were dependent on the properties of solvents [229].

Diblock and triblock copolymers composed of MMA-*co*-glycidyl methacrylate block and triblock MA copolymers and pendant POSS cubics were prepared by ATRP. Monofunctional and bifunctional P(MMA-*co*-GMA) copolymers, synthesized by ATRP, served as macroinitiators for ATRP of POSS-containing MA group. The obtained block copolymers were characterized by ^{1}H-NMR spectroscopy, SEC, and determination of their T_gs by DSC [230].

Silicone acrylates were prepared by the platinum-catalyzed hydrosilation of polyacryloxy functional monomers with PDMS-*co*-PMHS. The hydrosilylation of either 1,6-hexanediol diacrylate or trimethylpropane triacrylate (TMPTA) with PDMS-*co*-PMHS occured with one of two or three olefinic groups, respectively, as long as excess monomer was present. A release coating showing excellent cohesive and adhesive strength on paper was obtained by the electron beam (EB) curing of the product of hydrosilylation of TMPTA with the PDMS-*co*-PMHS containing 4 mol% methylhydrosiloxane units at a dose to cure of 2 Mrad, but the aged release force was excessively high. The poor aged release performance was probably due to dilution of the Me_2SiO unit content of the coating with excess monomer required to stabilize the silicone acrylate. These polysiloxane-acrylate copolymers can be used as possible EB-curable release coatings for paper [231].

Acrylate monomers were grafted onto commercial silicones by a two-step emulsion polymerization method. A high grafting ratio was achieved when the silicone content was more than 20% – it was confirmed in FT-IR spectra. SEM and energy-dispersive X-ray spectroscopic results showed the "sea–island" structure on the surface of release films that were formed from acrylate-modified silicone latexes. The silicones formed the continuous "sea" structure and the grafted acrylate polymers formed the segregated "island" structure. Improved properties of the release films were observed. The release force increased from 8.1 to 47 g/in, and surface tension of silicone release films increased from 19–21 mN/m (before acrylate modification) to 30 mN/m [232].

Core–shell silicone acrylic emulsions with 3-methacryloxypropyl(trimethoxy)silane (MATS) in the shell were prepared by radical polymerization in the presence of polymerizable maleate surfactant (MT). The incorporation of the surfactant into copolymer structure was confirmed by FT-IR spectroscopy. The small amount of octadecyl acrylate affected emulsion stability and caused a decrease in the particle size from 194.6 to 165.7 nm, while the particle size increased from 165.7 to 242.9 nm with the increase of MPTS concentration. Better water resistance was observed when MPTS concentration was higher than 2% and MT concentration was lower than 1.5%. The surface morphology of film depended on MATS concentration [233].

Three kinds of polysiloxane-functionalized acrylic latexes were prepared from ethyl acrylate and 2-ethylhexyl acrylate by emulsion polymerization. Molecular compositions of acrylic-siloxane copolymers were modified by using as comonomers MATS, octamethylcyclotetrasiloxane (D_4), and ethylene glycol dimethacrylate. After modification of acrylic latexes with polysiloxanes, microphase separation was observed by atomic force microscopy (AFM). Modification of polysiloxane structure resulted in decrease of the storage modulus and surface energy of latex films [234].

Aqueous dispersions of silicone resin nanoparticles were prepared through emulsion polymerization of silicone monomers in water, according to Weyenberg et al.'s procedure [235]. Octamethylcyclotetrasiloxane (D_4), methyltrimethoxysilane, and methacryloxy(trimethoxy)silane were used. After removal of water and drying, silicone-acrylic core–shell nanopowders (ca. 100 nm in size) were obtained. Next epoxy-polyester and polyester coatings were modified by addition of 3 wt% of these nanopowders. The modified coatings showed significantly higher CA values and significantly lower surface-free energy (SFE) values in comparison with unmodified coatings, presumably due to migration of silicone resin to their surface. The presence of silicone copolymer on the surface was confirmed by energy-dispersive spectroscopy (EDS), X-ray photoelectron spectroscopy (XPS), and AFM studies. Surface properties of the coatings were strongly dependent on the chemical structure of silicone resin [236].

Fluorine–silicone acrylic resins with different contents of 2,2,6,6-tetramethyl-4-piperidyl methacrylate (TMPM) having ability to form nitroxide and free-radical catching were prepared by solution copolymerization of different acrylate monomers, MMA, and vinyl(trimethoxy)silane. The obtained resins formed coatings with good thermal stability and hydrophobic properties (water CAs >100°). Their weather resistance showed a great improvement for the synergetic protection by bonded TMPM, organic fluorine, and silicone. The modified coatings saved excellent mechanical properties after aging tests, with an impact strength and flexibility of 400 Ncm and 2 mm, respectively [237].

Poly(vinylacetate-*b*-dimethylsiloxane-*b*-vinylcetate) triblock copolymer of well-defined microstructure was prepared by ATRP method from bis(bromoalkyl)-terminated PDMS macroinitiator (Br-PDMS-Br) and vinyl acetate telomers. A low T_g of PDMS segments in the microstructure of triblock copolymer was reponsible for its flexible properties, suggesting new potential applications [238].

7.2.2.2 Polystyrene–polysiloxane copolymers

Block copolymer poly(styrene-*b*-dimethylsiloxane), free of the parent homopolymers, was prepared by polymerization of hexamethylcyclotrisiloxane with "living" polystyrene (PS), which was obtained from styrene and alkyllithium. The AB block copolymers resembled surfactants due to the extreme differences in solubility between the two blocks. Films of the block copolymers when cast from solution showed a morphology determined by the nature of the solvent and the degree to which each block was solvated [239].

Low-polydispersity PDMS, end functionalized with hydroxyl-terminated PDMS prepared by a reversible addition fragmentation chain transfer (RAFT) method, using a carboxylic acid-functional RAFT agent, was further modified by the free radical polymerization with styrene and styrene derivatives such as monomers, giving block copolymers containing PDMS moieties. A thin film of these PDMS-*b*-PS copolymers showed a microphase separation [240]. Strong phase separation in these copolymers was evidenced by the high Flory–Huggins interaction parameter $\chi = 0.26$ [241]. PS–PDMS was used as a precursor for nanoporous materials [242], orientation-controlled self-assembled nanolithography [243], and nanoimprint lithography [244]. Owing to the excellent thermal and oxidative stability and good etch selectivity of PDMS this copolymer seems to be an attractive material for microelectronics [242, 243, 245–248].

Blends of poly(styrene-*block*-polydimethylsiloxane) (PS–PDMS, with MWs 31,000 and 15,000 g/mol, respectively, weight-average MW $M_w = 45.5000$ g/mol and polydispersity $M_w/M_n = 1.15$) with silicone MQ resin (which was obtained by trimethylsilylation of silicate nanoparticles containing $Me_3SiO_{1/2}$ (M) and $SiO_{4/2}$ (Q) units) showed similar properties to block copolymers with different PS/PDMS ratios. MQ particles that were localized in the PDMS phase caused an increase in the volume fraction in the block copolymer and controlled microdomain morphology. Thus, MQ resin played a role of a robust morphology modifier [249].

A series of polystyrene-*b*-poly(dimethylsiloxane)-*b*-polystyrene (PS-*b*-PDMS-*b*-PS) triblock copolymers were synthesized by ATRP. The products were characterized by FT-IR spectroscopy, GPC, TGA, CA, and SEM. The PS-*b*-PDMS-*b*-PS triblock copolymers showed good thermal stability, low surface tension, and microphase separation [250]. Hybrid nanocomposites (NCs) of PS and methacryl phenyl POSS were prepared by reactive melt blending with dicumyl peroxide (DCP) as a free radical initiator and styrene monomer as a chain transfer agent. The degree of POSS hybridization (α-POSS) increased with the POSS content, DCP/POSS ratio, and rotor speed. In the absence of styrene during processing of PS-POSS materials, an increase in the α-POSS content led to a reduction in the MW by PS chain scission, due to the free radical initiation. The addition of styrene reduced the steric hindrance in the hybridization reaction between POSS and PS, enhanced the degree of POSS hybridization, and avoided degradation of PS chains. The PS-POSS nanoclusters (NCs) contained POSS NCs and crystalline microparticle POSS aggregates. An improved

POSS dispersion and lower amounts of nonbound POSS were observed for PS-POSS with higher α-POSS values [251].

Core–shell PDMS/PS and PS composites in the form of microspheres were prepared by combined emulsion and suspension in situ polymerization. PDMS was grafted with MATS. The core–shell emulsion was broken by addition of calcium chloride electrolyte. PDMS soft particles were dispersed uniformly in the composites, which was proved by TEM. Diameters of the composite beads ranged from 0.5 to 4.0 mm, which were controlled by different hydroxyapatite loading. The properties of the latex film including water absorption ratio, CA, pendulum hardness, and transparency were studied. The properties of emulsion were dependent on the content of solids. The CA of the latex films reached 106.07° with the increasing of core–shell emulsion, while the absorption and pendulum hardness decreased to 0.37% and 322 S, respectively [252].

The ABC-type triblock copolymers: poly(styrene-*block*-acrylic acid) (PS-*b*-PAA) and poly-(styrene-*block*-3-trimethoxysilylpropyl methacrylate) (PS-*b*-PMSMA) were obtained by living polymerization. They formed porous films by blending with poly (methyl silsesquioxane) (PMSSQ), followed by spin coating and multistep baking. The chemical bonding between the above ABC copolymers and PMSSQ resin caused significant differences in the morphologies and affected properties of the hybrids and their porous derivatives. Macrophase separation was observed in the PMSSQ/PS-*b*-PAA hybrid due to intramolecular and intermolecular hydrogen bonding. The blend of PMSMA with PMSSQ did not show macrophase separation. Results of modulated DSC also suggested a significant difference in the miscibility of the two hybrid systems. The pore size of the nanoporous thin film prepared from the PMSSQ/PS-*b*-PMSMA hybrid, estimated by TEM, was less than 15 nm. The dielectric constant and refractive index of the prepared porous films decreased from 2.60 to 1.84 and from 1.35 to 1.23, respectively, when the PS-*b*-PMSMA content increased from 0 to 50 wt% [253].

Amphiphilic diblock copolymers of PEG and poly(pentafluorostyrene) (PPFS) were prepared by ATRP method, using modified PEG as a macroinitiator. Different amounts of aminopropylisobutyl POSS were grafted onto the PPFS block of the PEG-*b*-PPFS via aromatic nucleophilic substitution reactions of the fluorine atoms in *para* position of the aromatic rings. The self-assembly of the resultant amphiphilic PEG-*b*-PPFS and PEG-*b*-PPFS-*g*-POSS in aqueous media was studied using laser light scattering, fluorescence, and TEM techniques. The grafted POSS played a role of a plasticizer: the T_g of the PPFS block in the copolymer substantially decreased with increased POSS content [254].

Diblock copolymers, polydimethylsiloxane-*b*-poly-2,5-bis(4-methoxyphenyl)oxycarbonyl styrene (PDMS-*b*-PMPCS), prepared by controlling PMPCS length and thermal annealing conditions, formed lamellae, hexagonally packed cylinders, and double gyroid phases. Their T_g and T_g breadth of PDMS were dependent on the detailed confining geometry – the nanostructure and confining size [255].

7.2.2.3 Copolymers of polysiloxanes with polyurethanes

Segmented thermoplastic polyurethane (TPU)–siloxane copolymers (PU-S) were prepared from 4,4′-methylenediphenyl diisocyanate (MDI) and 1,4-butanediol as a hard segment and 40–90 wt% of α,ω-dihydroxy-terminated poly(propylene oxide)-*b*-poly(dimethylsiloxane)-*b*-poly(propylene oxide) (PPO-PDMS) as the soft segment. The structure, MWs, and crystallinity of these copolymers were analyzed by FT-IR, ^{1}H- and ^{2}D-NMR spectroscopy, and GPC and DSC methods. Surface free energy analysis revealed highly hydrophobic nature to the obtained polyurethane (PU) films, due to the presence of hydrophobic (siloxane) groups on the surface. The increase in the hydrophobic PPO-PDMS segment content led to the decrease in the percentage of absorbed water in copolymers. SEM and AFM analyses showed that copolymers with lower content of PPO-PDMS segments have higher microphase separation between segments. The obtained PU-S based on PPO-PDMS exhibited good surface and morphological properties and should find a variety of applications as hydrophobic coatings in biomedicine [256].

Waterborne epoxy-resin-based polyol dispersions modified with PDMS units were prepared by the reaction of 2,4-toluene diisocyanate with 2,2′-bis(hydroxymethyl) propionic acid, bisphenol A epoxy-resin-based polyol, and carbofunctional hydroxypropyl-terminated PDMS (HOPr-PDMS). The HOPr-PDMS-modified polyol dispersions showed a small particle size and an excellent dispersion stability. From the HOPr-PDMS-modified polyol dispersion and a hydrophilic-modified polyisocyanate, a two-component waterborne PU was prepared. The tensile strength, modulus, and pencil hardness values of the films decreased with increasing HOPr-PDMS content, while the thermal stability of the cross-linked waterborne PU films increased with increasing HOPr-PDMS loading. PDMS segments migrated to the surface of the films. The water absorption ratio of the films decreased from 6.6% to 5.0% and the CA of the films increased from 71° to 96°, when the HOPr-PDMS content in the films increased from 0% to 10%. Thus, the water resistance of the films was enhanced by the incorporation of HOPrT-PDMS into the PU networks [257].

PUs and PU-S copolymers were often used for biomedical applications in the past few decades, for example, in drug delivery systems, hard and soft tissue regeneration, and scaffolding [258].

Waterborne PU-containing siloxane segments (WS-PU) were fabricated by polyaddition reaction of poly(tetramethylene oxide)glycol, PEG, and α,ω-aminopropyl PDMS (APDMS) (both as soft segments), 2,2-di(hydroxylmethyl)propionic acid (as a hydrophilic chain extender), 1,4-butanediol to isophorone diisocyanate (as the hard segment). The thermal properties of WS-PU films were analyzed by DSC, TGA, and DMA. The mechanical properties of WS-PU film were also studied. The DSC and DMA results showed that there was a microphase separation in the WS-PU film. As a result of the incorporation of APDMS into PU made the thermal stability of hard segment decreased while that of soft segment increased. The elasticity of WS-PU was improved when the APDMS content not exceeded 10%; first, the vapor permeability of coatings increased and then decreased with increasing content of APDMS, which resulted from the hydrophilicity change and microstructure change of membrane. With 10% APDMS content in the WS-PU, the water vapor permeability of coated fabric was 2,130.15 g/(m^2 · 24 h), and the water resistance reached 30.0 kPa [259].

Novel PU copolymers were prepared by direct transurethanetion reaction of a commercial PU with carbofunctional PDMS containing monohydroxyethyl-oxypropyl end-groups with different MWs (PDMS, MWs 1,000, 5,000, and 10,000 g/mol). Transurethanetions with different mass ratios of PDMS to PU chains were carried out in solution at 65 and 100 °C. Dibutyltin dilaurate catalyzed bond cleavage in the PU chain, accelerated the reaction between hydroxyl end groups of PDMS, and led to recovery of NCO groups of PU. The chemical structures of the prepared copolymers were characterized by ^{1}H, ^{13}C, and ^{29}Si NMR. The content of PDMS ranged between 3 and 16 wt%. The GPC results showed that MWs of the PU-PDMS copolymers were lower than those of the starting PU. These copolymers showed lower values of T_g when compared with T_g of the starting PU. According to ATR (Attenuated Total Reflectance) FT-IR spectroscopy, the surface of the copolymer films was enriched with siloxane groups and the electron microscopy showed that PU-PDMS copolymers were textured with microspheres. The static CAs for copolymer films measured with water ranged from 94° to 117° [260].

A series of dihydroxy-terminated polyether-polydimethylsiloxane-polyether [α,ω-dihydroxy-(PE-PDMS-PE)] triblock copolymers were prepared from Si–H-terminated oligodimethylsiloxanes ($HMe_2SiO(Me_2SiO)_xSiMe_2H$, $M^HD_xM^H$, $x = 8$, 15, 20, 30, 40, and 50) and methacryloxy-functionalized PPO. The one-step hydrosilylation reaction was carried out with chloroplatinic acid at 80 °C. These ABA oligomers, terminated by hydroxy groups, were characterized via ^{1}H-NMR, ^{13}C-NMR, ^{29}Si-NMR, FT-IR, and GPC. The ABA triblock oligomers were used for two-step bulk synthesis of TPUs without any solvents. The effect of triblock oligomers impact on TPUs' mechanical properties, thermal performance, surface water repellency, and morphology performance was studied by Instron material tester, DSC, TGA, water CAs, SEM, and TEM. DSC and TGA showed that PE-PDMS-PE–modified TPUs had T_g under −120 °C and the temperature of 50% weight loss was improved from 280 to 340 °C. PE-PDMS-PE–modified TPU did not show the reduction on mechanical properties than TPU without PDMS. Tensile strength was kept at 13 MPa and elongation at break at 300%. According to SEM and TEM results, these copolymers showed a pseudo-three-phase separation. The PE-PDMS-PE–modified TPU had significantly improved hydrophobic properties [261].

Novel hydroxyalkyl carbamate and dihydroxyalkyl carbamate-terminated PDMS oligomers and their carbamate-linked block copolymers with poly(ε-caprolactone) (PCL) were used for preparation of many kinds of PU-S coatings (Schemes 7.16 and 7.17).

They were cross-linked with triisocyanurate resin of isophorone diisocyanate and branched commercial polyol. Eight coatings were further characterized, using DSC, DMA, XPS, and surface energy analysis. Two coatings showed good performance in algal (*Ulva*), bacterial (*Cytophaga lytica*, *Halomonas pacifica*), and barnacle (*Balanus*

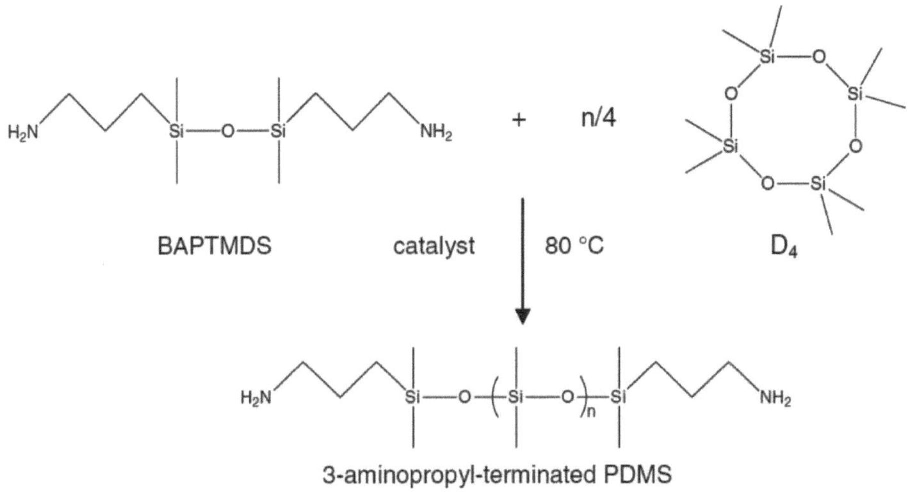

Scheme 7.16: Reaction scheme for the synthesis of 3-aminopropyl-terminated PDMS oligomers. Reprinted with permission from [262]. Copyright 2007, Springer Nature.

amphitrite) laboratory screening assays and were potential candidates for underwater marine applications [262].

Segmented poly(urethane-urea-siloxane)s (Scheme 7.18) were prepared using MDI and ethylene diamine as the hard segment components (38–65 wt%) and HOPr-PDMS as the soft segment component – by a two-step polymerization in a mixture of polar solvents: tetrahydrofuran and *N*-methylpyrrolidone. The microphase-separated copolymers with high tensile strength were obtained [263].

These copolymers showed high water resistance and became more hydrophobic with increasing weight fraction of PDMS. Formation of globular superstructures in the copolymer films was observed by SEM and AFM [263].

Properties of waterborne PU cationomers, prepared by the reaction of 4,4′-methylene-bis(phenyl isocyanate) (MDI), PEG 2000, *N*-methyldiethanolamine, and HCOOH, were modified with different amounts of α,ω-di(hydroxylethylenoxypropyl)-PDMS (3.0–7.3 wt%). FT-IR spectra confirmed the structure of obtained PU-S copolymers. CA measurements increased that with increasing content of the polysiloxane segments in the hybrid films [264].

7.2.2.4 Copolymers of polysiloxanes with polyimides

An aromatic PI containing Si atoms and pendant aryl rings was prepared by polycondensation of a silicon-containing aromatic diamine with 4,4′-(hexafluoroisopropylidene)diphthalic anhydride. It was copolymerized in situ with different loading of colloidal SiO_2 in the presence of 3-aminopropyl(triethoxy)silane (APTES) as a coupling agent, giving PI-silica nanohybrids. The APTES was partially hydrolyzed and

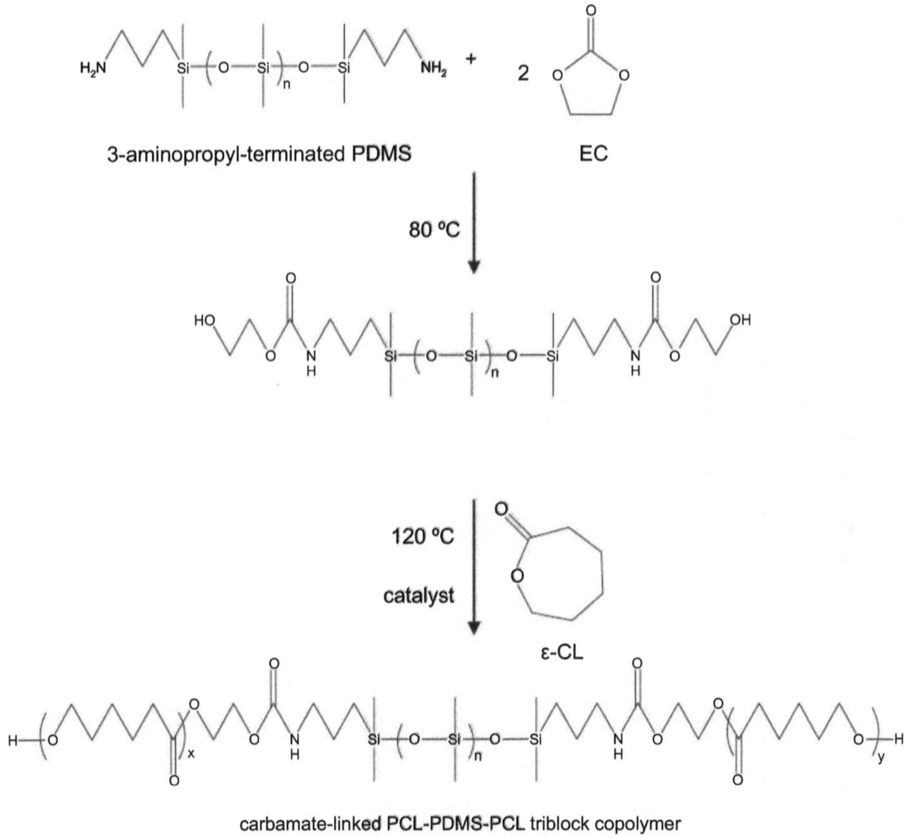

Scheme 7.17: The synthesis of hydroxyalkyl carbamate-terminated PDMS oligomers and carbamate-linked PCL-PDMS-PCL triblock copolymers. Reprinted with permission from [262]. Copyright 2002, Springer Nature.

condensed in situ, followed by grafting onto silica surface. A modified silica network was chemically bound with the PI matrix and homogeneously dispersed throughout the PI matrix with 50–70 nm size range. The best miscibility of polymer and silica phases in the nanohybrids was obtained when up to 40 wt% nanoparticles were incorporated into the backbone of PI matrix [265].

A novel type of high-performance cyanate ester (CE) resins was obtained by copolymerization of 2,2′-bis(4-cyanatophenyl) *iso*-propylidene with hyperbranched polysiloxane (HbrPSi). The reactive HbrPSi was prepared via a hydrolytic condensation of 3-(trimethoxysilyl)propyl methacrylate (Scheme 7.19). The incorporation of HbrPSi effectively promoted the curing reaction of CE, increased the apparent free volume fraction of the cured network, and also significantly affected the properties of the final resins. The impact strength of modified CE resin with 15 wt% of HbrPSi was 19.6 kJ/m^2, which was more than 2 times higher than for pure CE resin. The CE

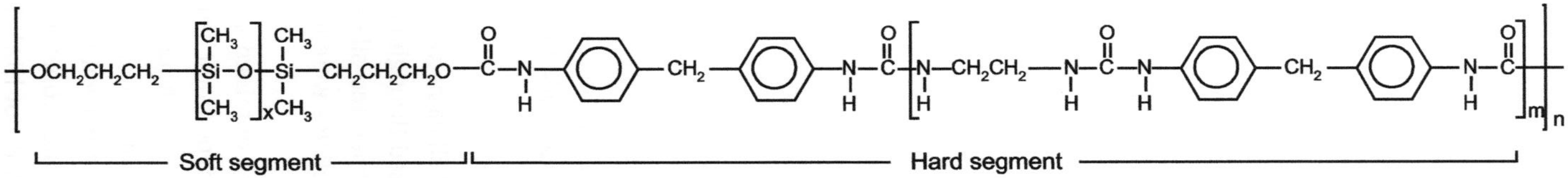

Scheme 7.18: Chemical structure of the poly(urethane-urea-siloxane) (PUUS) copolymers. Reprinted with permission from [263]. Copyright 2013, Springer Nature.

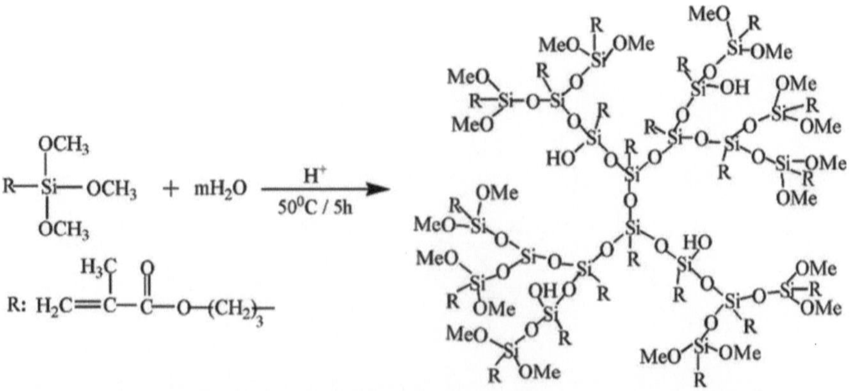

Scheme 7.19: Scheme of a partial hydrolytic condensation of 3-(trimethoxysilyl)propyl methacrylate. Reprinted with permission from [266]. Copyright 2011, Springer Nature.

resin modified with silicone oligomers also showed better thermal stability, moisture resistance and dielectric properties than starting CE resin [266].

HbrPSi containing amino groups and silanol functional groups was synthesized by the hydrolytic polycondensation of (γ-aminopropyl)methyl(diethoxy)silane with phenyl(trimethoxy)silane (PTMS) (Scheme 7.20) in *N,N′*-dimethylacetamide (DMAc) solution. Its hyperbranched topology of HbrPSi was confirmed by ^{29}Si-NMR spectroscopy. The HbrPSi was incorporated into PI chains through copolycondensation reactions. This new material showed good mechanical strength, satisfactory break elongation, excellent thermal stability, high optical transparency, and outstanding atomic oxygen (AO) erosion resistance. Addition of 29.7 wt% of HbrPSi significantly increased resistance of PI copolymer to AO, which showed only 7.7% weight loss to that of pristine PI after 22 h AO exposure. A formation of an ablation silica protective layer was responsible for the "self-healing" surface properties of this new PI. The passivation layer prevented the underlying polymer from additional erosion [267].

PI/silica (PI/SiO_2) hybrid films containing imidazole moieties were prepared from 3,3′,4,4′-biphenyltetracarboxylic dianhydride and 2,2′-di(*p*-aminophenyl)-5,5′-dibenzimidazole, in DMAc solution, by sol–gel method and thermal imidization. A homogeneous dispersion of silica nanoparticles in the PI matrix was stated by SEM and TEM techniques. The hybrid film containing up to 40% of silica was transparent. The hybrid film PI/SiO_2 with 15% SiO_2 showed good mechanical properties: a tensile modulus of 5.66 GPa, a tensile strength of 222 MPa, and an elongation at break of 12% [268].

A series of novel superhydrophobic-segmented PI-siloxane electrospun fibers were prepared from 4,4′-oxydianiline and 3,3′,4,4′-benzophenone tetracarboxylic dianhydride as hard segments and changing contents of aminopropyl-terminated PDMS (APPS) (10 or 20 mol%) as a soft segment, as well as the MW of APPS (1,000 or 2,500 g/mol, respectively). Electrospinning of the polyamic acids was followed by thermal imidization which gave the electrospun PI-siloxane materials. They were

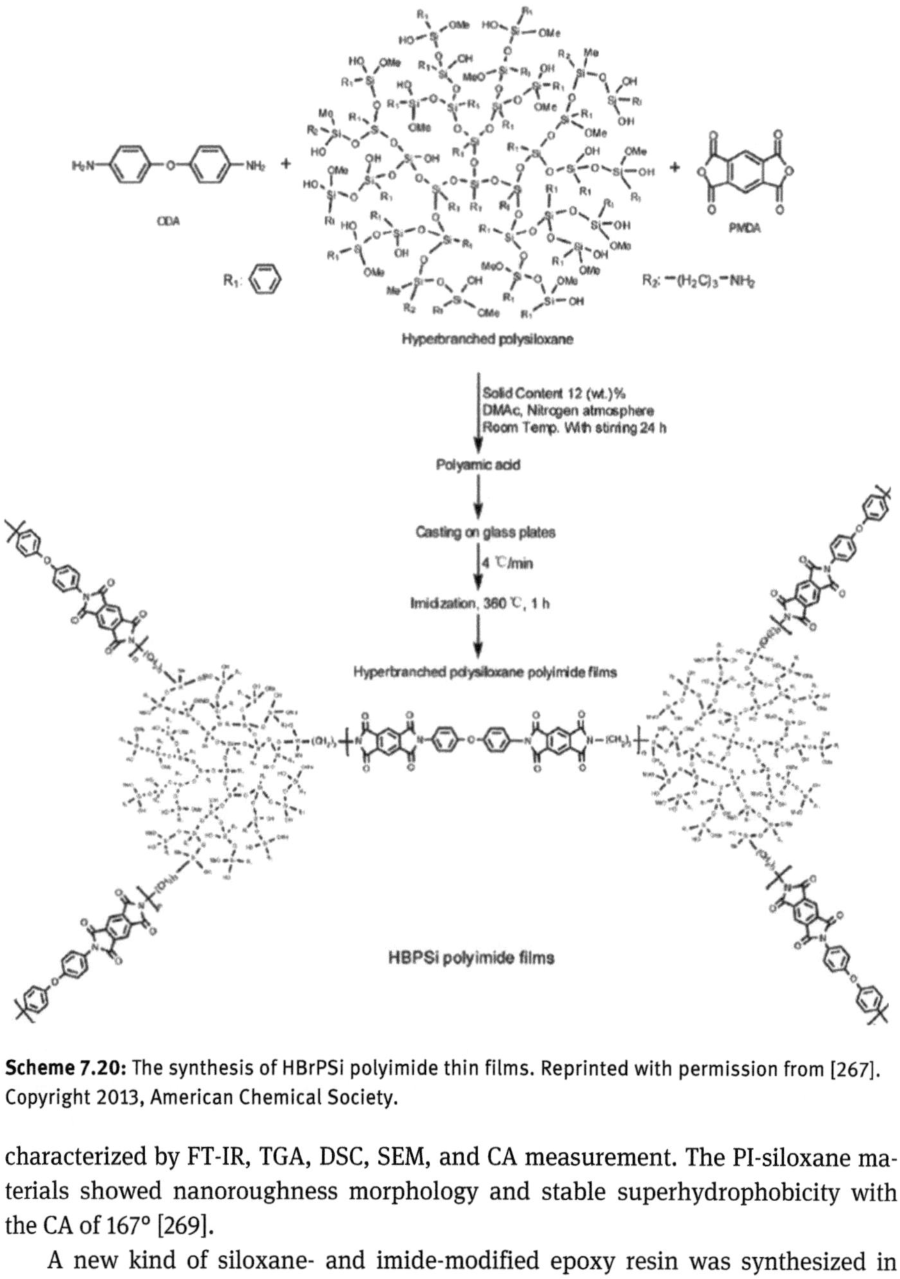

Scheme 7.20: The synthesis of HBrPSi polyimide thin films. Reprinted with permission from [267]. Copyright 2013, American Chemical Society.

characterized by FT-IR, TGA, DSC, SEM, and CA measurement. The PI-siloxane materials showed nanoroughness morphology and stable superhydrophobicity with the CA of 167° [269].

A new kind of siloxane- and imide-modified epoxy resin was synthesized in three steps. By hydrosilylation of nadic anhydride with 1,1,3,3-tetramethyldisiloxane, 5,5′-(1,1,3,3-tetramethyl-1,1,3,3-disiloxanediayl)-bisnorbornane-2,3-dicarboxylic anhydride (I) was obtained, which reacted with 4-aminophenol to give *N,N′*-bis(4-hydroxyphenyl)-5,5′-(1,1,3,3-tetramethyl-1,1,3,3-disiloxanediallyl)-bisnorbornane-2,3-dicarboximide (II). Then this intermediate product reacted with epichlorohydrin to give siloxane- and imide-modified epoxy (i.e., *N,N′*-diglycidylether-bis-(4-phenyl)-5,5′-(1,1,3,3-tetramethyl-1,1,3,3-disiloxanediallyl)-bis-norbornane-2,3-dicarboximide (III) (Scheme 7.21). This new resin was blended with commercial epoxy resin based on bisphenol F (1:1 and 1:0.8, wt/wt) and cured with various equivalent ratios (1:1 and 1:0.8) with siloxane-modified dianhydride (I). TGA thermograms revealed that in air atmosphere the cured samples were stable up to approximately 300 °C, while in nitrogen atmosphere their thermal stability was only slightly better. Thermal and dynamic mechanical properties were measured with thermal mechanical analysis and DMA. All new materials were cured nearly completely and showed good water-repellency and excellent chemical resistance to alkali and acid [270].

Two novel benzocyclobutene (BCB)-functionalized oligomers were prepared via imidization of siloxane containing dianhydride with 4-aminobenzocyclobutene, which was obtained by hydrosilylation of Si–H-terminated PDMS oligomers with 5-norbornene-2,3-dicarboxylic anhydride (nadic anhydride) (Scheme 7.22). The BCB oligomers were soluble in various organic solvents. They were cross-linked through ring opening and the following Diels–Alder reaction at elevated temperature. The BCB polymers showed low water absorption, excellent dielectric properties, good thermal stability, and good planarization. The mechanical and thermal properties of BCB resins were controlled by the length of the siloxane unit. The BCB resins with a shorter siloxane chain showed higher modulus, higher T_g, and lower coefficient of thermal expansion than BCB resins with longer chains [271].

Polysiloxane-modified PIs are important materials, which have outstanding mechanical properties and thermal stability. The synthesis, structure, and properties of polyimide-*b*-polysiloxane copolymers has been reviewed by L. Feng et al. The processability, toughness, and adhesion of PIs were substantialy improved by incorporation of polysiloxanes. Copolymerization of polyimide-*b*-linear polysiloxane copolymers with ladder-like polysiloxanes having excellent thermal stability and mechanical properties led to further improvement of their initial modulus and thermal stability [272].

Two series of borosiloxane-containing copolyimides were prepared by the reaction of 3,3′,4,4′-biphenyltetracarboxylic dianhydride and 2,3′,3,4′-biphenyltetracarboxylic dianhydride with *p*-phenylenediamine, 4,4′-oxydianiline, and different borosiloxane diamine monomers (BOSiA). The obtained copolyimides containing borosiloxane units in the main chain and in the side chain exhibited better solubility than borosiloxane-free copolyimides and showed quite large elongation at break (10–14%), high T_g (320–360 °C), excellent thermal stability (570–620 °C for T_{10}), and a low coefficient of thermal expansion (14–24 ppm/°C). The CAs of copolyimides increased to 96° for 5% content of borosiloxane in the main chain of the copolymer and up to 107° for 10% of BOSiA in the side chain of the copolymer as compared to 72° for neat copolyimide [273].

Modified PI aerogels were prepared by reacting the polyamic acid oligomer with different cross-linking agents: cage octa(aminophenyl) silsesquioxane, cyclic ladder-like poly(aminophenyl) silsesquioxanes and (1,3,5-triaminophenoxy)benzene (TAB).

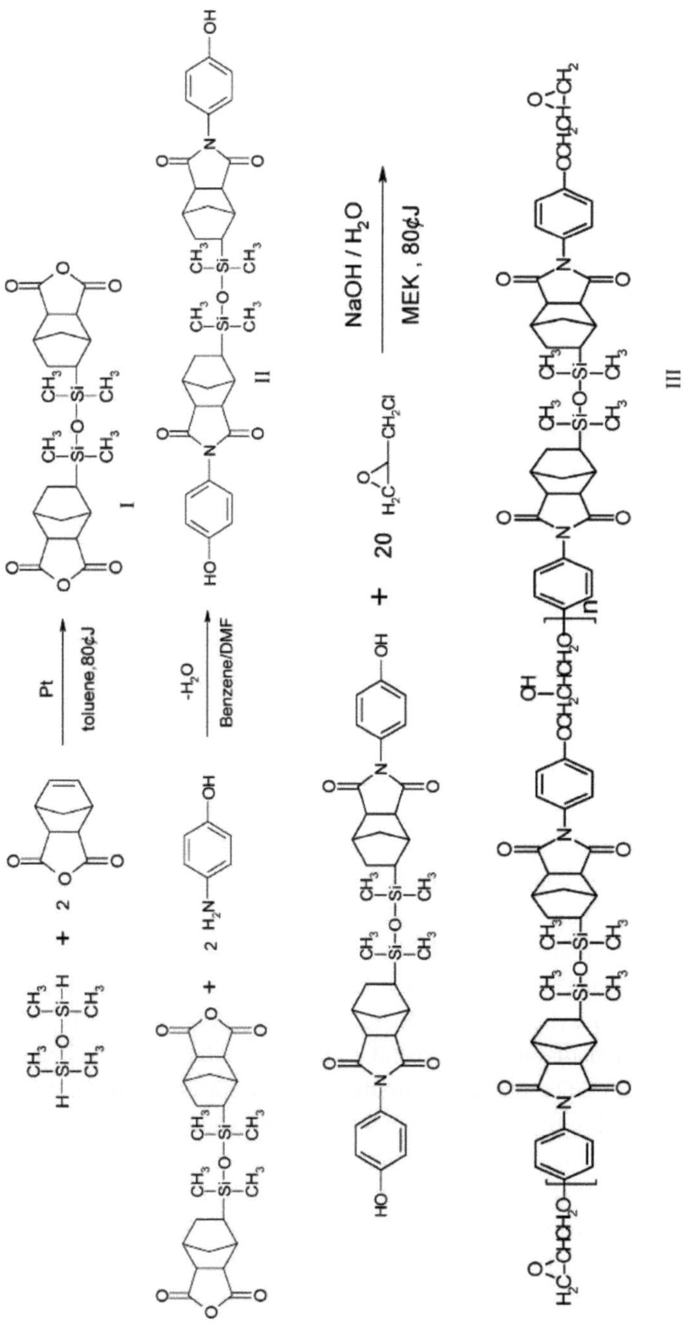

Scheme 7.21: Synthesis of siloxane containing dianhydride I, intermediate compound II, and final epoxy resin III. Reprinted with permission from [270]. Copyright 2005, Springer Nature.

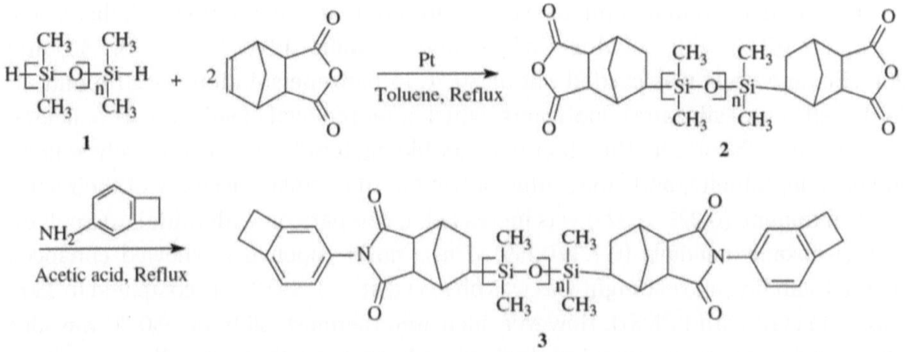

Scheme 7.22: The synthetic route of benzocyclobutene (BCB)-based siloxane-containing monomers. Reprinted with permission from [271]. Copyright 2016, John Wiley and Sons.

The PI–POSS aerogels formed more stable and stronger structures than with the TAB, based on the density, porosity, shrinkage, and compression measurements. The incorporation of POSS molecules gave stable cross-linking network structures with improved physical properties. The thermal properties of the PI aerogels were tested by DMA, TGA, and transient hot wire method. The PI–POSS aerogels showed strong mechanical properties, much better thermostability, lower thermal conductivities, and better thermal insulation. These aerogels may find important application as thermally resistant materials [274].

Novel 5,11,14,17-tetraanilinooctaphenyl double-decker silsesquioxane (DDSQ), tetraamino-substituted POSS, was used for the preparation of PI NCs. The hybrid PI NCs formed nanostructures in which the POSS component was aggregated into spherical microdomains with a diameter of 40–80 nm. In comparison to unmodified PI, the PI NCs containing POSS units exhibited improved thermal stability and surface hydrophobicity. Due to the incorporation of POSS microdomains, the dielectric constants of the PI NCs were significantly decreased as compared to plain PI [275].

7.2.2.5 Polysiloxane–polysilazane and polysiloxazane copolymers

An important group of silicon-containing copolymers are polysiloxazanes, which have Si–O and Si–N bonds in macromolecules. They exhibit intermediate properties and are very promising for potential applications [276–278], for example, as precursors of coatings and ceramic materials [276, 277]. Polysiloxazanes have been synthesized either by copolymerization of cyclosiloxanes and cyclosilazanes [276, 279] or by homopolymerization of cyclosiloxazanes [278, 280–285]. However, the polymerization of various cyclosiloxazanes required specific conditions (in bulk at high temperature) but without any control since, besides the polymer, a large amount of various cyclic compounds were formed [280, 283–285].

Polycondensation of a mixture of chloro-terminated permethylcyclodisilazane oligomers (CDS) with polydimethylsiloxane-α,ω-diols (PDMSD) of 100, 150, and 200 siloxane units was carried out at 80 °C in benzene solution for 20 h and afforded siloxane–silazane copolymers, which after removal of solvents were heated at 200 °C/0.2 Tr for 2 h. The obtained wax-like materials were completely soluble in benzene, toluene, and some other solvents. An intrinsic viscosity of their solutions in toluene (0.225–0.325) was increased in comparison with initial α,ω-polydimethylsiloxane-α,ω-diols (0.113–0.159). These novel copolymers showed enhanced thermal stability: a 15% weight loss was observed at 430–510 °C, as compared to 350–360 °C to start with PDMSD. However, increased thermostability by ~50 °C was also observed for copolymer samples obtained with 3% (wt/wt) of CDS oligomer, which was 25% of equimolar amount. On the other hand, cross-linking and formation of a gel were observed during polycondensation of equivalent amounts of CDS oligomers with octamethyltetrasiloxane-1,7-diol, when a rubber-like product was prepared [283, 286].

An anionic ROP of several six-membered *N*-phenyl substituted cyclosilazoxanes with potassium PDMS-diolate as an initiator and DMSO as an activator at 120–160 °C within 20–60 h gave only an equilibrium mixture of oligomeric cyclic and linear silazoxanes. After quenching living polymer chains with an excess of Me_3SiCl liquid oligomers of average MWs ranging from 1,000 to 6,500 g/mol (M_n, osmometric) were obtained. Similarly proceeded polymerizations of eight-membered *N*-phenyl-substituted cyclosilazoxanes, although anionic copolymerization of the cyclosilazoxanes with hexamethylcyclotrisiloxane (D_3) in various ratios provided high MW polysilazoxanes (M_n: 16,000–80,000 g/mol, depending on molecular ratio of comonomers). These polymers were viscous liquids of high thermal stability (up to 500 °C) [283, 284]. However, in copolymerization of eight-membered *N*-phenylcyclosilazoxanes with octamethylcyclotetrasiloxane (D_4), MWs (M_n) did not exceed 7,200 g/mol [284].

A. Soum et al. synthesized polysiloxazanes through controlled equilibrium ring-opening anionic polymerization of the heptamethyl-1,3-dioxa-5-aza-2,4,6-trisilacyclohexane (D_2D^{NMe}) [287]. The polymers with a regular structure with a very small amount of a backbiting product, nonamethyl-1,3,5-trioxa-7-aza-2,4,6,8-tetrasilacyclooctane (D_3D^{NMe}), were obtained. The anionic ROP of heptamethyl-1,3-dioxa-5-aza-2,4,6-trisilacyclohexane was initiated with organolithium compounds and was carried out in toluene, mostly at 30 °C. With the addition of dimethylformamide (DMF) as an activator, the polymerization was controlled, backbiting reactions were limited, and gave a low content of unique specific cyclic compound. Thermodynamic and kinetic studies confirmed that polymerization was equilibrated. The active species were ion pairs externally solvated by DMF in equilibrium with unreactive aggregated ion pairs. Yields of copolymers precipitated from acetonitrile (and dried under vacuum) were 18–43% (whithin 3–97 h of reaction time), and their number average MWs ranged from 6.6 to 48.5 kDa, and PDI = 1.1–1.5 [287].

A polysiloxane containing *N,N′*-bis(diphenylsilyl)tetraphenylcyclodisilazane segments was prepared by anionic ROP of D_3 with a lithium salt of *N,N′*-bis-(chlorodiphenylsilyl)tetraphenylcyclodisilazane as an initiator (Scheme 7.23), in toluene solution at 80 °C for 24 h, using THF and DMSO as promoters.

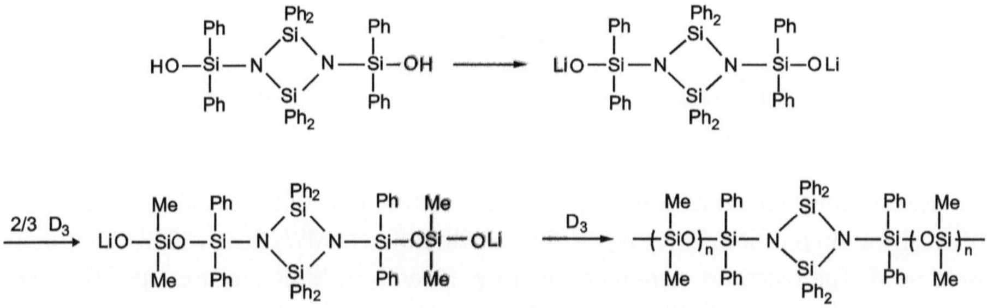

Scheme 7.23: Reprinted with permission from [288]. Copyright 2000, Springer Nature.

The prepared polysiloxazane copolymer was characterized by ^{1}H- and ^{29}Si-NMR, GPC, and intrinsic viscosity. The number average MW (M_n) of the polymer was 32,000 g/mol and its viscosity average MW (M_v) was 39,000 g/mol [288, 289].

A series of new cross-linked poly[(methylsiloxane)-*co*-(dimethylsilazane)] copolymers were prepared by cationic ring opening polymerization (CROP) of 1,3,5,7- tetramethylcyclotetrasiloxane and 2,2′,4,4′,6,6′-hexamethylcyclotrisilazane in the presence of triflic acid in methylene chloride within 10 days at 30 °C. The ROP reactions were terminated with Et_3N in methanol, copolymers were washed with CH_2Cl_2 and dried under vacuum. Yields ranged from 70% to 78% and MWs, determined by osmometry – from 9.9 to 27.1 kDa. Their structures were composed of random linear segments and cross-linked segments and were confirmed by FT-IR studies, and thermal stability was strongly dependent on the concentration of methylsiloxane units in the copolymers. Thermal and UV-curing processes were applied. The pyrolysis of the copolymers, carried out under argon atmosphere at 1,200 °C and in air at 1,000 °C, afforded $Si_xC_yN_z$ ceramic [290].

Ladder-like polysiloxazanes with a maximum weight average MWs 70,000 g/mol were synthesized with high yields (88–99%) by ammonolysis of 1,1′,3,3′-tetraisocyanato-1,3-dimethyl-1,3-disiloxane (Scheme 7.24):

This reaction was carried out in THF, diethyl ether, or 1,4-dioxane at −80 or −60 °C, with a formation of cyanuric acid as a byproduct, followed by aging within 2–9 days. After filtration, polymers in a form of white powder were isolated from filtrate by concentrating to ~30 wt%, precipitation from hexane, next filtration, and drying. They were characterized by FT-IR, ^{1}H, ^{13}C, ^{29}Si-NMR, and GPC methods and were further used for fabrication of SiCNO ceramics by pyrolysis under ammonia or nitrogen atmosphere at progressing temperatures of up to 900 °C [291].

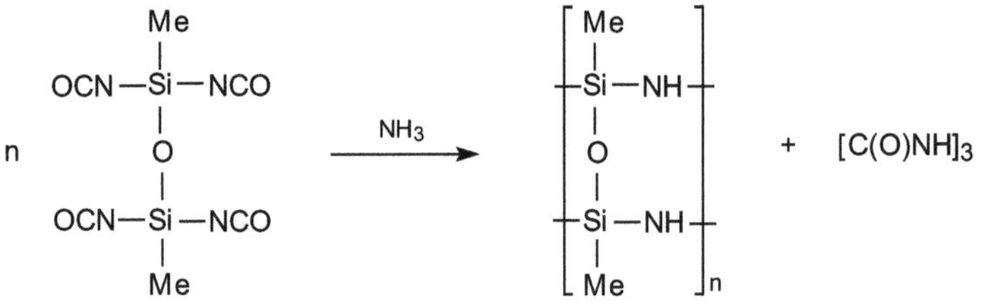

Scheme 7.24: Synthesis of ladder-like polysiloxazanes by ammonolysis of 1,1′,3,3′-tetraisocyanato-1,3-dimethyl-1,3-disiloxane.

New coating materials were prepared by grafting PEO onto polysilazane (PSZ) by hydrosilylation reaction. Three types of PEO with different MWs (350, 750, 2,000 g/mol) were used. The reactions were monitored by ^{1}H- and ^{13}C-NMR spectroscopy. The PEO grafting density onto PSZ increased with a reduction of the Si–H to allyl groups ratio and a decrease of the PEO chain length. The PEO-*graft*-PSZ (PSZ-PEO) hybrid coatings were cured by moisture at room temperature and can prevent the adhesion of marine bacteria on surfaces. The antiadhesion properties, and the antifouling activity, of the coatings against three marine bacteria species were tested. They were dependent on the grafting density and the chain length of PEO. The best antifouling activity showed shortest copolymer PEO(350 g/mol)-*graft*-PSZ having the highest graft density. As the density of grafted PEO(750 g/mol) and PEO(2,000 g/mol) chains onto the PSZ surface was approximately equal, the relative effectiveness of these two types of PEO was controlled by the length of the PEO chain. The bacterial antiadhesion properties of PEO(2,000 g/mol)-*graft*-PSZ coatings were more efficient than the PEO(750 g/mol)-*graft*-PSZ coatings [292].

7.2.2.6 Miscellaneous polysiloxane-organic copolymers

PDMS-PCL block copolymers of A-B-A architecture (PDMS-*b*-PCL-*b*-PDMS) were synthesized by ROP of ε-caprolactone with hydroxybutyl-terminated polydimethylsiloxane, catalyzed in bulk (at 140 °C) by stannous octanoate. The PDMS-*b*-PCL-*b*-PDMS ranged from viscous liquids to hard, crystalline solids, depending on their molecular composition. They were next used for the preparation of polysiloxane-PU/(urea) copolymers and modifications of epoxy resin and polyamide (Nylon 6) [293].

Polycarbonate–polysulfone and polycarbonate–PDMS block copolymers were prepared by J.E. McGrath, J.S. Riffle, and coworkers [294–296]. Some of them were obtained via an interfacial process [297]. Silicone–polycarbonate copolymers having PDMS segments or heptamethyltrisiloxane units were prepared by the interfacial polymerization of chloroformate-functionalized PDMS and trisiloxane. The latter were obtained from Eugenol terminated PDMS and heptamethyltrisiloxane derivative, which were synthesized through the hydrosilylation reaction of eugenol with $M^{H}{}_{2}D_{n}$ or $MD^{H}M$ toward Karstedt's catalyst. Flexibility and wettability of the synthesized polymers increased with increasing silicone content. These new copolymers showed good thermooxidative stability and transparency, similar to polycarbonate based on bisphenol-A. Their weight average MW (M_w) and PDI (M_w/M_n), determined by GPC, ranged from 45,000 to 55,000 g/mol and 2.2–2.0, respectively [298].

Well-defined octa-polyisobutylene-arm radial star polymers with polysiloxane cores were prepared by hydrosilylation of allyl-terminated polyisobutylene with octa (dimethylsiloxy)octasilsesquioxane ($Q_8M^H{}_8$) in the presence of Karstedt's catalyst at room temperature [299].

Two cycloalkylene–siloxane linear copolymers were prepared from dicyclopentadiene and silacyclopentene, and the thermal stability of the raw polymers was evaluated by DSC and TGA. The DSC data in nitrogen indicated that both polymer systems had excellent thermal stability. In air, these polymers began to oxidize around 150 °C, while their very intensive oxidation occurred at about 400 °C [300].

A homologous series of octasilsesquioxane–PEG copolymers were prepared by the hydrosilylation of allyl-terminated PEGs having various chain lengths with octa(dimethylsiloxy)silsesquioxane ($Q_8M^H{}_8$) or octahydridosilsesquioxane ($T^H{}_8$) (Scheme 7.25).

The products of grafting were chemically characterized by FT-IR and ^{1}H-, ^{13}C-, and ^{29}Si-NMR spectroscopy, and thermal properties of the POSS hybrids were characterized by DSC and TGA. The T_gs of the HO-functionalized PEGs-POSS macromonomers were dependent on the chain length of PEGs and concomitant suppression of crystallization [301].

New transparent and flexible copolymers (DDSQ-BZ-PDMS) (Scheme 7.26) were prepared by hydrosilylation of a bis-allyl benzoxazine derivative of double-decker silsesquioxane (DDSQ-BZ) with Si–H-terminated PDMS $M^H{}_2D_n$. The char yield (73 wt%) after pyrolysis of DDSQ-BZ-PDMS was much higher than in the case of a typical polybenzoxazine. Moreover, DDSQ-BZ-PDMS saved high flexibility and transparency after thermal curing. The thermal decomposition temperature of DDSQ-BZ after thermal curing was increased up to 500 °C, and the char yield was 80 wt% [302].

Hybrid NCs of unsaturated polyester (UP) with POSS spherosilicate were prepared through the reaction between maleimide groups of octa(maleimido phenyl) silsesquioxane (OMPS) as a nanocross-linker, and olefinic reactive sites (maleimide and styrenic units) present in the UP resin system – through free radical polymerization, initiated by benzoyl peroxide. The hybrid molecular structure of NCs was analyzed by FT-IR spectroscopy. The homogeneous morphology and nanoscale dispersion of OMPS into the polyester hybrid NCs were confirmed by XRD, SEM, and TEM analysis data. The incorporation of octamaleimide-functionalized POSS into UP systems significantly improved the values of T_g, the thermal properties, char yield, and storage modulus of the hybrid NCs according to varying contents of POSS (1, 3, 5, and 10 wt%). The hybrid UP-OMPS resin NCs can be used as coatings, adhesives, sealants, and matrices for the fabrication of advanced composite ingredients for high-performance applications [303].

Block and random tercopolymers of POSS with fluorinated poly(aryl ether sulfones) (PAESs) (Scheme 7.27) were prepared by different synthetic methods.

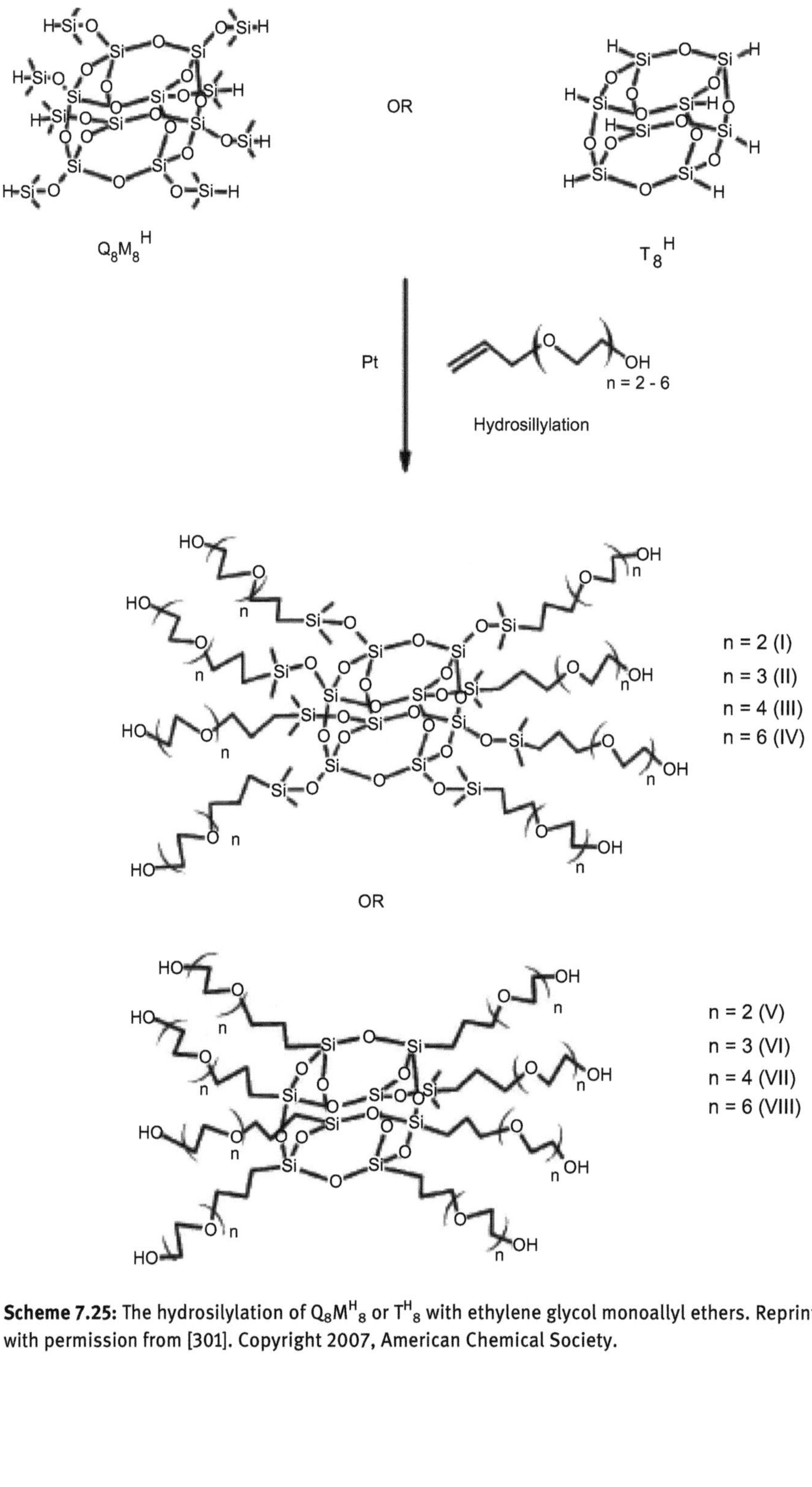

Scheme 7.25: The hydrosilylation of $Q_8M^H{}_8$ or $T^H{}_8$ with ethylene glycol monoallyl ethers. Reprinted with permission from [301]. Copyright 2007, American Chemical Society.

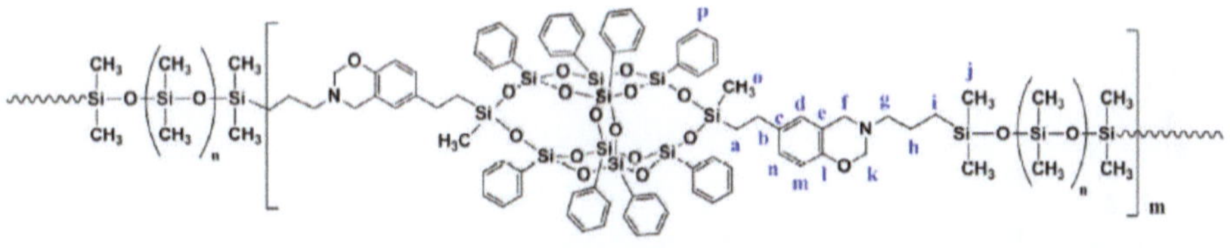

Scheme 7.26: The chemical structure of copolymers DDSQ-BZ-PDMS. Reprinted with permission from [302]. Copyright 2017, American Chemical Society.

In order to investigate the effect of POSS content and the structure–function in relation to the dielectric property and hydrophobicity, three kinds of PAESs random terpolymers with different chemistry structure at variable POSS content in the main chain were prepared. The structures of PAESs were characterized by IR, NMR, and wide-angle X-ray diffraction (WXRD) spectra. The results show that the dielectric constant initially increased and then decreased to 2.68 at 100%-DDSQ-PAES (molar content of DDSQ = 100%) at 1 MHz. The CA increased to 97.5° at 100%-DDSQ-PAES, while the chemical structure of organic chains also played an important role on thermostability, dielectric, and hydrophobic properties. The results were discussed and interpreted in detail.

These copolymers exhibited increased water CAs and T_g values and decreasing k values with growing content of POSS. The block copolymers had higher T_g values (up to 187 °C) than random copolymers (173 °C) under the same POSS molar content, owing to their different sequence distribution. The dielectric constants of the terpolymers achieved low values (2.71 at 1 MHz) and were strongly reduced due to the presence of POSS and fluorine. The sequence distribution did not affect its surface properties and dielectric properties. All the copolymers were slightly soluble in various common organic solvents. Due to the presence of POSS cages and fluorine atoms in the main chains, the copolymers were highly hydrophobic and possessed very low dielectric constants. The char yields of these terpolymers were significantly improved by the ceramic formation from POSS moieties during thermal decomposition [304].

Three other kinds of PAESs random terpolymers (Scheme 7.28) with different chemistry structure were also prepared from bishydroxy derivative of DDSQ (at variable POSS content in the main chain) and from bisphenol A, bisphenol S, and diphenyl bisphenol. The chemical structure of organic chains affected thermostability, and dielectric and hydrophobic properties of these copolymers. The thermal stability of copolymers was improved with increasing content of DDSQ and the CA increased to 97.5° at 100% content of DDSQ [305].

Hydrosilylation of α,ω-oligo(dimethylsiloxanes) with α-cyclodextrins (α-CD) afforded terminally substituted oligo(dimethylsiloxanes) containing α-CD moieties, which served as host end groups for the cyclopentadienyl rings of ferrocene. The

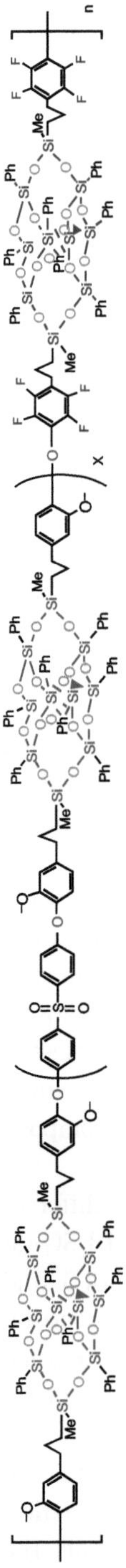

Scheme 7.27: The structure of block polyhedral oligomeric silsesquioxane/fluorinated poly aryl ether sulfone tricopolymers. Reprinted with permission from [304]. Copyright 2017, J. Wiley & Sons Inc.

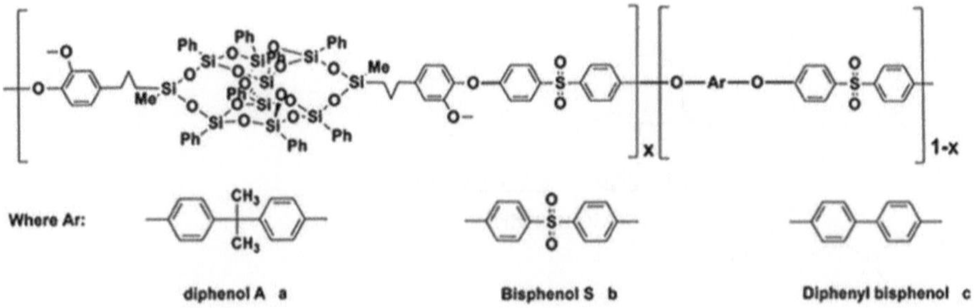

Scheme 7.28: The structure of hybrid copolymers DDSQ–PAESs. Reprinted with permission from [305]. Copyright 2017, J. Wiley & Sons Inc.

double complexation of unsubstituted ferrocene led to a supramolecuar formation of the siloxane networks [306].

Poly(ferrocenyldimethylsilane)-*b*-polyvinylsiloxane copolymers were functionalized with *N*-hydroxysuccinimidyl (NHS) ester. The NHS ester-functionalized polymers were further modified by facile reaction with primary amines, which enabled the incorporation of a range of functionalities through standard thiol-ene "click" reactions [307].

Poly(lactide) (PLA) NCs were prepared by solution blending of commercial poly (L-lactide) (PLLA) and biodegradable core–shell particles, which were synthesized via ring-opening copolymerization of a mixture of ε-caprolactone and L-lactide to form poly(ε-caprolactone-*co*-lactide) (PCLLA), as rubbery core, initiated by octa-[3-hydroxy-3-methylbutyl(dimethylsiloxy)]silsesquioxane (POSS), followed by sequential polymerization of D-lactide to form poly(D-lactide) (PDLA) as outer shell (Scheme 7.29). The outer PDLA layer could facilitate strong interactions between core–shell rubber particles and PLLA matrix via formation of stereocomplex. The random microstruxture of PCLLA and the subsequent grafting of PDLA were monitored by NMR. The rubbery characteristic of PCLLA was confirmed by DSC, which showed a T_g of ~–7 °C. Formation of stereocomplexes between PLLA and POSS-rubber-PDLA polymer, denoted as POSS-rubber-D was confirmed using FT-IR spectroscopy, DSC, and XRD. The obtained biodegradable NCs exhibited a 10-fold increase of elongation at break but retained other mechanical properties: a tensile strength and Young's modulus. XRD, light scattering, SEM, and TGA studies suggested that strong stereocomplex matrix/rubber interactions, good particle dispersion, rubber-initiated crazing, and low rubber content were responsible for these properties [308].

Silicone-cellulose membranes were obtained from PMHS and (hydroxypropyl) cellulose (Scheme 7.30). Their hydrolytic stability in water and in the presence of a cellulase enzyme did not increase with PMHS content but was rather optimal for an appropriate polysiloxane: polysaccharide ratio. The biopolymer was sensitive to enzymatic degradation and water solubilization [309].

PDMS films were modified with poly(2-hydroxyethyl methacrylate) (PHEMA) and collagen by sequential method of IPNs. Collagen was grafted through methyl sulfonyl

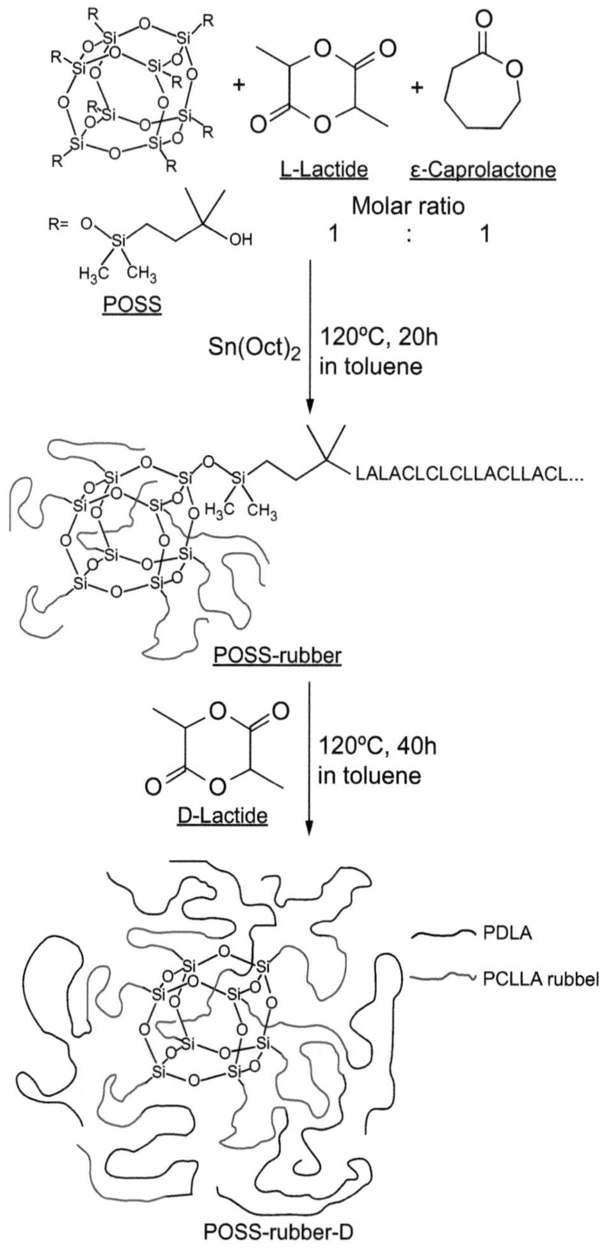

Scheme 7.29: The synthesis of core–shell POSS-rubber-D particles by ROP. Reprinted with permission from [308]. Copyright 2013, American Chemical Society.

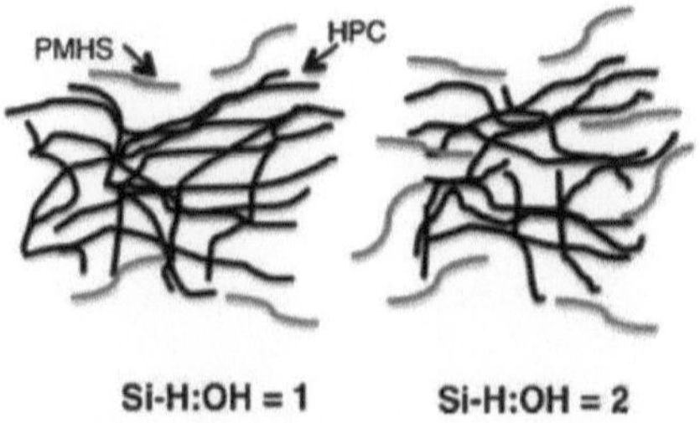

Scheme 7.30: Proposed structures of (hydroxypropyl)cellulose (HPC)/PMHS composite films. Reprinted with permission from [309]. Copyright 2012, Springer Nature.

chloride and covalently linked protein layers, which was confirmed using ATR-FT-IR and XPS. The hydrophilicity of modified samples increased as the PHEMA loading increased. The collagen-modified surfaces showed significant cell adhesion and growth in comparison with unmodified PDMS samples. The biocompatibility of the PDMS surface was confirmed by culture of fibroblast cells (L929) [310].

A series of hybrid gelatine/silica materials, containing 30–80 wt% of gelatine, was prepared by a sol–gel technique. They showed a diverse structure: a part of silica formed a homogeneous composite with gelatine, while another part formed microspherical particles of complex structure. These gelatine–silica hybrids were crosslinked with branched polyalkoxysiloxanes [311]. Gelatine–siloxane hybrid materials were also prepared in the reaction of gelatine with carbofunctional siloxanes containing epoxy- or both epoxy- and fluoroalkyl groups. The samples modified with fluoroalkyl-substituted siloxanes or with mixed siloxanes showed higher thermostability than gelatine. T_g of the hybrids occurred at lower values than for gelatine, exhibiting the plasticizing effect of the siloxanes used. Enthalpy values of the hybrids decreased, indicating the changes in the helical structure of gelatine caused by the incorporation of siloxane units. Similarly, gelatine–siloxane hybrids were prepared in the reaction of gelatine and functional siloxane containing epoxy- and polysiloxane containing both epoxy- and fluoroalkyl substituents. The hybrids were characterized by FT-IR and ^{1}H NMR spectroscopies, and SEM. Their thermal properties were determined by TGA and DSC measurements and were dependent on the content of organosilicon compounds.[312].

A novel comb-like polysiloxane copolymer (GPPDMS), containing guanidyl- and phosphoric acid groups, was prepared by grafting P_2O_5 and dicyandiamide from siloxane oligomers which were obtained by hydrolytic condensation of APTES (Scheme 7.31).

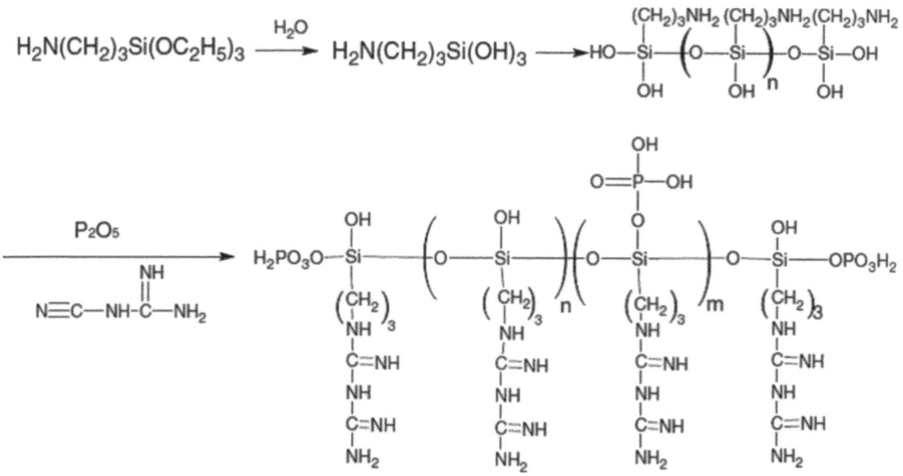

Scheme 7.31: Synthetic route to GPPDMS and its proposed chemical structure. Reprinted with permission from [313]. Copyright 2012, Springer Nature.

This copolymer was used as flame-retardant agent for a cotton fabrics which showed a LOI value of 31.9 and a highest reduction in heat of combustion when the textile was treated with 18.6% (add-on) of GPPDMS. Results of EDS analysis showed that after combustion substantial amount of Si and P elements were present on the surface of fibers and caused synergic effect, enhancing the flame retardancy of cotton fabric. GPPDMS inhibited the degradation of cotton fabrics and the formation of volatile species, but favored the formation of char [313, 314]. Morover, GPPDMS showed good antimicrobial activity of the modified cotton fabrics against *Escherichia coli* and *Staphylococcus aureus* (97% and 96%, respectively) [314].

New synthetic routes to alkyl silicates and oganosiloxanes were elaborated by G.B. Goodwin and M.E. Kenney. Tetraethoxysilane ($Si(OEt)_4$) was prepared from γ-Ca_2SiO_4, Ca_3SiO_4O, and Portland cement. $(EtO)_3SiOSi(OEt)_3$ was obtained from $Ca_2ZnSi_2O_7$, and $(n\text{-}PrO)_3SiO(n\text{-}PrO)_2Si(On\text{-}Pr)_3$ was obtained from $Ca_3Si_3O_9$. Isomers of $(EtO)_{10}Si_6O_7$ were prepared from $Ca_6Si_6O_{18}\cdot 6H_2O$ and from $Na_2Ca_4Si_6O_{18}$, while isomers of $Me_{10}Si_6O_7$ from isomers of $(EtO)_{10}Si_6O_7$ [315].

References

[1] J.D. Summers, C.S. Elsbernd, P.M. Sormani, P.J.A. Brand, C.A. Arnold, I. Yilgor, J.S. Riffle, S. Kilie and J.E. McGrath, Chapter 14: *Recent advances in organosiloxane copolymers*, in *Inorganic and Organometallic Polymers*, M. Zeldin, K.J. Wynne and H.R. Allcock (Eds.), *ACS Symposium Series*, 1988, Vol. 360, 180–198.

[2] C. Burger and F.-H. Kreuzer, Charter 3: *Polysiloxanes and polymers containing siloxane groups*, in *Silicon in Polymer Synthesis*, H.R. Kricheldorf (Ed), Springer, 1996, 113–219.

[3] C. Eaborn, *Organosilicon Compounds*, Butterworth Sci. Publ., London, 1960, and referencies cited therein.
[4] W. Noll, *Chemie und Technologie der Silikone*, 2 Aufl, Verlag Chemie, GmbH, Weinheim, 1968, and referencies cited therein.
[5] B. Hardman and A. Torkelson, Chapter: *Silicones*, in *Encyclopedia of Polymer Science and Engineering, Second Edition*, Vol. 15, J. Wiley & Sons, New York, NY, 1989, 271–289.
[6] J. Chruściel, *Progress* in polymethylhydrosiloxane chemistry Polimery, 1999, 44(7–8), 462. and referencies cited therein. In Polish.
[7] P. Rościszewski, *Zastosowanie silikonów*, WNT, Warszawa, 1964, In Polish.
[8] P. Rościszewski and M. Zielecka, *Silikony, właściwości i zastosowanie*, WNT, Warszawa, 2002, In Polish.
[9] J.W. White, Chapter 22: *Recent achievements in industrial silicon polymer science*, in *organosilicon chemistry, Proceedings of the Xth International Symposium, Poznań, 1993*, B.Marciniec and J. Chojnowski (Eds.), Gordon & Breach Scientific Publishers, 1995, 363–386.
[10] H.R. Kricheldorf (Ed.), *Silicon in Polymer Synthesis*, Springer-Verlag, Berlin-Heidelberg-New York, 1996.
[11] M.A. Brook, *Silicon in Organic, Organometallic and Polymer Chemistry*, J. Wiley & Sons, New York – Chichester – Weinheim – Brisbane – Singapor – Toronto, 2000, and referencies cited therein.
[12] R.O. Sauer, *US Patent* 2 595 890, 1952; *Chemical Abstracts*, 1952, 46, 7820.
[13] N.N. Sokołow, *Metody syntezy poliorganosiloksanów*, PWN, Warszawa, 1961, In Polish.
[14] *British Patent* 632 954, 1949; *Chemical Abstracts*, 1950, 44, 7866.
[15] P. Rościszewski, E. Jagielska and K. Bartosiak, *Patent PL* 92 926, 1974.
[16] P. Rościszewski, R. Witlib and M. Książek, *Patent PL* 92 637, 1975.
[17] R.D.C. Richards, J. Hollinghurst and J.A. Semlyen, Cyclic polysiloxanes: 5. Preparation and characterization of poly(hydrogenmethylsiloxane), *Polymer*, 1993, 34(23), 4965.
[18] V.O. Reikhsfeld and A.G. Ivanova, *Vysokomolekularnye Soedineniya*, 1962, 4, 30.
[19] T. Graczyk and Z. Lasocki, *Bulletin de L'Academie Polonise des Sciences, Série* des *Sciences Chimiques*, (a), 1978, 36(12), 917. (b) 1979, 37(3), 181; (c) 1979, 37(3), 185.
[20] T.C. Williams, *US Patent* 5 354 830, 1994; *Chemical Abstracts*, 1995, 122, 267860y.
[21] H. Hara and M. Ikeno, *Japanese Patent* JP 0616942 (1994); *Chemical Abstracts*, 1995, 121, 11004x.
[22] R.K. Harris and B.J. Kimber, ^{29}Si nuclear magnetic resonance studies of some oligo- and polymeric siloxanes, *Journal of Organometallic Chemistry*, 1974, 70(1), 43.
[23] J.V. Crivello, J.L. Lee and D.A. Conlon, Chapter: *A new free radical approach to the synthesis of polydimethylsiloxane-vinyl monomer block polymers*, in *Advances in Elastomers and Rubber Elasticity*, J. Lal and J.E. Mark (Eds.), Plenum Press, New York & London, 1986, 157–183.
[24] E. Husemann and G. Greber, *German Patent* 1 140 348, 1962; *Chemical Abstracts*, 1963, 58, 12600.
[25] M.R. Stober, M.C. Musolf and J.L. Speier, The Addition of Silicon Hydrides to Olefinic Double Bonds. XI. Exchange of Methyl and Trimethylsiloxy Groups in Bistrimethylsiloxymethylsilane, *Journal of Organic Chemistry*, 1965, 30(5), 1651.
[26] R. Hager, *XI International Symposium on Organosilicon Chemistry, Universite Montpellier II*, France, 1996, *Abstracts*, LD5.
[27] R. Hager and J. Weis, *German Patent* 4 317 909, 1993; *Chemical Abstracts*, 1995, 123, 10862e.
[28] K.A. Andrianov, L.A. Gavrikova and E.F. Rodionova, *Vysokomolekularnye Soedineniya*, 1971, 13A, 937.

[29] K.A. Andrianov, I.I. Abhazova, L.M. Khananashvili, G.D. Bogratishvili, T.P. Dogsopulc and V.A. Chagelishvili, *Zhurnal Obshchei Khimii*, 1971, 41, 606.

[30] Y. Zhang, Z. Zhang, Q. Wang and Z. Xie, Synthesis of Well-Defined Difunctional Polydimethylsiloxane with an Efficient Dianionic Initiator for ABA Triblock Copolymer, *Journal of Applied Polymer Science*, 2007, 103(1), 153.

[31] A. Saxena, S. Rajaraman and M. Leatherman, Synthesis of Narrowly Dispersed Bis-Hydride-Capped Polydimethylsiloxane Using Difunctional Anionic Initiator Based on 1,3-Diisopropenylbenzene, *Macromolecules*, 2007, 40(3), 752.

[32] J. Chruściel and Z. Lasocki, Reduction of α,ω-dichloropoly(dimethylsiloxanes) with lithium aluminium hydride *Polish Journal of Chemistry*, 1979, 53(6), 1383.

[33] J. Beckmann, D. Dakternieks, A. Duthie and R. Foitzik, The use of Pearlman's catalyst for the oxidation of Si–H bonds. Synthesis, structures and acid-catalysed condensation of novel α,ω-oligosiloxanediols $HOSiMe_2O(SiPh_2O)_nSiMe_2OH$ (n = 1–4), *Silicon Chemistry*, 2003, 2(1–2), 27.

[34] D.F. Wilcock and T.H. Dallas, *US Patent* 2 547 678, 1951; *Chemical Abstracts*, 1951, 45, 8029.

[35] R. Müller, R. Kohne and S. Śliwiński, Über Silikone. XLVII. Ein definiertes Siloxan aus Siliciumchloroform, *Journal für praktische Chemie*, 1959, 9(1–2), 71.

[36] G. Greber and L. Metzinger, Über oligomere siliciumverbindungen mit funktionellen gruppen. 2. Mitt. Über die darstellung von polysiloxanhydriden und deren addition an ungesättigte verbindungen. *Die Makromolekulare Chemie: Macromolecular Chemistry and Physics*, 39(1), 189–216. https://doi.org/10.1002/macp.1960.020390114

[37] G. Greber and L. Metzinger, *German Patent* 1 085 875, 1960; *Chemical Abstracts*, 1961, 55, 27057.

[38] D.F. Wilcock, *US Patent* 2 491 843, 1949; *Chemical Abstracts*, 1950, 44, 22887.

[39] J.I. Gallagher, *British Patent* 1 042 784, 1966; *Chemical Abstracts*, 1967, 66, 3026n.

[40] Y. Yamamoto and T. Matsuda, *Patent EP* 435 654, 1991; *Chemical Abstracts*, 1991, 115, 183573z.

[41] T. Nennendal, E. Aguirre-Kaufmann, W.H. Binder, S. Knaus and H. Gruber, Equilibration control of poly(dimethylsiloxane)-co-(hydromethylsiloxane), *Zeszyty Naukowe Politechniki Śląskiej, Seria: Chemia*, 1999, 140, 115.

[42] M. Lefort and N. Parasko, *French Patent* 1 438 344, 1966; *Chemical Abstracts*, 1967, 66, 3525a.

[43] L.N. Lewis, D.S Johnson and J.D. Rich, The chemistry of the Si-H bond with acid treated clays, *Journal of Organometallic Chemistry*, 1993, 448(1–2), 15.

[44] M. Pratt, J. Smid and I.M. Khan, *Polymer Preprints*, 1994, 35(2), 643.

[45] J.S. Razzano, P.P. Anderson and R.J. Perry, *US Patent 5670596; Chemical Abstracts*, 1997, 127, 278623.

[46] S.A. Nye, D.A.Riccio and B.S.Wutzer, *US Patent* 5698654; *Chemical Abstracts*, 1998, 128, 75824.

[47] W.P. Liao and S.A. Nye, *European Patent Application EP 838547; Chemical Abstracts*, 128, 295432 (1998).

[48] R.J. Perry, *US Patent* 6,043,388, 2000.

[49] R.J. Perry, *Japanese Patent* JP 2000191791, 2000.

[50] C.L. Lee, Polymerization of cyclosiloxanes mechanism of cross-linkage in the basecatalyzed polymerization of 2,4,6,8-tetramethylcyclotetrasiloxane, *Journal of Organometallic Chemistry*, 1966, 6(6), 620.

[51] J.K. Paulasaari and W.P. Weber, Preparation of Highly Regular Poly(1-hydrido-1,3,3,5,5-pentamethyltrisiloxane) and Its Chemical Modification by Hydrosilylation, *Macromolecules*, 1999, 32(20), 6574.

[52] K. Yoshino and A. Kawamata (Kao Corp.), *Japanese Patent* JP 02306980, 1990.
[53] J. Chojnowski, S. Rubinsztajn, W. Fortuniak and J. Kurjata, Oligomer and polymer formation in hexamethylcyclotrisiloxane (D3)-hydrosilane systems under catalysis by tris (pentafluorophenyl) borane, *Journal of Inorganic and Organometallic Polymers and Materials*, 2007, 17(1), 173.
[54] J. Chojnowski, W. Fortuniak, J. Kurjata, S. Rubinsztajn and J.A. Cella, Oligomerization of Hydrosiloxanes in the Presence of Tris (pentafluorophenyl) borane, *Macromolecules*, 2006, 39(11), 3802.
[55] J. Chruściel, J. Kulpiński and Z. Romanowski, New Block Polymethylhydrosiloxanes, *Zeszyty Naukowe Politechniki Śląskiej, Seria: Chemia*, 1999, 140, 109.
[56] P. Cancouët, E. Daudet, G. Hélary, M. Moreau and G. Sauvet, Functional polysiloxanes.I. Microstructure of poly(hydrogenmethylsiloxane-*co*-dimethylsiloxane)s obtained by cationic copolymerization, *Journal of Polymer Science Part A: Polymer Chemistry*, 2000, 38(5), 826.
[57] P. Cancouët, S. Pernin, G. Hélary and G. Sauvet, Functional polysiloxanes. II. Neighboring effect in the hydrosilylation of poly(hydrogenmethylsiloxane-*co*-dimethylsiloxane)s by allylglycidylether, *Journal of Polymer Science Part A: Polymer Chemistry*, 2000, 38(5), 837.
[58] M.A. Grunlan, N.S. Lee, F. Mansfeld, E. Kus, J.A. Finlay, J.A Callow, M.E. Callow and W.P. Weber, *Journal of Polymer Science Part A: Polymer Chemistry*, 2006, 44(8), 2551.
[59] J.J. Chruściel, M. Fejdyś and W. Fortuniak, Synthesis, Characterization and Microstructure of New Liquid Poly(methylhydrosiloxanes) Containing Branching Units $SiO_{4/2}$, *Polymers*, 2018, 10(5), 484. DOI: 10.3390/polym10050484.
[60] J. Chruściel, unpublished results.
[61] L. Wilczek, Sun Y., *XXVII Organosilicon Symposium, 18–19.03.1994., Troy, New York, USA, Lecture Abstracts*, B-30.
[62] L. Wilczek, *US Patent* 5233071, 1993.
[63] B. Marciniec, J. Guliński, W. Urbaniak, T. Nowicka and J. Mirecki, Catalysis of hydrosilylation part XVIII. $Pt(PPh_3)_2(CH_2{=}CH_2)$ – A versatile catalyst for hydrosilylation of olefins, *Applied Organometallic Chemistry*, 1990, 4(1), 27.
[64] B. Marciniec, J. Guliński, L. Kopylowa, H. Maciejewski, M. Grunwald-Wyspiańska and M. Lewandowski, Catalysis of hydrosilylation: Part XXXI. Functionalization of poly(methylhydro) siloxanes via hydrosilylation of allyl derivatives, *Applied Organometallic Chemistry*, 1997, 11 (10–11), 843.
[65] Z.M. Michalska and K. Strzelec, Rhodium(I) complex catalysts immobilized on polyamides having a pyridine moiety: Effect of the polymer structure, *Journal of Molecular Catalysis A: Chemical*, 2001, 177(1), 89.
[66] R. West, K. Oka., M.A. Miller, H. Takahashi. and R.L. Skrupky, *XXVI Silicon Symposium, 26–27.03.1993., Indianapolis, Indiana, USA, Lecture Abstracts*, C-1
[67] West R., Oka K., Takahashi H., Miller M.A., Gunji T., Chapter 9: *Some silicon polymers of C60 (Buckminstefullerene)*, in *inorganic and organometallic polymers II, advanced materials and intermediates*, in *American Chemical Society Symposium Series*, Vol. 572, P. Wisian-Neilson, H.R. Allcock and K.J. Wynne (Eds.), USA, Washington DC, 1994, 92–101.
[68] A. Bachari, G. Belorgey, G. Helary and G. Sauvet, Synthesis and characterization of multiblock copolymers poly[poly(L-lactide)-*block*-polydimethylsiloxane], *Macromolecular Chemistry and Physics*, 1995, 196(1), 411.
[69] K. Matyjaszewski, P.J. Miller, J. Pyun, G. Kickelbick and S. Diamanti, Synthesis and characterization of star polymers with varying arm number, length, and composition from organic and hybrid inorganic/organic multifunctional initiators, *Macromolecules*, 1999, 32(20), 6526.
[70] R. Muskus and T. Ganicz, *XIII Polish Organosilicon Symposium*, Chmielno, Poland, 2007, *Abstracts*, p. 38 (oral communication). In Polish.

[71] K. Łęszczak, K. Szubert and B. Marciniec, *XIII Polish Organosilicon Symposium*, Chmielno, Poland, 2007, *Abstracts*, p. 78 (poster). In Polish.
[72] K. Pancer, H. Maciejewski and B. Marciniec, *XIII Polish Organosilicon Symposium*, Chmielno, Poland, 2007, *Abstracts*, p. 79 (poster). In Polish.
[73] R. Spindler and D.F. Shriver, Synthesis, NMR characterization, and electrical properties of siloxane-based polymer electrolytes, *Macromolecules*, 1988, 21(3), 648.
[74] I.M. Khan, Y. Yuan, D. Fish, E. Wu and J. Smid, Comblike Polysiloxanes with Oligo (oxyethylene) Side Chains. Synthesis and Properties, *Macromolecules*, 1988, 21(9), 2684.
[75] E. Wu, I.M. Khan, and J. Smld, Comb polysiloxanes with ω-hydroxyoligo(oxyethylene) side chains, Polymer Bulletin, 1988, 20(5), 455–461. https://doi.org/10.1007/BF01153437
[76] G.-B. Zhou, I.M. Khan and J. Smid, Solvent-Free Cation-Conducting Polysiloxane Electrolytes with Pendant Oligo(oxyethylene) and Sulfonate Groups, *Macromolecules*, 1993, 26(9), 2202.
[77] S. Garin, L. Lecamp, B. Youssef and C. Bunel, New polysiloxanes bearing heterocyclic groups - synthesis and curability, *European Polymer Journal*, 1999, 35(3), 473.
[78] Z. Zhang and S. Fang, Novel network polymer electrolytes based on polysiloxane with internal plasticizer, *Electrochimica Acta*, 2000, 45(13), 2131.
[79] M.K. Mapes, D. Schumacher, L.J. Lyons, D. Moline, R. Hooper and R. West, *Polymer Preprints*, 2000, 41(1), 309.
[80] R. Hooper, L.J. Lyons, D.A. Moline and R. West, Novel siloxane polymers as polymer electrolytes for high energy density lithium batteries, *Silicon Chemistry*, 2002, 1(2), 121.
[81] S.S. Abed-Ali, B.J. Brisdon and R. England, Poly(organosiloxanes) containing crown ether functionalities, *Macromolecules*, 1989, 22(10), 3969.
[82] J. Kang and S. Fang, *Polymer Bulletin*, 2002, 49(2–3), 127.
[83] R. Zhang, Z. Zhang, K. Amine and R. West, Modification of polymethylhydrosiloxane by dehydrocoupling reactions catalyzed by transition metal complexes: Evidence for the preservation of linear siloxane structures, *Silicon Chemistry*, 2003, 2(5), 271.
[84] Z. Zhang, L.J. Lyons, K. Amine and R. West, *Polymer Preprints*, 2004, 45(1), 583.
[85] Z. Zhang, N.A.A. Rossi, L.J. Lyons, K. Amine and R. West, *Polymer Preprints*, 2004, 45(1), 700.
[86] Z. Zhang, L.J. Lyons, J.J. Jin, K. Amine and R. West, Synthesis and ionic conductivity of cyclosiloxanes with ethyleneoxy-containing substituents, *Chemistry of Materials*, 2005, 17 (23), 5646.
[87] Z. Zhang, L.J. Lyons, K. Amine and R. West, Network-Type Ionic Conductors Based on Oligoethyleneoxy-Functionalized Pentamethylcyclopentasiloxanes, *Macromolecules*, 2005, 38(13), 5714.
[88] N.A.A. Rossi, Z. Zhang, Y. Schneider, K. Morcom, L.J. Lyons, Q. Wang, K. Amine and R. West, Synthesis and Characterization of Tetra- and Trisiloxane-Containing Oligo(ethylene glycol)ss Highly Conducting Electrolytes for Lithium Batteries, *Chemistry of Materials*, 2006, 18(5), 1289.
[89] Z. Zhang, L.J. Lyons, K. Amine and R. West, Synthesis and ionic conductivity of mixed substituted polysiloxanes with oligoethyleneoxy and cyclic carbonate substituents, *Silicon Chemistry*, 2005, 3(5), 259.
[90] Y. Karatas, N. Kaskhedikar, M. Burjanadze and H.-D. Wiemhöfer, Synthesis of Cross-Linked Comb Polysiloxane for Polymer Electrolyte Membranes, *Macromolecular Chemistry and Physics*, 2006, 207(4), 419.
[91] W.-J. Liang, Y.-P. Chen, C.-P. Wu and P.-L. Kuo, Solid Polymer Electrolytes. XI. Preparation, Characterization and Ionic Conductivity of New Plasticized Polymer Electrolytes Based on Chemical-Covalent Polyether–Siloxane Hybrids, *Journal of Applied Polymer Science*, 2006, 100(2), 1000.
[92] A. Synytska, E. Biehlig and L. Ionov, Adaptive PEG–PDMS brushes: Effect of architecture on adhesiveness in air and under water, *Macromolecules*, 2014, 47(23), 8377.

[93] E. Martinelli, E. Guazzelli, C. Bartoli, M. Gazzarri, F. Chiellini, G. Galli, M.E. Callow, J.A. Callow, J.A. Finlay and S. Hill, *Journal of Polymer Science, Part A: Polymer Chemistry*, 2015, 53(10), 1213.

[94] V. Shibaev and N. Boiko, Chapter 10: *Liquid crystalline silicon-containing dendrimers with terminal mesogenic groups*, pp 237–283, in *Silicon-Containing Dendritic Polymers. Advances in Silicon Science*, Vol. 2, P.R. Dvornic and M.J. Owen (Eds.), Springer, Dordrecht, 2009. Part of the *Advances in Silicon Science* book series (ADSS).

[95] S. Boileau and D. Teyssie, Synthesis and properties of liquid crystalline polysiloxanes, *Journal of Inorganic and Organometallic Polymers*, 1991, 1(3), 247.

[96] A.P. Polishchuk, T.V. Timofeeva, N.N. Makarova, M.Yu. Antipin and T. Struchkov, A new structural type of liquid crystal Crystal structures and thermotropic mesophases of dihydroxytetraalkyldisiloxanes: Columnar phases with hydrogen bonded assemblies, *Liquid Crystals*, 1991, 9(3), 433.

[97] Y.K. Godovsky and V.S. Papkov, Thermotropic mesophases in element-organic polymers, *Advances in Polymer Science*, 1989, 88, 129–177.

[98] Y.K. Godovsky and V.S. Papkov, Thermotropic mesophases in linear polysiloxanes, *Macromolecular Chemistry, Macromolecular Symposia*, 1986, 4(1), 71.

[99] K.J. Miller, J. Grebowicz, J.P. Wesson and B. Wunderlich, Conformations of poly (diethylsiloxane) and its mesophase transitions, *Macromolecules*, 1990, 23(3), 849.

[100] J.M. Pochan, M.F. Frois, A.J. Goedde, C.L. Beatty and O.F. Pochan, *Liquid Crystals, the Fourth State of the Matter*, F.D. Saeva (ed.), Marcel Dekker, 1979, 41.

[101] C.L. Beatty and F.E. Karasz, Transitions in poly(diethyl siloxane), *Journal of Polymer Science, Polymer Physics Edition*, 1975, 13(5), 971.

[102] Y.K. Godovsky, Rubber-elastic mesophase siloxane systems, *Angewandte Makromoleculare Chemie*, 1992, 202–203, 187.

[103] T. Ganicz and W.A. Stańczyk, Side-chain Liquid Crystal Polymers (SCLCP): Methods and Materials. An Overview, *Materials*, 2009, 2(1), 95.

[104] J.-S. Hu, B. Zhang, L.-M. Liu and F.-B. Meng, Synthesis, Structures, and Properties of Side-Chain Cholesteric Liquid-Crystalline Polysiloxanes, *Journal of Applied Polymer Science*, 2003, 89(14), 3944.

[105] J. Zhi, B. Zhang and G. Shi, Synthesis and Properties of Photochromic Cholesteric Liquid Crystalline Polysiloxane Containing Chiral Mesogens and Azobenzene Photochromic Groups, *Journal of Applied Polymer Science*, 2002, 85(10), 2155.

[106] Y. Jia, B. Zhang, Z. Feng and Y. Guan, Synthesis and characterization of side-chain liquid crystalline polymer containing dichroic dye monomer, *European Polymer Journal*, 2003, 39 (8), 1701.

[107] Y. Wang, B. Zhang and X. He and J.-W. Wang, Side-chain cholesteric liquid crystalline polymers containing menthol and cholesterol–synthesis and characterization, *Colloid Polymer Science*, 2002, 285(10), 2155.

[108] T. Ganicz, T. Pakuła, W. Fortuniak and E. Białecka-Florjańczyk, Linear and hyperbranched liquid crystalline polysiloxanes, *Polymer*, 2005, 46(25), 11380.

[109] T. Ganicz, T. Pakuła W.A. Stańczyk, Novel liquid crystalline resins based on MQ siloxanes, *Journal of Organometallic Chemistry*, 2006, 691(23), 5052.

[110] T. Ganicz and W.A. Stańczyk, Synthesis of novel tri-podal mesogenic alkenes and side-chain polysiloxanes, *Journal of Organometallic Chemistry*, 2004, 689(16), 2606.

[111] S.H. Goh and Y.H. Lai, Effect of donor-acceptor interaction on thermal stability of supramolecular side chain liquid crystalline polymers based on poly(3-carboxypropylmethylsiloxane), *Liquid Crystals*, 2002, 29(5), 675.

[112] S.H. Goh, Y.H. Lai and S.X. Cheng, Supramolecular side chain liquid crystalline polymers assembled via hydrogen bonding between carboxylic acid-containing polysiloxane and azobenzene derivatives, *Liquid Crystals*, 2001, 28(10), 1527.
[113] X. Li, S.H. Goh and Y.H. Lai, Effects of the H-bond donor structure on the properties of supramolecular side chain liquid crystalline polymers based on poly(3-carboxypropylmethylsiloxane- *co*-dimethylsiloxane)s and stilbazoles, *Liquid Crystals*, 2003, 30(7), 811.
[114] X. Li, S.H. Goh and A.T.S. Wee, Miscibility of carboxyl-containing polysiloxane/poly (vinylpyridine) blends, *Polymer*, 2000, 41(17), 6563.
[115] G.H. Mehl and I.M. Saez, Polyhedral liquid crystal silsesquioxanes, *Applied Organometallic Chemistry*, 1999, 13(4), 261–272.
[116] E. Białecka-Florjańczyk and J.T. Sołtysiak, Liquid crystalline silicon-containing oligomers, *Journal of Organometallic Chemistry*, 2010, 695(15–16), 1911.
[117] T. Makowski, T. Ganicz, W. Zajączkowski, W. Pisula, W.A. Stańczyk and A. Tracz, Synthesis and thermotropic behavior of side chain polysiloxane bearing triphenylene moiety, *eXPRESS Polymer Letters*, 2015, 9(7), 636.
[118] C. Racles, A. Airinei, V. Cozan, M. Cazacu and I.E. Sajo, New Arylidene–Siloxane Polyethers: Liquid-Crystalline and Photosensitive Properties, *Journal of Applied Polymer Science*, 2003, 90(11), 3093.
[119] C. Racles and M. Cazacu, Siloxane-Containing Liquid-Crystalline Supramolecular Polymers: Preparation and Study of Thermotropic Behavior, *Journal of Applied Polymer Science*, 2008, 109(6), 4000.
[120] S.A. Ponomarenko, N.I. Boiko, V.P. Shibaev, R.M. Richardson, I.J. Whitehouse, E.A. Rebrov and A.M. Muzafarov, Carbosilane Liquid Crystalline Dendrimers: From Molecular Architecture to Supramolecular Nanostructures, *Macromolecules*, 2000, 33(15), 5549.
[121] X.-M. Zhu, N.I. Boiko, E.A. Rebrov, A.M. Muzafarov, M.V. Kozlovsky, R.M. Richardson and V.P. Shibaev, Carbosilane liquid crystalline dendrimers with terminal chiral mesogenic groups: structure and properties, *Liquid Crystals*, 2001, 28(8), 1259.
[122] K.L. Genson, J. Holzmueller, I. Leshchiner, E. Agina, N. Boiko, V.P. Shibaev and V.V. Tsukruk, Organized Monolayers of Carbosilane Dendrimers with Mesogenic Terminal Groups, *Macromolecules*, 2005, 38(19), 8028.
[123] V.P. Shibaev and A.Y. Bobrovsky, Liquid crystalline polymers: development trends and photocontrollable materials, *Russian Chemistry Reviews*, 2017, 86(11), 1024.
[124] V.P. Shibaev and N.I. Boiko, Chapter 10: *Liquid crystalline silicon-containing dendrimers with terminal mesogenic groups*, in *Silicon-Containing Dendritic Polymers*, P. Dvornik, M. Owen (Eds), London, Springer, 2009, 237–283.
[125] D. Grande, T. Ganicz, W.A. Stańczyk, N. Lacoudre and S. Boileau, Novel oligocarbosilazanes bearing promesogenic end groups, *eXPRESS Polymer Letters*, 2009, 3(7), 446.
[126] H. Finkelmann, H.-J. Kock and G. Rehage, Liquid crystalline elastomers – a new type of liquid crystalline material, *Makromolekulare Chemie, Rapid Communications*, 1981, 2(4), 317.
[127] W. Meier and H. Finkelmann, Liquid Crystal Elastomers with Piezoelectric Properties, *MRS Bulletin*, 1991, 16, 29.
[128] W. Gleim and H. Finkelmann, Thermoelastic and photoelastic properties of crosslinked liquid-crystalline side chain polymers, *Makromolekulare Chemie*, 1987, 188(6), 1489.
[129] K. Hammerschmidt and H. Finkelmann, Stress-optical and thermomechanical measurements on liquid crystalline elastomers, *Makromolekulare Chemie*, 1989, 190(5), 1089.
[130] W. Meier and H. Finkelmann, Piezoelectrocity of Cholesteric Elastomers, *Makromolekulare Chemie, Rapid Communications*, 1990, 11(12), 599.

[131] R. Zentel and G. Reckert, Liquid crystalline elastomers based on liquid crystalline side group main chain and combined polymers, *Makromolekulare Chemie*, 1986, 187(8), 1915.
[132] K.-H. Hanus, W. Pechhold, F. Soergel, B. Stoll and R. Zentel, Phase behavior and elastic properties of a slightly crosslinked liquid crystalline main-chain polymer *Colloid and Polymer Science*, 1990, 268(3), 222.
[133] R. Zentel, G. Reckert, S. Bualek and H. Kapitza, Liquid-crystalline elastomers with cholesteric and chiral smectic C* phases, *Makromolekulare Chemie*, 1989, 190(11), 2869.
[134] R. Zentel, Liquid crystalline elastomers, *Angewandte Chemie*, 1989, 101(10), 1437.
[135] S.U. Vallerien, F. Kremer, E.W. Fischer, H. Kapitza, R. Zentel and H. Poths, Experimental proof of piezoelectricity in cholesteric and chiral smectic C*-phases of LC-elastomers, *Makromolekulare Chemie, Rapid Communications*, 1990, 11(12), 593.
[136] J. Küpfer and H. Finkelmann, Nematic liquid single crystal elastomers, *Makromolekulare Chemie, Rapid Communications*, 1991, 12(12), 717.
[137] F.-H. Kreuzer, R. Maurer and J. Stohrer, Abstracts of the 14th International Liquid Crystal Conference, Pisa, 1991, p. 936.
[138] R.D. Joyner and M.E. Kenney, Phthalocyaninosiloxane, *Inorganic Chemistry*, 1962, 1(3), 717.
[139] E. Orthmann and G. Wegner, Catalysis of the polycondensation of dihydroxysiliconphthalocyanine, *Makromolekulare Chemie, Rapid Communications*, 1986, 7(5), 243.
[140] T. Sauer and G. Wegner, Rigid rod polymers from liquid crystalline phthalocyanines, *Makromolekulare Chemie, Macromolecular Symposia*, 1989, 24(1), 303.
[141] B.J. Exsted and M.U. Urban, *American Chemical Society, Polymer Preprints*, 1990, 31, 663.
[142] K.E. Polmanteer, T.D. Talcott and D.N. Willing, *German Patent* 2 041 633, 1971; *Chemical Abstracts*, 1971, 75, 7123z.
[143] R.C. McAfee, *US Patent* 4 558 109, 1985; *Chemical Abstracts*, 1986, 104, 131220k.
[144] P.S. Chang, T.S. Hughes, Y. Zhang, G.R. Webster, Jr., D. Poczynok and M. Buese, Synthesis and characterization of oligocyclosiloxanes via the hydrosilation of vinylsilanes and vinylsiloxanes with heptamethylcyclotetrasiloxane, *Journal of Polymer Science Part A: Polymer Chemistry*, 1993, 31(4), 891.
[145] M.J. Michalczyk, W.E. Farneth and A.J. Vega, High temperature stabilization of crosslinked siloxanes glasses, *Chemistry of Materials*, 1993, 5(12), 1687.
[146] T. Matsushita and Y. Shigehisa, *Patent EP* 590 595, 1994; *Chemical Abstracts*, 1995, 121, 302784e.
[147] H. Hara, M. Ikeno and T. Okami, *Japanese Patent* JP 06107947, 1994; *Chemical Abstracts*, 1995, 121, 302248h.
[148] K. Hashimoto, K. Noda, T. Hiiro, M. Isurugi and K. Yonezawa, *Japanese Patent* JP 06256634, 1994; *Chemical Abstracts*, 1995, 122, 21554g.
[149] K. Hashimoto, N. Kazumasa, H. Koji, I.Tomoki and K. Yonezawa, *Japanese Patent* JP 06256640, 1994; *Chemical Abstracts*, 1995, 122, 190149p.
[150] M. Zielecka and P. Rościszewski, *Patent PL* 164 929, 1994; *Chemical Abstracts*, 1996, 125, 277562d.
[151] *German Patent* 3 416 694, 1985; *Chemical Abstracts*, 1986, 104, 110815t.
[152] *Japanese Patent* JP 60 163 809, 1985; *Chemical Abstracts*, 1986, 104, 10402n.
[153] S.F. Reutsch, *Patent EP* 529 847, 1993; *Chemical Abstracts*, 1993, 118, 240473e.
[154] J. Chruściel and T. Graczyk, Preparation and properties of elastomeric silicone anti-adhesive layers on paper and parchment, *Polimery*, 2006, 51(3), 192. and referencies cited therein. In Polish.
[155] J. Chruściel, Crosslinking of polydimethylsiloxane-a,w-diols with polymethylhydrosiloxanes and some mechanical properties of resulting silicone rubbers, *Polimery*, 1999, 44(9), 586. and references cited therein. In Polish.

[156] M. Fejdyś-Kaczmarek, J. Chruściel and A. Miazga, Mechanical properties of silicone elastomers and rubbers cross-linked with polymethylhydrosiloxanes, *Polimery*, 2006, 51(1), 48. In Polish.
[157] J. Chojnowski, Silicon preceramic polymers, *Polimery*, 1987, 32(5), 167. In Polish.
[158] R. West, The polysilane high polymers, *Journal of Organometallic Chemistry*, 1986, 300(1–2), 327. and references cited therein.
[159] K. Matyjaszewski, Chapter 7: *Polimery nieorganiczne i organometaliczne*, in *Chemia polimerów*, vol. III, pp. 120–134, in Oficyna Wydawnicza Politechniki Warszawskiej, Z. Florjańczyk and S. Penczek (Eds.), Warszawa, 1998. In Polish.
[160] R.M. Laine and F. Babonneau, Preceramic polymer routes to silicon carbide, *Chemistry of Materials*, 1993, 5(3), 260.
[161] H. Finkelman, Liquid Crystalline Polymers, *Angewandte Chemie, International Edition in English*, 1987, 26(9), 816.
[162] I. Śledzińska, J. Sołtysiak and W. Stańczyk, Liquid crystal oligo(cyclosiloxanes), *Journal of Inorganic and Organometallic Polymers and Materials*, 1994, 4(2), 199.
[163] E. Białecka-Florjańczyk, J. Sołtysiak and W. Stańczyk, Tire effect of the length of the flexible spacer on the mesomorphic properties of oligo(cyclosiloxanes), *Polimery*, 1995, 40, 461. In Polish.
[164] V. Percec and C.S. Wang, *Journal of Macromoleculecular Science-Chemistry*, (a) Synthesis and characterization of poly(methylsiloxane)s containing 4-[s(-)-2-methyl-1-butoxy1]-4'-[p-(ω-alkan-l-yloxy)benzoyloxy]-α-methyl-stilbene side groups 1991, A28(8), 687. (b) Synthesis and characterization of polymethacrylates, polyacrylates, and poly(methylsiloxane)s containing 4-[s(–)-2-methyl-1-butoxy]-4'-(ω-alkanyl-1-OXY)-α-methylstilbene side groups, 1992, A29(2), 99.
[165] G.P.C. Chien, J.F. Kuo and C.Y. Chen, Synthesis and characterization of terminally carboxyl ethylene glycol monomethyl ethers-substituted side-chain liquid-crystalline polysiloxanes, *Journal of Polymer Science Part A: Polymer Chemistry*, 1993, 31(10), 2423.
[166] P.J. Madec and E. Maréchal, *Polymer Preprints*, 1993, 34(1), 814.
[167] S. Ruder, D.L. Embrey, S.D. Allen and G.E. Wnek, *Polymer Preprints*, 1998, 39(1), 294.
[168] P. Xie, Y. Wan, B. Zhou, J. Hou, D. Dai, Z. Li, D. Liu and R. Zhang, Synthesis and mesomorphic properties of fishbone-like liquid crystalline polysilsesquioxanes, 3. Fishbonelike, azo-based liquid crystalline polysilsesquioxane, *Macromoleculecular Chemistry and Physics*, 1996, 197 (2), 745.
[169] H. Uchida, Y. Kabe, K. Yoshino, A. Kawamata, T. Tsumuraya and S. Masamune, General strategy for the systematic synthesis of oligosiloxanes. Silicone dendrimer, *Journal of American Chemical Society*, 1990, 112(19), 7077.
[170] A. Morikawa, M. Kakimoto and Y. Imai, Synthesis and Characterization of New Polysiloxane Starburst Polymers, *Macromolecules*, 1991, 24(12), 3469.
[171] A.M. Caminade and J.P. Majoral, Main group elements-based dendrimers, *Main Group Chemistry News*, 1995, 3(1), 14. and referencies cited therein.
[172] G. de Marignan, D. Teyssié, S. Boileau, J. Malthête, C. Noël, Liquid crystalline side-chain polymers. 1. Polysiloxanes with a carbonate group in the spacer, *Polymer*, 29, 7, 1318–1322.
[173] M.S. Ahsan, S. Sasaki and Y. Kawakami, Liquid crystalline compounds with various rigid siloxane frameworks, *Reactive and Functional Polymers*, 2007, 67(11), 1200.
[174] A.W. Van Der Made and P.W.N.M. Van Leewen, Silane dendrimers, *Journal of Chemical Society, Chemical Communications*, 1992, 0, 1400.
[175] A. Morikawa, M. Kakimoto and Y. Imai, Convergent synthesis of siloxane starburst dendrons and dendrimers via hydrosilylation, *Macromolecules*, 1992, 25(12), 3247.
[176] L.L. Zhou and J. Roovers, *Macromolecules*, 1993, 26(5), 963.
[177] S. Rubinsztajn and J. Stein, A silane functionalized styrene monomer and its polymerization, *Journal of Inorganic and Organometallic Polymers and Materials*, 1995, 5(1), 43.

[178] J.S. Shim, S. Asthana, N. Omura and J.P. Kennedy, *Polymer Preprints*, 1998, 39(1), 196.
[179] G. Kickelbick, P.J. Miller and K. Matyjaszewski, *Polymer Preprints*, 1998, 39(1), 284.
[180] R. Kollefrath, O. Nuyken and B. Voit, *Polymer Preprints*, 1996, 37(2), 407.
[181] A. Bachrach and A. Zilkha, *European Polymer Journal*, 1984, 20(5), 493.
[182] K. Ruehlmann and I. Jansen, *Chapter: Olefin-Cyclosiloxan-Copolymere mit Stabilisatorgruppen, in Wybrane zagadnienie chemii krzemu, B. Marciniec (Ed.)*, A. Mickiewicz Univ. Press, Poznań, 1985, 37–58. In German.
[183] M. Zielecka, K. Cyruchin and D. Sobkowiak, *Patent PL* 177972, 2000.
[184] S. Boileau, L. Bouteiller, R.B. Khalifa, Y. Liang and D. Teyssie, *Polymer Preprints*, 1998, 39(1), 457.
[185] B. Oh, D. Vissers, Z. Zhang, R. West, H. Tsukamoto and K. Amine, New interpenetrating network type poly(siloxane-g-ethylene oxide) polymer electrolyte for lithium battery, *Journal of Power Sources*, 2007, 119–121, 442.
[186] P. Hron, J. Šlechtová, K. Smetana, B. Dvořankova and P. Lopour, Silicone rubber-hydrogel composites as polymeric biomaterials. IX. Composites containing powdery polyacrylamide hydrogel, *Biomaterials*, 1997, 18(15), 1069.
[187] L.S. Miller, A.L. Rhoden, N. Byrne, J. Heptinstall and D.J. Walton, Polysiloxane-based biomembranes, *Materials Science and Engineering*, 1995, C3(15), 187.
[188] M.A. Brook, J. Jiang, P. Heritage, B. Underdown and M.R. Mc Dermott, Silicone-modified starch/protein microparticles: Protecting biopolymers with a hydrophobic coating, *Colloids Surfaces B: Biointerfaces*, 1997, 9(6), 285.
[189] M. Manzano, D. Arcos, M.R. Delgado, E. Ruiz, F.J. Gil and M. Vallet-Regi, Bioactive star gels, *Chemistry of Materials*, 2006, 18(24), 5696.
[190] N.A. Chekina, V.N. Pavlyuchenko, V.F. Danilichev, N.A. Ushakov, S.A. Novikov and S.S. Ivanchev, A new polymeric silicone hydrogel for medical applications: synthesis and properties, *Polymers for Advanced Technologies*, 2006, 17(11–12), 872.
[191] G. Greber and S. Jäger, Über oligomere Siliciumverbindungen mit funktionellen Gruppen. 12. Mitt. Über Herstellung von oligomeren siliciumorganischen Diolen und Diaminen und ihre Umsetzungen mit organischen Diisocyanaten, *Die Makromolekulare Chemie*, 1962, 57(1), 150.
[192] A. Voliotis, C.M. Liauw, G.C. Lees, R.N. Rothon and B. Thomas, *Macromolecular Symposia*, 2003, 194(1), 335.
[193] V. Tertykh, V. Yanishpolskii and Y. Bolbukh, *Macromolecular Symposia*, 2003, 194(1), 141.
[194] T. Cosgrave, C.A. Prestige and B. Vincent, Chemisorption of linear and cyclic polymethylsiloxanes on alumina studied by Fourier-transform infrared spectroscopy, *Journal of Chemical Society, Faraday Transactions*, 1990, 86(9), 1377.
[195] K. Reihs, R.A. Colom, S. Gleditzsch, M. Deimel, B. Hagenhoff and A. Benninghoven, Chemisorption of poly(methylhydrogensiloxane) on oxide surfaces: A quantitative investigation using static SIMS, *Applied Surface Science*, 1995, 84(1), 107.
[196] P.M. Greco, F.C. Stedile and J.H. Zimnoch Dos Santos, Influence of PMHS loading on the silica surface, on catalyst activity and on properties of resulting polymers, *Journal of Molecular Catalysis A: Chemical*, 2003, 197(1–2), 233.
[197] J. Bik, W.M. Rzymski and J. Chruściel, Modification of hydrogenated styrene-butadiene rubber with polymethylhydrogen siloxane, *Elastomery*, 2003, 7(6), 9.
[198] J. Chruściel, G. Janowska, P. Rybiński and L. Ślusarski, Effect of the top layer modification of polymers on their thermostability and flammability, *Journal of Thermal Analysis and Calorimetry*, 2006, 84(2), 339.
[199] D. Bieliński, P. Głąb and J. Chruściel, Modification of styrene-butadiene rubber with polymethylsiloxanes. Part I. Interactions of the modifier with rubber, *Polimery*, 2007, 52(3), 195. In Polish.

[200] D. Bieliński, P. Głąb and J. Chruściel, Modification of styrene-butadiene rubber with polymethylsiloxanes. Part II. Systems filled with silica, *Polimery*, 2007, 52(4), 268. In Polish.
[201] J.F. Miravet and J.M.J. Frechet, *Polymeric materials science and engineering, fall meeting, sept. 8–11, 1997, Las Vegas (Nevada), Proceeding of the American Chemical Society, Division of PMSE*, 1997, 77, 141.
[202] Y. Kobayashi, *Japanese Patent* 10 087 530, 1998; *Chemical Abstracts*, 1998, 128, 257190.
[203] R.M. Lopez and G.C. Fu, A mild, convenient, and inexpensive method for converting imines into amines: Tin-catalyzed reduction with polymethylhydrosiloxane (PMHS), *Tetrahedron*, 1997, 53(48), 16349.
[204] J. Lipowitz and S.A. Bowman, Use of polymethylhydrosiloxane as a selective, neutral reducing agent for aldehydes, ketones, olefins, and aromatic nitro compounds, *Journal of Organic Chemistry*, 1973, 38(1), 162.
[205] H. Fritzsche. U. Hasserodt and F. Korte, Reduktion organischer Verbindungen des fünfwertigen Phosphors zu Phosphinen, I. Reduktion tertiärer Phosphinoxyde zu tertiären Phosphinen mit Silanen, *Berichte*, 1964, 97(7), 1988.
[206] H. Fritzsche, U. Hasserodt and F. Korte, Reduktion organischer Verbindungen des fünfwertigen Phosphors zu Phosphinen, II. Reduktion tertiärer Phosphinoxyde zu tertiären Phosphinen mit Trichlorsilan, *Berichte*, 1965, 98(1), 171.
[207] S.J. Clarson and J.A. Semlyen, *Siloxane Polymers*, Prentice Hall, New Jersey, USA, 1993.
[208] R.D. Archer, *Inorganic and Organometallic Polymers*, John Wiley & Sons, Inc., New York – Chichester – Weinheim – Brisbane – Singapore – Toronto, 2001.
[209] V. Chandrasekhar, Chapter 6: *Polysiloxanes*, in *Inorganic and Organometallic Polymers*, Springer-Verlag, Berlin-Heidelberg, 2005, 219–248.
[210] J.J. Kennan, *Siloxane Polymers*, S.J. Clarson and J.A. Semlyen (Eds.), Prentice Hall, Englewood Clifs, 1993, 72–134.
[211] M.J. Ziemelis and J.C. Saam, Sequence distribution in poly(dimethylsiloxane-co-methylvinylsiloxanes), *Macromolecules*, 1989, 22(5), 2111.
[212] K.R. Rózga-Wijas, J. Chojnowski, T. Zundel and S. Boileau, Controlled Synthesis of Siloxane Copolymers Having an Organosulfur Group by Polymerization of Cyclotrisiloxanes with Mixed Units, *Macromolecules*, 1996, 29(8), 2711.
[213] W.P. Weber and G. Cai, Anionic and Cationic Ring-Opening Polymerization of 2,2-Divinyl-4,4,6,6-tetramethylcyclotrisiloxane. Synthesis of Regular Poly[2,2-bis(perfluoroalkyl-1‘, *H*1‘*H*,2‘*H*,2‘*H*-dimethylsilylethyl)-4,4,6,6- tetramethyltrisiloxanes] by Chemical Modification of Regular Poly(2,2-divinyl-4,4,6,6-tetramethyltrisiloxane), *Macromolecules*, 2001, 34(13), 4355.
[214] J. Bauer, N. Hüsing and G. Kickelbick, Preparation of functionalized block copolymers based on a polysiloxane backbone by anionic ring-opening polymerization, *Journal of Polymer Science, Part A: Polymer Chemistry*, 2002, 40(10), 1539.
[215] K.-L. Xie, K. Hu and Y. Chen, Synthesis of Amphiphilic Polysiloxanes Modified with Multi-cationic Groups to Improve Wettability of Polyester Materials, *Iranian Polymer Journal*, 2010, 19(6), 447.
[216] X. Wang, J. Wu, Y. Li, C. Zhou and C. Xu, Pyrolysis kinetics and pathway of polysiloxane conversion to an amorphous SiOC ceramic, *Journal of Thermal Analysis and Calorimetry*, 2014, 115(1), 55.
[217] J. Wu, Y. Li, L. Chen, Z. Zhang, D. Wang and C. Xu, Simple fabrication of micro/nano-porous SiOC foam from polysiloxane, *Journal of Materials Chemistry*, 2012, 22(14), 6542.
[218] B.A. Cavic and M. Thompson, Protein adsorption to organosiloxane surfaces studied by acoustic wave sensor, *Analyst*, 1998, 123(10), 2191.
[219] T. Gunji, T. Shioda, K. Tsuchihira, H. Seki, T. Kajiwara and Y. Abe, Preparation and properties of polyhedral oligomeric silsesquioxane–polysiloxane copolymers, *Applied Organometallic Chemistry*, 2010, 24(8), 545.

[220] T.C. Loh, C.M. Ng, R.N. Kumar, H. Ismail and Z. Ahmad, Improvement of thermal ageing and transparency of methacrylate based poly(siloxane–silsesquioxane) for optoelectronic application, *Journal of Applied Polymer Science*, 2017, 134(37), 45285.
[221] H. Zhou, Y. Su, X. Chen, J. Luo and Y. Wan, High-performance PDMS membranes for pervaporative removal of VOCs from water: The role of alkyl grafting, *Journal of Applied Polymer Science*, 2016, 133(31), 43700.
[222] D. Lin, Q. Zhao and M. Yan, Surface modification of polydimethylsiloxane microfluidic chips by polyamidoamine dendrimers for amino acid separation, *Journal of Applied Polymer Science*, 2016, 133(25), 43580.
[223] K.D. Belfield, X.Z. Lin and I. Cabasso, Hydrosilylation of α, β-unsaturated esters: Application for the synthesis of functional polysiloxane graft copolymers and macromonomers, *Journal of Polymer Science, Part A: Polymer Chemistry*, 1991, 29(7), 1073.
[224] X.W. He, J.M. Widmaier, J.E. Herz and M.C. Meyer, Polydimethylsiloxane/poly(methyl methacrylate) interpenetrating polymer networks: 2. Synthesis and properties, *Polymer*, 1992, 33(4), 866.
[225] J.-F. Stumbé, F. Calderara and G. Riess, Synthesis of poly(methyl methacrylate)-g-poly (dimethyl siloxane) graft copolymers via a miniemulsion process, *Polymer Bulletin*, 2001, 47 (3–4), 277.
[226] H. Inoue and K. Matsukawa, Synthesis and Gas Permeability of Cyclotetrasiloxane-Containing Methacrylate Copolymers, *Journal of Macromolecular Science – Pure and Applied Chemistry*, 1992, A29(6), 415.
[227] J. Kurjata, J. Chojnowski, C.-T. Yeoh, N.A.A. Rossi and S.J. Holder, Synthesis of poly [dimethylsiloxane-block-oligo(ethylene glycol) methyl ether methacrylate]: an amphiphilic copolymer with a comb-like block, *Polymer*, 2004, 45(18), 6111.
[228] C.P. Xing, T. He, Y.J. Peng and R.H. Li, Synthesis and characterization of poly(methyl methacrylate)/polysiloxane composites and their coating properties, *Journal of Applied Polymer Science*, 2018, 135(51), 46358.
[229] Y. Sun and W. Liu, Synthesis and properties of triblock copolymers containing PDMS via AGET ATRP, *Polymer Bulletin*, 2012, 68(7), 1815.
[230] M. Janata, A. Sikora, P. Látalová, E. Čadová, V. Raus, L.r Matějka and P. Vlček, Synthesis of defined polyhedral oligosilsesquioxane-containing diblock and triblock methacrylate copolymers by atom transfer radical polymerization, *Journal of Applied Polymer Science*, 2013, 128(6), 4294.
[231] B.J. Kokko, Silicone acrylates through the hydrosilation of polyacryloyloxy functional monomers with copolymers of dimethyl and hydrogen methylsiloxanes, *Journal of Applied Polymer Science*, 1993, 47(7), 1309.
[232] S.-Q. Wei, Y.-P. Bai and L. Shao, A novel approach to graft acrylates onto commercial silicones for release film fabrications by two-step emulsion synthesis, *European Polymer Journal*, 2008, 44(8), 2728.
[233] H.H. Wang, X.R. Li, G.Q. Fei and J. Mou, Synthesis, morphology and rheology of core-shell silicone acrylic emulsion stabilized with polymerisable surfactant, *eXPRESS Polymer Letters*, 2010, 4(11), 670.
[234] S. Bas and M.D. Soucek, Comparison of Approaches to Prepare Polysiloxane-Functionalized Acrylic Latexes, *Silicon*, 2013, 5(2), 139.
[235] D.R. Weyenberg, D.E. Findlay, J. Cekada Jr. and A.E. Bey, Anionic emulsion polymerization of siloxanes, *Journal of Polymer Science, Part C*, 1969, 27(1), 27.
[236] I. Ofat and J. Kozakiewicz, Modification of epoxy-polyester and polyester powder coatings with silicone-acrylic nanopowders – effect on surface properties of coatings, *Polimery*, 2014, 59(9), 643.

[237] H. Lei, D. He, Y. Guo, Y. Tang and Y. Lu, Modification of a fluorine–silicone acrylic resin with a free-radical-catching agent, *Journal of Applied Polymer Science*, 2018, 135(25), 46385.
[238] M.A. Semsarzadeh and S. Amiri, Synthesis and characterization of PVAc-b-PDMS-b-PVAc Triblock copolymers by atom transfer radical polymerization initiated by PDMS macroinitiator, *Journal of Inorganic and Organometallic Polymers and Materials*, 2013, 23(3), 553.
[239] J.C. Saam, D.J. Gordon and S. Lindley, Block Copolymers of Polydimethylsiloxane and Polystyrene, *Macromolecules*, 1970, 3, 1(1).
[240] M.L. Wadley and K.A. Cavicchi, Synthesis of Polydimethylsiloxane-Containing Block Copolymers via Reversible Addition Fragmentation Chain Transfer (RAFT) Polymerization, *Journal of Applied Polymer Science*, 2010, 115(2), 635.
[241] T. Nose, Coexistence curves of polystyrene/poly(dimethylsiloxane) blends, *Polymer*, 1995, 36 (11), 2243.
[242] S. Ndoni, M.E. Vigild and R.H. Berg, Nanoporous Materials with Spherical and Gyroid Cavities Created by Quantitative Etching of Polydimethylsiloxane in Polystyrene–Polydimethylsiloxane Block Copolymers, *Journal of the American Chemical Society*, 2003, 125(44), 13366.
[243] Y.S. Jung and C.A. Ross, Orientation-Controlled Self-Assembled Nanolithography Using a Polystyrene–Polydimethylsiloxane Block Copolymer, *Nano Letters*, 2007, 7(7), 2046.
[244] P. Choi, P. Fu and L.J. Guo, Siloxane Copolymers for Nanoimprint Lithography, *Advanced Functional Materials*, 2007, 17(1), 65.
[245] H. Ahn, K.J. Lee and W.R. Childs, Micron and submicron patterning of polydimethylsiloxane resists on electronic materials by decal transfer lithography and reactive ion-beam etching: Application to the fabrication of high-mobility, thin-film transistors, *Journal of Applied Physics*, 2006, 100(8), 084907.
[246] M.A. Hartney, A.E. Novembre and F.S. Bates, Block Copolymers as Bilevel Resists, *Vacuum Science & Technology B: Microelectronics and Nanometer Structures Processing, Measurement, and Phenomena*, 1985, 3(5), 1346.
[247] J.M. DeSimone, G.A. York, J.E. McGrath, A.S. Gozdz and M.J. Bowden, Synthesis, bulk, surface and microlithographic characterization of poly(1-butene sulfone)-g-poly(dimethylsiloxane), *Macromolecules*, 1991, 24(19), 5330.
[248] Y.S. Jung and C.A. Ross, Solvent-Vapor-Induced Tunability of Self-Assembled Block Copolymer Patterns, *Advanced Materials*, 2009, 21(24), 2540.
[249] G. Kim, S. Swier, H.-J. Lee and C. Wang, Morphology Control of Poly(styrene-block-dimethylsiloxane) by Simple Blending with Trimethylsilylated Silicate Nanoparticles, *Macromolecules*, 2016, 49, 7370.
[250] Y. Jiang, H. Shi, M. Cai, Y. Liang, B. Li, H. Zhang and R. Song, Synthesis and Characterization of PS-b-PDMS-b-PS Triblock Copolymer, *Journal of Applied Polymer Science*, 2013, 129(1), 247.
[251] O. Bianchi, L.G. Barbosa, G. Machado, L.B. Canto, R.S. Mauler and R.V.B. Oliveira, Reactive melt blending of PS-POSS hybrid nanocomposites, *Journal of Applied Polymer Science*, 2013, 128(1), 811.
[252] C. Hu, Y. Zhang, W. Li and Q. Gu, A Novel Method for Preparing Composites of PDMS/PS Core–Shell Emulsion and Polystyrene, *Journal of Applied Polymer Science*, 2007, 105(5), 2571.
[253] Y. Chang, C.-Y. Chen and W.-C. Chen, Poly(methylsilsesquioxane)/Amphiphilic Block Copolymer Hybrids and Their Porous Derivatives: Poly(styrene-*block*acrylic acid) and Poly (styrene-*block*-3-trimethoxysilylpropyl methacrylate), *Journal of Polymer Science, Part B: Polymer Physics*, 2004, 42(24), 4466.
[254] H. Hussain, B.H. Tan, K.Y. Mya, Y. Liu, C.B. He and T.P. Davis, Synthesis, micelle formation, and bulk properties of poly(ethylene glycol)-b-poly(pentafluorostyrene)-g-polyhedral

oligomeric silsesquioxane amphiphilic hybrid copolymers, *Journal of Polymer Science, Part A; Polymer Chemistry*, 2010, 48(1), 152.
[255] Z. Huang, J. Jiang, L. Shi, X. Wang, G. Xue, L. Li, Z. Shen and D. Zhou, Dependences of Confining Size and Interfacial Curvature on the Glass Transition of Polydimethylsiloxane in Self-Assembled Block Copolymers, *Macromolecular Chemistry and Physics*, 2018, 219(3), 1700518.
[256] I.S. Stefanović, D. Godevac, M. Špírková, P. Jovančić, V. Tešević, V. Milačić and M.V. Pergal, Impact of the poly(propylene oxide)-b-poly(dimethylsiloxane)-b-poly(propylene oxide) macrodiols on the surface-related properties of polyurethane copolymers, *Hemijska Industrija*, 2016, 70(6), 725.
[257] D. Liu, G.-M. Wu and Z.-W. Kong, Preparation and characterization of a polydimethylsiloxane-modified, epoxy-resin-based polyol dispersion and its crosslinked films, *Journal of Applied Polymer Science*, 2017, 134(1), 44342.
[258] R. Aeinehvand, S.K. Rahimi and M.H. Kafshgari, Chapter: *Polyurethanes and silicone polyurethane copolymers* in *biomedical applications*, in *Encyclopedia of Biomedical Polymers and Polymeric Biomaterials*, M. Mishra (Ed.), Taylor and Francis Group, 2016. DOI: 10.1081/E-EBPP-120050583.
[259] H. Lei, Y. Luo, Z. Ge, X. Li and S. Wang, Preparation and properties of waterborne siloxane-containing polyurethane for moisture permeable textile coating, *e-Polymers*, 2010, 069.
[260] F. Joki-Korpela and T.T. Pakkanen, Incorporation of polydimethylsiloxane into polyurethanes and characterization of copolymers, *European Polymer Journal*, 2011, 47(8), 1694.
[261] Z. Dai, K. Yang and Q. Dong, Synthesis and characterization of hydroxy-terminated polyetherpolydimethylsiloxane-polyether (PE-PDMS-PE) triblock oligomers and their use in the preparation of thermoplastic polyurethanes, *Journal of Applied Polymer Science*, 2015, 132(38), 42521.
[262] A. Ekin, D.C. Webster, J.W. Daniels, S.J. Stafslien, F. Casse, J.A. Callow and M.E. Callow, Synthesis, formulation, and characterization of siloxane–polyurethane coatings for underwater marine applications using combinatorial high-throughput experimentation, *Journal of Coatings Technology and Research*, 2007, 4(4), 435.
[263] M. Balaban, V. Antic, M. Pergal, D. Godjevac, I. Francolini, A. Martinelli, J. Rogan and J. Djonlagic, Influence of the chemical structure of poly(urea-urethane-siloxane)s on their morphological, surface and thermal properties, *Polymer Bulletin*, 2013, 70(9), 2493.
[264] B. Król, K. Pielichowska, P. Król and P. Chmielarz, Polyurethane cationomers modified by polysiloxane, *Polymers for Advanced Technology*, 2017, 28(11), 1366.
[265] S. Babanzadeh, S. Mehdipour-Ataei and A.R. Mahjoub, Preparation and Characterization of Novel Polyimide/SiO2 Nano-hybrid Films by In Situ Polymerization, *Journal of Inorganic and Organometallic Polymers and Materials*, 2012, 22(6), 1404.
[266] C. Zhou, A. Gu, G. Liang, L. Ji and L. Yuan, Preparation and properties of novel resins based on cyanate ester and hyperbranched polysiloxane, *Journal of Polymer Research*, 2011, 18(1), 139.
[267] X.F. Lei, Y.Chen, H.P. Zhang, X.J. Li, P. Yao and Q.Y. Zhang, Space Survivable Polyimides with Excellent Optical Transparency and Self-Healing Properties Derived from Hyperbranched Polysiloxane, *ACS Appllied Materials and Interfaces*, 2013, 5(20), 10207.
[268] X. Ma, X. Ma, X. Qiu, F. Liu, R. Jin, C. Kang, H. Guo and L. Gao, Preparation and Properties of Imidazole-containing Polyimide/Silica Hybrid Films, *Chemical Research in Chinese Universities*, 2014, 30(6), 1047.
[269] B. Oktay, R.D. Toker and N. Kayaman-Apohan, Superhydrophobic behavior of polyimide–siloxane mats produced by electrospinning, *Polymer Bulletin*, 2015, 72(11), 2831.
[270] H.-T. Li, M.-S. Lin, H.-R. Chuang and M.-W. Wang, Siloxane- and Imide-modified Epoxy Resin Cured with Siloxane-containing Dianhydride, *Journal of Polymer Research*, 2005, 12(5), 385.

[271] J. Yang, Y. Cheng, Y. Jin and F. Xiao, Synthesis and properties of novel benzocyclobutene-functionalized siloxane thermosets, *Polymer International*, 2013, 62(12), 1684.
[272] L. Feng and J.O. Iroh, Polyimide-*b*-polysiloxane Copolymers: Synthesis and Properties, *Journal of Inorganic and Organometallic Polymers and Materials*, 2013, 23(3), 477.
[273] Y. Lei, Y. Shuc and J. Huo, Synthesis and characterization of novel borosiloxane-containing aromatic copolyimides with excellent thermal stability and low coefficient of thermal expansion, *Polymer International*, 2017, 66(8), 1173.
[274] Y.-W. Wu, M.-F. Ye, W.-C. Zhang and R.-J. Yang, Polyimide aerogels crosslinked through cyclic ladder-like and cage polyamine functionalized polysilsesquioxanes, *Journal of Applied Polymer Science*, 2017, 134(37), 45296.
[275] J. Qiu, S. Xu, N. Liu, K. Wei, L. Li and S. Zheng, Organic–inorganic polyimide nanocomposites containing a tetrafunc tional polyhedral oligomeric silsesquioxane amine: synthesis, morphology and thermomechanical properties, *Polymer International*, 2018, 67(3), 301.
[276] J. Lipowitz, J.A. Rabe and I.M. Carr, Chapter 12: NMR Characterization of Polymethyldisilylazane. A Precursor to Si-C-N-O Ceramics, in Inorganic and Organometallic Polymers, edited by M. Zeldin, K.J. Wynne and H.R. Allcock, *ACS Symposium Series*, 1988, vol. 360, p. 156–165.
[277] A. Lukacs and G.J. Knasiak (KION Corporation), *Patent WO* 02090108, 2002.
[278] M. Bouquey, C. Brochon, S. Bruzaud, A.F. Mingotaud, M. Schappacher and A. Soum, Ring-opening polymerization of nitrogen-containing cyclic organosilicon monomers, *Journal of Organometallic Chemistry*, 1996, 521(1–2), 21.
[279] M. Bouquey and A. Soum, Anionic Ring-Opening Copolymerization of Cyclodisilazanes, *Macromolecular Chemistry and Physics*, 2001, 202(7), 1232.
[280] C.R. Krüger and E.G. Rochow, Mixed Cyclic and Linear Compounds with Siloxane and Silazane Linkages, *Inorganic Chemistry*, 1963, 2(6), 1295.
[281] J.G. Murray, (a) *French Patent* 1 357 957, 1964; (b) *US Patent* 3 239 550, 1966.
[282] J.G. Murray, *US Patent* 3 271 261, 1966.
[283] Z. Lasocki, B. Dejak, J. Kulpiński, E. Leśniak, S. Piechucki and M. Witekowa, Chapter 13: Silicon-Nitrogen-Containing Rings and Polymers, in *Inorganic and Organometallic Polymers*, in Inorganic and organometallic polymers, M. Zeldin, K.J. Wynne and H.R. Allcock (Eds.), *ACS Symposium Series*, 1988, 360, 166.
[284] Z. Lasocki and M. Witekowa, Polymerization of N-Phenylcyclosilazoxanes, *Journal of Macromolecular Science, Part A: Pure and Applied Chemistry*, 1977, A11(3), 457.
[285] Z. Lasocki and M. Witekowa, Synthesis of N-phenylcyclosilazoxanes, *Synthesis and Reactivity in Inorganic and Metal-Organic Chemistry*, 1974, 4(3), 231.
[286] J. Kulpiński, Z. Lasocki and E. Leśniak, *Patent PL* 145905, 1989.
[287] C. Brochon, A.F. Mingotaud, M. Schappacher and A. Soum, Equilibrium Anionic Ring-Opening Polymerization of a Six-Membered Cyclosiloxazane, *Macromolecules*, 2007, 40(10), 3547.
[288] N. Zhou, Z.-J. Zhang, C.-H. Xu and Z.-M. Xie, *Chinese Journal of Polymer Science*, 2000, 18(6), 551.
[289] N. Zhou, Z.-J. Zhang, C.-H. Xu and Z.-M. Xie, Synthesis and Properties of Polysiloxane Containing N,N-Bis(diphenylsilyl)tetraphenylcyclodisilazane, *Journal of Applied Polymer Science*, 2001, 82 (4), 929.
[290] M. Rodriguez-Baeza, A.C. Neira and C.J. Aguilera, Synthesis of poly[(methylsi1oxane)-co-(dimethylsilazane)] copolymers as precursors of ceramic materials, *Polymer Bulletin*, 2003, 50(1–2), 25.
[291] T. Gunji, Y. Taniguchi and Y. Abe, *Journal of Ceramic Society of Japan*, 2006, 114(6), 492.
[292] T.D.H. Nguyen, F.-X. Perrin and D.L. Nguyen, New hybrid materials based on poly (ethyleneoxide)-grafted polysilazane by hydrosilylation and their anti-fouling activities, *Beilstein Journal of Nanotechnology*, 2013, 4(1), 671.

[293] J.S. Riffle, W.P. Steckle, Jr., K.A. White and R.S. Ward, *Polymer Preprints*, 1985, 26, 251.
[294] J.E. McGrath, D.W. Dwight, J.S. Riffle, T.F. Davidson, D.C. Webster and R. Viswanathan, *Polymer Preprints*, 1979, 20(2), 528.
[295] D.C. Webster, P.J. Andolino, J.S. Riffle, F.L. Keohan and J.E. McGrath, *Polymer Preprints*, 1983, 24(1), 161.
[296] P.J. Andolino-Brandt, D.C. Webster and J.E. McGrath, *Polymer Preprints*, 1984, 25(2), 91.
[297] J.S. Riffle, A.K. Banthia, D.C. Webster and J.E. McGrath, *Organic Coatings and Plastics Preprints*, 1980, 42(1), 122.
[298] M.S.I. Mollah, Y.-D. Kwon, M.M. Islam, D.-W. Seo, H.-H. Jang, Y.-D. Lim, D.-K. Lee and W.-G. Kim, Synthesis and characterization of polycarbonates containing terminal and chain interior siloxane, *Polymer Bulletin*, 2012, 68(6), 1551.
[299] I. Majoros, T.M. Marsalkó and J.P. Kennedy, Synthesis and characterization of novel well-defined stars consisting of eight polyisobutylene arms emanating from an octa(dimethylsiloxy)octasilsesquioxane core *Polymer Bulletin*, 1997, 38(1), 15.
[300] Y.K. Kim, D.B. Bourrie and O.R. Pierce, Polycycloalkylene-siloxane polymers: Synthesis and thermal study, *Journal of Polymer Science, Polymer Chemistry Edition*, 1978, 16(2), 483.
[301] E. Markovic, M. Ginic-Markovic, S. Clarke, J. Matisons, M. Hussain and G.P. Simon, Poly(ethylene glycol)-Octafunctionalized Polyhedral Oligomeric Silsesquioxane: Synthesis and Thermal Analysis, *Macromolecules*, 2007, 40(8), 2694.
[302] Y.-T. Liao, Y.-C. Lin and S.-W. Kuo, Highly Thermally Stable, Transparent, and Flexible Polybenzoxazine Nanocomposites by Combination of Double-Decker-Shaped Polyhedral Silsesquioxanes and Polydimethylsiloxane, *Macromolecules*, 2017, 50(15), 5739.
[303] S. Jothibasu, A. Chandramohan, A.A. Kumar and M. Alagar, Polyhedral oligomeric silsesquioxane (POSS) reinforced-unsaturated polyester hybrid nanocomposites: Thermal, thermomechanical and morphological properties, *Journal of Macromolecular Science, Part A: Pure and Applied Chemistry*, 2018. DOI: https://doi.org/10.1080/10601325.2018.1453264.
[304] Q. Jiang, J. Hao, Y. Wei and J. Mu, Synthesis and characterization of block and random POSS/fluorinated PAES tricopolymers, *Polymers for Advanced Technology*, 2017, 28(6), 658.
[305] Y. Wei, Q. Jiang, J. Hao and J. Mu, Y. Wei, Q. Jiang, J. Hao and J. Mu, POSS-based poly(aryl ether sulfone)s random terpolymer linked POSS to the main chain: Effect of chemical structure and POSS content on properties, *Polymers for Advanced Technology*, 2017, 28(10), 516.
[306] H. Ritter, B. Knudsen and V. Durnev, Linkage of α-cyclodextrin-terminated poly(dimethylsiloxanes) by inclusion of quasi bifunctional ferrocene, *Beilstein Journal of Organic Chemistry*, 2013, 9, 1278.
[307] C.E. Boott, D.J. Lunn and I. Manners, Versatile and Controlled Functionalization of Polyferrocenylsilane-b-Polyvinylsiloxane Block Copolymers Using a N-hydroxysuccinimidyl Ester Strategy, *Journal of Polymer Science, Part A: Polymer Chemistry*, 2016, 54(2), 245.
[308] Y. Sun and C. He, Biodegradable "Core–Shell" Rubber Nanoparticles and Their Toughening of Poly(lactides), *Macromolecules*, 2013, 46(24), 9625.
[309] N. Etteyeb, V. Jaouen, N. Steunou, N. Gharbi and T. Coradin, Elaboration, Stability and Enzymatic Degradation of Hydroxypropylcellulose/Polysiloxane Biocomposite Membranes, *Silicon*, 2012, 4(1), 79.
[310] S.M. Rezaei and Z.A.M. Ishak, Grafting of collagen onto interpenetrating polimer networks of poly(2-hydroxyethyl methacrylate) and poly(dimethyl siloxane) polymer films for biomedical applications, *Express Polymers Letters*, 2014, 8(1), 39.
[311] P. Pietras, R. Przekop and H. Maciejewski, New approach to preparation of gelatine/SiO2 hybrid systems by the sol-gel process, *Ceramics – Silikáty*, 2013, 57(1), 58.

[312] P. Wojciechowska, P. Pietras and H. Maciejewski, Synthesis, Characterization, and Thermal Properties of Organic–Inorganic Hybrids Based on Gelatin and Organomodified Silicones, *Advances in Polymer Technology*, 2014, 33(51), 21459.
[313] C. Dong, Z. Lu, P. Zhu, F. Zhang and X. Zhang, Combustion behaviors of cotton fabrics treated by a novel guanidyl- and phosphorus-containing polysiloxane flame retardant, *Journal of Thermal Analysis and Calorimetry*, 2015, 119(1), 349.
[314] C. Dong, Z. Lu and F. Zhang, Preparation and properties of cotton fabrics treated with a novel guanidyl- andphosphorus-containing polysiloxane antimicrobial and flame retardant, *Materials Letters*, 2015, 142(1), 35.
[315] G.B. Goodwin and M.E. Kenney, Chapter 18: *A new approach to the synthesis of alkyl silicates and organosiloxanes*, in *Inorganic and Organometallic Polymers*, Vol. 360, M. Zeldin, K.J. Wynne and H.R. Allcock (Eds.), *ACS Symposium Series*, 1988, 238–248.

Chapter 8
Ceramics derived from silicon polymers

The relatively new field of *preceramic polymer science* has been developed mainly within a recent three decades. The term *preceramic polymer* (resin) concerns oligomeric and polymeric materials which contain both organic and inorganic molecular segments and can be thermally transformed into inorganic ceramics during high-temperature pyrolysis or in fire conditions. Especially silicon and nitrogen atoms, as inorganic ingredients, can interact with organic carbon, liberating gaseous by-products. Cross-linking, pyrolysis, and crystallization processes lead to ceramic materials with unique properties [1].

Si-based preceramic polymers mainly serve as precursors for the fabrication of advanced (*high-tech*) ceramics, which are often called as *polymer-derived ceramics* (PDCs). A variety of functional properties of PDCs was achieved, especially in the recent 20 years. The highly interdisciplinary scientific and technological development of PDCs has been conducted by thousands of chemists, physicists, materials scientists, engineers, and mineralogists around the world as the background of micro- and nanoscience and technology. The investigations of transformation processes of polymers into ceramics, decomposition, crystallization, phase separation, and creep processes enabled significant technological progress in the ceramic science and technology, for instance in the development of ceramic special coatings and fibers or ceramics stable at ultrahigh temperatures (up to 2,000 °C). The preceramic polymers have been used as reactive binders in production processes of *high-tech* technical ceramics. They allowed the formation of ordered pores in the mesoporous materials and have been used for joining advanced ceramic components, and have been processed into macroporous or bulk materials. Novel studies of PDCs provided insights into their structure at the nanoscale level and much better basic understanding of the various unique and useful features of PDCs, mainly related to their high chemical durability, semiconducting properties, or high creep resistance. Thus, possible fields of applications of PDCs have been significantly extended in different engineering fields, including hard materials, high-temperature-resistant materials (automotive, aerospace, energy materials, etc.), functional materials in electrical engineering, in micro- and nano-electronics, or chemical engineering (catalyst support, food- and biotechnology, etc.) [2–8].

Production and availability of the polymeric preceramic precursors substantially increased over the past two decades. The PDCs have special microstructural features. Unusual materials properties of the PDCs result from their unique nano-sized microstructure which makes preceramic polymers of great interest in many disciplines. Usually fabrication of ceramics from preceramic polymers requires special processing technologies. An advantage of the special characteristics of the preceramic polymers leads to their numerous new applications [2–9].

https://doi.org/10.1515/9783110643671-008

Different structural and functional properties of silicon-containing ceramic nanocomposites and preparative strategies leading to polymer-derived ceramic nanocomposites (PDC-NCs) were reviewed by Ionescu et al. [10, 11].

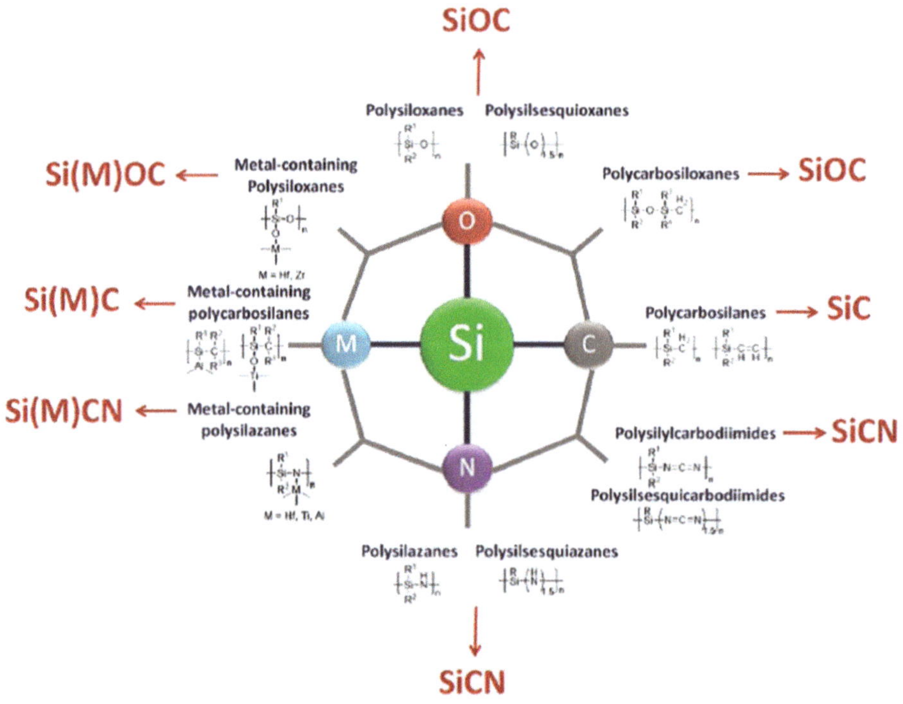

Figure 8.1: Organosilicon polymers applied as precursors for ceramics. Reprinted with permission from [11] under the terms and conditions of the Creative Commons Attribution license (CC BY 4.0 rule). Copyright 2015, MDPI, Switzerland.

Perspective applications of the PDC-NCs may include materials for heterogeneous catalysis, highly thermally stable materials, anode materials for secondary ion batteries, membranes for separation of hot gases, and nanomaterials for hydrogen storage applications. Many preceramic polymers are often low-cost materials which are attractive for the manufacture of ceramic fibers and matrix materials in *ceramic matrix composites* (CMCs) [11, 12].

The most important preceramic polymers include polysilanes, polycarbosilanes (PCSs), polycarbosiloxanes, polysilazanes (PSZs), polysilylcarbodiimides, poly(silsesquioxane-carbodiimides), and other polymers (Fig. 8.1) [6, 11, 13–15].

8.1 Polymer-derived silicon carbide ceramics

Yajima et al. elaborated thermal degradation of previously non-attractive, non-meltable poly(dimethylsilane) $-(Me_2Si)_n-$ into silicon carbide fibers (β-SiC). This process involved thermal rearrangement of poly(dimethylsilane), at ~450 °C, into of fiber forming poly(methylhydrocarbosilane) with an idealized (hypothetical) structure $-(MeHSiCH_2)_n-$, which was further pyrolyzed at ~1,100 °C into β-SiC (see Chapters 4 and 5) [16–20].

The direct pyrolysis of poly(methylsilane) $-[(H)CH_3Si]_n-$ (PMS, $n > 30$) gave high yield of the silicon carbide [15]. Similarly, a low viscous liquid polycarbosilane (LPCS) was prepared by a thermal rearrangement of poly(dimethylsilane) $-(Me_2Si)_n-$, which was carried out in an autoclave at 400 °C for 5 h and showed a high content of Si–H bonds. The pyrolysis of LPCS gave ceramic residue of 34% at 900 °C in argon atmosphere. The formation of SiC during pyrolysis of LPCS at 1,500 °C was confirmed by X-ray diffraction (XRD) method. The LPCS resin was used for the fabrication of the composite containing C and SiC particles, without any interfacial coating over the carbon fabric reinforcement. The carbon fabric showed density of 1.36 g/cm^3 and a flexural strength value of 64 MPa after six infiltration and pyrolysis cycles. Scanning electron microscope (SEM) analysis indicated the formation of nano- to submicron wires in the pores of the hybrid composite. A mechanical strength was further improved by addition of reinforcement fillers: boron nitride (BN) or carbon fabric [21].

The pyrolysis of PCS which was obtained from alkenylsilanes led to relatively high char yields. Formation of the crystalline silicon carbide (β-SiC) was optimum for a copolymer of alkylsilane and alkenylsilane with a silane to carbosilane backbone ratio of 85:15 and a C:Si ratio of 1.3:1 [22]. G.D. Soraru et al. studied the pyrolysis of PCS into a mixture of microcrystalline β-SiC and α-SiC ceramic at 1,700 °C by MAS ^{29}Si- and ^{13}C-NMR (nuclear magnetic resonance), transmission electron microscopy (TEM) and XRD analysis [23]. PCS prepared from alkenylsilanes gave high char yields in comparison with polyalkylsilane precursors (except for methylsilane) which provided low ceramic yields. At higher temperature, PCS precursors derived from vinylsilane pyrolyzed to a carbon-rich material with amorphous morphology. A stoichiometric SiC ceramics was fabricated by hydrosilylation of vinylsilane with methylsilane [24, 25]. A maximum weight loss, during pyrolysis of PMS and polyvinylsilane (PVS) into SiC ceramic, was observed below 650 °C. Copolymers of PMS with PVS formed the stoichiometric SiC. However, polymers of PMS easily oxidized and were pyrophoric sometimes [26].

Silicon carbide fibers were prepared with 65–70% ceramic yield by cross-linking of a low MW PCS with vinylsilazane monomer in the presence of dicumyl peroxide, followed by pyrolysis of the spun fibers at 1,000 °C [27]. PCS ceramic precursors were useful as ceramic binders in the preparation of powder metallurgy (PM) of functionally graded materials (FGMs). An Al/SiC composite can find aerospace applications and Cu/SiC can be used for dynamic seal applications. The two main advantages of using ceramic precursors for PM preparation of FGMs are: (1) elimination of the commercially used debinding step and (2) shrinkage control of the individual layers of the composite [28]. Melt-spinnable polyalumino-carbosilane (PACS) served as the precursor of Si–C–Al–O ceramic fibers, which can be used for the preparation of SiC fibers with high tensile strength and good thermal stability. Low-softening-point polycarbosilane (LPCS) was obtained by pyrolysis of polydimethylsilane and was applied to prepare PACS precursors with changing aluminum content by the reaction with aluminum(III) acetylacetonate. The composition and structure of the PACS precursors were analyzed by GPC, ^{1}H NMR, UV–Vis, FT-IR, ^{29}Si NMR, ^{27}Al MAS NMR, thermogravimetric analysis (TGA), and elemental analysis. The Si–C–Al–O fibers were formed by melt spinning, followed by curing, and pyrolysis of the precursors. This method can be also applied for fabrication other melt-spinnable polycarbosilanes containing metals (Zr, Ti, Fe, Co, etc.) of high ceramic yield and adjustable metal content by reacting LPCS with other metal-containing compounds [29].

The pyrolysis of a high-yield precursor to the stoichiometric SiC, a silicon analog of polyethylene, a linear poly(silylenemethylene) $–(SiH_2CH_2)_n–$ (PSM), with a regular polycarbosilane structure, was studied by Interrante et al. by different techniques (solid-state NMR and IR spectroscopies, TGA, and analysis of evolved gases). An elimination of H_2 bound to silicon was a key step in the cross-linking process. This conclusion was based on observed evolution of D_2 from $(SiD_2CH_2)_2$ as the primary gaseous product of the thermal cross-linking process occurring in the range of 250–400 °C. A reaction mechanism of the pyrolysis and cross-linking of PSM was proposed in which both 1,1-H_2 elimination and intramolecular H-transfer reactions led to highly reactive silylene intermediates. Their insertion into Si–H bonds of neighboring polymer chains forming Si–Si bonds took place, followed by the rapid rearrangement to Si–C bonds at these temperatures and formation of Si–C interchain cross-links. The cross-links prevented extensive fragmentation of the PCS network and at the increasing temperature to the range (>420 °C) homolytic bond cleavage took place, leading to free radical species. At higher temperatures (>475 °C), the free radical processes were probably responsible for the rearrangement of the SiC network structure, which was confirmed by a solid-state NMR spectroscopy. Prolonged heating of PCS to 1,000 °C led to the formation of the silicon carbide SiC in high yield (~85%) [30]. Polyphenylcarbosilane $–(PhHSiCH_2)_n–$ (M_w = 2,500 Da), prepared via Kumada rearrangement of $–(PhMeSi)_n–$ at 350 °C, decomposed at 1,200 °C into β-SiC ceramic and the ceramic yield was ~60% based on TGA results [31]. Cages (cubes) and NPs of β-SiC were prepared by a modified solvent-free Yajima process via vapor–liquid reactions of methylchlorosilanes (Me_2SiCl_2 or $MeSiCl_3$) with Na. Linear or rigid cross-linked intermediate PCS precursors of SiC (containing NaCl) were formed, respectively [32].

A thermooxidative stability of multiwalled carbon nanotubes (MWCNTs) was improved by coating with the silicon carbide (SiC) using PCS as the precursor. The SiC coatings on MWCNTs were formed from PCS coatings which were heated to ~1,300 °C under an inert atmosphere. The formation of SiC on the surface of MWCNTs was confirmed by XRD, energy dispersive X-ray analysis, and SEM. The tubular structure of the MWCNTs was retained as it was confirmed by TEM. The thermooxidative stabilities of coated and virgin MWCNTs was also studied by TGA. Sonication studies showed that the mechanical strength of the SiC coated MWCNTs was increased [33].

Partially allyl-substituted poly(hydridocarbosilane) (containing 5 mol% of allyl groups) (APHCS) underwent self-crosslinking at lower temperatures without any addition of cross-linking agents. In argon atmosphere, APHCS gave ceramic residue of 72% at 900 °C and 70% at 1,500 °C. At 1,500 °C, the SiC with the particle size of 3–4 nm was prepared. The APHCS was used as a matrix resin for the fabrication of C/SiC composite with density of 1.7 g/cm^3 and a flexural strength of 74–86 MPa after four infiltration and pyrolysis cycles [34].

Different preceramic polymer routes to silicon carbide were reviewed by R.M. Laine [5]. The poly(silmethylene) prepared by ring-opening polymerization of 1,3-disilacyclobutane with Pt catalyst was molded and shaped into fibers, followed by the pyrolysis process, which was carried out at temperatures of at least 900 °C [35]. The β-SiC fine fiber composite was prepared by force spinning technology combined with microwave energy processing [36].

Another preceramic polymer was prepared by hydrosilylation of polymethylsilane $-(HMeSi)_n-$ with 1,3,5,7-tetramethyl-1,3,5,7-tetravinylcyclotetrasiloxane $(MeViSiO)_4$ ($D^{Vi}{}_4$) and showed outstanding thermal and adhesive properties. It exhibited excellent thermal and bonding properties. At 1,200 °C, the ceramic yield was about 81% (in argon) and 90.6% (in air). The shear strength was 14.9 MPa at room temperature, and increased to 31.7 MPa after heating for 2 h in air at 1,000 °C. Reinforcement of the preceramic polymer with B_4C powder improved its shear strength up to 50.8 MPa after annealing at 1,200 °C under air atmosphere. The obtained adhesive can serve as a binder for SiC ceramics especially in high-temperature aeronautical and astronautical applications [37]. The β-SiC finds many other applications at high temperatures as composite reinforcements or refractory filtration systems having valuable properties like: low density, thermal stability, oxidation, and wear resistance. The nonwoven fine β-SiC fibers were prepared by spinning from a solution of polystyrene (PS) and poly (carbomethylsilane) (PCMS) as the precursor materials under nitrogen to control and reduce oxygen content. The long continuous nonwoven fine fibers with diameters ranging from 270 nm to 2 μm (depending on the selected processing parameters) were obtained with high yields. The fine fiber materials showed formation of the highly crystalline β-SiC fine fibers after microwave irradiation. They were characterized by SEM, XRD, and FTIR spectroscopy [36]. The SiC-based nanomaterials exhibited unique chemical, mechanical and physical (luminescent, electrical, thermal) properties and may find wide potential applications in composites, microwave absorbers, supercapacitors, as catalysts, and bioimaging probes [38]. The silicon carbide has also been recognized as a third generation semiconductor because of its valuable properties, for example, exceptional chemical inertness, outstanding mechanical behavior, high thermal stability, and conductivity. Thus, it has unique advantage and can be used under high-power/high-temperature/high-voltage harsh conditions. One-dimensional (1D) SiC nanostructures form numerous morphologies (nanorods, nanoneedles, nanochains, nanosprings, nanobelts, nanoprisms, nanocables, nanoarrays, etc.). Several techniques, such as pyrolysis of polymeric precursors, carbothermal reduction, chemical vapor deposition (CVD), and thermal evaporation, were used for the fabrication of SiC nanostructures. The 1D SiC can find applications in supercapacitors, field-effect transistors, field emitters, photocatalysts, pressure sensors, nano-electromechanical devices, microwave absorbers, superhydrophobic coatings, and so on [38, 39].

The SiC-based ceramic fibers were also derived from PCS or polymetallocarbosilane precursors. The first generations were Si–C–O (Nicalon type) fibers and Si–Ti–C–O fibers contained more than 10 wt% oxygen due to oxidation during curing – it caused a strong decrease of strength at temperatures exceeding 1,300 °C. Their maximum use temperature was 1,100 °C. The second-generation fibers were SiC (Hi-Nicalon) fibers and Si–Zr–C–O fibers. The oxygen content of these fibers was reduced to less than 1 wt% by electron beam irradiation curing in He. Their thermal stability was improved up to 1,500 °C, but their creep resistance was limited to a maximum of 1,150 °C since their C:Si atomic ratio resulted in excess carbon. The third-generation fibers were stoichiometric SiC fibers. They exhibited improved thermal stability and creep resistance up to 1,400 °C and fulfilled many of the requirements for the use of ceramic matrix composites (CMCs) for high-temperature application. The $SiBN_3C$ fibers derived from polyborosilazane also remained in the amorphous state up to 1,800 °C, and showed good high-temperature creep resistance [40].

The SiC–Ti ceramics was obtained from a hybrid precursor of titanium-containing PCS which was prepared from hyperbranched polycarbosilane (HBPCS), followed by cross-linking with tetrabutyl titanate (TBT) at 160 °C and pyrolysis at high temperatures. The cross-linking reaction of HBPCS–TBT hybrid precursor was studied by FTIR, solid-state ^{29}Si MAS NMR, and GPC. It proceeded through condensation reaction between the Si–H bond of HBPCS and butoxy group in TBT with the formation of Si–O–Ti units. The thermal properties and the structure of cross-linked hybrid precursor, the crystallization behavior, and composition of final ceramics were analyzed by FTIR, Raman spectroscopy, TGA, XRD, and energy dispersive elemental analysis. The ceramic yield of SiC–Ti was significantly increased by incorporation of TBT and it reached 83% at 1,400 °C for HBPCS–TBT (95:5) (based on TGA results). The SiC–Ti ceramic was amorphous at 900 °C. The characteristic peaks of β-SiC and TiC appeared until 1,600 °C. The growth of SiC crystals was inhibited by the formation of TiC [41].

Ceramic composite coatings containing ZrO_2 were prepared on steel substrates by dip-coating zirconia powder and glass particles dispersions (in di-*n*-butyl ether) with commercial organosilazane polymers: perhydropolysilazane (PHPS) and poly (methylvinylhydrosilazane) (PMVHS) (Figure 8.2), followed by drying and laser pyrolysis. Laser irradiation enabled the transformation of these composites into ceramic coatings. Reactions of the glasses with the monoclinic ZrO_2 fillers led to the formation of stabilized dendritic tetragonal ZrO_2 crystals. The resulting dense semicrystalline ceramic coatings showed a significantly different morphology compared to the same coating system pyrolyzed in a furnace due to different forming mechanisms. They were studied by ATR-FTIR, SEM with energy-dispersive X-ray spectroscopy (EDS), and XRD. The thermal stability of the coating ingredients was analyzed by TGA [42].

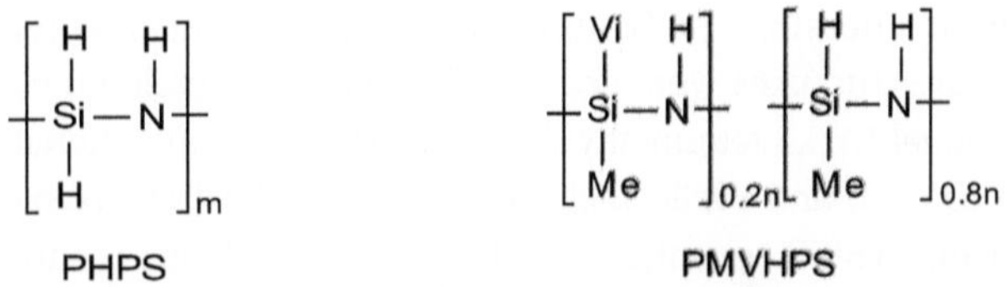

Figure 8.2: Chemical structures of the organosilazane polymers (a) PHPS and (b) PMVHPS [42].

The condensation of $(Me_3Si)_2NLi$ with $ZrCl_4$ gave di[bis(trimethylsilyl)amino](dichloro) zirconium $[(Me_3Si)_2N]_2ZrCl_2$, which was copolymerized by Würtz–Fittig method (with lithium in tetrahydrofuran (THF)) with a mixture of chlorosilane monomers ($MeHSiCl_2$, $MeViSiCl_2$, and Me_3SiCl), using molar ratio 1:4:1:2. The obtained new polyzirconosilane oligomer containing Si–H, Zr–Si–N–Si_2, and Si–Vi units was used as the precursor for preparation of SiC–ZrC ceramics. It was thermally cross-linked at 110–200 °C and next pyrolyzed at 1,000–1,400 °C giving SiC–ZrC ceramics with ~65 wt% yield. It showed extremely high thermooxidation resistance up to 1,200 °C [43].

8.2 Polymer-derived Si–C–O ceramics

Preceramic polysiloxane (PS) gels containing Si–H bonds were pyrolyzed at 1,000 °C into silicon oxycarbides (SiOC) (Si–C–O ceramics). The Si–H bonds played an active role in the incorporation of C atoms within the silica network. It was proposed that dehydrocarbocondensation reaction between Si–CH_3 and Si–H groups, proceeding at 600 °C, was responsible for the cross-linking process, leading to the polymer-to-glass transformation in the Si–C–O system and usually involved cross-linking reactions at the C sites, which transformed Si–CH_3 sites into $C(Si)_4$ structures and Si–C–O networks. The formation of Si–CH_2–Si bridges during the early stages of the pyrolysis of the PS containing Si–CH_3 and Si–H groups was postulated based on studies by MAS-NMR. The TGA coupled with mass spectrometry (TG/MS) was also applied [44].

Various Si–O–C- and Si–C-based porous ceramics were prepared from PS precursors by different processing strategies, including cross-linking reactions of PSs. Their properties can be tailored for a specific applications and were highly dependent on the processing method (replica, sacrificial template, direct foaming, and reaction techniques), but involved different cross-linking methods. The production of porous ceramics from PS precursors has some advantages: simple processing methodology, low processing cost, and easy control of porosity and other properties of the formed ceramics [45]. Silicone resins are also good precursors to the SiOC ceramics. They are useful binding agents which form protective silica layers on material surfaces beyond 1,200 °C in an oxidative atmosphere [46].

The pyrolytic conversion of vinyl-terminated poly(methylhydrosiloxane-*co*-vinylmethylsiloxane) into SiOC ceramic was a complex process, which involved many reactions occurring with liberation of H_2 and CH_4 [47]. Two linear vinyl functionalized PSs with regular structures were cross-linked with hydrosiloxanes of different structures. Difunctional 1,1,3,3-tetramethyldisiloxane ($M^H{}_2$), tetrafunctional star octamethylpentasiloxane ($QM^H{}_4$), and cyclic 1,3,5,7-tetramethylcyclotetrasiloxane ($D^H{}_4$) were used as new precursors to SiCO ceramics. Functionality of both kinds of substrates affected the thermal properties of the cross-linked PSs. Their thermal transformation under argon into ceramics was followed by FTIR spectra. The samples formed at 1,000 °C were a mixture of Si–C–O ceramic and a free carbon phase and did not contain pores [48].

The SiOC fibers with different chemical compositions were prepared by electrospinning a mixture of poly(methylsilsesquioxane) (PMSQ) or poly(methylphenylsilsesquioxane) (PMPSQ) preceramic polymers with polyvinylpyrrolidone (PVP), which were next cross-linked and pyrolyzed at 1,000 °C in argon. The morphology of the produced fibers was dependent on the processing procedure (cross-linking catalyst, additives, solvent selection). The 20% volume content of *N,N*-dimethylformamide in PMSQ/isopropanol system enabled the decrease of the diameter of the as-spun fibers from 2.7 µm to 1.7 µm. The SiOC fibers prepared from PMSQ and PMPSQ resins had uniform morphology with an average diameter of 0.98 µm and 1.08 µm, respectively. These silsesquioxane-derived SiOC ceramic fibers should find different potential applications [49]. SiOC (SiC_xO_y) thin films and nanowires, prepared by chemical methods showed interesting light-emitting properties and are widely used in the Si semiconductor industry as passivation layers, low-k dielectrics, and etch-stop layers. Moreover, SiC_xO_y may find numerous prospective applications in other technological fields (energy, lighting, biological applications, etc.) for instance as gas sensors, anode materials for lithium batteries, white light-emitting materials, hydrogen storage materials, and biomedical devices. The SiC_xO_y materials doped with erbium showed intensive luminescence properties in a broad emission spectral range (from the ultraviolet, through the visible, up to the near-infrared spectrum) [50].

Micro- and nanoporous SiOC foams with narrow pore size distribution were obtained by pyrolysis at different temperatures (850, 1,000, 1,250, and 1,450 °C) of polymeric gels prepared from mixtures of functional polymethyl(vinyl, hydro)siloxane (PMVHS) and polypermethylsiloxane (PDMS). The pore sizes and porosities of the porous SiOC materials changed in the range from 10 nm to 3 mm, and from 20% to 90%, respectively, by changing the viscosity of PDMS or its content. The preceramic PS containing both Si–H and Si–vinyl groups was cross-linked and pyrolyzed into SiOC ceramics in high yield, while PDMS, which completely decomposed during pyrolysis, was used as a porogen. The PMVHS was cross-linked via self-hydrosilylation reaction which led to gelation. The microphase separation was observed. The pyrolysis of the gel gave porous SiOC materials. Pore size, pore volume, and morphology were controlled by choosing viscosity of PDMS and by changing the PMVHS to PDMS ratio [51].

SiOC and SiCN membranes based on commercial PS and PSZ precursors were deposited as thin layers on porous ceramic substrates by coating a polymeric precursor solution, followed by cross-linking and pyrolysis above 600 °C in inert atmosphere. High ideal permselectivities of these membranes were observed at 300 °C. The SiOC membrane exhibited nearly 10 times higher permeances of hydrogen to CO_2 and achieved molecular sieving properties in separation of H_2/CO_2 mixtures. The SiOC and SiCN membranes had excellent physical and chemical properties: high mechanical strength, high temperature resistance, and chemical inertness [52].

SiOC glasses were prepared by inert atmosphere pyrolysis at 1,000 °C of gel precursors which were synthesized by cohydrolysis of triethoxysilane, $HSi(OEt)_3$, and methyldiethoxysilane, $HMeSi(OEt)_2$. The oxycarbide structures were characterized by means of ^{29}Si-MAS-NMR and Raman spectroscopies, XRD, and chemical analysis. Depending on experimental conditions and the composition of the starting gels, a pure oxycarbide phase was formed or it contained some carbon or silicon phase. By increasing the temperature up to 1,500 °C, the SiOC glasses showed compositional and weight stability; however, the amorphous network underwent structural rearrangements which led to the precipitation of nano-sized β-SiC crystallites into amorphous silica. Crystallization of silicon was also observed at 1,500 °C [53]. The SiOC glasses were also prepared by the pyrolysis of a preceramic PS, which was obtained by cross-linking a linear PMHS with 1,3,5,7-tetravinyl-1,3,5,7-tetramethylcyclotetrasiloxane (D^{Vi}_4) or divinylbenzene (DVB) as cross-linkers, via a Pt-catalyzed hydrosilylation reaction, with addition from 18 to 60 vol% of free carbon. Polymeric preceramic films with a thickness in the range 100–200 μm were pyrolyzed for 1 h at 1,200 °C in argon atmosphere. SiOC samples with low content of free C were obtained by cross-linking PHMS with D^{Vi}_4, while for C-rich composition PHMS was cross-linked with DVB. The SiOC glasses were characterized by nanoindentation method. Their elastic modulus decreased linearly from ~106 to ~80 GPa when the C-free content increased from ~18 vol.% to ~60 vol.%, while the hardness decreased from ~11 to ~8 GPa [54, 55].

Hybrid gels, obtained by the hydrolytic polycondensation of (diethoxy)dimethylsilane and zirconium propoxide mixtures with different Si:Zr ratio were pyrolyzed into ceramic materials containing even 50% of ZrO_2 [56]. Fibers composed of SiOC and titanium dioxide ($SiOC/TiO_2$) were prepared by reaction of poly(methylhydrosiloxane) with TBT, followed by electrospinning and pyrolysis at 1,000 °C. The $SiOC/TiO_2$ fibers showed hydrophobic properties with water contact angle of 130°. X-ray photoelectron spectroscopy, XRD, and TEM were used for characterization on the poly(titanosiloxane) gel fibers and the ceramic fibers. The gel fibers were converted into ceramic fibers composed of amorphous silicon and titanium SiOC glass during pyrolysis, and next decomposed to cristobalite–SiO_2, brookite–TiO_2 and trace of TiC nanoparticles incorporated in amorphous phase at 1,300 °C [57].

8.3 Polymer-derived Si_3N_4, Si–C–N, and Si–C–N–O ceramics

PSZs are polymers containing silicon, nitrogen, hydrogen, and carbon. They include either "inorganic" (PHPSs) or "organic" structures (organopolysilazanes and polycarbosilazanes). PSZs are precursors for the preparation of fibers, coatings, and 3D continuous fiber-reinforced ceramic matrix composites that often cannot be prepared by traditional ceramic-processing methods [58].

Ceramic materials containing Si, C, N, and O as the main elements were also prepared from rice husks, as a source of naturally existing silica and C. The various ceramic materials containing silicon were prepared by the thermal decomposition of rice husks and further heat treatment at temperatures 1,200–1,450 °C under a nitrogen atmosphere in the presence of $FeSO_4$. The formation of various SiCNO ceramics and silicon nitride (Si_3N_4) were investigated with different concentrations of $FeSO_4$ (4–10%). Morphology and surface properties of the ceramics were studied by SEM. Elemental analysis of the whiskers was analyzed by means EDX. The formation of different phases, including whiskers, was confirmed by XRD and FTIR spectra [59].

The direct preparation (on a large scale) of silica nanospheres with controlled morphology and high purity is still a difficult task. Nano- and submicron amorphous silica spheres were prepared by pyrolysis of an amorphous PSZ preceramic powder with $FeCl_2$ as a catalyst. The obtained microspheres had diameter of 600–800 nm and high purity. Their surfaces were smooth and clean without any flaws [60].

Micro- and macro-cellular SiCN and SiOCN foams were prepared from a commercially available poly(methylvinyl)silazane as the preceramic polymer. A mixture of partially cross-linked PSZ and poly(methylmethacrylate) microspheres (used as a sacrificial filler) was pressed and next pyrolyzed to form microcellular foams. Alternatively, in a one-step process, a mixture of liquid PSZ with a blowing agent (azodicarbonamid) was cured and next pyrolyzed into macrocellular ceramics. The prepared foams showed a mostly interconnected porosity ranging from ~60 to 80 vol% and a compressive strength of ~1–11 MPa. Some foams containing fillers were slightly contaminated with oxygen, probably due to the adsorbed moisture on their surface. The SiCN-based foams with tailored pore architecture and properties may find high temperature applications [61].

Oligosilazanes $(MeHSiNH)_n$ (obtained by ammonolysis of $MeHSiCl_2$) were transformed in reaction with KH into oligosilazanes with MW 600–1,800 Da, which upon pyrolysis afforded high yields (80–85%) of ceramic materials composed of mixtures of silicon nitride Si_3N_4, silicon carbide SiC, and free C [62]. Similar thermal treatment of PSZ (obtained by ammonolysis of $MeHSiCl_2$ and Me_2SiCl_2), which proceeded at temperature 1,300 °C via dehydrocondensation between Si–H and N–H bonds, led to amorphous Si_3N_4/SiC ceramic fibers [63]. By pyrolysis of $Si[N(CH_2CH_3)_2]_4$ in argon atmosphere, silicon carbonitride (SiCN) ceramics was formed [64].

Alternatively, silicon nitride coatings were very often prepared by CVD techniques from mixtures of gases: SiH_4/NH_3 or $SiCl_4/NH_3$. Their mechanical, chemical, and electrical properties might be applied as passivation layers and dielectric coatings. A pyrophoric monomer, trisilylamine $[(H_3Si)_3N]$, was also used as a silicon nitride precursor [65]. More often, preceramic polymers were applied as PSZs and non-volatile resinous sesquisilazanes which were not suitable for the fabrication of thin films of silicon nitride. However, PSZs not containing carbon substituents were hydrolytically unstable during storage [66], while some organosilazane derivatives gave high yields of silicon carbide on pyrolysis [67–69].

The liquid methylcyclosilazanes $-(MeHSiNMe)_3-$ and $[(MeHSiNMe)_2(MeHSiNH)_2]-$ were pyrolyzed into silicon nitride with 50–85% yield, while 30–40% yield was observed for linear poly(1,1-dimethylsilazane) $-(Me_2SiNH)_n-$ [70]. Pyrolysis of polyhydrosilazanes (PHSZ) containing units $-(RHSiNR')_n-$ and $-[(H)SiNR')]_n-$ (PHSZ, R=H; R'=H, *t*-Bu), which were formed by amminolysis (with *t*-$BuNH_2$) or by ammonolysis of chlorosilane mixtures (H_2SiCl_2 and $HSiCl_3$) in THF solution, afforded high yields of ceramic residues upon pyrolysis at 800–1,820 °C. The PHSZ were used as preceramic binders and sintering agents for Si_3N_4 powder during further thermal treatment at high temperatures. After pyrolysis processes, the silicon nitride existed in α and β crystalline forms [71]. Microporous Si_3N_4 membranes were prepared by pyrolysis (at 1,800 °C for 4 h) of a water suspension of α-Si_3N_4 powder coated with the liquid PSZ precursor (with average MW 1,250 g/mol) which was infiltrated into pores. Their top ceramic layers were formed by dip coating from diluted PSZ solutions in toluene and fired at 800 °C. The mechanically strong composite Si_3N_4 membranes contained 42 vol.% pores with diameter between 0.4 and 0.52 μm. The permeation of different gases (He, N_2, and CO_2) through these membranes was pressure-independent. The separation factors of 4.7 for He/N_2 were determined [72]. Porous and dense Si_3N_4/nano-SiC NCs were prepared by pyrolysis of polycarbosilazane with a powdered Si_3N_4 as a filler. However, Si_3N_4/nano-SiC decomposed at 1,800 °C into β-SiC and N_2 [73].

PSZs with different substituents at silicon and nitrogen, prepared by dehydrocoupling of Si–H and N–H bonds, catalyzed by $Ru_3(CO)_{12}$, were pyrolyzed into amorphous, silicon based ceramics. The relationships between the structure and chemical content of polymers and their pyrolyzed ceramic compositions and yields were discussed. A possible reaction mechanism during pyrolysis was proposed. The decomposition product patterns at different temperatures and the compositions of the final ceramics suggested specific kinetically and thermodynamically controlled thermolysis pathways. Amorphous ceramic products were heated at 800 °C and crystallized at 1,600 °C [74].

A composition of 40–70 wt% of low MW PSZ, 15–35 wt% medium MW polysilazane, and 5–30 wt% of methylvinylcyclosilazane or other unsaturated organic or organosilicon compound containing at least two alkenyl groups was used as the binder for ceramic powders: silicon carbide or/and silicon nitride [75]. Pyrolysis of PHPS under an ammonia atmosphere gave nearly stoichiometric Si_3N_4, while pyrolysis under nitrogen produced silicon-rich material. Pyrolytic products were composed of microcrystals of α-Si_3N_4, β-Si_3N_4, and silicon. Majority of the char was crystalline at ~1,270 °C, and the entire char was crystallized at 1,400 °C [76]. Pyrolysis of linear and branched PSZs containing Si–Vi or Si–H functional groups (Figure 8.3) under inert atmospheres led to amorphous Si–C–N–(H) ceramics at 1,000 °C. Further heating caused the transformation into the thermodynamically stable crystalline phases. The structural changes associated with cross-linking, pyrolysis, and crystallization were studied at 300 °C and 1,600 °C by means of ^{29}Si and ^{13}C solid-state NMR and FTIR spectroscopy, XRD, TGA, and density measurements. A correlation of the polymer architecture with the structure of the amorphous ceramic material was also studied [77].

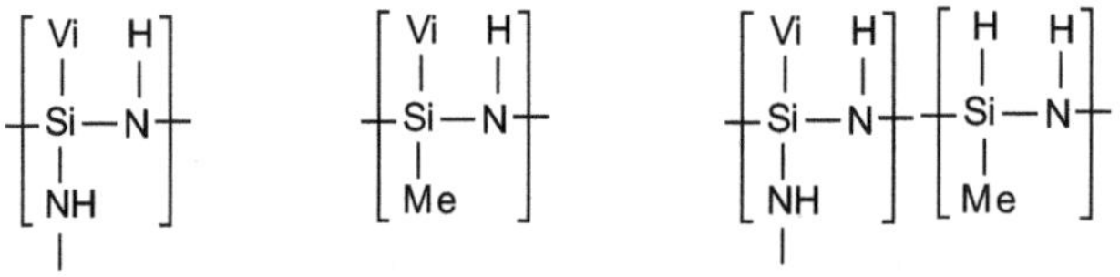

Figure 8.3: Examples of chemical structures of some polysilazane precursors [77].

Many other PSZ precursors of the different structures (e.g., **1**, Figure 8.4.) were used for preparation of SiOC, SiCN, and SiOCN ceramics [7, 78].

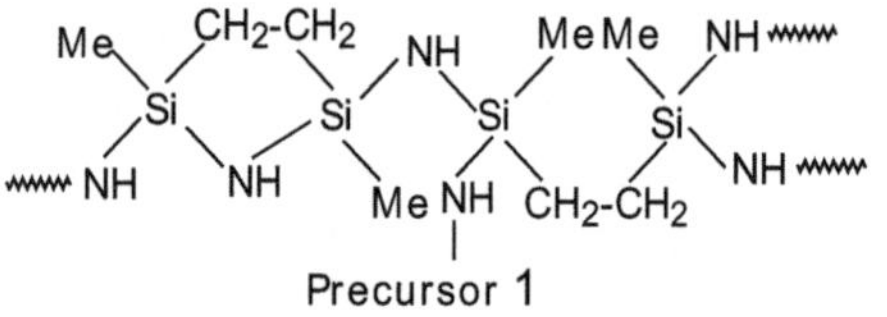

Figure 8.4: Chemical structure of the polysilazane precursor **1** [78].

The Si_3N_4 thin films were also obtained by plasma-enhanced chemical vapor deposition process (PECVD) of SiH_4/N_2 mixtures. Mechanical properties and corrosion resistance of Si_3N_4 films were determined. They may be applied in mechanical and optical devices [79]. The hardest transparent spinel ceramic, polycrystalline cubic silicon nitride (c-Si_3N_4) with a grain size of ~150 nm, was prepared by sintering commercially available α-Si_3N_4 powder under high pressure at temperature 1,600–1,800 °C. This material, which is often called γ-Si_3N_4, showed an intrinsic optical transparency over a wide range of wavelengths below its band-gap energy (258 nm) [80]. The cubic Si_3N_4 is one of the hardest materials next to diamond and cubic boron nitride (c-BN), exhibiting the high temperature metastability of c-Si_3N_4 in air. The transparent c-Si_3N_4 ceramic can serve as a window under extremely severe conditions: to protect optical sensors and detectors. The c-Si_3N_4 showed excellent thermal properties in air [81, 82], at least up to 1,400 °C, which is superior as compared to diamond, c-BN, and other hard materials [83].

Oligosilazanes obtained on two different methods:

(a) from reaction of a mixture of three equivalents of dichlorosilane, H_2SiCl_2, with one equivalent of methyldichlorosilane ($MeSiHCl_2$) with ammonia of a structure $[(SiH_2NH)_3(MeSiHNH)]_n$ ($n = 12–14$) (**2**);

(b) from reaction of dichlorosilane with methylamine and ammonia in a 3:1 ratio of the structure $[(SiH_2NH)_3(MeSiHNMe)]_n$ (**3**)

were cross-linked through *n*-BuLi-catalyzed dehydrocoupling reactions of Si–H and N–H bonds into products **2a** and **3a**, respectively. The molecular structures of **2** and **3** were analyzed by means of high resolution ^1H–, ^{13}C–, ^{29}Si–NMR in C_6D_6 solution, and by FTIR spectroscopy. The insoluble cross-linked products **2a** and **3a** were characterized by elemental analysis and solid-state ^1H–, ^{13}C–, and ^{29}Si CP-MAS NMR in combination with FTIR spectroscopy. Pyrolysis in an argon atmosphere up to 1,400 °C of the cross-linked products gave Si_3N_4/SiC ceramics in 94% yield. The absence of "free" carbon phase was confirmed by TG-MS and neutron wide-angle scattering. A one-step decomposition in the 250–700 °C range with predominant liberation of hydrogen took place. An elimination of methane and ammonia (from **2**) or methane and methylamine (from **3**) was below 0.5 mass%. High-temperature properties of Si_3N_4/SiC ceramics were studied by different techniques: high temperature TGA, TEM of annealed ceramic samples, XRD, and EDX. The onset of crystallization in both materials was around 1,500 and 1,300 °C (1 atm N_2), respectively, while α- and β-Si_3N_4 were formed besides α/β-SiC. Above 1,900 °C, ceramics derived from both **2a** and **3a** decomposed to α/β-SiC/β-Si_3N_4/α-Si composites [84].

By reacting an organosilazane oligomer or polymer containing Si–H and/or Si–N functional groups, especially of the structure $(MeHSiNH)_n$, with borane compounds in an organic solvent, were prepared preceramic polymers. The molar ratio of the organosilicon polymer repeat units to borane was ≤15. Borosilazane prepolymers were soluble in organic solvents (e.g., in toluene) and can serve as binders for other ceramic powders (Si_3N_4, BN, SiC, etc.) and can be used for preparation of Si_3N_4/BN fibers after treatment at high temperature. They were pyrolyzed at 1,000–1,500 °C in argon stream, providing black ceramic residue, while pyrolysis at 1,000 °C in a stream of ammonia allowed to remove most of the carbon and gave a white ceramic material of borosilicon nitride. These samples heated under argon to 1,500 °C contained more boron and became darker (grey or dark grey, depending on the B content) [85].

Recently low-temperature deposition of Si_3N_4 (at 250 °C) on Si(111) wafers and HO-terminated silica, using di-*sec*-butylaminosilane $(s\text{-}Bu)_2NSiH_3$ (DSBAS) and hydrazine (N_2H_4) by atomic layer deposition (ALD), was described [86]. Although at 250 °C, using DSBAS and N_2H_4 on HO-terminated SiO_2, no film growth was observed, the addition of a trimethylaluminum (TMA) pulse prior to ALD led to observable ligand exchange and measurable Si_3N_4 film deposition (~1 nm), although the growth was limited to ~1 nm. The addition of a second TMA pulse after 40 cycles of DSBAS and N_2H_4 allowed an increased film growth. Density functional theory calculations showed that the initial strong adsorption of N_2H_4 to aluminum increased reactivity of DSBAS with N_2H_4.

The hydrosilylation reaction was found to be an efficient and fast thermal cross-linking reaction for the preparation of Si–C–N precursors at relatively low temperature (120 °C). Moreover, hydrosilylation led to the formation of carbosilane bridges, Si–C–Si or Si–C–C–Si, that were not affected by the main depolymerization reactions (exchange of Si–N bonds and transamination reactions) which were responsible for low ceramic yields. After pyrolysis, these carbosilane bridges led to high carbon content in the ceramic residue. It was concluded that silazane oligomers containing Si–vinyl and Si–H groups might provide a good compromise between ceramic yield and final carbon content [87].

The thermal cross-linking and the pyrolysis of various oligosilazanes under argon were studied by R. Corriu et al. using TGA/MS method. The ceramic yield, the ceramic composition, and the gaseous evolutions were strongly dependent on the number and the nature of the functional groups attached to the silicon atom. The cross-linking occurred through complex reactions: hydrosilylation of vinyl groups with Si–H bonds, dehydrogenation of Si–H groups, dehydrocoupling of Si–H and N–H groups, transamination, and polymerization of vinyl groups (Scheme 8.1) [87].

The thermal degradation and fire performance of coatings obtained from polymethylsilazane-bearing Si–H and vinylsilane functionalities (PSZ, Figure 8.5) were studied by S. Duquesne et al. The thermal degradation of the coating occurred in three steps and led to the formation of Si_3N_4 and char formation at high temperature. The flame retardancy of PSZ coating was improved by addition of a fire retardant [88].

PSZs are excellent precursors for silicon nitride Si_3N_4 and silicon carbide SiC. Vinyl functional PSZ were thermoset below 200 °C with peroxide initiators by different methods. Commercial liquid peroxides and silyl and silazane peroxides were used. They were prepared in reactions of various silazanes with hydroperoxides. Peroxides promoted crosslinking of PSZ, especially containing both $CH_2{=}CH{-}Si$ and Si–H groups. Peroxide-curable PSZ were found to be very useful binders for advanced ceramic injection molding [89].

A controlled atmosphere pyrolysis process (in argon, NH_3 and a 10% NH_3/argon mixture) was used for preparation of Si_3N_4/SiC composites from a polyureasilazane (PUSZ). A volume fraction of SiC in a Si_3N_4 matrix was in 0–30% range, accompanied

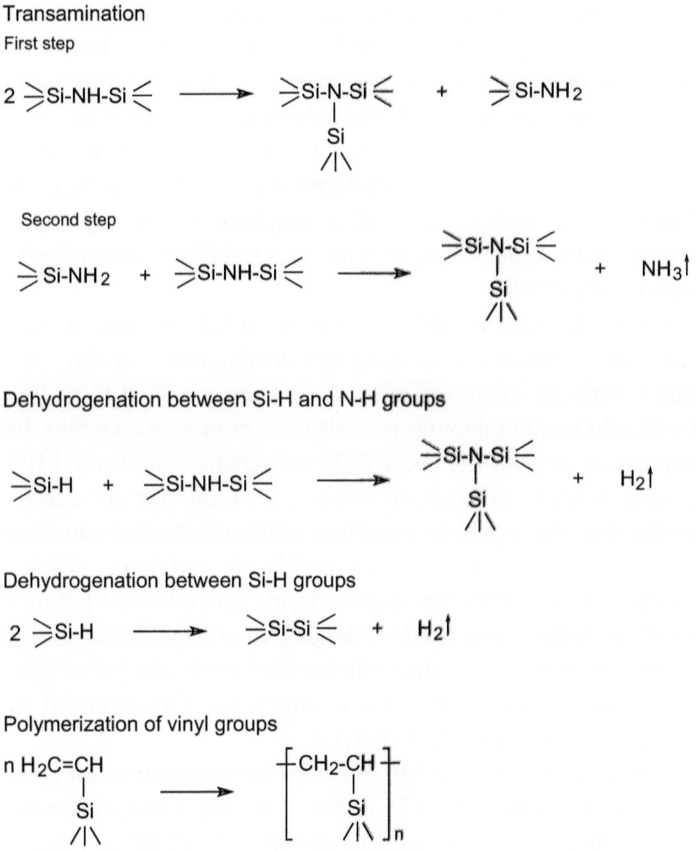

Scheme 8.1: Possible cross-linking reactions during pyrolysis of oligovinylsilazane [87].

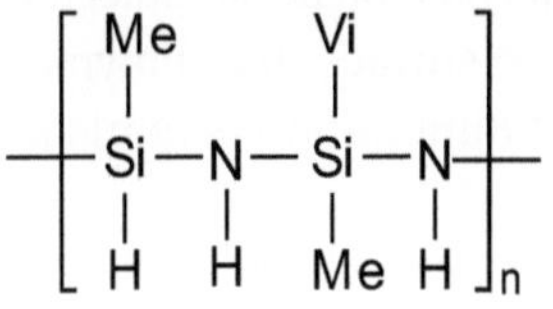

Figure 8.5: The chemical structure of poly(methylhydro-*co*-methylvinyl)silazane [88].

by a free carbon phase was present up until 1,850 °C. It was also found that carbon existed in samples that did not have observable XRD peaks of SiC. This novel process could be used for any type of PSZ polymer to produce SiCN powders for Si_3N_4/SiC nanocomposites with varying SiC content [90].

Newer insulating materials were prepared from PSZs which were cured with an isocyanate, an epoxy resin, or their mixtures in the presence of gas-generating compounds (water, an alcohol, an amine, or their combinations) [91].

A polysiloxazane precursor for SiCON ceramics was prepared by the partial hydrolysis of chlorosilanes ($MeViSiCl_2$, $MeHSiCl_2$, and $MeSiCl_3$) at −20 °C, followed by ammonolysis reaction of the obtained dichloro-terminated siloxane oligomers with NH_3. The structure and thermal properties of the polysiloxazane precursor were studied by means of FTIR, ^{1}H- and ^{29}Si-NMR spectra, gel permeation chromatography (GPC), and TGA. The polysiloxazane was pyrolyzed at 1,500 °C under N_2 or Ar atmosphere, providing SiCON ceramics. Under N_2 atmosphere α-Si_3N_4 crystalline phase was mainly formed, while under argon atmosphere, crystalline phases of α/β-SiC and/or α-Si_3N_4 were detected [92].

A new preceramic polyvinylsilazane (PVSZ) substituted with ethoxy groups, was prepared through the ammonolysis of (trichloro)vinylsilazane in THF, followed by reaction with controlled amount of ethanol. End cap modified PVSZ (EC-PVSZ) was prepared with addition of (chloro)trimethylsilane, in order to stabilize the polymer without incorporating oxygen. The above PSZs were characterized with FTIR-ATR, ^{1}H-NMR, GPC, TGA, residual gas analysis (RGA), and X-ray powder diffraction (XRD) of the ceramic char after pyrolysis in various atmospheres. Both modified systems demonstrated improved shelf-life. CVD was used for deposition of ZnO/SiO_2 interfacial coatings onto commercial woven ceramics fabric (Nextel-440™). A new low pressure CVD (LP-CVD) furnace was used for deposition of SiO_2. The effects of precursor amounts, temperature, and coating thicknesses were studied to optimize the mechanical strength of the coated fabric, which was characterized by SEM, scanning Auger microscopy (SAM), and XRD [12].

Hexamethyldisilazane (HMDSz) ($(Me_3Si)_2NH$) was used as the precursor for preparation of amine-functionalized polymer films by means of plasma assisted chemical vapor deposition (PA-CVD) technique. It was characterized by FTIR, field emission scanning electron microscopy (FESEM) and AFM techniques. The surface of the film contained large amount of NH_2 functional groups and was nanometrically smooth. The optical band gap of the film was ~3 eV. It can serve as a convenient biomolecule sensing material by immobilizing different enzymes on the surface [93]. Polymer films deposited by plasma method were quite stable and durable in comparison with chemically synthesized polymer films.

Composites of PSZs reinforced with single-walled carbon nanotubes (SWCNTs) were fabricated by two different approaches. First, SWCNTs covalently functionalized by disilazane groups were used for improvement of interfacial interaction between the nanotube surface and the PSZ matrix. Alternatively, an electrical field was applied in the cross-linking step during the nanocomposite synthesis, which allowed a good simultaneous dispersing and aligning the SWCNTs in the Si-C-N matrix [94]. The silicon carbonitride and graphite (SiCN–graphite) composite was also prepared by pyrolysis of the mixture of polysilylethylenediamine-derived amorphous SiCN and graphite powder at a temperature of 1,000 °C for 1 h in argon. This SiCN-graphite material, was used as anode active material in a lithium ion battery and showed excellent electrochemical properties. The SiCN-graphite anode exhibited a high initial and steady specific discharge capacity (975.6 and 425.0 $mAhg^{-1}$, respectively) after 30 charge–discharge cycles at a current density of 40 mAg^{-1}. Both these values were higher than that of polymer-derived SiCN and commercial graphite under the same charge–discharge condition [95].

A novel amorphous ceramic was prepared from mixtures of precursors: PCS and PSZ [PSZ/PCS=2 (w/w)] which were pyrolyzed at 1,200–1,500 °C in nitrogen. During formation of the Si–N–C ceramics, crystallization of microcrystalline α-Si_3N_4 and nanocrystalline SiC occurred. The obtained Si–N–C ceramics were characterized by ceramic yield, density, porosity, XRD, and SEM. The ratio of PSZ/PCS and the annealing temperature strongly affected the crystallization behavior and microstructure. Non-oxide Si–N–C ceramics showed homogeneous elemental distribution, better oxidation resistance and high thermal stability, and excellent properties, attractive for technological materials. Thus, it finds numerous practical applications in various branches of technology [96]. The SiCN ceramics, obtained by pyrolysis of a commercial poly(ureasilazane) at 1,200 °C, was applied for fabrication of pressure sensors, consisted of a thick blocking element and slender sensitive element. The thickness and dimensions of the sensors films depend on their application, and especially if they work in extremely high temperature environments, for example, in gas turbine engines [97].

The bulk and solvent copolymerization of styrene with 40 wt% of a commercial, vinyl-functionalized PSZ –$(MeViSiNH)_{0.2}(MeHSiNH)_{0.8}$– (HTT 1800) was carried out in the presence of dicumyl peroxide as a radical initiator at high temperature under argon atmosphere. The PSZ HTT 1800, as a flame retardant additive, was also incorporated on PS matrix during radical polymerization reaction in (1-methylethyl)benzene solution. The solvent was next removed by vacuum distillation. Chemical and physical properties of the hybrid organic–inorganic copolymer were analyzed by FTIR, TGA, and DSC. A higher increase of glass transition temperature for PS/HTT 1800 obtained by bulk copolymerization (101.7 °C), than for the solvent copolymerization (91.0 °C) was observed. This behavior was probably caused by cross-linking reactions between organic and inorganic monomers giving material with enhanced thermal stability. TGA analyses showed a substantial decrease of weight loss of PS when polysilazane was added. However, this hybrid material passed only a V-1 flammability test UL94 [98].

SiCN films were also prepared from the 1,1-dimethylhydrazine silyl derivatives: dimethyl(2,2′-dimethylhydrazino)silane and dimethyl-bis-(2,2′-dimethylhydrazino)silane, by the chemical vapor deposition (CVD) process modified with remote plasma. The obtained ceramic films contained different types of composite compounds having Si–N, Si–C, and C–N bonds. The films prepared below 400 °C were partially hydrogenated. Electron microscopic studies showed that nanocrystals 50–200 nm in size were formed in the amorphous matrix of films. The formation of crystals and their crystalline forms were independent of the substrate temperature. Thus, it was concluded that the nanocrystals were formed in the gas phase, or on a surface with growing film. The films exhibited high thermal stability up to 1,000 °C [99].

An amorphous hydrogenated SiCN (a-Si:C:N:H) films were prepared by the remote hydrogen plasma chemical vapor deposition (RP-CVD) from (dimethylamino) dimethylsilane Me_2NSiMe_2H at temperature 30–400 °C. A chemical composition, structure and physical, optical, and mechanical properties of the deposited films were studied by different methods. Their density, refractive index, hardness, elastic modulus, friction coefficient, adhesion to a substrate, optical and tribological properties, and resistance to wear were determined [100]. SiCN films were also prepared by the remote nitrogen plasma chemical vapor deposition from tetramethyldisilazane $(HMe_2Si)_2NH$ at substrate temperature 30–400 °C. Structure-property relationships of SiCN films were determined. It was found that a friction coefficient μ value decreased with increasing content of the Si–N and Si–C bonds in the film network. A minimum value of the friction coefficient ($\mu = 0.01$) was observed for the optimum film composition determined by the atomic ratios N/Si = 0.4 and C/Si = 0.8–0.9. Owing to a relatively high hardness ($H = 14.3$ GPa), low friction coefficient ($\mu = 0.02$, against stainless steel), and high refractive index ($n = 2.0$) which was found for SiCN film deposited at the substrate temperature 400 °C, this material may be used for special coatings [101]. Plasma polymer NPs have great potential for their use in photonics, nanomedicine, and other applications and may provide a valuable addition to the field of nanoscale-dispersed polymers. However, studies on the mechanisms of nanoparticle formation in plasma from organosilicon substrates are still in progress and open new horizons in precise tuning of NPs size, shape, chemical composition, surface charge, and wettability [102].

The oxidation behavior of SiCN–ZrO_2 fibers (which were prepared from alkoxide-modified silazane) and SiCN ceramics at 1,350 °C were compared. The lower oxidation rate of these fibers was a result of the lower carbon content in the fiber material [103]. A commercially available ceramic precursor $-(MeViSiNH)_{0.2}(MeHSiNH)_{0.8}-$ was reacted with the aminopyridinato copper complex $[Cu_2(ApTMS)_2]$ (ApTMSH=(4-methylpyridin-2-yl)trimethylsilylamine) which occurred with elimination of aminopyridine (ApTMSH) and formation of copper-modified poly(carbosilazane) (PCSZ-Cu). Further cross-linking of the PCSZ-Cu and subsequent pyrolysis gave the copper-containing ceramics (Cu-SiCN). During the pyrolysis step at 1,000 °C, the metallic copper was formed, as it was confirmed by solid-state ^{65}Cu-NMR spectroscopy, EDS, SEM pictures, and powder diffraction experiments. The Cu-SiCN ceramics showed catalytic activity toward the oxidation of cycloalkanes in air, which increased with increasing copper content, while the catalysts were recyclable [104].

Microporous silicon oxycarbonitride ceramics with accessible and tailored micropores was prepared by an NH_3-assisted thermolysis of mixtures of polysiloxane, PCS, and PSZ $-(MeViSiNH)_{0.2}(MeHSiNH)_{0.8}-$. The ^{29}Si-MAS NMR showed that the incorporation of nitrogen gave structures containing substantial amounts of SiN_4 and SiO_2N_2 building units. The samples derived from PCS remained non-porous: for such a C-rich and N-bearing phase, the NMR, TEM, and a pair distribution function analysis results suggested a Si network exhibiting domains dominated by either Si–N or Si–C bonds. ^{13}C NMR showed a presence of primarily "carbidic" CSi_4 in the C-rich phases and the formation of an amorphous *sp*^{2}-hybridized carbon phase, presumably both were crucial for the formation of micropores [105].

Aluminum nitride (AlN) was coated with amorphous SiOC and silicon oxynitrocarbide (SiONC) ceramic films, which were prepared by dip-coating method using preceramic PSZ, followed by pyrolysis at 700 °C in different atmospheres (air, Ar, N_2, and ammonia). Thus, PSZ was converted into SiOC in air and into SiONC ceramic in Ar, N_2, and NH_3. The characteristics of amorphous SiOC and SiONC ceramic films on AlN surface was studied by FTIR, XRD, and XPS. The interfacial adhesion between silicone rubber and AlN was significantly improved after the incorporation of amorphous SiOC and SiONC ceramic films on AlN surface. The morphology of SiOC–air and SiONC–NH_3 ceramic films on AlN surface (which was more flat and smooth than SiONC–N_2 and SiONC–Ar ceramic films) was confirmed from AFM observations. Moreover, the enhancement of the thermal conductivity of silicone rubber composites was observed owing to the decrease in the surface roughness of SiOC and SiONC ceramic films on the surface of thermally conductive filler (AlN) [106].

Silicon-based ceramics on aluminum nitride (Si/AlN) matrix was also prepared by pyrolysis of PSZ (see Figure 8.6) coated AlN at 1,600 °C in air, Ar, N_2, and NH_3 atmospheres. The presence of crystalline phases SiO_2 in Si/AlN$_{air}$, SiC whisker in Si/AlN$_{Ar}$, SiC whisker/Si_3N_4 grain in Si/AlN$_{N2}$, and Si_3N_4 ceramics in Si/AlN$_{NH3}$ on AlN particles was found from XRD, XPS, and SEM/EDS studies. The Si/AlN hybrid fillers were thermally conductive. The improvement of the thermal conductivity of silicone rubber filled with Si/AlN hybrid fillers which were fabricated in Ar and NH_3 was observed. This was explained by the presence of nanoscale SiC whiskers and a thermal conductive layer of Si_3N_4 on Si/AlN$_{air}$ and Si/AlN$_{NH3}$ surface, respectively [107].

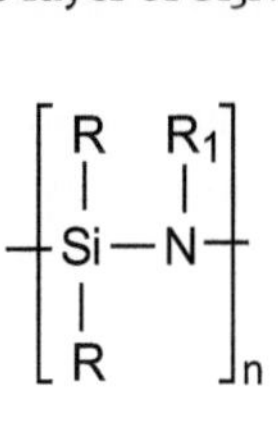

(R = H, Me, Et, NR_1; R_1 = H, Me, SiR_3)

Figure 8.6: Chemical structures of polysilazane precursors [107].

Pyrolysis of a polytitanosilazane which was synthesized from PHPS and $Ti(NMe_2)_4$ at 1,000 °C in NH_3, followed by heat treatment at 1,800 °C in N_2 gave Si_3N_4-TiN ceramics. TiN formed particles with diameter smaller than 100 nm [108]. Perhydropolytitanosilazane (PHPTiS) of the chemical composition $(Si_{1.1}Ti_{0.4}C_{2.2}N_{1.1}H_{6.2})_n$ with a molecular weight of 4,200 g/mol was obtained in reaction between PHPS and tetrakis (dimethylamino)titanium in a 2.5:1 molar ratio. Ammonolysis of PHPTiS at −40 °C caused cross-linking. Pressing at 110 °C and pyrolysis under ammonia (at 1,000 °C) and nitrogen (at 1,300 °C) gave bulk Si_3N_4–TiN nanocomposites in 69.6% ceramic yield. TiN nanocrystallites with an average diameter of 3.1 nm were homogeneously incorporated in an amorphous Si_3N_4 matrix and showed a high Vickers hardness of 25.1 GPa. A low molecular weight TiN ceramics was also mixed with a polymeric Si_3N_4 precursor in controlled molar ratio and was further processed by ammonolysis, warm pressing, and controlled nanocrystal growth, giving nanocomposites with the expected properties and very high hardness [109]. An inexpensive semiconductor, the powdered titanium disilicide $TiSi_2$, showed unusual optoelectronic properties and it was found to catalyze splitting of water into hydrogen and oxygen, under sunlight irradiation. Hydrogen was liberated readily and oxygen was adsorbed on a surface of $TiSi_2$, enabling the separation of these two gases. Oxygen was desorbed at increased temperature (above 100 °C) [110].

By the pyrolysis of commercial PSZ precursors deposited on polydivinylbenzene microspheres was prepared a macro- and microporous SiCN hybrid material, which showed good selectivity and adsorption properties for organic dyes, high adsorption capacity and good regeneration and recycling ability for the dyes having triphenyl structure (malachite green, methyl blue, acid fuchsin, and basic fuchsin). However, this hierarchically porous SiCN hybrid material did not adsorb the dyes containing azobenzene structures (methyl red, methyl orange, and congo red) and may find applications in the treatment of waste water containing complex organic pollutants [111].

The conductivity of the microwave-treated sample of a polymer-derived SiCN ceramic was about 40 times higher than that of the conventional heat-treated one at the same temperature and dwell time conventionally. The XRD patterns showed that both samples were amorphous without obvious crystallization. X-ray photoelectron spectra revealed that there was a significant sp^3-to-sp^2 transition of free carbon in the microwave-treated sample, which contributed to the distinct increase of the conductivity of the sample [112].

Copolymers of ferrocene with hexamethyldisilane and HMDSz were obtained by PA-CVD coupled with chemical transport reaction. They were further pyrolyzed and sintered at 1,200 °C under argon/nitrogen atmospheres giving nanostructured magnetic ceramics. The morphology of preceramic polymers and resulted ceramic, their crystallinity and composition were characterized with SEM, XRD, and Raman spectroscopy to gain information of their morphology, crystallinity and composition which were dependent on the deposition parameters. Magnetic properties of these ceramic materials were also evaluated [113].

Homogeneous nanocomposites of silicon nitride (Si_3N_4) with different contents of reduced-graphene oxide (rGO) were prepared upon pyrolysis of a graphene oxide (GO)-filled PSZ (which was synthesized in situ), in Ar atmosphere. Hot-pressing of the obtained nanocomposite powders gave monolithic rGO/Si_3N_4. The rGO phase was homogeneously dispersed within an amorphous or crystalline Si_3N_4 matrix.

An increasing amount of rGO in the NCs gradually suppressed the crystallization of the silicon nitride matrix into α-Si_3N_4. Different dielectric properties and tunable electromagnetic waves (EMW) behavior were observed depending on the volume fraction of the graphene phase in the ceramic NCs [114].

In the past 15 years, a particular effort has also been made on the forming methods and control of the mesoporosity and the shape of preceramic PDC with focus on the processing of tailor-made polymeric precursors to mesoporous materials, and to the characterization of their properties. Mesoporous PDCs can be also widely used as electrodes in fuel and solar cells [115]. The development of the catalytic activity of mesoporous PDC (for application in various catalyst-assisted reactions) was achieved by two strategies: that is, deposition of metal nanoparticles on the mesoporous network or in-situ generation of the catalytically active phase in the mesoporous PDC matrix. Novel solid catalysts based on mesoporous PDC had a narrow and uniform pore size, pore interconnectivities, and high surface areas. Owing to the possibility to tailor the preceramic polymer chemistry and architecture, these materials exhibited adjustable phase compositions and microstructures useful as supports of active catalysts (i.e., metal nanoparticles) or directly as catalysts. The mesoporous PDC functionalized with metal NPs gave many useful catalysts. A direct incorporation of catalytically active species during the formation of the mesoporous PDC matrix was often applied, especially for the modification of preceramic organosilicon precursors with metals. Organosilicon polymers in most cases easily dissolve in polar and non-polar solvents and have pendant functional groups, that is, Si–H, N–H, and Si-CH=CH_2 or Si-CH_2-CH=CH_2, bound to the polymer backbone which can easily react with organometallic compounds to give metal/PDC nanocomposite precursors.

The first mesoporous PDC using silica templates were made of SiC. The synthesis of mesoporous SiC materials was achieved by using mesoporous silica (SBA-15 and KIT-6) as hard templates and a PCS as a precursor [116]. Many mesoporous carbon templates were often used as well. For example, they were impregnated with a polyvinyl-silazane diluted in toluene (60 wt%), followed by the pyrolysis at 1,400 °C under nitrogen (into Si–C–N ceramics) and template removal under air at 650 °C [117].

Mesoporous PDC were used as support materials of catalytically active metallic nanoparticles (NPs). Most of the work done on metal NPs supported mesoporous PDC was based on ceramics derived from PSZs and on the catalytic hydrolysis of sodium borohydride ($NaBH_4$) [118–120]. Alternatively, the catalytically active mesoporous phases were incorporated in PDC and were directly used as catalysts. For example, Pt/Si–C–N ceramics was obtained by combining micromolding and two-component colloidal self-assembly with cooperative assembly of five compounds [solvent, amphiphilic block copolymer (PI-b-PDMAEMA), radical initiator (dicumyl peroxide, DCP), preceramic polymer poly(ureamethylvinyl)silazane and the dimethyl-(1,5-cyclooctadiene) platinum catalyst] [121].

Self-assembly was found an important tool for “bottom-up” fabrication of ordered mesoporous ceramics via microphase separation mechanism [122–124]. It was well developed for sol–gel method in a one-step process. The PDCs with a well-defined mesoporosity was achieved by using a soluble structure-directing agent (SDA) plying a role of a template (SDA, amphiphilic molecules, which were either cationic, anionic or non-ionic surfactants as block copolymers). The soft-template approach requires the compatibility between the SDA and the preceramic polymer [125].

8.3.1 The Si–C–N and Si–C–N–O ceramics derived from PSZ and PUSZ

Commercial linear and cyclic liquid PSZs –$(MeViSiNH)_{0.2}(MeHSiNH)_{0.8}$– and PUSZ, with chemical structures presented in a Figure 8.7, easily reacted to form a solid preceramic material and then were pyrolyzed to either silicon carbide or silicon nitride-containing high performance ceramics for *Ceramic Matrix Composites* (CMCs), *Metal Matrix Composites* (MMCs) [as ceramic fibers, ceramic precursor infiltrant, and polymer infiltration/pyrolysis (PIP)] [126].

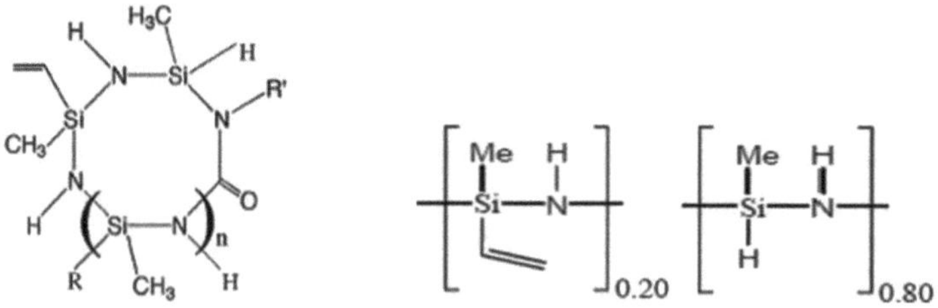

Figure 8.7: The chemical structure of commercial cyclic and linear polyureasilazanes [126].

Low viscosity liquid thermosetting PUSZ and PSZ resins are versatile reagents, which contain repeat units in which silicon and nitrogen atoms are bonded in an alternating sequence. The PUSZ have a small content of urea functionality, while PSZ contain fewer low molecular weight silazane oligomers. The PUSZ and PSZ are thermoset, which were cured to a solid materials by heating up to 180–200 °C with a free radical initiators or by exposure to UV radiation in the presence of a UV sensitizer. Both polymers convert to silicon carbide or silicon nitride ceramics at high temperatures [127, 128].

Thin (~40–60 µm) and thick (~2–3 mm) ceramic SiCN layers were prepared from liquid poly(ureasilazane) precursor for fabrication of pressure sensors of desirable shape, especially useful for high temperatures applications. It was formed on gelatin layer, which was easily dissolved in hot water before cross-linking and pyrolysis for preparation of very thin SiCN films. Fabrication of SiCN films includes thermosetting, cross-linking, and pyrolysis. The best mechanical properties (the hardness was 23 GPa) were reached for pure commercial polyureasilazane, which was thermally set at a temperature of 240 °C, cross-linked at 13.8 MPa and pyrolyzed at 1,200 °C. An addition of photo initiator 2,2-dimethoxy-2-phenyl-acetophenone enabled photo-lithographical patterning of the preceramic polymer by UV lithography. Mesoporous 3*C*-SiC hollow fibers were prepared through electrospinning of PUSZ and PVP solution followed by high-temperature pyrolysis. The concentration of PUSZ seemed to play a determined role on the formation of hollow fibers. The obtained fibers were characterized by field emission scanning electron microscopy (FESEM), high-resolution transmission electron microscopy, XRD, and N_2 adsorption. The 3*C*-SiC hollow fibers with mesoporous walls had uniform diameters (~34 nm) and a low surface area of ~22 m^2/g [129].

Many other examples of preparation of SiCN ceramics by different experimental techniques were described in a literature [130–135]. Highly stable SiCN ceramic patterns on Si substrates with submicron dimensions were prepared with a liquid commercial PSZ ceramic precursor and nanoscale CD and DVD soft lithography techniques, modified imprint lithography and micromolding in capillaries. The soft or hard PDMS mold was used to give the cured polymer patterns at 70–90 °C, which were next pyrolyzed at 800 °C in a nitrogen atmosphere. They were characterized by AFM and SEM [136].

Carbon-rich SiCN and SiOC ceramics showed significantly improved resistance against crystallization. The formation of crystalline phases was hindered or retarded by the presence of free carbon phase which acted as a diffusion barrier [137].

8.3.2 The carbon-rich Si–C–N ceramics derived from polysilylcarbodiimides

Another group of valuable precursors for the synthesis of SiCN-based ceramics are polysilylcarbodiimides [4, 5, 138]. The scientific interest on polysilylcarbodiimides increased in the end of twentieth century as it was reported by Riedel et al. [139–145]. Their thermal transformation to SiCN ceramics gave different microstructures with thermal stabilities which were dependent on the branching of chains [146]. The decomposition of this highly branched polymer led to the first ternary crystalline phases in the Si–C–N system, namely SiC_2N_4 and Si_2CN_4 [145]. Poly(phenylsilylcarbodiimides) $-[PhRSi\text{-}NCN]_n-$, (R=H, Me, Vi, Ph) were found to be good precursors for carbon-rich nanostructured SiCN ceramics [4, 5, 147]. The processing route significantly affected on the nanostructure of a bulk carbon-rich SiCN and SiBCN PDCs which were prepared by pressing of poly(phenylvinylsilylcarbodiimide) and poly(borophenylsilylcarbodiimide) at 1,100 and 1,400 °C. They showed an improved thermal stability against crystallization in comparison to their powder analogues [78].

Hybrid silsesquicarbodiimides of the structure $[(NCN)_{1.5}Si(CH_2)_xSi(NCN)_{1.5}]_n$ (where $x = 2$, 6, and 8) were prepared by sol–gel polycondensation reactions of bis(trichlorosilyl)alkanes and bis(trimethylsilyl)carbodiimide. They were composed of flexible organic chains within an inorganic network. The presence of the NCN groups in xerogel structures was identified by FTIR spectra. The composition and molecular structures were characterized by solid-state ^{13}C CP MAS- and ^{29}Si CP MAS-NMR spectroscopies, XRD, and elemental analysis. Their morphology was studied by SEM and TEM methods, and the pore structure of the materials was examined by the gas adsorption (BET) method. The surface area decreased with increasing length of the alkylene spacing group [146].

The carbon-rich SiCN ceramics was synthesized by the thermal decomposition of poly(phenylvinylsilylcarbodiimide) under argon atmosphere at five temperatures, namely 1,100 °C, 1,300 °C, 1,500 °C, 1,700 °C, and 2,000 °C. The SiCN was used for preparation of anode material for lithium-ion batteries, without any conducting additives. The free carbon phase was identified as *soft carbon* [147]. The thermal decomposition of poly(diphenylsilylcarbodiimide) precursor gave micro- and mesoporous SiCN ceramics with high specific surface area (~570 m^2/g). High-resolution TEM studies confirmed that the pores were embedded only in the carbon phase [148].

8.4 Polymer-derived Si–B–C–O ceramics

Carborane-based polymers are mainly obtained by polycondensation of bis(methoxydimethylsilyl)-*m*-carborane with bis(chlorosilyl)-*m*-carborane derivatives, alkylchlorosilanes, or alkylchlorosiloxanes in the presence of a catalyst [149–153]. Poly(carboranesiloxane)s find applications as heat resistant materials, for example, for GC phases [154] or as high temperature elastomers [155]. Linear inorganic–organic hybrid polymers containing at least one alkynyl group and at least one bis(silyl or siloxanyl)carboranyl unit within their backbone were polymerized into oxidatively stable thermosets and at temperature 600–1,000 °C formed ceramics [156].

Polyorganoborosiloxane ceramics precursors were synthesized by the hydroboration reaction of 1,3,5,7-tetramethyl-1,3,5,7-tetravinylcyclotetrasiloxane with borane dimethylsulfide. They were pyrolized at 1,100 °C in argon atmosphere giving black and amorphous SiBCO ceramics, thermally stable up to 1,200 °C. At temperatures exceeding 1,300 °C, decomposition of the SiBCO matrix took place with formation of crystalline β-SiC. Crystalline silica phases were not formed during annealing of SiBCO up to 1,500 °C [157].

Borosilicate gels were prepared by the sol–gel method of mixtures of methyl(triethoxy)silane and dimethyl(diethoxy)silane (DMDES) and boric acid. The use of $B(OH)_3$ allowed the hydrolytic condensation of ethoxysilanes in the absence of water or catalyst. Difunctional $-Me_2SiO-$ units promoted the formation of linear oligomers which facilitated fiber drawing before gelation. The SiBOC glasses were prepared by pyrolysis of the borosilicate gels in argon atmosphere at 1,000 °C. Results of TG-DTA showed that the ceramic yield decreased by increasing the amount of DMDES. Gel fibers were prepared from partially aged solutions by hand drawing. Their pyrolysis under argon atmosphere at 1,000 °C gave thin homogeneous SiBOC glass fibers [158].

Borosiloxane gels with B/Si ratios of 0.2 and 0.5, prepared by hydrolytic condensation reactions of alkoxysilanes with boric acid, had absorption bands at 880 cm^{-1} (in FTIR spectra), corresponding to B–O–Si moiety, due to the incorporation of the cross-linker trigonal units of BO_3 into the polymeric network. XRD analyses showed on an amorphous morphology of the resulting PDCs obtained by pyrolysis up to 1,000 °C under inert atmosphere. CMCs, composed of a SiBCO matrix and unidirectional carbon fiber rods as a reinforcement phase, were prepared by pyrolysis of carbon fiber rods coated with polysiloxane or poly(borosiloxane) (PBS) matrices. The C/SiBCO composites exhibited better thermal stability than the C/SiOC materials. Good adhesion between the carbon fiber and the ceramic phase was observed by SEM microscopy [159, 160].

Poly(borosiloxanes) (PBS), with a B/Si molar ratio of 0.2, were also prepared by a dehydrocondensation reaction of 1,3,5,7-tetramethyl-1,3,5,7-tetracyclotetrasiloxane (D^H_4) with $B(OH)_3$ or by hydrolytic polycondensation of vinyltriethoxysilane with $B(OH)_3$, followed by a hydrosilylation reaction with D^H_4. Their pyrolysis at 1,000 °C, in argon atmosphere gave black and amorphous boron and SiOC glasses (SiBCO). The PBS and SiBCO glasses were characterized by TGA, XRD, FTIR, ^{29}Si, and ^{11}B MAS NMR spectroscopies [161].

Polycondensation of vinyltriethoxysilane with $B(OH)_3$ in 2:1, 1.5:1, and 1:1 mole ratio in diglyme at 83–87 °C for 3 h was catalyzed by hydrochloric acid and gave vinyl-functionalized borosiloxane oligomers soluble in the reaction medium. After removal of ethanol, the by-product, and diglyme the obtained borosiloxane oligomers were characterized by FTIR and TGA and pyrolyzed to ceramics at 900 °C, 1,500 °C, and 1,650 °C in argon atmosphere. The ceramics obtained were characterized by IR, Raman, ^{13}C–, and ^{29}Si-MAS NMR and XRD. SiOC/SiBOC glass was formed at 900 °C, while β-SiC at 1,500 °C. Further heating at 1,650 °C led to a mixture of α- and β-SiC and diamond-like carbon phases [162].

SiBOC thin films were prepared by the sol–gel method using a mixture of triethoxysilane (as a source of T^H units) and methyldiethoxysilane (as a source of D^H units) with a T^H/D^H molar ratio of 2 giving SiOC with traces of free C. Triethylborate $B(OEt)_3$ (TEB), was used as the source of boron (in B/Si ratio of 0.1). Ethanol was used as a solvent and the hydrolysis was carried out with acidified water. Thin borosiloxane films containing $T^HD^H_2$ units were deposited on Si and SiO_2 substrates by spin coating and then pyrolyzed in a carbon furnace under Ar flow at temperatures in a range of 800–1,200 °C. Results were compared with boron free SiOC films, containing $T^HD^H_2$ units, prepared from the same sol–gel precursor solution without addition of TEB [163].

Alternatively, by controlled pyrolysis of sol–gel derived precursors SiBOC films were prepared. It was shown that the addition of B into SiOC ceramics changed drastically the properties of the film with superior performances for light emitting applications. The SiBOC film showed a lower thermal shrinkage compared to SiOC, which allowed an easy processing or shaping of the film. A complex structure of the SiBOC film resulted in a wide distribution of emitting centers, which led to a high efficient, strong broadband tunable visible emission, observed from SiBOC ceramic thin films. Thus, SiBOC thin films seemed to be very promising materials for phosphor and electroluminescence applications [164].

The shrinkage during pyrolysis of a gel precursor as thin film and as bulk sample were compared. The hybrid silica gel, precursor for SiOC glasses, contained Si–CH_3 and Si–H groups. The shrinkage of bulk samples was measured with conventional dilatometry, while for thin films it was studied for the first time with in situ dilatometry allowing to measure the thickness and the refractive index during pyrolysis. Both type of samples continued to shrink well above the temperature (800 °C) at which the weight losses were complete according to TGA data. Compared to bulk samples, thin films (1) showed a higher shrinkage (15–20 vol%) and (2) the onset of the pyrolytic transformation was shifted 100–150 °C toward lower temperatures [165].

SiOC glasses find broad applications. Its thermostability was improved by incorporation of boron. SiBOC gel was prepared from single-source polyborosiloxanes which were synthesized from polymethylethoxysiloxane and $B(OH)_3$ by a sol–gel process. SiBOC ceramic fibers, obtained by electrospinning, showed excellent thermostability [166]. SiBCO composites were prepared through densification process of porous C/C composites from preceramic polymers: PCS, phenylborosiloxane, and vinylborosiloxane [167].

8.5 Polymer-derived Si–B–C–N ceramics

Poly(borosilazanes) were synthesized via hydroboration, from (trivinyltrimethyl)cyclotrisilazane $[CH_2{=}CH(CH_3)SiNH]_3$, and trimethylamineborane complex: $(CH_3)_3N{:}BH_3$. The poly(borosilazanes) (with Si/B molar ratios: of 3 or 9) were pyrolyzed up to 1,000 °C under nitrogen atmosphere, giving amorphous silicon boron carbonitride glasses (SiBCN). The high-temperature behavior of SiBCN glasses was studied by XRD and IR spectroscopy. The glasses containing the larger amounts of B remained amorphous to higher temperatures in comparison with the B-free SiCN glass, which, at 1,500 °C, crystallized in SiC. At 2,000 °C, the ceramics obtained from SiBCN formed BN + SiC composite [168].

New liquid polycarbo(boro)silazanes (PCBSZ, $M_w = 2{,}511$; $M_n = 1{,}655$) were prepared via hydroboration polymerization of borazine with (divinyl)tetramethyldisilazane in THF, without the use of any catalyst or the formation of any by-products. The soluble liquid PCBSZ was used as a precursor of Si–B–C–N ceramics and changed into an insoluble solid due to α- or β-vinyl addition cross-linking reactions of the unreacted vinyl groups to the B–C alkyl bridges between the borazine ring and vinylsilazane, which were carried out at 200 °C to 600 °C. It was changed to an insoluble ceramic with the composition $Si_{1.5}C_{4.2}B_{1.0}N_{2.4}$ after post-treatment at 170 °C. Next, it was pyrolyzed at 1,400 °C to SiCN and BN amorphous structures with 70–86% ceramic yields. These processes were monitored by FTIR, ^{1}H, ^{11}B, ^{13}C, and ^{29}Si NMR spectroscopy, and TGA [169].

A liquid viscous linear carborane–siloxane–acetylene hybrid copolymers were prepared by T. Barton et al. [170, 171], by reacting 1,7-bis(chlorotetramethyl)disiloxyl)-*m*-

carborane with dilithiobutadiyne. It was soluble in most organic solvents and was used as thermoset polymeric precursor, which was thermally or photochemically cured through the triple bonds of the acetylenic moieties at 300–400 °C, followed by pyrolysis at 1,000 °C. Thus, it was converted in high yield into a shaped ceramic material, which showed high thermal and oxidative stability at high temperature [172].

Excellent precursors for ceramics are decaborane-based polymers. The high boron content in carboranes and decaborane was important for formation of boron carbide and boron nitride ceramics. Carborane and decaborane structures stabilize these polymers against degradation without cross-linking. On pyrolysis, poly(carboranesiloxane)s form mixed non-oxide ceramics B_4SiC, which can be used especially as protective coatings for oxidizable substrates. However, the high cost of these polymers limits their practical applications [154].

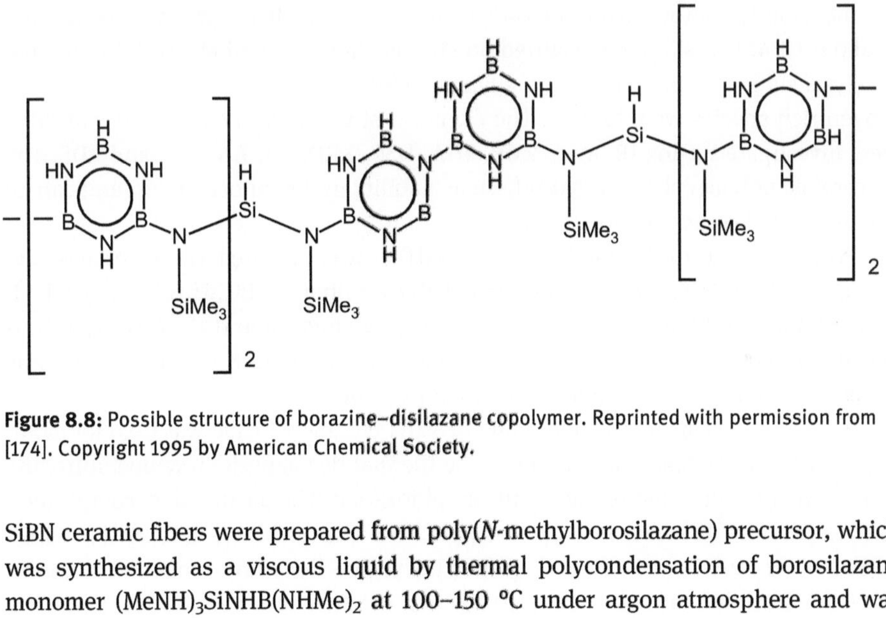

Figure 8.8: Possible structure of borazine–disilazane copolymer. Reprinted with permission from [174]. Copyright 1995 by American Chemical Society.

SiBN ceramic fibers were prepared from poly(*N*-methylborosilazane) precursor, which was synthesized as a viscous liquid by thermal polycondensation of borosilazane monomer $(MeNH)_3SiNHB(NHMe)_2$ at 100–150 °C under argon atmosphere and was further processed by melt-spinning [173]. The copolymers containing borazine and disilazane units in a backbone (Figure 8.8) were prepared by the thermal dehydrocondensation of borazine ($B_3N_3H_6$) with tris(trimethylsilylamino)silane $(Me_3SiNH)_3SiH$ (**TTS**) and 1,1,3,3,5,5-hexamethylcyclotrisilazane $(Me_2SiNH)_3$ (**HCT**) [175]. Formation of trimethylsilane as a side product (0.25 mol for each mole of borazine) was observed, presumably due to a Si–N cleavage reaction. Ring-opening reactions of **HCT** with borazine proceeded with formation of borazine–disilazane segments and side products (volatile Me_2SiH_2 and $Me_2SiH–NH–B_3N_3H_5$). It was assumed that both series of copolymers contained borazine and silazane-nitrogen moieties. Elemental analyses of the **TTS** copolymers were consistent with an approximate $\{[(\text{-}B_3N_3H_4)–NH]_3SiH\}_x$

structural unit, while the **HCT** copolymers had compositions ranging from $(B_3N_3H_4)_{1.00}N_{1.81}(SiMe_2)_{1.57}H_{1.6}$ to $(B_3N_3H_4)_{1.00}N_{1.07}(SiMe_2)_{1.28}H_{0.9}$ having large polydispersities of MWs, and relatively low intrinsic viscosities of their solutions, showing on highly branched structures. These two series of copolymers gave different types of ceramic materials upon pyrolysis. The **TTS** copolymers gave amorphous BNSiC ceramics up to 1,400 °C. Heating up to 1,800 °C resulted in further loss of silicon leading to ceramics with compositions ranging from $B_{1.00}N_{0.90}Si_{<0.01}C_{<0.01}$ to $B_{1.00}N_{0.91}Si_{0.14}C_{0.07}$. The DRIFT spectra of both kinds of copolymers showed the presence of boron nitride. The ceramics obtained from the **HCT** copolymers at 1,400 °C were also amorphous and had more silicon and carbon, with typical compositions of $B_{1.0}N_{1.5}Si_{0.4}C_{0.2}$. Pyrolyzes of the **HCT** copolymers up to 1,800 °C gave materials with compositions ranging from $B_{1.00}N_{2.58}Si_{0.94}C_{0.30}$ to $B_{1.00}N_{0.86}Si_{0.16}C_{0.19}$. Their XRD spectra showed crystalline phases of β-Si_3N_4, β-SiC, and Si, but no crystalline boron-containing species [174].

Another borosilazane preceramic polymer (PBSZ) was synthesized from trichlorosilane ($HSiCl_3$), boron trichloride (BCl_3), and hexamethydisilazane, at a molar ratio of about 1:1:4. Cross-linking occurred due to the elimination of Me_3SiCl. In the final pyrolysis stage exceptionally thermally stable SiBCN-based ceramics, CH_4, and hydrogen-rich species were formed. The chemical structure and composition of PBSZ were investigated using IR, MAS-NMR, TGA, DSC, XRD, EA, AA, ICP, and EDS. The preceramic polymer showed hydrothermal stability under autoclave test and can be useful for nuclear application [175].

Polyorgano(boro)silazane $[B(C_2H_4SiMeNH)_3]_n$ was prepared via the ammonolysis (in THF at 80 °C) of tris(dichloromethylsilylethyl)borane $B(C_2H_4SiMeCl_2)_3$, which was obtained by hydroboration of dichloromethylvinylsilane $(CH_2{=}CH)CH_3SiCl_2$. It was pyrolyzed at 1,300 °C in argon atmosphere into Si–B–C–N ceramic, which was characterized by XRD and Raman spectroscopy [176].

A ceramic precursor tris[(chlorodimethylsilyl)ethane]borazine was used for modification of novolac resin and improved the thermal decomposition temperature, the char yield of novolac and promoted the graphitization. The maximum decomposition rate temperature was increased up to ~575 °C. The char yield was increased to 75.5% at 800 °C and 68.7% at 1,200 °C, respectively. The B_3N_3 ring remained stable at 1,400 °C. During heat treatment a fixed siloxane structure SiO_x was formed. The molecular structure and thermal properties of borazine modified novolac were characterized by FTIR, Raman spectra, XPS, NMR, and TGA. The weight loss and gas evolution of modified novolac during the thermal degradation was evaluated by DTA/TG/MS test [177].

A great number of other examples of *high-tech ceramics* derived from organosilicon polymers have been published in the world literature in recent decades [2–11].

References

[1] R. Lee, Carbosilanes: Reactions & Mechanisms of SMP-10 Pre-Ceramic Polymers, Marshall Space Flight Center, Jacobs ESTS Group / ICRC, July 2009.

[2] D. Seyferth, Chapter: *Organosilicon Polymers as Precursors for Silicon-Containing Ceramics*, R.M. Laine, Ed., *Transformation of Organometallics into Common and Exotic Materials: Design and Activation*, NATO ASI Series (Series E: Applied Sciences), Vol. 141, Springer, Dordrecht, 1988. DOI: https://doi.org/10.1007/978-94-009-1393-6.

[3] R.M. Laine and F. Babonneau, Preceramic Polymer Routes to Silicon Carbide, *Chemistry of Materials*, 1993, 5(3), 260–279.

[4] R. Riedel, G. Mera, R. Hauser and A. Klonczynski, Silicon-Based Polymer-Derived Ceramics: Synthesis, Properties and Applications – A Review, *Journal of the Ceramic Society of Japan*, 2006, 114(6), 425–444.

[5] P. Colombo, G. Mera, R. Riedel and G.D. Sorarů, Polymer-Derived Ceramics: 40 Years of Research and Innovation in Advanced Ceramics, *Journal of American Ceramic Society*, 2010, 93(7), 1805–1837.

[6] S. Somiya, *Handbook of Advanced Ceramics: Materials, Applications, Processing, and Properties* (e-Book Google), Academic Press, 2013.

[7] M. Birot, J.-P. Pilot and J. Dunogues, Comprehensive Chemistry of Polycarbosilanes, Polysilazanes, and Polycarbosilazanes as Precursors of Ceramics, *Chemical Reviews*, 1995, 95(5), 1443–1477.

[8] S. Fu, M. Zhu and Y. Zhu, Organosilicon polymer-derived ceramics: An overview, *Journal of Advanced Ceramics*, 2019, 8(4), 457–478. ISSN 2226-4108. DOI: https://doi.org/10.1007/s40145-019-0335-3.

[9] S. Packirisamy, K. Sreejith, D. Deepa and B. Swaminathan, Chapter: Polymer-Derived Ceramics and Their Space Applications, Handbook of Advanced Ceramics and Composites, M. Yashwant and J. Roy, Eds., Springer Intern. Publ, 2020, 1–107.

[10] G. Mera and E. Ionescu, Chapter: *Silicon-Containing Preceramic Polymers, Encyclopedia of Polymer Science and Technology*, J. Wiley & Sons, Inc., Hoboken, NJ, USA, 2013.

[11] G. Mera, M. Gallei, S. Bernard and E. Ionescu, Ceramic Nanocomposites from Tailor-Made Preceramic Polymers, *Nanomaterials*, 2015, 5. 468–540. DOI: 10.3390/nano5020468.

[12] G.C. Richards, Development of New Matrix and Interfacial Materials for Ceramic Matrix Composites, Doctoral Dissertation, University of Connecticut, Storrs, USA (2014).

[13] T.-H. Yoon, L.-Y. Hong and D.-P. Kim, Chapter 21: *Fabrication of SiC-based Ceramic Microstructures from Preceramic Polymers with Sacrificial Templates and Softlithography Techniques, Lithography*, Ed, M. Wang, 427–446, February 2010, INTECH, Croatia. ISBN 978-953-307-064-3.

[14] F.I. Hurwitz, P.J. Heimann, J.Z. Gyekenyesi, J. Masnovi and X.Y. Bu, Polymeric Routes to Silicon Carbide and Silicon Oxycarbide CMC, NASA Technical Memorandum 104372 (1991).

[15] Z. Zhang, F. Babonneau, R.M. Laine, Y. Mu, J.F. Harrod and J.A. Rahn, Poly(methylsilane) – A High Ceramic Yield Precursor to Silicon Carbide, *Journal of the American Ceramic Society*, 1991, 74(3), 670–673.

[16] S. Yajima, J. Hayashi and M. Omori, Continuous Silicon Carbide Fiber of High Tensile Strength, *Chemistry Letters*, 1975, 4(9), 931–934. DOI: http://dx.doi.org/10.1246/cl.1975.931.

[17] S. Yajima, K. Okamura and J. Hayashi, Structural analysis in continuous silicon carbide fiber of high tensile strength, *Chemistry Letters*, 1975, 4(12), 1209–1212. DOI: https://doi.org/10.1246/cl.1975.1209.

[18] S. Yajima, M. Omori, J. Hayashi, K. Okamura, T. Matsuzawa and C.F. Liaw, Simple Synthesis of the Continuous SiC Fiber with High Tensile Strength, *Chemistry Letters*, 1976, 5(4), 551–554. DOI: https://doi.org/10.1246/cl.1976.551.

[19] S. Yajima, Y. Hasegawa, J. Hayashi and M. Imura, Synthesis of continuous silicon carbide fibre with high tensile strength and high Young's modulus, *Journal of Materials Science*, 1978, 13(12), 2569–2576. DOI: https://doi.org/10.1007/PL00020149.

[20] S. Yajima, J. Hayashi, M. Omori and K. Okamura, Development of a silicon carbide fibre with high tensile strength, *Nature*, 1976, 261(8), 683–685. DOI: https://doi.org/10.1038/261683a0.

[21] B. Swaminathan, A. Painuly, P.V. Prabhakaran and S. Packirisamy, Liquid polycarbosilane-derived C/C–SiC composites with improved mechanical strength for high temperature applications, *Ceramics International*, 2015, 41(3), 3574–3577.

[22] J.F. Harrod, Chapter 7: "*Polymerization of Group 14 Hydrides by Dehydrogenative Coupling*", *Inorganic and Organometallic Polymers*, M. Zeldin, K.J. Wynne and H.R. Allcock, Eds., ACS Symposium Series, Vol. 360, 1988, and references therein, 89–100.

[23] G.D. Soraru, F. Babonneau and J.D. Mackenzie, Structural evolutions from polycarbosilane to SiC ceramic, *Journal of Materials Science*, 1990, 25(9), 3886–3893.

[24] J. Masnovi, X.Y. Bu, P. Conroy, A.H. Andrist, F.I. Hurwitz and D. Miller, Better Ceramics Through Chemistry IV, Brinker, D.E. Clark, D.R. *Ulrich* and B.J.J. Zelinsky, Eds., Materials Research Society Symposium Proceedings, Vol. 180, 23 Nov. 1990, 779–784.

[25] J. Masnovi, X. Bu, K. Beyene, P.J. Heimann, T.A. Kacik, A.H. Andrist and F.I. Hurwitz, Syntheses, structures and properties of polycarbosilanes formed directly by polymerization of Alkenyl-silanes, *Technical Report*: NASA-CR-191927, USA, Jan. 1, 1993.

[26] F.I. Hurwitz, T.A. Kacik, X.-Y. Bu, J. Masnovi, P.J. Heimann and K. Beyene, Pyrolytic conversion of methyl- and vinylsilane polymers to Si-C ceramics, *Journal of Materials Science*, 1995, 30 (6), 3130–3136.

[27] G.J. Choi, Silicone Carbide Fibers from Copolymers of Commercial Polycarbosilane and Silazane, *Journal of Industrial and Engineering Chemistry*, 1997, 3, 223–228.

[28] P. Czubarow and D. Seyferth, Application of poly(methylsilane) and NicalonT polycarbosilane precursors as binders for metal/ceramic powders in preparation of functionally graded materials, *Journal of Materials Science*, 1997, 32(4), 2121–2130.

[29] Y. Gou, H. Wang, K. Jian, Y. Wang, J. Wang, Y. Song and Z. Xie, Facile synthesis of melt-spinnable polyaluminocarbosilane using low-softening-point polycarbosilane for Si–C–Al–O fibers, *Journal of Materials Science*, 2016, 51, 8240–8249.

[30] Q. Liu, H.-J. Wu, R. Lewis, G.E. Maciel and L.V. Interrante, Investigation of the Pyrolytic Conversion of Poly(silylenemethylene) to Silicon Carbide, *Chemistry of Materials*, 1999, 11(8), 2038–2048.

[31] Y.J. Lee, J.H. Lee, S.R. Kim, W.T. Kwon, H. Oh, J.-H.P. Klepeis, S.J. Teat and Y.H. Kim, Synthesis and characterization of novel preceramic polymer for SiC, *Journal of Materials Science*, 2010, 45(4), 1025–1031. DOI: https://doi.org/10.1007/s10853-009-4034-2.

[32] C.-H. Wang, Y.-H. Chang, M.-Y. Yen, C.-W. Peng, C.-Y. Lee and H.-T. Chiu, Synthesis of Silicon Carbide nanostructures via a Simplified Yajima process – Reaction at the Vapor-Liquid Interface, *Advanced Materials*, 2005, 17(4), 419–422.

[33] R.K. Gupta, R. Mishra, K. Mukhopadhyay, R.K. Tiwari, A. Ranjan, A.K. Saxena and A. New, Technique for Coating Silicon Carbide Onto Carbon Nanotubes Using a Polycarbosilane Precursor, *Silicon*, 2009, 1(2), 125–129.

[34] R. Sreeja, B. Swaminathan, A. Painuly, T.V. Sebastian and S. Packirisamy, Allylhydridopolycarbosilane (AHPCS) as matrix resin for C/SiC ceramic matrix Composites, *Materials Science and Engineering: B*, 2010, 168(1–3), 204–207.

[35] T.L. Smith Jr., U.S. Pat. 4,631,179 (1986).

[36] A. Salinas, M. Lizcano and K. Lozano, Synthesis of β-SiC Fine Fibers by the Forcespinning Method with Microwave Irradiation, *Journal of Ceramics*, 2015, Article ID 217931. DOI: https://doi.org/10.1155/2015/217931.

[37] X. Wang, J. Wang and H. Wang, Synthesis of a novel preceramic polymer (V-PMS) and its performance in heat-resistant organic adhesives for joining SiC ceramic, *Journal of the European Ceramic Society*, 2012, 32, 3415–3422. DOI: http://dx.doi.org/10.1016/j.jeurceramsoc.2012.03.032.

[38] R. Wu, K. Zhou, C.Y. Yue, J. Wei and Y. Pan, Recent progress in synthesis, properties and potential applications of SiC nanomaterials, *Progress in Materials Science*, 2015, 72, 1–60.

[39] S. Chen, W. Li, X. Li and W. Yang, One-dimensional SiC nanostructures: Designed growth, properties, and applications, *Progress in Materials Science*, 2019, 104, 138–214.

[40] H. Ichikawa, Polymer-Derived Ceramic Fibers, *Annual Review of Materials Research*, 2016, 46, 335–356. DOI: 10.1146/annurev-matsci-070115-032127.

[41] Z. Yu, J. Zhan, C. Zhou, L. Yang, R. Li and H. Xia, Synthesis and Characterization of SiC(Ti) Ceramics Derived from a Hybrid Precursor of Titanium-Containing Polycarbosilane, *Journal of Inorganic and Organometallic Polymers*, 2011, 21, 412–420.

[42] K. Tangermann-Gerk, G. Barroso, B. Weisenseel, P. Greil, T. Fey, M. Schmidt and G. Motz, Laser pyrolysis of an organosilazane-based glass/ZrO_2 composite coating system, *Materials Design*, 2016, 109, 644–651.

[43] M. Wang, C. Huang and Z. Wang, Polyzirconosilane Preceramic Resin as Single Source Precursor of SiC–ZrC Ceramics, *Journal of Inorganic and Organometallic Polymers*, 2016, 26, 24–31.

[44] V. Gualandris, D. Hourlier-Bahloul and F. Babonneau, Structural Investigation of the First Stages of Pyrolysis of Si-C-O Preceramic Polymers Containing Si–H Bonds, *Journal of Sol-Gel Science and Technology*, 1999, 14(1), 39–48.

[45] B.V.M. Kumar and Y.-W. Kim, Processing of polysiloxane-derived porous ceramics: a review, *Science and Technology of Advanced Materials*, 2010, 11(4), 044303. DOI: 10.1088/1468-6996/11/4/044303.

[46] M. Narisawa, Silicone Resin Applications for Ceramic Precursors and Composites, *Materials*, 2010, 3, 3518–3536. DOI: 10.3390/ma3063518.

[47] X. Wang, J. Wu, Y. Li, C. Zhou and C. Xu, Pyrolysis kinetics and pathway of polysiloxane conversion to an amorphous SiOC ceramic, *Journal of Thermal Analysis and Calorimetry*, 2014, 115, 55–62.

[48] A. Nyczyk-Malinowska, M. Wójcik-Bania, T. Gumuła, M. Hasik, M. Cypryk and Z. Olejniczak, New precursors to SiCO ceramics derived from linear poly(vinylsiloxanes) of regular chain composition, *Journal of the European Ceramic Society*, 2014, 34, 889–902.

[49] A. Guo, M. Roso, M. Modesti, J. Liu and P. Colombo, Preceramic Polymer-Derived SiOC Fibers by Electrospinning, *Journal of Applied Polymer Science*, 2014, 131, 39836.

[50] S. Gallis, V. Nikas and A.E. Kaloyeros, Chapter 14: *Silicon Oxycarbide Thin films and Nanostructures: Synthesis, Properties and Applications*, InTech Open Science (2017); dx.doi.org/10.5772/66992

[51] J. Wu, Y. Li, L. Chen, Z. Zhang, D. Wang and X. Caihong, Simple fabrication of micro/nano-porous SiOC foam from polysiloxane, *Journal of Materials Chemistry*, 2012, 22(14), 6542–6545.

[52] Y. Jüttke, H. Richter, I. Voigt, R.M. Prasad, M.S. Bazarjani, A. Gurlo and R. Riedel, Polymer Derived Ceramic Membranes for Gas Separation, *Chemical Engineering Transactions*, 2013, 32, 1891–1896. DOI: 10.3303/CET1332316.

[53] G.D. Sorarù, G. D'Andrea, R. Campostrini, F. Babonneau and G. Mariotto, Structural Characterization and High-Temperature Behavior of Silicon Oxycarbide Glasses Prepared from Sol-Gel Precursors Containing Si-H Bonds, *Journal of the American Ceramic Society*, 1995, 78(2), 379–387.
[54] Y. Blum, G.D. Sorarù, A.P. Ramaswamy, D. Hui and S.M. Carturan, Controlled mesoporosity in SiOC via chemically bonded polymeric "spacers", *Journal of the American Ceramic Society*, 2013, 96, 2785–92.
[55] G.D. Sorarù, L. Kundanati, B. Santhosh and N. Pugno, Influence of free carbon on the Young's modulus and hardness of polymer-derived silicon oxycarbide glasses, *Journal of the American Ceramic Society*, 2019, 102(3), 907–913. DOI: https://doi.org/10.1111/jace.16131.
[56] S. Dire, R. Campostrini and R. Ceccato, Pyrolysis Chemistry of Sol-Gel-Derived Poly (dimethylsiloxane)-Zirconia Nanocomposites. Influence of Zirconium on Polymer-to-Ceramic Conversion, *Chemistry of Materials*, 1998, 10(1), 268–278.
[57] D. Su, L. Wang, K. Liang, F. Zhang, D. Lin, Y. Hu, H. Ji, X. Li, C. Shunquan Chenb and X. Yan, Silicon oxycarbide/titanium dioxide fibers with wrinkle-like surface by electrospinning, *Materials Letters*, 2016, 172, 202–206.
[58] A. Lukacs, III, Polysilazane Precursors to Advanced Ceramics, *American Ceramic Society Bulletin*, 2007, 86(1), 9301–9306.
[59] M. Sarangi, Effect of an Iron Catalyst and Process Parameters on Si-Based Ceramic Materials Synthesized From Rice Husks, *Silicon*, 2009, 1(2), 103–109.
[60] F. Gao, Z. Peng and X. Fu, One-Step Synthesis and Characterization of Silica Nano-/ Submicron Spheres by Catalyst-Assisted Pyrolysis of a Preceramic Polymer, *Journal of Nanomaterials*, 2013, Article ID 843570.
[61] C. Vakifahmetoglu, I. Menapace, A. Hirsch, L. Biasetto, R. Hauser, R. Riedel and P. Colombo, Highly porous macro- and micro-cellular ceramics from a polysilazane precursor, Ceramics Internship, 2009, 35, 3281–3290.
[62] D. Seyferth and G.H. Wiseman, High-yield synthesis of Si_3N_4/SiC ceramic materials by pyrolysis of a novel polyorganosilazane, *Journal of the American Ceramic Society*, 1984, 67(7), C132–C133.
[63] Y.C. Song, Y. Zhao, C.X. Feng and Y. Lu, Synthesis and pyrolysis of polysilazane as precursor of Si_3N_4/SiC ceramic, *Journal of Materials Science*, 1994, 29(21), 5745–5756.
[64] D.M. Narsavage, L.V. Interrante, P.S. Marchetti and G.E. Maciel, Condensation polymerization of tetrakis(ethylamino)silane and its thermal-decomposition to Si_3N_4/SiC ceramics, *Chemistry of Materials*, 1991, 3(4), 721–730.
[65] A.R. Reinberg, Single Component Monomer for Silicon Nitride Deposition, U.S. Patent 4,200,666 (1980).
[66] D. Seyferth, G.H. Wiseman and C. Prud'homme, A liquid silazane precursor to silicon nitride, *Journal of the American Ceramic Society*, 1983, 66, C13.
[67] B.A. Mazaev, B.N. Talanov and M.M. Eleen, *Inorganic Materials*, 1983, 19, 2044.
[68] C.L. Beatty, *Ultrastructure Processing of Advanced Structural and Electronic Materials*, Ed, L. Hench, Noyes Data, 1984.
[69] D.Ya. Zhinkin, E.A. Semenova, M.V. Sobolevskii, and K.A. Andrianov, *Plastitscheskie Massy*, 1963, 12, 16.
[70] B. Arkles, Silicon Nitride from Organosilazane Cyclic and Linear Prepolymers, *Journal of the Electrochemical Society*, 1986, 133(1), 233–234.
[71] S.T. Schwab, Polysilazane precursors for silicon nitride and resultant products, US Pat. 5,294,425 (1994).

[72] H. Mori, S. Mase, N. Yoshimura, T. Hotta, K. Avama and J.I. Tsubaki, Fabrication of Si_3N_4 supported membranes using the pyrolysis of liquid polysilazane precursor, *Journal of Membrane Science*, 1998, 147(1), 23–33.
[73] U. Degenhardt, F. Stegner, C. Liebscher, U. Glatzel, K. Berroth, W. Krenkel and G. Motz, Sintered silicon nitride/nano-silicon carbide materials based on preceramic polymers and ceramic powder, *Journal of the European Ceramic Society*, 2012, 32(9), 1893–1899.
[74] Y.D. Blum, K.B. Schwartz and R.M. Laine, Preceramic Polymer Pyrolysis I. Pyrolytic Properties of Polysilazane, *Journal of Materials Science*, 1989, 24(5), 1707–1718. DOI: https://doi.org/10.1007/BF01105695.
[75] P. Hogan, Pat. WO 92/01732 (1992).
[76] C.R. Blanchard and S.T. Schwab, X-ray Diffraction Analysis of the Pyrolytic Conversion of Perhydro-polysilazane into Silicon Nitride, *Journal of the American Ceramic Society*, 1994, 77 (7), 1729–1739.
[77] S. Traβl, D. Suttor, G. Motz, E. Rőssler and G. Ziegler, Structural characterisation of silicon carbonitride ceramics derived from polymeric precursors, *Journal of the European Ceramic Society*, 2000, 20, 215–225.
[78] Y. Gao, G. Mera, H. Nguyen, K. Morita, H.J. Kleebe and R. Riedel, Processing route dramatically influencing the nanostructure of carbon-rich SiCN and SiBCN polymer-derived ceramics. Part I: Low temperature thermal transformation, *Journal of the European Ceramic Society*, 2012, 32, 1857–1866.
[79] L. Liu, W.-G. Liu, N. Cao and C.-L. Cai, Study on The Performance of PECVD Silicon Nitride Thin Films, *Defence Technology*, 2013, 9, 121–126.
[80] N. Nishiyama, R. Ishikawa, H. Ohfuji, H. Marquardt, A. Kurnosov, T. Taniguchi, B.-N. Kim, H. Yoshida, A. Masuno, J. Bednarcik, E. Kulik, Y. Ikuhara, F. Wakai and T. Irifune, Transparent polycrystalline cubic silicon nitride, *Scientific Reports*, 2017, 7(44755) DOI: 10.1038/srep44755.
[81] M.G. Schwarz, M.A. Zerr, E. Kroke, B.T. Poe, H. Fuess and R. Riedel, Spinel-Si_3N_4: multi-anvil press synthesis and structural refinement, *Advanced Materials*, 2000, 12(12), 883–887.
[82] J.Z. Jiang, K. Kragh, D.J. Frost, K. Ståhl and H. Lindelov, Hardness and thermal stability of cubic silicon nitride, *Journal of Physics: Condensed Matter*, 2001, 13(22), L515–L520.
[83] A. Zerr, R. Riedel, T. Sekine, J.E. Lowther, W.Y. Ching and I. Tanaka, Recent advances in new hard high-pressure nitrides, *Advanced Materials*, 2006, 18(22), 2933–2948.
[84] M. Hőrz, A. Zern, F. Berger, J. Haug, K. Müller, F. Aldinger and M. Weinmann, Novel polysilazanes as precursors for silicon nitride/silicon, carbide composites without "free" carbon, *Journal of the European Ceramic Soc*, 2005, 25, 99–110.
[85] D. Seyferth, Preceramic organosilicon-boron polymers, *Patent EP Applied*, 1990, 0(424), 082.
[86] A. Dangerfield, C.E. Nanayakkara, A. Mallikarjunan, X. Lei, R.M. Pearlstein, A. Derecskei-Kovacs, J. Cure, A. Estève and Y.J. Chabal, Role of Trimethylaluminum in Low Temperature Atomic Layer Deposition of Silicon Nitride, *Chemistry of Materials*, 2017, 29(14), 6022–6029.
[87] N.S.C.K. Yive, R.J.P. Corriu, D. Leclercq, P.H. Mutin and A. Vioux, Silicon Carbonitride from Polymeric Precursors: Thermal Cross-Linking and Pyrolysis of Oligosilazane Model Compounds, *Chemistry of Materials*, 1992, 4(1), 141–146.
[88] B. Gardelle, S. Duquesne, C. Vue and S. Bourbigot, Thermal degradation and fire performance of polysilazane-based coatings, *Thermochimica Acta*, 2011, 519(1–2), 28–37.
[89] J.M. Schwark and A. Lukacs, *Chapter 5: Polysilazane Thermosets as Precursors for silicon carbide and silicon nitride, Inorganic and Organometallic Polymers II, Advanced Materials and Intermediates*, P. Wisian-Neilson, H.R. Allcock and K.J. Wynne, Eds., ACS Symp. Series, Vol. 572, 1994, 43–54.

[90] K.T. Strong Jr., S.A. Arreguin and R.K. Bordia, Controlled atmosphere pyrolysis of polyureasilazane for tailored volume fraction Si_3N_4/SiC nanocomposites powders, *Journal of the European Ceramic Society*, 2016, 36, 3663–3669.
[91] R.S. Larson and M.E. Fuller, Insulating materials comprising polysilazane, methods of forming such insulating materials, and precursor formulations comprising polysilazane, *US Patent*, 2013, 8(598), 245.
[92] J. Luo, Z. Zhang, W. Liu, X. Wang, Z. Peng, Y. Luo and C. Xu, The Synthesis of SiCON Ceramics Through Precursor Method, *Journal of Applied Polymer Science*, 2012, 126(3), 853–859.
[93] S. Das, S. Jana, D. De, S. Garain, S. Ray and U. Gangopadhyay, Synthesis of Ammine (NH_2)-Functionalized Silazane Based Plasma Polymer by rf-PACVD Technique as a Promising Biosensing Film, *International Journal: Adamas Technical Review*, 2014, 1(1), 49–53.
[94] L. Fernandez, Y. Li, M. Burghard, Z. Burghard, P. Gerstel, J. Bill and F. Aldinger, Dispersion of carbon nanotubes in polysilazanes for the preparation of reinforced Si-C-N composites, *Key Engineering Materials*, 2009, 403, 231–234.
[95] Y. Feng, -N.-N. Feng and G.-X. Du, Preparation and Electrochemical Performance of Polymer-derived SiCN-graphite Composite as Anode Material for Lithium Ion Batteries, *International Journal of Electrochemical Science*, 2012, 7, 3135–3140.
[96] X. Bao, S. Li, X. Tang and Y. Zhang, Synthesis of Si-N-C Ceramic Composites by Pyrolysis of Polysilazane and Polycarbosilane, *Key Engineering Materials*, 2012, 512–515, 306–309.
[97] A. Leo, S. Andronenko, I. Stiharu and R.B. Bhat, Characterization of Thick and Thin Film SiCN for Pressure Sensing at High Temperatures, *Sensors*, 2010, 10, 1338–1354. DOI: 10.3390/s100201338.
[98] S.A.C. Abarca, O. Flores, A.L.G. Prette, G.S. Barroso, T. Coana, G. Motz and R.A.F. Machado, Synthesis and Thermal Characterization of Silicon-based Hybrid Polymer, *Chemical Engineering Transactions*, 2013, 32, 1621–1626.
[99] T.P. Smirnova, A.M. Badalyan, L.V. Yakovkina, N.H. Sysoeva, I.P. Asanov, V.V. Kaichev, V.I. Bukhtiyarov, A.N. Shmakov, V.I. Rakhlin and A.N. Fomina, Silicon Carbonitride Films as a Promising Material Synthesized from New Sources, *Chemical Sustainable Development*, 2001, 9, 23–29.
[100] I. Błaszczyk-Łężak, A.M. Wróbel and D.M. Bieliński, Remote hydrogen microwave plasma chemical vapor deposition of silicon carbonitride films from a (dimethylamino)dimethylsilane precursor: Compositional and structural dependencies of film properties, *Diamond & Related Materials*, 2006, 15, 1650–1658.
[101] I. Błaszczyk-Łężak, A.M. Wróbel and D.M. Bieliński, Remote nitrogen microwave plasma chemical vapor deposition from a tetramethyldisilazane precursor. 2. Properties of deposited silicon carbonitride films, *Thin Solid Films*, 2006, 497, 35–41.
[102] A. Choukourov, P. Pleskunov, D. Nikitin, V. Titov, A. Shelemin, M. Vaidulych, A. Kuzminova, P. Solař, J. Hanuš, J. Kousal, O. Kylián, D. Slavínská and H. Biederman, Advances and challenges in the field of plasma polymer nanoparticles, *Beilstein Journal Nanotechnology*, 2017, 8, 2002–2014.
[103] A. Saha, S.R. Shah and R. Raj, Oxidation Behavior of SiCN–ZrO_2 Fiber Prepared from Alkoxide-Modified Silazane, *Journal of American Ceramic Society*, 2004, 87(8), 1556–1558.
[104] G. Glatz, T. Schmalz, T. Kraus, F. Haarmann, G. Motz and R. Kempe, Copper-Containing SiCN Precursor Ceramics (Cu@SiCN) as Selective Hydrocarbon Oxidation Catalysts Using Air as an Oxidant, *Chemistry – A European Journal*, 2010, 16, 4231–4238.
[105] S.C. Turdean-Ionescu, C. Bazarjani, M.S. Tai, C.-W. Li, D. Fasel, C. Donner, W. Shen, J. Riedel, R. Gurlo and A.E. Mattias, Silicon oxycarbonitrides synthesized by ammonia-assisted thermolysis route from polymers: A total X-ray scattering, solid-state NMR, and TEM structural study, *Journal of the European Ceramic Society*, 2016, 35, 979–989.

[106] H.T. Chiu, T. Sukachonmakul, C.H. Wang, K. Wattanakul, M.T. Kuo and Y.H. Wang, Effect of pyrolysis atmospheres on the morphology of polymer-derived silicon oxynitrocarbide ceramic films coated aluminum nitride surface and the thermal conductivity of silicone rubber composites, *Applied Surface Science*, 2014, 292, 319–327.

[107] H.-T. Chiu, T. Sukachonmakul, C.-H. Wang, K. Wattanakul, M.-T. Guo and Y.-H. Wang, Fabrication and characterization of silicon-based ceramic/aluminum nitride as thermally conductive hybrid filler in silicone rubber composite, *Materials Chemistry and Physics*, 2014, 147, 11–16.

[108] Y. Iwamoto, K. Kikuta and S. Hirano, Synthesis of Poly-Titanosilazanes and Conversion into Si_3N_4-TiN Ceramics, *Journal of the Ceramic Society of Japan*, 2000, 108(4), 350–356.

[109] M.C. Bechelany, V. Proust, C. Gervais, R. Ghisleni, S. Bernard and P. Miele, In Situ Controlled Growth of Titanium Nitride in Amorphous Silicon Nitride: A General Route Toward Bulk Nitride Nanocomposites with Very High Hardness, *Advanced Materials*, 2014, 26, 6548–6553.

[110] P. Ritterskamp, A. Kuklya, M.-A. Wüstkamp, K. Kerpen, C. Weidenthaler and M. Demuth, A Titanium Disilicide Derived Semiconducting Catalyst for Water Splitting under Solar Radiation – Reversible Storage of Oxygen and Hydrogen, *Angewandte Chemie International Edition*, 2007, 46, 7770–7774.

[111] L. Meng, X. Zhang, Y. Tang, K. Su and J. Kong, Hierarchically porous silicon–carbon–nitrogen hybrid materials towards highly efficient and selective adsorption of organic dyes, *Scientific Reports*, 2015, 5(7910), 1–16. DOI: 10.1038/srep07910.

[112] G. Shao, W. Peng, C. Ma, W. Zhao, J. Guo, Y. Feng, H. Wang, R. Zhang and L. An, Enhanced electric conductivity of polymer-derived SiCN ceramics by microwave post-treatment, *Journal of the American Ceramic Society*, 2017, 100, 842–847.

[113] A. Dumitru, V. Ciupina, I. Stamatin, G. Prodan, A. Morozan and C. Mirea, Plasma polymerization of ferrocene with silane and silazane monomers for design of nanostructured magnetic ceramics, *Journal of Optoelectronics and Advanced Materials*, 2006, 8, 50–54.

[114] X. Wang, G. Mera, K. Morita and E. Ionescu, Synthesis of polymer-derived graphene/silicon nitride-based nanocomposites with tunable dielectric properties, *Journal of the Ceramic Society of Japan*, 2016, 124(10), 981.

[115] A. Lale, M. Schmidt, M.D. Mallmann, A.V.A. Bezerra, E.D. Acosta, R.A.F. Machadob, U.B. Demircic and S. Bernard, Polymer-Derived Ceramics with engineered mesoporosity: From design to application in catalysis, *Surface & Coatings Technology*, 2018, 350, 569–586.

[116] Y.F. Shi, Y. Meng, D.H. Chen, S.J. Cheng, P. Chen, T.F. Yang, Y. Wan and D.Y. Zhao, Highly ordered mesoporous silicon carbide ceramics with large surface areas and high stability, *Advanced Functional Materials*, 2006, 16, 561–567.

[117] J. Yan, A. Wang and D.-P. Kim, Preparation of ordered mesoporous SiCN ceramics with large surface area and high thermal stability, *Microporous Mesoporous Mater.*, 2007, 100, 128–133.

[118] C. Salameh, A. Bruma, S. Malo, U.B. Demirci, P. Miele and S. Bernard, Monodisperse platinum nanoparticles supported on highly ordered mesoporous silicon nitride nanoblocks: Superior catalytic activity for hydrogen generation from sodium borohydride, *RSC Advances*, 2015, 5, 58943–58951.

[119] O. Majoulet, C. Salameh, M.E. Schuster, U.B. Demirci, Y. Sugahara, S. Bernard and P. Miele, Synthesis of periodic mesoporous silicon-aluminum-carbon-nitrogen frameworks surface-decorated with Pt nanoparticles, *Chemistry of Materials*, 2013, 25, 3957–3970.

[120] A. Lale, A. Wasan, R. Kumar, P. Miele, U.B. Demirci and S. Bernard, Organosilicon polymer-derived mesoporous 3D silicon carbide, carbonitride and nitride structures as platinum supports for hydrogen generation by hydrolysis of sodium borohydride, Intern. *Journal of Hydrogen Energy*, 2016, 41, 15477–15488.

[121] M. Kamperman, A. Burns, R. Weissgraeber, N. Van Vegten, S.C. Warren, S.M. Gruner, A. Baiker and U. Wiesner, Integrating structure control over multiple length scales in porous high temperature ceramics with functional platinum nanoparticles, *Nano Letters*, 2009, 9, 2756–2762.
[122] J. Liu, A.Y. Kim, L.Q. Wang, B.J. Palmer, Y.L. Chen, P. Bruinsma, B.C. Bunker, G.J. Exarhos, G.L. Graff, P.C. Rieke, G.E. Fryxell, J.W. Virden, B.J. Tarasevich and L.A. Chick, *Advances in Colloid and Interface Science*, 1996, 69, 131–180.
[123] Y. Wan, Y. Shi and D. Zhao Designed synthesis of mesoporous solids via nonionicsurfactant-templating approach, *Chemical Communications*, 2007, 897–926.
[124] N. Pal and A. Bhaumik, Soft templating strategies for the synthesis of mesoporousmaterials: inorganic, organic-inorganic hybrid and purely organic solids, *Advances in Colloid and Interface Science*, 2013, 189–190, 21–41.
[125] Y. Shi, Y. Wan and D. Zhao, Ordered mesoporous non-oxide materials, *Chemical Society Reviews*, 2011, 40, 3854–3878.
[126] http://www.matweb.com/search/datasheettext.aspx?matguid=d401532d736946d69341fb28a58e1b98
[127] (a) U.S. Pat. 4,929,704 (1990); (b) U.S. Pat. 5,001,090 (1991); (c) U.S. Pat. 5,021,533 (1991); (d) U.S. Pat. 5,032,649 (1991); (e) U.S. Pat. 5,155,181 (1992); (f) U.S. Pat. 6,329,487 (2001).
[128] R.L.K. Matsumoto, Crystallization of pyrolyzed polysilazanes, *Materials Research Society Symposium Proceedings*, 1990, 180, 797–800.
[129] Y. Liu, H. Hou, X. He and W. Yang, Mesporous 3*C*-SiC Hollow Fibers, *Scientific Reports*, 2016, 7(1893) DOI: 10.1038/s41598-017-02147-8.
[130] D. Bahloul, M. Pereira and C. Gerardin, Pyrolysis chemistry of polysilazane precursors to silicon carbonitride Part 1. Thermal degradation of the polymers, *Journal of Materials Science*, 1997, 7, 109–116.
[131] C. Haluschka, H.J. Kleebe, R. Franke and R. Riedel, Silicon carbonitride ceramics derived from polysilazanes Part I. Investigation of compositional and structural properties, *Journal of the European Ceramic Society*, 2000, 20, 1355–1364. DOI: https://doi.org/10.1016/S0955-2219(00)00010-8.
[132] C. Haluschka, C. Engel and R. Riedel, Silicon carbonitride ceramics derived from polysilazanes Part II. Investigation of electrical properties, *Journal of the European Ceramic Society*, 2000, 20, 1365–1374. DOI: 10.1016/S0955-2219(00)00009-1.
[133] Y.L. Li, E. Kroke, R. Riedel, C. Fasel, C. Gervais and F. Babonneau, Thermal cross-linking and pyrolytic conversion of poly(ureamethylvinil) silazanes to silicon-based ceramics, *Appl. Organometal. Chem.*, 2001, 15, 820–832.
[134] C.A. Lewinson, P. Colombo, I. Reimanis and O. Unal, Stress occurring during joining of ceramics using preceramic polymers, *Journal of the American Ceramic Society*, 2001, 84, 2240–2244.
[135] N. Fainer, M. Kosinova, Y. Rumyantsev, E. Maksimovskii, F. Kuznetsov, V. Kesler, V. Kirienko, H.B. Shan and L. Cheng, Synthesis and physicochemical properties of nanocrystalline silicon carbonitride films deposited by microwave plasma from organoelement compounds, *Glass Physics and Chemistry-Engl. Tr.*, 2005, 31, 427–433.
[136] D.H. Lee, K.H. Park, L.Y. Hong and D.P. Kim, SiCN ceramic patterns fabricated by soft lithography techniques, *Sensors and Actuators A: Physical*, 2007, 135, 895–901.
[137] G. Mera, A. Navrotsky, S. Sen, H.-J. Kleebe and R. Riedel, Polymer-derived SiCN and SiOC ceramics–structure and energetics at the nanoscale, *Journal of Materials Science A*, 2013, 1, 3826–3836.
[138] G. Mera, R. Riedel, F. Poli and K. Müller, Carbon-rich SiCN ceramics derived from phenyl-containing poly(silylcarbodiimides), *Journal of the European Ceramic Society*, 2009, 29, 2873–2883.

[139] R. Riedel, E. Kroke, A. Greiner, A.O. Gabriel, L. Ruwisch, J. Nicolich and P. Kroll, Inorganic solid-state chemistry with main group element carbodiimides, *Chemistry of Materials*, 1998, 10, 2964–2979.
[140] A.O. Gabriel, R. Riedel, S. Storck and W.F. Maier, Synthesis and thermally induced ceramization of a non-oxidic poly(methylsilsesquicarbodiimide) gel, *Applied Organometallic Chemistry*, 1997, 11, 833–841.
[141] Y. Iwamoto, W. Volger, E. Kroke, R. Riedel, T. Saitou and K. Matsunaga, Crystallization behavior of amorphous silicon carbonitride ceramics derived from organometallic precursors, *Journal of the American Ceramic Society*, 2001, 84, 2170–2178.
[142] A. Kienzle, J. Bill, F. Aldinger and R. Riedel, Nanosized Si–C–N-powder by pyrolysis of highly crosslinked silylcarbodiimide, *Nanostructured Materials*, 1995, 6, 349–352.
[143] R. Riedel, A. Greiner, G. Miehe, W. Dressier, H. Fuess, J. Bill and F. Aldinger, The first crystalline solids in the ternary Si–C–N system, *Angewandte Chemie International Edition*, 1997, 36, 603–606.
[144] W. Dressler and R. Riedel, Progress in silicon-based non-oxide structural ceramics, *International Journal of Refractory Metals and Hard Materials*, 1997, 15, 13–47.
[145] S. Widgeon, G. Mera, Y. Gao, E. Stoyanov, S. Sen, A. Navrotsky and R. Riedel, Nanostructure and energetics of carbon-rich SiCN ceramics derived from polysilylcarbodiimides: Role of the nanodomain interfaces, *Chemistry of Materials*, 2012, 24(6), 1181–1191. DOI: https://doi.org/10.1021/cm2038238.
[146] S. Nahar-Borchert, E. Kroke, R. Riedel, B. Boury and R.J.P. Corriu, Synthesis and characterization of alkylene-bridged silsesquicarbodiimide hybrid xerogels, *Journal of Organometallic Chemistry*, 2003, 686(1–2), 127.
[147] M. Graczyk-Zajac, G. Mera, J. Kaspar and R. Riedel, Electrochemical studies of carbon-rich polymer-derived SiCN ceramics as anode materials for lithium-ion batteries, *Journal of the European Ceramic Society*, 2010, 30, 3235–3243.
[148] R.M. Prasad, G. Mera, K. Morita, M. Müller, H.-J. Kleebe, A. Gurlo, C. Fasel and R. Riedel, Thermal decomposition of carbon-rich polymer-derived silicon carbonitrides leading to ceramics with high specific surface area and tunable micro- and mesoporosity, *Journal of the European Ceramic Society*, 2012, 32, 477–484.
[149] T.L. Heying, S. Papetti and O.G. Schaffling, U.S. Pat., 3,388,090 (1968).
[150] S. Papetti, B.B. Schaeffer, A.P. Grey and T.L. Heying, *Journal of Polymer Science Part A*-l, 1966, 4(6), 1623–1636.
[151] H. Schroeder, Chapter: Polymer chemistry of boron cluster compounds, *Inorganic Macromolecules Review*, 1970, 1, 45.
[152] T.L. Heying, S. Papetti and O.G. Schaffhng, U.S. Pat., 3,388,091 (1968).
[153] T.L. Heying, S. Papetti and O.G. Schaffling, U.S. Pat., 3,388,092 (1968).
[154] S. Packkirisamy, Decaborane(14)-basd polymers, *Progress in Polymer Science*, 1996, 21, 707–773. DOI: https://doi.org/10.1016/0079-6700(96)00002-0.
[155] P.R. Dvomic and R.W. Lenz, *High Temperature Siloxane Elastomers,* 233–305, Hitthig and Wepf Verlag, Base1, 1990.
[156] T.M. Keller and L.J. Henderson, Carborane-(siloxane or silane)-unsaturated hydrocarbon based polymers, US Pat. 5,272,237 (1993).
[157] V. Liebau, R. Hauser and R. Riedel, Amorphous SiBCO ceramics derived from novel polymeric precursors, *Comptes Rendus Chimie*, 2004, 7, 463–469.
[158] R. Pena-Alonso and G.D. Soraru, Synthesis and characterization of hybrid borosiloxane gels as precursors for Si–B–O–C fibers, *Journal of Sol-Gel Science and Technology*, 2007, 43, 313–319.

[159] R.L. Siqueira, I.V.P. Yoshida, L.C. Pardini and M.A. Schiavon, Poly(borosiloxanes) as Precursors of Ceramic Matrix in Carbon Fiber Composites p.186, IIBCC 2006-10th Int. Inorganic-Bonded Fiber Composite Conference, Sao Paulo, Brazil. October 15–18, 2006. Universidade de Sao Paulo & University of Idaho: Sao Paulo, 2006.
[160] R.L. Siqueira and I.V.P. Yoshida, Poly(borosiloxanes) as Precursors for Carbon Fiber Ceramic Matrix Composites, *Materials Research*, 2007, 10(2), 147–151.
[161] M.A. Schiavon, N.A. Armelin and I.V.P. Yoshida, Novel poly(borosiloxane) precursors to amorphous SiBCO ceramics, *Materials Chemistry and Physics*, 2008, 112, 1047–1054.
[162] D. Devapal, K.J. Sreejith, B. Swaminathan, S. Chinthalapalli, S. Bhuvaneswari and S. Packirisamy, Influence of Heat Treatment Temperature on the Microstructure Evolution of Poly(vinylborosiloxane) Derived Ceramics, *Journal of Inorganic and Organometallic Polymers and Materials*, 2020, 30, 2224–2233.
[163] G.D. Sorarù, G. D'Andrea, R. Campostrini, F. Babonneau and G. Mariotto, Structural characterization and high-temperature behavior of silicon oxycarbide glasses prepared from sol-gel precursors containing Si-H bonds, *Journal of the American Ceramic Society*, 1995, 78, 379.
[164] A. Karakuscu, R. Guider, L. Pavesi and G.D. Sorarù, Broad-band tunable visible emission of sol–gel derived SiBOC ceramic thin films, *Thin Solid Films*, 2011, 519, 3822–3826.
[165] G.D. Sorarù, A. Karakuscu, C. Boissiere and F. Babonneau, On the shrinkage during pyrolysis of thin films and bulk components: The case of a hybrid silica gel precursor for SiOC glasses, *Journal of the European Ceramic Society*, 2012, 32, 627–632.
[166] X. Li, G. Wen, T. Zhang, L. Xia, B. Zhong and S. Fan, Large-scale Synthesis of SiBOC Fibers Produced by Electrospinning, *Physics Procedia*, 2012, 25, 189–193.
[167] B. Swaminathan, A. Painuly, S.K. Manwatkar, S. Packirisamy, A.N.I.L. Painuly and S.K. Manwatkar, *Preceramic Polymer-Derived C/C-SiC and C/C-SiBCO Composites for High Temperature Applications, High Temperature Ceramic Materials and Composites,* W. Krenkel and J. Lamon, Eds., VISO Verlagsgegelschaft mbH, Berlin (Germany), 2010, 724–729.
[168] M.A. Schiavon, G.D. Soraru and I.V.P. Yoshida, Poly(borosilazanes) as precursors of SiABACAN glasses: synthesis and high temperature properties, *Journal of Non-Crystalline Solids*, 2004, 348, 156–161.
[169] Q.D. Nghiem and D.-P. Kim, Polymerization of Borazine with Tetramethyldivinyldisilazane as a New Class SiCBN Preceramic Polymer, *Journal of Industrial and Engineering Chemistry*, 2006, 12(6), 905–910.
[170] S. Ijadi-Maghsoodi, Y. Pang and T.J. Burton, Efficient, "one-pot" synthesis of silylene–acetylene and disilylene–acetylene preceramic polymers from trichloroethylene, *Journal of Polymer Science Part A: Polymer Chemistry*, 1990, 28, 955.
[171] S. Ijadi-Maghsoodi and T.J. Burton, Synthesis and study of silylene-diacetylene polymers, *Macromolecules*, 1990, 23, 4485.
[172] L.J. Henderson Jr. and T.M. Keller, Chapter 21: *Poly(carborane-siloxane-acetylene) as Precursor to High Temperature Termoset and Ceramic, Inorganic and Organometallic Polymers II, Advanced Materials and Intermediates*, P. Wisian-Neilson, H.R. Allcock and K.J. Wynne, Eds., *ACS Symp. Series*, Vol. 572, 1994, 416–426.
[173] Y.Q. Peng, K.Q. Han, X. Zhao, Z.H. Wang, Z.H. Deng, Z.Q. Liu, J. Zhang, X.D. Li, Z.Y. Sun and M.H. Yu, Preparing continuous SiBN ceramic fiber from Precursor Polymer of N-methylpolyborosilazanes, 18th Intern. Conference on Composite Materials, 21–26 August, 2011, Jeju Island, Korea.
[174] T. Wideman, K. Su, E.E. Remsen, G.A. Zank and L.G. Sneddon, Synthesis, Characterization, and Ceramic Conversion Reactions of Borazine/Silazane Copolymers: New Polymeric Precursors to SiNCB Ceramics, *Chemistry of Materials*, 1995, 7, 2203–2212.

[175] J. Lee, D.P. Butt, R.H. Baney, C.R. Bowers and J.S. Tulenko, Synthesis and pyrolysis of novel polysilazane to SiBCN ceramic, *Journal of Non-Crystalline Solids*, 2005, 351, 2995–3005.
[176] A.B. Kousaalaya, R. Kumar and S. Packirisamy, Characterization of free carbon in the as-thermolyzed Si–B–C–N ceramic from a polyorganoborosilazane precursor, *Journal of Advanced Ceramics*, 2013, 2, 325–332.
[177] Y. Wang, L. Chen, T. Xu, Y. Yan, J. Gu, J. Yun and J. Feng, High char yield novolac modified by Si-B-N-C precursor: Thermal stability and structural evolution, *Polymer Degradation and Stability*, 2017, 137, 184–196. DOI: https://doi.org/10.1016/j.polymdegradstab.2017.01.013.

Chapter 9
Polycrystalline silicon, silicon nanoparticles and nanowires

9.1 Polycrystalline silicon and silicon nanoparticles

The metallurgical-grade silicon should have at least 98% purity, while the solar-grade silicon has purity of 6N class (99.9999%). The crystalline silicon is a very important element in the semiconductor industry. The very pure crystalline silicon (at least 99.9999999%) has the 9N class of purity and it is a basic electronic material for a production of computer chips and other high-technology semiconductor electronic devices [1, 2]. Single silicon crystals are most often prepared by Czochralski's method [3].

Polycrystalline silicon is commonly called as *polysilicon* and it is most often purified by so called Siemens process, which involves distillation of volatile silicon compounds. The obtained pure monosilane (SiH_4) is further decomposed into silicon at high temperatures [4, 5]. Polysilicon is the purest synthetic material on the market and it is the basic material for large-scale photovoltaic (PV) applications as solar cells. For many current optoelectronic applications, high-purity silicon is required, but for PV applications, it is not necessary, and therefore alternate processes and materials are used for fabrication of PV-grade silicon [6–13].

Crystalline silicon (c-Si) and amorphous silicon (a-Si) are the most fundamental, fully inorganic materials used for microelectronics, optoelectronics, and photonics, which are realized through the industrial lithographic and *p-n* doping processes. These materials are also produced in vacuum and vapor-phase deposition processes and mechanical slicing and polishing techniques of Si-wafers. An amorphous silicon (a:Si) is produced by glow discharge of SiH_4 or Si_2H_6. It can be applied in thin-film solar cells [14, 15]. Polysilicon is mostly used for the semiconductor industry. Solar energy is newer application for polycrystalline silicon [16].

Silicon nanoparticles (NPs) were prepared by a new combined vapor phase and solution phase process. Pyrolysis of monosilane (SiH_4), induced with CO_2 laser, was used for fabrication of Si NPs (with a very small average diameter 5 nm) at rather high rates (20–200 mg/h). Silicon NPs were prepared directly by this vapor phase (aerosol) synthesis and were etched with mixtures of hydrofluoric acid (HF) and nitric acid (HNO_3) leading to reduction of the particles size and passivation of their surfaces. After acidic etching, they exhibited bright visible photoluminescence (PL) at room temperature. The wavelength of maximum PL intensity was tuned from above 800 nm to below 500 nm by controlling the conditions and etching time. NPs with blue emission (maximum PL intensity at 420 nm) were prepared by rapid thermal oxidation of orange-emitting particles. The silicon NPs were characterized by FTIR

https://doi.org/10.1515/9783110643671-009

spectroscopy, transmission electron microscopy (TEM), X-ray diffraction (XRD) technique, and X-ray photoelectron spectroscopy (XPS). The stability of the PL properties with time was studied as well [17].

A method for efficient grafting of organic molecules onto photoluminescent silicon NPs was described by F. Hua et al. [18]. ^{1}H-NMR was used for monitoring the photoinitiated hydrosilylation reaction on the surface of silicon. High coverage grafting was realized by use of the modified etching process, which led to H-Si-terminated surface on the NPs contaminated with traces of oxygen. It was necessary to avoid a presence of oxygen during the grafting process. Thus, it was possible to:
(1) prepare air-stable green-emitting silicon NPs,
(2) prepare stable dispersions of grafted silicon NPs in a variety of organic solvents from which particles were readily precipitated by addition of nonsolvent, dried, and redispersed,
(3) separate these NPs by size (and therefore, emission color) using conventional chromatographic methods,
(4) protect the particles from chemical reagents and PL quenching, and
(5) introduce functional groups on the particle surface for further derivatization.

Graphene-wrapped Si NPs (G/Si NPs) supported in carbon composite nanofibers containing 5–20 wt% G/Si, were fabricated by electrospinning and subsequent thermal treatment. They can be used as an anode material for lithium-ion batteries. The graphene wrapping the Si NPs helped to prevent agglomeration and ensured a good dispersion of Si NPs inside the carbon nanofibers (CNF) matrix [19]. Silicon NPs, prepared by reduction of nanosilica with Mg NPs, were converted with very high yield of 89% into an alkoxyl-functionalized hydrophobic colloid by reacting with 1-pentanol. Its dispersion in ethanol was mixed with ethanolic solution of oligomeric phenolic precursor resin and was used for preparation surface carbon coating. The Si NPs coated with carbon exhibited excellent lithium storage performance comparable to the Si-based anode materials for lithium-ion batteries [20].

Recently a porous poly(methyl methacrylate) (PMMA) film on a Si surface with a novel dipping method was prepared [21]. The Si substrate was directly dipped into acidic aqueous media at 10 °C. This process was initiated by spontaneous reduction of 4-nitrobenzene diazonium tetrafluoroborate at the Si surface which led to the grafting of aryl radicals onto the Si surface. Thus, the polymerization of methyl methacrylate (MMA) was initiated by aryl radicals, and the PMMA chains, terminated with radical, formed in the solution were also grafted onto the silicon surface. The grafted PMMA chains underwent aggregation and formed a porous polymer film. The porosity of the film was dependent on the grafting time and the MMA concentration. Excellent dielectric properties of PMMA film on silicon support may lead to many industrial applications.

The preparation of SiO_2-embedded silicon nanocrystals (Si-NCs) from the thermal processing of silsesquioxane polymers prepared from trichlorosilane ($HSiCl_3$)

was described by E.J. Henderson et al. [22]. In reaction of water with $HSiCl_3$ a cross-linked hydrogen silsesquioxane $(HSiO_{1.5})_n$ was formed, which was a precursor of bulk quantities of SiO_2-embedded Si-NCs. Methyl-modified silsesquioxane copolymers $(HSiO_{1.5})_n(CH_3SiO_{1.5})_m$ ($m \ll n$, $m + n = 1$) were also prepared, with a low content of methyl groups, which acted as a polymer network modifier and affected the formation of Si-NCs during thermal process. The Si-NC size was readily tuned by controlled variations of processing temperature for $(HSiO_{1.5})_n$ and composition (n and m) for $(HSiO_{1.5})_n(CH_3SiO_{1.5})_m$. The obtained Si-NCs exhibited size-dependent PL. Freestanding Si-NCs were obtained through chemical etching of the oxide matrix and exhibited tunable PL throughout the visible spectrum.

Silicon NPs are useful materials for solar cells, optoelectronic devices, and biological markers. The synthesis of air-stable Si NPs with tunable optoelectronic properties is a very important process. The mechanochemical synthesis of silicon NPs through high-energy ball milling gave different covalently bonded surfaces, depending on the nature of the organic liquid used in the milling process. The use of the C_8 substrates: 1-octanol, 1-octaldehyde, octanoic acid, and 1-octene resulted in surfaces functionalized with strong Si–C bonds or strong Si–O bonds. The surfaces of the NPs were characterized by IR and NMR spectroscopy. The grafted Si NPs were remarkably stable against air oxidation and agglomeration and were soluble in organic solvents. However, the luminescence and optical properties of the NPs were very sensitive to the nature of their modified surface [23].

Silicon (Si) plays dominating role in the PVs industry and some new attempts were introduced in order to reduce its fabrication cost. In recent years cyclohexasilane (Si_6H_{12}) was used by P. Boudjouk and coworkers as a new liquid precursor to amorphous silicon and silicon-containing electronic materials: amorphous silicon (a-Si:H) field-effect transistors (FETs) and rectifying diodes [24, 25]. Spin coating of Si_6H_{12}-based inks with subsequent polymerization in UV light during spin coating led to the formation of a-Si:H films, which were crystallized via thermal and laser annealing. The n-type and p-type a-Si were formed which were used in heterojunction structures (i.e., thin a-Si films on heavily doped Si wafers) [25]. The possibility of generating 10 lm wide lines of a-Si materials by collimated Aerosol Beam deposition Direct-Write (CAB-DWTM) of Si_6H_{12}-based inks was reported. A new printing approach (i.e., collimated aerosol beam direct write, CAB-DW) allowed the deposition of silane-based features with line widths <10 μm [26]. The silicon-based materials with good electrical properties, prepared at low cost, may be applied as semiconductor devices. While a-Si nanowires (NWs) obtained by electrospinning method from Si_6H_{12}-polymer inks can be used for lithium ion battery anodes [27].

Through thermal and laser annealing polysilane films were crystallized. Hydrogenation in plasma was carried out at atmospheric pressure which caused further structural changes in these films. The transformation from amorphous to microcrystalline phase was studied using scanning electron microscopy (SEM), atomic force microscopy, Raman and impedance spectroscopy. A four-decade increase in the electrical conductivity was attributed to a disorder–order transition in a bonded Si network. These results showed a potentially attractive use of a solution process followed by ambient postprocessing for preparation of crystalline thin silicon films [28, 29].

An application of a commercially available, high purity (semiconductor grade, >99.99%) pyrophoric trisilane (Si_3H_8), for the preparation of a printable silicon ink was reported by A.P. Cádiz Bedini et al. [30]. This synthesis was carried out in solution at room temperature (or below) in nitrogen atmosphere, at ambient pressure, and involved an initial sonication step, followed by irradiation with ultraviolet light. The production of higher order silanes was accelerated by ultrasounds. Formation of a highly branched silicon hydride polymer was achieved by a combined sonication and photolysis methods, which were monitored by a gas chromatography and NMR spectroscopy. Silicon NPs were analyzed by SEM. The obtained silicone ink was used in the field of printable electronics for preparation of amorphous silicon thin films by spin coating and atmospheric pressure chemical vapor deposition. Optoelectronic properties of the Si NPs were similar to those obtained by plasma-enhanced chemical vapor deposition (PECVD) material.

Pristine-soluble hydrogen polysilane $H(SiH_2)_nH$ molecules were used as a precursor material for the solution-based deposition of hydrogenated amorphous Si (a-Si:H) thin-films for less expensive PV devices. Three different synthetic methods were applied: (1) hydrogenation of Si anionic compound, (2) Würtz-type reductive coupling reaction, and (3) the dehydrocoupling reaction. The Würtz-type reaction of diiodosilane as a direct synthetic route to hydrogenated polysilane was carried out under air-free conditions, which were necessary in order to avoid the formation of siloxane bonds from the spontaneous reaction between silane-based molecules and moisture from air. The hydrogenation of $CaSi_2$ and the dehydrocoupling reaction of phenylsilane gave more stable polysilane [31].

A synthesis of colloidal silicon nanocrystals (Si NCs), with widely tunable average diameter, from less than 3 to 90 nm, was described by C.M. Hessel et al. [32]. The synthesis was carried out at the high decomposition temperature (>1,100 °C) of hydrogen silsesquioxane (HSQ) to form Si quantum dots with good crystallinity and a narrow size distribution with tunable size embedded in SiO_2. The silica matrix was etched with HF in the dark, followed by thermal hydrosilylation with alkenes, giving solvent-dispersible free Si nanocrystals exhibiting bright PL. The prepared Si NCs showed peak PL from visible wavelengths to the bulk band gap of Si at 1,100 nm. Their properties and relationship between PL energy and size were confirmed by TEM, small-angle X-ray scattering (SAXS), and XRD [32].

A dual channel fluorescence system which combined the optical properties of nanocomposites containing soluble polysilane, silicon quantum dots, and a fluorescent cytotoxic agent 2-(4-chlorophenyl)-6-(thiophen-2-yl)pyridine, was obtained by L. Sacarescu et al. [33].

A highly stable colloidal dispersion of silicon quantum dots–polysilane nanocomposite was prepared by one-pot synthetic route based on microwave-assisted Würtz

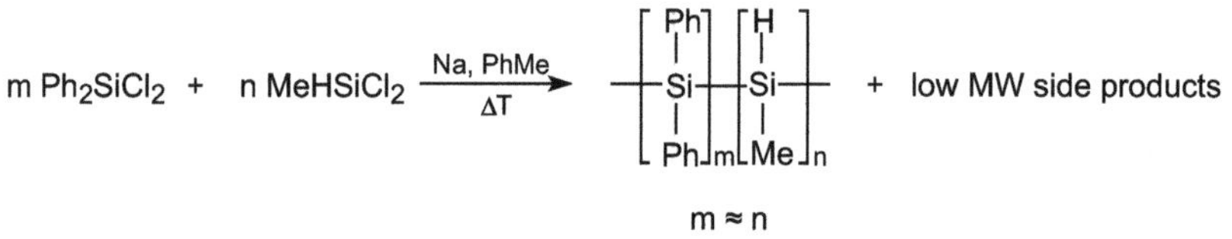

Figure 9.1: Synthesis of $(Ph_2Si)_m$-*co*-$(MeHSi)_n$ [33].

coupling of equimolar amounts of diphenyl(dichloro)silane Ph_2SiCl_2 and methyl-(dichloro)silane $MeHSiCl_2$ with a dispersion of sodium in toluene, at temperature 120 °C within 10 min. (Figure 9.1). Elimination and insertion reactions of silylenes (MeHSi: and H_2Si:), as well elimination of hydrogen, were responsible for the formation of silicon quantum dots. The size and shape of the silicon quantum dots which were stabilized within the polysilane thin film were studied by TEM. The colloidal dispersions were investigated by SAXS, which proved that polysilane prevented aggregation of Si NPs, which alternatively emitted signals at two different wavelengths by excitation at a single wavelength. UV-vis spectrophotometry of the silicon quantum dots–polysilane NCs in the presence of 2-(4-chlorophenyl)-6-(thiophen-2-yl)pyridine was used to define the active wavelength range and establish the fluorescence detection method [33].

Nanocrystalline Si-SiO_2 (*nc*-Si/SiO_2) nanocomposites were prepared in bulk through reductive thermal annealing of a well-defined molecular precursor, HSQ, in 4% H_2/96% N_2 atmosphere, within 1 h (see Scheme 9.1). The presented method gave large quantities and quantitative yields of composite powders. Freestanding, hydride-surface-terminated silicon NCs showed PL in the visible light, after etching *nc*-Si/SiO_2 NC powders in ethanol-water solutions of HF acid [34].

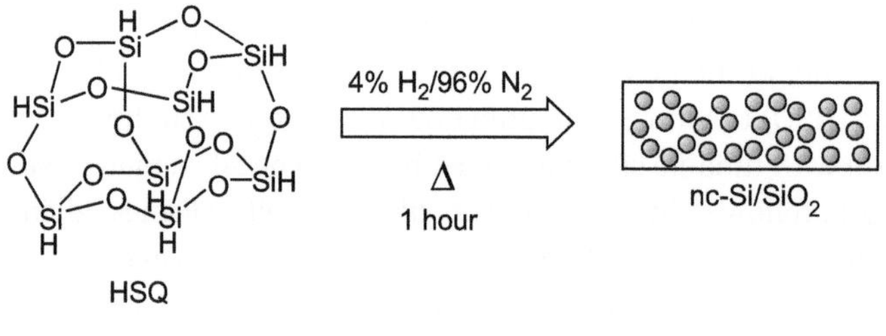

Scheme 9.1: Thermal decomposition of hydrosilsesquioxane (HSQ) into silicon nanoparticle/SiO_2-like (*nc*-Si/SiO_2) nanocomposites. Reprinted (adapted) with permission from [34]. Copyright 2006, by American Chemical Society.

Freestanding particles and composites were characterized by means of FTIR, PL spectroscopy, TEM, XRD, XPS, selected area electron diffraction, and thermogravimetric analysis [34].

Si-nanocrystals (Si NCs) were also prepared by annealing Si-rich oxide multilayer films in N_2 and Ar atmosphere. The XPS results showed that the Si-rich oxide (SRO) and SiO_2 layers were composed of $SiO_{1.1}$ and SiO_2, respectively. The FTIR results showed that phase separation between Si and SiO_2 occurred after annealing treatment, while the TEM images proved that Si NCs were formed, exhibiting interesting photoluminescent properties [35].

An organic–inorganic hybrid solar cell was prepared by addition of freshly etched silicon nanocrystals (Si-NCs) to poly(3-hexylthiophene) blend containing [6,6]-phenyl-C_{61}-butyric acid methyl ester. The incorporation of the Si-NCs into the bulk heterojunction structure stimulated the coupling of optical excitation between the polymers and Si-NCs, and caused an extended optical excitation response and improved power conversion efficiency of the hybrid solar cell [36].

Silicon is not only the most important material in modern semiconductor industry, but it is also one of the most widely studied materials used in light-emitting diodes and as antireflective coatings, thus it was widely used for minimization of reflection loss and good compatibility with silicon PVs (in solar cells). For the fabrication of silicon nanostructure arrays (NSAs) several effective strategies were adopted: vapor-phase growth or deposition and different multistep etching techniques (plasma, dry, and wet etching). Most silicon NSA coatings, with graded refractive index, were prepared by the etching methods due to the simple processing and low cost [37].

Pseudospherical alloy nanocrystals with a mixed composition $Si_{0.45}Ge_{0.55}$ were prepared through the thermally induced simultaneous disproportionation reactions of $H_8Si_8O_{12}$ (HSQ) and GeI_2 (see equations (9.1)–(9.3)) at temperatures 350–500 °C, leading to *nc*-$Si_{0.45}Ge_{0.55}$. The presence of mixed nanocrystals *nc*-$Si_{0.45}Ge_{0.55}$ was confirmed by Raman spectroscopy, TEM, XRD, and EDX. FTIR spectra indicated termination of NPs surfaces with M-H bonds (M = Si or Ge) after HF etching [38]:

$$H_8Si_8O_{12} \rightarrow 6SiO_2 + 2SiH_4 \quad (9.1)$$

$$SiH_4 \rightarrow Si + 2\ H_2 \quad (9.2)$$

$$2GeI_2 \rightarrow Ge + GeI_4 \quad (9.3)$$

The silicon–germanium nanocrystalline alloys Si_xGe_{1-x} ($64 < x < 100$) (NCs) were prepared by mixing GeI_2-trialkylphosphine adducts with HSQ. The homogeneous coprecipitation of GeI_2 and HSQ was observed, with improved control over the size and composition of the resulting Si_xGe_{1-x} NCs. The oxide-embedded and freestanding materials were obtained with near-infrared PL comparable in quantum efficiency to Si NCs. The formation of bimodal kinds of Si-rich and Ge-rich NCs was observed, with homogeneous distribution of Ge within both kinds of NCs. The effect of precursor stoichiometry and annealing temperature and time, on particle size, composition of NCs, and on the near-IR PL was studied. These materials were analyzed using FTIR and Raman spectroscopy, PL spectroscopy, high-resolution transmission electron microscopy (HRTEM), powder XRD, and spatially resolved energy-dispersive X-ray (EDX) spectroscopy [39].

The Au/*n*-Si metal–semiconductor (MS) type Schotthy diodes with and without interfacial ($Ca_{1.9}Pr_{0.1}Co_4O_x$) layer were prepared on P-doped (*n*-Si) single Si crystal [40, 41].

Hydroxylated surfaces of silicon nanostructures were modified by a ring opening reaction with heterocyclic silanes containing Si–N or Si–S bonds in the ring, to generate –SH or –NH functional surfaces, grafted through O–Si bonds. The reaction was much faster at 25 °C than hydrolytic condensation of (trialkoxy)silanes on similar hydroxy-substituted surfaces, and retained PL of the quantum-confined Si nanostructures. These surfaces can trap protein lysozyme, which showed its enzymatic activity upon release from the modified porous nanostructure [42].

Water-soluble silicon NPs were prepared by grafting PEG polymers onto functionalized silicon NPs with distal alkyne or azide moieties [43].

Semiconducting PV materials and photoelectrodes, made of silicon are useful for light absorption, charge separation, and charge transfer. Amorphous, microcrystalline, and nanocrystalline silicon has been often used (via complex cost effective technology) for preparation of crystalline and thin-film silicon-based photoelectrodes with enhanced performance, which can be immersed in aqueous solution for photoelectrochemical (PEC) production of hydrogen from water electrolysis. They were also applied in PEC CO_2 reduction and PEC regeneration of discharged species in redox flow batteries. Silicon-based photoelectrodes were recommended for storage of solar energy in chemical bonds. Thus, silicon was considered as "an ideal material for the cost-effective production of solar chemicals through the PEC methods" [44].

9.2 Silicon nanowires

Silicon nanowires (SiNWs) are important components for future electronic and sensor nanodevices [1]. SiNWs are also useful in many different fields of science (and especially in materials science) and the chemical technology. A huge number of publications has been devoted to fabrication methods and characterization of SiNWs, and studies concerning their applications in many fields [45–51].

SiNWs are one-dimensional (1D) quasi-structures with a diameter of less than 100 nm. They are prepared by different methods, depending on their future applications [52]. Owing to the unique quasi-1D electronic structure and high surface to volume silicon ratio, many devices based on SiNWs have better properties than their traditional counterparts. Electronic devices fabricated from SiNWs are a development of silicon metal insulator semiconductor transistors. The high surface to volume ratio leads to a very high sensitivity of the sensor devices. Solar cells and anodes in Li-ion batteries were improved by exploiting the quasi-one-dimensionality as well [53].

SiNWs are promising ingredients of nanoelectronic devices: solar cells, photonic and optoelectronic devices, field effect transistors *FETs), and chemical or biological sensors. Their chemical, mechanical, thermal, electronic, and optical properties are different than for bulk silicon, which is an indirect gap material. SiNWs are useful 1D nanostructures which change their physical properties from indirect to direct band gap [54–56]. They are prepared by many different methods: chemical vapor deposition (CVD) (thermal CVD, PECVD and hot-wire CVD (HWCVD)), electron beam evaporation, laser ablation, template assisted, solution growth, and so on.

SiNW growth on glass substrates using HWCVD takes place on the Au nanotemplate during passing the gaseous monosilane (SiH_4) over a hot filament heated up to 1,600 °C, causing dissociation of the SiH_4 into Si and H_2. The Si species react with Au NPs of the template and form liquid eutectic droplets [55]. It was proposed that the growth of SiNWs proceeded in the presence of metal catalysts according to the so called vapor–liquid–solid mechanism (VLS) [57–59]. At high temperature the gas monosilane (SiH_4), in vapor phase, is catalytically dissociated by metal-Si liquid alloy droplets. Then Si atoms dissolve in the alloy liquid droplets, followed by precipitation of SiNWs from the supersaturated alloy droplets. Schematic of catalytic growth of SiNWs via VLS mechanism are presented in Figures 9.2 and 9.3.

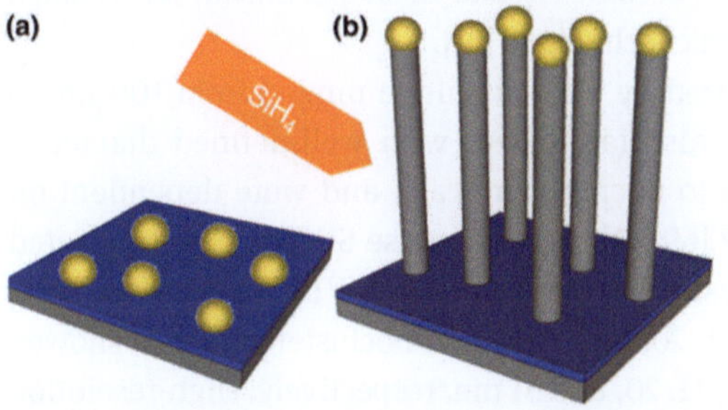

Figure 9.2: Schematic growth of silicon nanowires through VLS mechanism: (a) gold particles formed on the growth substrate; (b) VLS growth from monosilane as silicon precursor. Reprinted (adapted) with permission from [53]. Copyright 2015, by Springer Nature.

Other growth mechanisms for the SiNW synthesis were also proposed, such as solid–liquid–solid (SLS), vapor–solid–solid (VSS), oxide-assisted growth and stress mechanism. Yazawa, Westwater, and Ozaki [60–65] prepared SiNWs with VLS growth induced by Au metal layer on a Si surface. Yu et al. [66] elaborated oven-laser ablation method through simple physical evaporation approach for fabrication of very pure ultrafine freestanding SiNWs. Recently the wet chemical etching of SiNWs into a thin silicon layer on glass or single crystal silicon wafer substrates were described [47]. Their optical, optoelectronic, and PV properties were also studied.

SiNWs were prepared by the Si/SiGe epitaxy technique, combined with reactive ion etching and selective chemical etching. A high-quality $Si/Si_{1-x}Ge_x/Si$ heteroepitaxial

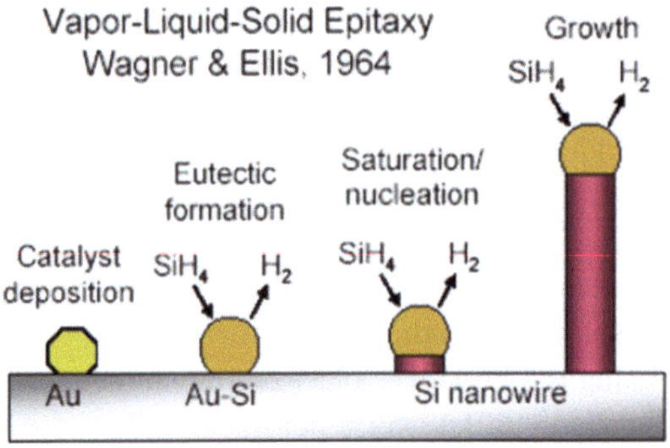

Figure 9.3: Schematic of catalytic growth of silicon nanowire growth according to VLS mechanism [54, 57].

film on a Si substrate was grown by very low-pressure chemical vapor deposition, followed by lithography and reactive ion etching to form new structures. The selective chemical wet etching with a solution of HNO_3, CH_3COOH, and diluted HF allowed to remove the $Si_{1-x}Ge_x$ layer and led to pentagon-shaped silicon wires with linewidth ~300 nm. The NWs were oxidized in wet oxygen at 750 °C or 850 °C and characterized by SEM. The samples oxidized at 850 °C became circular in shape. Silicon NWs coated with silica may be used in microelectronic technology [67].

The diameters of first NWs prepared by VLS technique ranged from 100 nm to 100 μm [57, 68]. Thinner NWs were also fabricated, with well-defined diameters ranging between few nanometers up to micrometer scale, and were dependent on the diameter of the metal catalyst NP [69–71]. Monodisperse SiNWs were fabricated by using gold nanoclusters as catalysts for 1D growth via a VLS mechanism. TEM studies of the SiNWs grown from 5, 10, 20, and 30 nm nanocluster catalysts showed that the NWs had mean diameters of 6, 12, 20, and 31 nm, respectively. High-resolution TEM (HRTEM) revealed that the NWs had single-crystal silicon cores coated with 1–3 nm of amorphous silica and that the cores remained highly crystalline for diameters as small as 2 nm [69].

SiNWs (up to 10 mm in length and 30–80 nm in diameter) were synthesized, with growth rates of up to 1 mm/min, by resistive heating of microline resistors and vapor-deposition in a room-temperature chamber [72]. The diameter of SiNWs affects the PL intensity [46]. Most SiNWs are quasi 1D structures with a diameter below 100 nm, which can be prepared with high precision using bottom-up techniques, and are useful as nanoelectronic materials. In recent years classical lithography techniques have been also applied for fabrication of SiNWs. Different electronic devices can strongly benefit from the high surface to volume ratio. The high surface to volume ratio has the potential to design very sensitive sensors, especially biosensors, which detect very small amounts of analytes [53].

SiNWs with single-crystal structures and controllable hole and electron doping were prepared by thermal annealing and surface passivation of oxide defects using chemical modification. Their average transconductance was increased from 45 to 800 nS and average mobility – from 30 to 560 cm^2/Vs with maximum values of 2,000 nS and 1,350 cm^2/Vs, respectively. The obtained SiNWs can be used as building blocks for nanoelectronics devices, for example, field effect transistors (FETs) [73].

Triangular-shaped SiNWs were also prepared from boron-doped silicon wafers by electrochemical method in aqueous HF and $AgNO_3$ solution. A series of SEM observations showed the detailed growth process of the SiNWs. The SiNWs and intermediate undetached Si nanotubes (NTs) were considered as promising materials for applications for future optoelectronic and nanoelectronic devices [74].

Gold was very often used as the catalyst for synthesis of Si NWs through the VLS mechanism [75–78]. However, gold traps electrons and holes in Si and causes a serious contamination problem for Si complementary metal oxide semiconductor (CMOS) processing. Other catalysts were also attempted for Si NW growth at high temperatures [79–83], but these new catalysts were not compatible with CMOS requirements. The use of aluminum for the epitaxial growth of Si NWs was reported for the first time by Wang et al. [82]. It was suggested that growth of SiNWs proceeded through a VSS rather than a VLS mechanism.

SiNWs were also prepared on silicon substrates through the thermal chemical vapor deposition system, reaction catalyzed by nickel, under N_2 atmosphere. Nickel catalyst (0.3–15.5 wt%) was deposited onto the silicon substrates by electroless nickel plating technique. The lengths, diameters, and growth densities of SiNWs were controlled by the content of the Ni layer on the silicon substrate [84].

SiNWs conductors with high aspect ratio (and width of 1.6 nm) were prepared by evaporation of silicon and its deposition on Ag surfaces [85].

A continuously growing need for rechargeable lithium batteries with higher energy capacity and longer cycle life is observed in the world – for applications in different electric vehicles, portable electronic devices, and implantable medical devices [86]. Silicon was found to be an attractive anode material for lithium batteries due to the highest known theoretical charge capacity (4,200 mAh/g [87]). Although silicon anodes are more than ten times better than graphite anodes and much larger than various nitride and oxide materials [88, 89], so far they have found limited applications [77], because silicon's volume changes by 400% upon insertion and extraction of lithium, resulting in pulverization and efficiency decrease [90]. In recent years it has been found that SiNW battery electrodes can accommodate large strain without pulverization, providing good electronic contact and conduction, and display short lithium insertion distances. For this purpose single-crystalline SiNWs were grown, by means the VLS growth method, on very thin stainless steel plates (0.002″ thick) coated with Au catalysts, inside a tube furnace. A Li foil as both reference and counter-electrodes was used, without any binders or conducting carbon additives. These substrates were heated to 530 °C and reacted in a chamber with 2% monosilane

SiH_4 (in Ar), at low pressure of (30 Torr). The α-$FeSi_2$ (up to ~25 wt%) was also formed. Their electrochemical properties were determined. The prepared SiNWs were used for lithium battery electrodes. C.K. Chan showed that SiNW battery electrodes accommodated large strain without pulverization, provided good electronic contact and conduction, and displayed short lithium insertion distances. The theoretical discharge capacity for silicon anodes close to 75% of maximum was achieved [91].

A new type of transistor with no junctions and no doping concentration gradients was fabricated from SiNWs. These devices showed extremely low leakage currents, near-ideal subthreshold slope, and less degradation of mobility with gate voltage and temperature than classical transistors, where semiconductor junctions were formed by introducing dopant atoms into the semiconductor material [92].

SiNWs were grown on gold-coated Si surfaces by chemical vapor deposition, via the vapor liquid solid growth, from a gaseous precursor, monosilane SiH_4, and gold as catalyst, in a stream of hydrogen H_2 as a carrier gas, in a low pressure chemical vapor deposition reactor, at constant pressure 20 mbar. Silicon pieces (10 × 10 mm^2) were first cleaned in a HF solution. A 2 nm thick Au layer was deposited onto these surfaces under a high vacuum pressure (10^{-6} Pa), at 850 °C for 10 min, under a H_2 flow. SiH_4 was introduced after cooling down the substrate to the growth temperature (450–650 °C). The length of Si NWs was dependent on their diameter. The length of NWs with diameter under 100 nm increased when diameter increased, because of an increase in the growth velocity. For the thicker diameter of SiNWs ($d > 100$ nm), length decreased when diameter increased. SiNWs were analyzed by SEM [93]. The diameter Si NWs were controlled by exploiting the difference in Au condensation coefficient on Si and SiO_2 surfaces at elevated temperature, in the range 520–700 °C. Au condensation was completely selective to Si [94].

SiNWs were also prepared at below Au–Si eutectic temperature from gaseous monosilane (SiH_4) and Au NP catalyst. SiNWs were grown onto Si (111) substrates, using very high frequency PECVD through a VSS mechanism at relatively low temperatures (363–230 °C). The morphology of the synthesized SiNWs was characterized by means of HRTEM, field-emission SEM equipped with EDX, XRD technique and Raman spectroscope [95].

Single-crystalline SiNWs were also synthesized on Si substrate coated with a SiO_x film in supercritical benzene, without nanocrystal metal catalysts. A 9 nm thick reactive SiO_x film on a silicon wafer was prepared by etching a Si substrate with boiling ultrapure water, followed by annealing at 1,100 °C for 30 min in argon atmosphere. SiNWs were prepared in supercritical fluid on the silicon substrate covered with SiO_x at temperatures 430–500 °C at 1,500 psi. A large amount of Si clusters with small diameters (2–3 nm) were formed on a surface of the reactive SiO_x layer, and these clusters most likely initiated growth of SiNWs [96].

Fibrous silicon NPs were synthesized from polished boron-doped silicon wafer targets, using a direct-diode-pumped Yb-doped fiber amplified femtosecond laser ablation with MHz pulse frequency at room temperature in air [97].

Thin film of NWs on the Si wafer were prepared from commercial Si wafers, which were first etched with a 5% HF aqueous solution for 5 min at room temperature. Then, the fresh Si surfaces that were Si–H terminated on the Si wafers were immediately dipped for 1 min into an Ag-coating solution containing 10% HF and 0.02 M $AgNO_3$. The wafers with a nonporous film of Ag were washed with water in order to remove any extra Ag^+ ions and were next immersed at 25 °C in an etching solution containing 10% HF and 0.6% H_2O_2. After 240 min of etching in the dark at room temperature, the wafers were washed with 10% HF solution in order to remove the oxide layer, and then were cleaned with isopropyl alcohol. The wafers were dried under a 6N-grade nitrogen flow. The vertically aligned SiNWs were characterized by SEM. The obtained SiNW thin film was used for fabrication of a high-performance SiNW FET [98].

Synthesis of SiNWs from trisilane (Si_3H_8) was elaborated by A.M. Chockla et al. – by a tin-seeded supercritical fluid–liquid–solid method. SiNWs were used as anodes in lithium batteries, carbon conductor, films with poly(vinylidene) fluoride (PVdF) or sodium alginate binder, and fluoroethylene carbonate-containing electrolyte – they gave reversible, high charge storage capacities of 1,800 mAhg. SiNWs also showed relatively good rate capability, with capacities of 400 mAhg [99].

SiNWs doped with ~10^{-16} cm^{-3} boron were prepared by the VLS growth technique. TEM analysis showed that these SiNWs contained almost entirely smooth Si cores (50 ± 5 nm in diameter), which were coated with ~5 ± 1 nm native SiO_2 layer. SiNWs were used for fabrication of gated SiNW gas sensors, which could be used for chemical and biological sensing applications. Their performance is usually accompanied by a "hysteresis" phenomenon [100].

Ultrasmall diameter core–shell SiNWs were prepared by dry thermal oxidation of 2 nm diameter (100) Si NWs at 300 and 1,273 K. The precise control of the Si-core radius of SiNWs and the SiO_x ($x \leq 2.0$) oxide shell was possible by controlling the growth temperature during the oxidation process. The core–shell SiNWs were recommended for fabrication of FETs and PV devices [101].

The growth of Si NWs in the vapor phase of an organic solvent directly on a variety of substrates (silicon, glass, and stainless steel) was described. Indium layer was evaporated during synthesis. Length and density of SiNWs were controlled by changing the reaction time. SiNWs were characterized by SEM, TEM, scanning transmission electron microscopy (STEM), XRD, EDX spectroscopy, and XPS. The indium-seeded SiNWs can serve as anode materials in lithium batteries [102].

A continuous progress on Si-based NWs and NTs as high capacity anode materials has been observed in recent years. It was found that one dimensional Si/Sn NWs and NTs have great potential for the next generation of advanced energy storage applications. These materials allow to achieve high energy density as well as long cycle life [103].

The VLS growth method was often used for SiNW epitaxial growth through decomposition of gaseous monosilane (SiH_4) or silicon tetrachloride ($SiCl_4$), under H_2

atmosphere, catalyzed by metallic eutectic particles, mostly gold (Au) in liquid phase of Au–Si droplet. Aluminum was also used as the catalyst for VLS growth [104].

The effect of aluminum-coating layers on the electrochemical properties of SiNWs was studied, using extensive SEM and TEM methods. An improved cycling performance in the SiNWs coated with 3 and 8 wt% aluminum was observed. A similar thickness alumina (Al_2O_3) coating was shown not to be as effective in reducing the long-term capacity loss. A synthetic procedure was further improved by application of electrically conducting TiN barrier layer. The SiNWs were used as negative electrodes in lithium ion battery half-cells [105].

The controlled growth of SiNWs in porous alumina templates was described. It can be further processed by classical microelectronic methods, in order to fabricate nanodevices. The amount of deposited a-Si (which could clogs the pores) was reduced by using atomic hydrogen instead of molecular hydrogen, due to its increase the etching rate of a-Si and the limited a-Si deposition through the formation of Si–H bonds [106].

SiNWs were also grown by the magnetron sputtering method using tin catalyst at temperatures ranging from 250 to 400 °C. As the growth temperature increases from 250 to 400 °C, the length and density of SiNWs first increased and then decreased. A mixed phase of amorphous and nanocrystalline silicon were formed in the synthesized SiNWs, and the crystallization degree of SiNWs increased rapidly with further increase in growth temperature. Sn NPs on the top of SiNWs were observed, which indicated that the VLS growth mechanism was responsible for SiNWs growth [107].

Hydrogen-functionalized silicon was often passivated with electroactive molecules, in order to obtain the most stable surfaces, which would be resistant to oxidation in air and in aqueous media. The silicon electrode stability depends on parameters such as molecular size, monolayer density, and crystal orientation of silicon surface. Direct surface passivation of silicon had several important advantages. It can be carried out at low voltages, while directly passivated SiNWs were much more effective biological sensors in high ionic strength solution [108, 109]. Many attempts were applied for surface functionalization of silicon materials. 3-Aminopropyl(triethoxy)silane, (3-aminopropyl) dimethyl(ethoxy)silane, and 3-(trimethoxysilyl)propyl aldehyde were the most widely used silane coupling agents (SCA) for modification of the Si surface covered with the silica layer [110–113]. The Si–OMe or Si–OEt groups of the SCA reacted with Si–OH groups on the surface and formed on the silicon oxide surface a monolayer functionalized with aldehyde or amine groups, which further reacted with amine or carboxylic acid groups of biological capture probes. For not oxidized Si surfaces two other methods were used to modify the surface for further biofunctionalization. The UV light was also applied for rapid photodissociation of the Si–H bond into Si radicals on the surface. These radicals reacted with terminal olefin groups of SCA molecules, forming stable Si–C bonds at the Si surface [114–116]. The SCA molecules usually have a protected terminal amine group, which can be used to attach biological probes after deprotection. Two-step chlorination/alkylation reactions were also applied for a formation of Si–C bonds on the Si surface [117, 118]. The Si–H surface was first chlorinated to form Si–Cl bond, followed by reaction with an allyl Grignard reagent. The allyl-functionalized surface can be used for further bioconjugation [119].

Surface-biofunctionalized silicon was often used for a fabrication of biosensors. Recently, in order to facilitate the diagnostics process in medicine, the concept of so-called point-of-care testing was developed. FET using nanomaterial as a kind of biosensors exhibited great characteristics for detection of a wide range of biomolecules due to their label-free and ultrasensitive properties. The working principles of surface functionalization, fabrication methods of such devices, and some current research trends in FET nanobiosensors were reviewed by M. Molaie [120]. The combination of an increasingly growing number of interesting (bio)receptors, improved electrical characterization procedures, sophisticated device fabrication methods, and advanced organic surface modification led to diverse applications of sensors in a variety of fields [45]. Selective electrochemical functionalization methods were applied for the surface modification of SiNWs [109].

Novel high-density amorphous silica NWs radially standing on a single-crystal silicon core were grown on a silicon wafer in grams quantity by a simple thermal evaporation process from SiO, in the presence of Sn, in a stream of hydrogen and argon (5:95), which was carried out at 1,320 °C and 350 mbar for 7 h. The column-like hierarchical Si-SiO_x nanostructure exhibited a single sharp PL peak at 378 nm and could be used in electronic devices and as an outstanding template for chemicals and biosensors due to its large surface-to-volume ratio [121].

Silicon nanomaterials with sphere or wire morphology were also prepared from solid microsized silicon powder by a nontransferred arc plasma technique. By changing the input power and gas flow rate, the mean diameter of synthesized silicon NPs (16.7–40.0 nm) was achieved [122].

Silicon has a low discharge potential, the highest known theoretical charge capacity, and is an attractive anode material for lithium batteries, which find applications in portable electronic devices, electric vehicles, and implantable medical devices [87,91,123]. SiNWs, prepared by rapid metal-catalyzed electroless etching, highly doped with different dopant atoms (e.g., As) were also used as a lithium-ion battery anode material. They gave better results than lightly doped SiNWs due to their highly conductive and highly porous nature and showed very high performance and cycle retention [123]. Facet-selective epitaxy of CdS semiconductors on SiNWs was achieved by M.N. Mankin et al. This method is a promising route to integration of compound semiconductors on Si. PL imaging and spectroscopy showed that the epitaxial shells displayed strong and clean band edge emission, confirming their high photonic quality [124]. Conformal layers of various materials (metals, metal oxides, and organic/inorganic semiconductors) onto high aspect ratio Si microwire and SiNW arrays were deposited by an electrochemical method [125].

Composite electrodes composed of SiNWs were mixed with amorphous carbon or carbon nanotubes (CNTs) and used as Li-ion battery anodes. The SiNWs coated with multiwalled carbon NTs as the conducting additive showed reversible capacities of 1,500 mAh/g for 30 cycles [126].

Sensor devices based on SiNWs were first applied by Cui and Lieber [49]. SiNW FET biosensors were used for detection of proteins [127]. Nowadays they serve as a general platform for ultra-sensitive, electrical detection of chemical and biological species. SiO_x coatings were used for the detection of protons [128] and gases [129]. The specific detection of other analytes, including ions and biomolecules became possible due to the presence of an affinity top layer (on the NW surface) which can interact with the analyte of interest.

A selective functionalization of the NWs is necessary to retain sensitivity of bioreceptors as the modification of the background oxide resulted in a reduced sensitivity [48]. In order to bind (bio)receptor molecules onto SiNW-based devices, most often the silanization on the NW oxide layer was applied (e.g., with APTES) [130, 131]. APTES, as a silane coupling agent (SCA), was the most often used for the subsequent immobilization of carboxylic acid- or aldehyde-terminated biomolecules. More frequently, further derivatization with glutaraldehyde was applied to chemically bind amine-terminated biomolecules. A disadvantage of silane compounds was their self-condensation and cross-linking between the alkoxy groups and Si-OH groups formed by hydrolysis, resulting in rough, unordered multilayers [48]. Alternatively, phosphonate monolayers were formed on SiNW-based devices. Apart from covalent functionalization of SiNW-based sensor devices also physicosorption was useful and allowed the detection of several compounds and ions [132].

Silicon *p-i-n* photodiode junction fibers fabricated via high pressure CVD can be used as optoelectronic devices such as solar power wires and high speed (1.8 GHz) photodetectors and have the potential for high-efficiency PV conversion and optoelectronic detection. Silicon layers over lengths of more than 10 m in fiber pores were deposited. Silane (SiH_4) was mixed with diborane B_2H_6 for deposition of *p*-type Si and SiH_4 with phosphine PH_3 for *n*-type Si. The dopant concentration, ranging from 10^{16} to 10^{20} cm^{-1}, was applied. At 400 °C, the deposition rate was <1 nm/min for intrinsic Si (*i*-Si) but increased dramatically to ~300 nm/min when ~1 wt% B_2H_6 was added. With ~1% PH_3 addition to SiH_4 at 500 °C, intrinsic Si deposited at 6–10 nm/min, while phosphorus doped Si deposited at 15–20 nm/min). At high pressure from homogeneous gas phase reactions of SiH_4 and/or PH_3 were formed intermediates (e.g., SiH_2, SiH_3PH_2). Uniform-thickness phosphorus-doped Si layers over a length of 1 m were deposited at a temperature of 525 °C and the precursor conversion efficiency was ~1.5 wt % per meter [133].

A phospholipid monolayer was covalently immobilized on SiO_x and Si–H surfaces, thus a second phospholipid layer was immobilized. Lipid bilayers provided a matrix for a virtually unlimited number of transmembrane proteins which accommodated different functionalities [134]. Organic monolayers on silicon surfaces were also prepared from alkenes. The UV-initiated methods were used for the chemical modification of oxide-free H-functionalized SiNW devices. The surface orientation of the NW played an important role [108]. A very good monolayer quality led to reduction of the number of interfacial defects [135]. The insulating properties and monolayer quality was improved by using alkynes instead of alkenes [136–139].

A selective functionalization of carbon–silicon (C-Si) alkyl and alkenyl monolayers covalently linked to all-(111) surface SiNW was used for preparation of biosensors. First, a silica layer was removed with 1% aqueous solution of HF forming Si–H surface, which was treated with 40% aqueous solution of NH_4F. Two kinds of C–Si monolayers on H–Si (111) surfaces were formed. By photochemical UV hydrosilylation reaction with 1-octadecene C_{18} alkyl monolayer was obtained, while with N-(5-hexynyl)phthalimide a phthalimide alkenyl monolayers were formed. Deprotected amine terminal groups were functionalized with biotin N-hydroxy-succinimide ester. The improvement in detection sensitivity of the alkyl- and alkenyl-passivated all-(111) SiNW biosensors, compared to conventional NW biosensor geometries and silicon dioxide passivation layers, was demonstrated [140].

In conclusion, a number of technical applications of SiNWs is still quickly growing, especially in the field of optoelectronics, photonics and PVs, catalysts, and different electronic devices (e.g., sensors and thermoelectric devices) [45, 47–49, 69, 75, 141–147]. Many other future potential applications of SiNWs seem possible, and a continuous progress in the SiNW technology should affect development of the existing silicon technology [53].

References

[1] A.M. Mitrašinović and T.A. Utigard, Refining silicon for solar cell application by copper alloying, *Silicon*, 2009, 1(4), 239–248.

[2] J. Wu, W. Ma, B. Yang, D. Liu and Y. Dai, Calculation and characterization of silicon-boron phases in metallurgical grade silicon, *Silicon*, 2012, 4(4), 289–295.

[3] *Wikipedia, the free encyclopedia.*

[4] B. Ceccaroli, E. Ovrelid and S. Pizzini, *Solar Silicon Processes: Technologies, Challenges, and Opportunities*, CRC Press, 2016, 258.

[5] https://en.wikipedia.org/wiki/Polycrystalline_silicon

[6] H. Yu, J.-Q. Zeng and Z.-R. Qiu, Chapter 7: *Silicon nanocrystals*, in *Crystalline Silicon – Properties and Uses*, S. Basu (Ed.), InTech, 2011, 121–134.

[7] M. Fujiki and G. Kwak, Chapter 1: *Amorphous and crystalline silicon films from soluble Si-Si network polymers*, in *Solar Silicon Processes: Technologies, Challenges, and Opportunities*, B. Ceccaroli, E. Ovrelid and S. Pizzini (Eds.), CRC Press, 2016, 1–22.

[8] H. Yu, J.-Q. Zeng and Z.-R. Qiu, Chapter 7: *Silicon nanocrystals*, in *Solar Silicon Processes: Technologies, Challenges, and Opportunities*, B. Ceccaroli, E. Ovrelid and S. Pizzini (Eds.), CRC Press, 2016, 121–134.

[9] S. Basu and J. Kanungo, Chapter 10: *Nanocrystalline porous silicon*, in *Solar Silicon Processes: Technologies, Challenges, and Opportunities*, B. Ceccaroli, E. Ovrelid and S. Pizzini (Eds.), CRC Press, 2016, 219–250.
[10] M.C. Arenas, M. Vega, O. Martínez and O.H. Salinas, Chapter 11: *Nanocrystalline porous silicon: Structural, optical, electrical and photovoltaic properties*, in *Solar Silicon Processes: Technologies, Challenges, and Opportunities*, B. Ceccaroli, E. Ovrelid and S. Pizzini (Eds.), CRC Press, 2016, 251–274.
[11] I. Rea, E. Orabona, I. Rendina and L. De Stefano, Chapter 12: *Porous silicon integrated photonic devices for biochemical optical sensing*, in *Solar Silicon Processes: Technologies, Challenges, and Opportunities*, B. Ceccaroli, E. Ovrelid and S. Pizzini (Eds.), CRC Press, 2016, 275–296.
[12] H. Takiguchi and K. Morita, Chapter 15: *Global flow analysis of crystalline silicon*, in *Crystalline Silicon – Properties and Uses*, S. Basu (Ed.), 329–344. InTech, Croatia, 2011. ISBN 978-953-307-587-7.
[13] K. Baert, E. Van Kerschaver and J. Poortmans, http://electroiq.com/blog/2009/08/crystalline-si-solar-cells-and-the-microelectronics-experience/
[14] M. Fujiki and G. Kwak, Chapter 1: *Amorphous and crystalline silicon films from soluble Si-Si network polymers*, in the book: *Crystalline Silicon – Properties and Uses*, http://www.intechopen.com/books/crystalline-silicon-propertiesand-uses
[15] D.L. Staebler and C.R. Wronski, Reversible conductivity changes in dischargeproduced amorphous Si, *Applied Physics Letters*, 1977, 31(4), 292–294. DOI: 10.1063/1.89674.
[16] J.-F. Tremblay, Solar market powers silicon, *Chemical and Engineering News*, October 2, 2006, 30–34.
[17] X. Li, Y. He, S.S. Talukdar and M.T. Swihart, Process for preparing macroscopic quantities of brightly photoluminescent silicon nanoparticles with emission spanning the visible spectrum, *Langmuir*, 2003, 19(20), 8490–8496.
[18] F. Hua, M.T. Swihart and E. Ruckenstein, Efficient surface grafting of luminescent silicon quantum dots by photoinitiated hydrosilylation, *Langmuir*, 2005, 21(13), 6054–6062.
[19] S.Y. Kim, K.S. Yang and B.-H. Kim, Improving the microstructure and electrochemical performance of carbon nanofibers containing graphene-wrapped silicon nanoparticles as a Li-ion battery anode, *Journal of Power Sources*, 2015, 273, 404–412.
[20] Z. Liu, X. Chang, T. Wang, W. Li, H. Ju, X. Zheng, X. Wu, C. Wang, J. Zheng and X. Li, Silica-derived hydrophobic colloidal nano-Si for lithium-ion batteries, *ACS Nano*, 2017, 11, 6065–6073.
[21] S. Zhang, J. Zhang, M. Li, W. Zhang and L. Cao, Grafting and properties of a porous poly(methyl methacrylate) film on a silicon surface by a one-step dipping method, *Journal of Applied Polymer Science*, 2017, 134(37), 44930.
[22] E.J. Henderson, J.A. Kelly and J.G.C. Veinot, Influence of HSiO1.5 sol-gel polymer structure and composition on the size and luminescent properties of silicon nanocrystals, *Chemistry of Materials*, 2009, 21(22), 5426–5434.
[23] A.S. Heintz, M.J. Fink and B.S. Mitchell, Silicon nanoparticles with chemically tailored surfaces, *Applied Organometallic Chemistry*, 2010, 24(3), 236.
[24] S.-B. Choi, B.-K. Kim, P. Boudjouk and D.G. Grier, Amine-promoted disproportionation and redistribution of trichlorosilane: Formation of tetradecachlorocyclohexasilane dianion, *Journal of American Chemical Society*, 2001, 123(33), 8117.
[25] S. Han, X. Dai, P. Loy, J. Lovaasen, J. Huether, J.M. Hoey, A. Wagner, J. Sandstrom, D. Bunzow, O.F. Swenson, I.S. Akhatov and D.L. Schulz, Printed silicon as diode and FET materials – Preliminary results, *Journal of Non-Crystalline Solids*, 2008, 354 (19–25), 2623.

[26] I.S. Akhatov, J.M. Hoey, O.F. Swenson and D.L. Schulz, Aerosol focusing in micro-capillaries: Theory and experiment, *Journal of Aerosol Science*, 2008, 39(8), 691.

[27] D.L. Schulz, J. Hoey, J. Smith, A. Elangovan, X. Wu, I. Akhatov, S. Payne, J. Moore, P. Boudjouk, L. Pederson, J. Xiao and J.-G. Zhang, Si_6H_{12}/Polymer Inks for electrospinning a-si nanowire lithium ion battery anode, *Electrochemical and Solid-State Letters*, 2010, 13(10), A143.

[28] G.R.S. Iyer, E.K. Hobbie, S. Guruvenket, J.M. Hoey, K.J. Anderson, J. Lovaasen, C. Gette, D.L. Schulz, O.F. Swenson, A. Elangovan and P. Boudjouk, Solution-based synthesis of crystalline silicon from liquid silane through laser and chemical annealing, *ACS Applied Materials and Interfaces*, 2012, 4(5), 2680.

[29] G.R.S. Iyer, S. Guruvenket, J.M. Hoey, K.J. Anderson, D.L. Schulz, O.F. Swenson, A. Elangovan, P. Boudjouk and E.K. Hobbie, The Smithsonian/NASA astrophysics data system solution-based synthesis of crystalline silicon thin films from liquid silane inks, *American Physical Society, APS March Meeting 2012*, February 27–March 2, 2012, abstract No. H33.008.

[30] A.P. Cádiz Bedini, S. Muthmann, J. Flohre, B. Thiele, S. Willbold and R. Carius, Sonophotolytically synthesized silicon nanoparticle-polymer composite ink from a commercially available lower silane, *Macromolecular Chemistry and Physics*, 2016, 217(15), 1655.

[31] S. Kim, C.-Y. Lee and M.H.-C. Jin, Fourier-transform infrared spectroscopic studies of pristine polysilanes as precursor molecules for the solution deposition of amorphous silicon thin-films, *Solar Energy Materials and Solar Cells*, 2012, 100, 61.

[32] C.M. Hessel, D. Reid, M.G. Panthani, M.R. Rasch, B.W. Goodfellow, J. Wei, H. Fujii, V. Akhavan and B.A. Korgel, Synthesis of ligand-stabilized silicon nanocrystals with size-dependent photoluminescence spanning visible to near-infrared wavelengths, *Chemistry of Materials*, 2012, 24(2), 393.

[33] L. Sacarescu, G. Roman, G. Sacarescu and M. Simionescu, Fluorescence detection system based on silicon quantum dots–polysilane nanocomposites, *eXPRESS Polymer Letters*, 2016, 10(12), 990.

[34] C.M. Hessel, E.J. Henderson and J.G.C. Veinot, Hydrogen silsesquioxane: A molecular precursor for nanocrystalline si–SiO2 composites and freestanding hydride-surface-terminated silicon nano-particles, *Chemistry of Materials*, 2006, 18(18), 6139.

[35] X. Wang, X. Yu, W. Yu, H. Feng, J. Wang, C. Yin, W. Lu and G. Fu, Effects of annealing ambient on the photoluminescence properties of Si-rich oxide/SiO2 multilayer films containing Si-nanocrystals, *Journal of Materials Science*, 2014, 49(3), 1353.

[36] S. Kim, K. Jeon, J.C. Lee, M.T. Swihart and M. Yang, Enhanced performance of a polymer solar cell upon addition of free-standing, freshly etched, photoluminescent silicon nanocrystals, *Applied Physics Express*, 2012, 5(2), 022302.

[37] J. Cai and L. Qi, Recent advances in antireflective surfaces based on nanostructure arrays, *Materials Horizons*, 2015, 2(1), 37.

[38] E.J. Henderson and J.G.C. Veinot, Synthesis of oxide encapsulated and freestanding hydride surface terminated Si1-*x*Ge*x* nanocrystals, *Chemistry of Materials*, 2007, 19(8), 1886.

[39] S.D. Barry, Z. Yang, J.A. Kelly, E.J. Henderson and J.G.C. Veinot, Synthesis of Si*x*Ge1–*x* Nanocrystals using hydrogen silsesquioxane and soluble germanium diiodide complexes, *Chemistry of Materials*, 2011, 23(22), 5096.

[40] S. Alialy, A. Kaya, E. Marıl, Ş. Altındal and İ. Uslu, Electronic transport of Au/(Ca1.9Pr0.1Co4Ox)/n-Si structures analysed over a wide temperature range, *Philosophical Magazine*, 2015, 95(13), 1448.

[41] A. Kaya, H.G. Cetinkaya, S. Altındal and I. Uslu, A comparative study on the electrical parameters of Au/*n*-Si Schottky diodes with and without interfacial ($Ca_{1.9}Pr_{0.1}Co_4O_x$) layer, *International Journal of Modern Physics B*, 2016, 30(16), 1650090.
[42] D. Kim, J.M. Zuidema, J. Kang, Y. Pan, L. Wu, D. Warther, B. Arkles and M.J. Sailor, Facile surface modification of hydroxylated silicon nanostructures using heterocyclic silanes, *Journal of American Chemical Society*, 2016, 138(46), 15106.
[43] Z. Xu, Y. Li, B. Zhang, T. Purkait, A. Alb, B.S. Mitchell, S.M. Grayson and M.J. Fink, Water-soluble PEGylated silicon nanoparticles and their assembly into swellable nanoparticle aggregates, *Journal of Nanoparticle Research*, 2015, 17, 56.
[44] D. Zhang, J. Shi, W. Zi, P. Wang and S.F. Liu, Recent advances in photoelectrochemical applications of silicon materials for solar-to-chemicals conversion, *Chemistry and Sustaiability Energy and Materials*, 2017, 10(22), 4324.
[45] V. Schmidt, J.V. Wittemann, S. Senz and U. Gösele, Silicon nanowires: A review on aspects of their growth and their electrical properties, *Advanced Materials*, 2009, 21(25–26), 2681.
[46] L. Ren, H. Li and L. Ma, Chapter 17: *The selective growth of silicon nanowires and their optical activation,* in *Nanowires – Implementations and Applications*, D.A. Hashim (Ed.), 375–416, and referencies cited therein; InTech, Croatia, 2011. ISBN: 978-953-307-318-7, Available from: http://www.intechopen.com/books/nanowires-implementations-and-applications/theselective-growth-of-silicon-nanowires-and-their-optical-activation.
[47] V. Sivakov, F. Voigt, B. Hoffmann, V. Gerliz and S. Christiansen, Chapter 3: *Wet – chemically etched silicon nanowire architectures: formation and properties*, pp. 45–80 (and referencies cited therein), in *Nanowires – Fundamental Research*, D.A. Hashim (Ed.), InTech, Croatia, 2011. ISBN: 978-953-307-327-9. Available from: http://www.intechopen.com/books/nanowires-fundamental-research/wet-chemically-etched-silicon-nanowirearchitectures-formation-and-properties.
[48] L.C.P.M. De Smet, D. Ullien, M. Mescher and E.J.R. Sudhölter, Chapter 13: *organic surface modification of silicon nanowire-based sensor devices*, pp. 267–288 (and referencies cited therein), in *Nanowires – Implementations and Applications*, D.A. Hashim (Ed.), InTech, Croatia, 2011, ISBN: 978-953-307-318-7. Available from: http://www.intechopen.com/books/nanowiresimplementations-and-applications/organic-surface-modification-of-silicon-nanowire-based-sensor-devices.
[49] Y. Cui and C.M. Lieber, Functional nanoscale electronic devices assembled using silicon nanowire building blocks, *Science*, 2001, 291(5505), 851.
[50] M. Shao, D.D.D. Ma and S.-T. Lee, Silicon nanowires – Synthesis, properties, and applications, *European Journal of Inorganic Chemistry*, 2010, 27, 4264.
[51] J. Ramanujam, D. Shiri and A. Verma, Silicon nanowire growth and properties: A review, *Materials Express*, 2011, 1(2), 105.
[52] T. Mikolajick, A. Heinzig, J. Trommer, S. Pregl, M. Grube, G. Cuniberti and W.M. Weber, *Physica Status Solidi – Rapid Research Letters*, 2013, 7(10), 793.
[53] T. Mikolajick and W.M. Weber, Chapter 1: *Silicon nanowires: Fabrication and applications*, pp. 1–25 (and referencies cited therein), in *Anisotropic Nanomaterials – Preparation, Properties, and Applications*, Q. Li (Ed.), 2015, ISBN: 978-3-319-18292-6; and referencies cited therein, http://www.springer.com/978-3-319-18292-6.
[54] N. Meshram, Job application paper: "Synthesis and characterization of silicon nanowire using Hot Wire Chemical Vapor Deposition"; available online in 2011.
[55] N.P. Meshram, A. Kumbhar and R.O. Dusane, Silicon nanowire growth on glass substrates using hot wire chemical vapor deposition, (a) *The Sixth International Conference on Hot Wire Chemical Vapor Deposition (Cat-CVD) Process, Palaiseau, France, Sept. 13–17, 2010* (*Poster HWCVD-6*); (b) *Thin Solid Films*, 2011, 519, 14, 4609.

[56] N. Meshram, A. Kumbhar and R. Dusane, Synthesis of silicon nanowires using tin catalyst by hot wire chemical vapor processing, *Materials Research Bulletin*, 2013, 48(6), 2254.
[57] R.S. Wagner and W.C. Ellis, Vapor-liquid-solid mechanism of single crystal growth, *Applied Physics Letters*, 1964, 4(89). DOI: https://doi.org/10.1063/1.1753975.
[58] A. Gómez-Martínez, F. Márquez, E. Elizalde and C. Morant, Microspheres for the growth of silicon nanowires via vapor-liquid-solid mechanism, *Journal of Nanomaterials*, 2014, Article ID 362798.
[59] J.D. Holmes, K.P. Johnston, R.C. Doty and B.A. Korgel, Control of thickness and orientation of solution-grown silicon nanowires, *Science*, 2000, 287(5457), 1471.
[60] M. Yazawa, M. Koguchi, A. Muto, M. Ozawa and K. Hiruma, Effect of one monolayer of surface gold atoms on the epitaxial growth of InAs nanowhiskers, *Applied Physics Letters*, 1992, 61 (25), 2051.
[61] K. Hiruma, M. Yazawa, K. Haraguchi, K. Ogawa, T. Katsuyama, M. Koguchi and H. Kakibayashi, GaAs free-standing quantum-size wires, *Journal of Applied Physics*, 1993, 74(5), 3162. DOI: https://doi.org/10.1063/1.354585.
[62] J. Westwater, D.P. Gosain and S. Usui, Control of the size and position of silicon nanowires grown via the vapor-liquid-solid technique, *Japanese Journal of Applied Physics*, 1997, 36(10), 6204.
[63] J. Westwater, D.P. Gosain, S. Tomiya, S. Usui and H. Ruda, Growth of silicon nanowires via gold/silane vapor-liquid-solid reaction, *Journal of Vacuum Science and Technology B*, 1997, 15(3), 554.
[64] N. Ozaki, Y. Ohno and S. Takeda, Silicon nanowhiskers grown on a hydrogen-terminated silicon {111} surface, *Applied Physics Letters*, 1998, 73(25), 3700. DOI: https://doi.org/10.1063/1.122868.
[65] S. Takeda, K. Ueda, N. Ozaki and Y. Ohno, Formation mechanism of nanocatalysts for the growth of silicon nanowires on a hydrogen-terminated Si {111} surface template, *Applied Physics Letters*, 2003, 82(6), 979. DOI: https://doi.org/10.1063/1.1541934.
[66] D.P. Yu, C.S. Lee, I. Bello, X.S. Sun, Y.H. Tang, G.W. Zhou, Z.G. Bai, Z. Zhang and S.Q. Feng, Synthesis of nano-scale silicon wires by excimer laser ablation at high temperature, *Solid State Communications*, 1998, 105(6), 403. DOI: https://doi.org/10.1016/S0038-1098(97)10143-0.
[67] J.L. Liu, Y. Lu, Y. Shi, S.L. Gu, R.L. Jiang, F. Wang and Y.D. Zheng, Fabrication of silicon nanowires, *Applied Physics A*, 1998, 66(5), 539.
[68] E.I. Givargizov, Fundamental aspects of VLS growth, *Journal of Crystal Growth*, 1975, 31(20). DOI: https://doi.org/10.1016/0022-0248(75)90105-0.
[69] Y. Cui, L.J. Lauhon, M.S. Gudiksen, J. Wang and C.M. Lieber, Diameter-controlled synthesis of single-crystal silicon nanowires, *Applied Physics Letters*, 2001, 78(15), 2214. DOI: https://doi.org/10.1063/1.1363692.
[70] T.I. Kamins, R.S. Williams, D.P. Basile, T. Hesjedal and J.S. Harris, Ti-catalyzed Si nanowires by chemical vapor deposition: Microscopy and growth mechanisms, *Journal of Applied Physics*, 2001, 89(2), 1008. DOI: https://doi.org/10.1063/1.1335640.
[71] Y. Xia, P. Yang, Y. Sun, Y. Wu, B. Mayers, B. Gates, Y. Yin, F. Kim and H. Yan, One-dimensional nanostructures: Synthesis, characterization, and applications, *Advanced Materials*, 2003, 15 (5), 353.
[72] O. Englander, D. Christensen and L. Lin, Local synthesis of silicon nanowires and carbon nanotubes on microbridges, *Applied Physics Letters*, 2003, 82(26), 4797.
[73] Y. Cui, Z. Zhong, D. Wang, W.U. Wang and C.M. Lieber, High performance silicon nanowirefield effect transistors, *Nano Letters*, 2003, 3(2), 149.

[74] T. Qiu, X.L. Wu, Y.F. Mei, G.J. Wan, P.K. Chu and G.G. Siu, Self-assembled growth and green emission of gold nanowhiskers, *Journal of Crystal Growth*, 2005, 277(1–4), 143. DOI: https://doi.org/10.1063/1.2138360.
[75] V. Schmidt, S. Senz and U. Gösele, Ferroelectric nanotubes fabricated using nanowires as positive templates, *Nano Letters*, 2005, 5(5), 931. DOI: https://doi.org/10.1063/1.2364260.
[76] D. Gao, R.H. He, C. Carraro, R.T. Howe, P.D. Yang and R. Maboudian, Selective growth of Si nanowire arrays via galvanic displacement processes in water-in-oil microemulsions, *Journal of American Chemical Society*, 2005, 127(13), 4574.
[77] J.B. Hannon, S. Kodambaka, F.M. Ross and R.M. Tromp, The influence of the surface migration of gold on the growth of silicon nanowires, *Nature*, 2006, 440(7080), 69.
[78] Y. Wu, Y. Cui, L. Huynh, C.J. Barrelet, D.C. Bell and C.M. Lieber, Controlled growth and structures ofmolecular-scale silicon nanowires, *Nano Letters*, 2004, 4(3), 433.
[79] A.M. Morales, C.M. Lieber and A. Laser Ablation, Method for the synthesis of crystalline semiconductor nanowires, *Science*, 1998, 279(5348), 208.
[80] J. Weyher, Some notes on the growth kinetics and morphology of VLS silicon crystals grown with platinum and gold as liquid-forming agents, *Journal of Crystal Growth*, 1978, 43(2), 235.
[81] V.A. Nebol'sin and A.A. Shchetinin, Role of surface energy in the vapor–liquid–solid growth of silicon, *Inorganic Materials*, 2003, 39(9), 899.
[82] E.iGivargizov, *Highly Anistropic Crystals*, Terra Scientific Publishing Company, Tokyo, Japan, 1987, 104–105.
[83] Y. Wang, V. Schmidt, S. Senz and U. Gösele, Epitaxial growth of silicon nanowires using an aluminium catalyst, *Nature Nanotechnology*, 2006, 1(3), 186.
[84] J.-F. Hsu and B.-R. Huang, The growth of silicon nanowires by electroless plating technique of Ni catalysts on silicon substrate, *Thin Solid Films*, 2006, 514(1–2), 20.
[85] P. De Padova, C. Quaresima, P. Perfetti, B. Olivieri, C. Leandri, B. Aufray, S. Vizzini and G. Le Lay, Growth of straight, atomically perfect, highly metallic silicon nanowires with chiral asymmetry, *Nano Letters*, 2008, 8(1), 271.
[86] G.A. Nazri and G. Pistoia (Eds.), *Lithium Batteries: Science and Technology*, Kluwer Academic/Plenum, Boston, 2004.
[87] B.A. Boukamp, G.C. Lesh and R.A. Huggins, All-solid lithium electrodes with mixed-conductor matrix, *Journal of Electrochemical Society*, 1981, 128(4), 725.
[88] T. Shodai, S. Okada, S. Tobishima and J. Yamaki, Study of Li3 − *x*M*x*N (M: Co, Ni or Cu) system for use as anode material in lithium rechargeable cells, *Solid State Ionics*, 1996, 86–88(2), 785.
[89] P. Poizot, S. Laruelle, S. Grugeon, L. Dupont and J.-M. Tarascon, Nano-sized transition-metal oxides as negative-electrode materials for lithium-ion batteries, *Nature*, 2000, 407(6803), 496.
[90] U. Kasavajjula, C. Wang and A.J. Appleby, Nano- and bulk-silicon-based insertion anodes for lithium-ion secondary cells, *Journal of Power Sources*, 2007, 163(2), 1003.
[91] C.K. Chan, H. Peng, G. Liu, K. McIlwrath, X.F. Zhang, R.A. Hugins and Y. Cui, High-performance lithium battery anodes using silicon nanowires, *Nature Nanotechnology*, 2008, 3(1), 31.
[92] J.-P. Colinge, C.-W. Lee, A. Afzalian, N.D. Akhavan, R. Yan, I. Ferain, P. Razavi, B. O'Neill, A. Blake, M. White, A.-M. Kelleher, B. McCarthy and R. Murphy, *Nature Nanotechnology*, 2010, 5(3), 225.
[93] F. Dhalluin, T. Baron, P. Ferret, B. Salem, P. Gentile and J.-C. Harmand, Silicon nanowires: Diameter dependence of growth rate and delay in growth, *Applied Physics Letter*, 2010, 96 (13), 133109. DOI: https://doi.org/10.1063/1.3373546.
[94] J. Westwater, D.P. Gosain and S. Usui, Control of the size and position of silicon nanowires grown via the vapor-liquid-solid technique, *Japanese Journal of Applied Physics*, 1997, 36(10), 6204.

[95] H. Hamidinezhad, Y. Wahab, Z. Othaman and A.K. Ismail, Synthesis and analysis of silicon nanowire below Si–Au eutectic temperatures using very high frequency plasma enhanced chemical vapor deposition, *Applied Surface Science*, 2011, 257(21), 9188.
[96] F.-W. Yuan and H.-Y. Tuan, Supercritical fluid– solid growth of single-crystalline silicon nanowires: An example of metal-free growth in an organic solvent, *Crystal Growth Design*, 2010, 10(11), 4741.
[97] B. Tan and K. Venkatakrishnan, Synthesis of fibrous nanoparticle aggregates by femtosecond laser ablation in air, *Optics Express*, 2009, 17(2), 1064.
[98] T.I. Lee, W.J. Choi, K.-J. Moon, J.-H. Choi, J.H. Park, U. Jeong, H.K. Baik and J.M. Myoung, A simple and rapid formation of wet chemical etched silicon nanowire films at the air–water interface, *Journal of Materials Chemistry*, 2011, 21(37), 14203.
[99] A.M. Chockla, K.C. Klavetter, C.B. Mullins and B.A. Korgel, Tin-seeded silicon nanowires for high capacity Li-ion batteries, *Chemistry of Materials*, 2012, 24(19), 3738.
[100] Y. Paska and H. Haick, Interactive effect of hysteresis and surface chemistry on gated silicon nanowire gas sensors, *ACS Applied Materials and Interfaces*, 2012, 4(5), 2604.
[101] U. Khalilov, G. Pourtois, A.C.T. Van Duin and E.C. Neyts, Self-limiting oxidation in small-diameter Si nanowires, *Chemistry of Materials*, 2012, 24(11), 2141.
[102] H. Geaney, T. Kennedy, C. Dickinson, E. Mullane, A. Singh, F. Laffir and K.M. Ryan, High density growth of indium seeded silicon nanowires in the vapor phase of a high boiling point solvent, *Chemistry of Materials*, 2012, 24(11), 2204.
[103] N.-S. Choi, Y. Yao, Y. Cui and J. Cho, One dimensional Si/Sn-based nanowires and nanotubes for lithium-ion energy storage materials, *Journal of Materials Chemistry*, 2011, 21(27), 9825.
[104] Y. Civale, L.K. Nanver, P. Hadley and E.J.G. Goudena. *Aspects of silicon nanowire synthesis by aluminum-catalyzed vapor-liquid-solid mechanism*, Proceedings of 7th Annual Workshop on Semiconductor Advances for Future Electronics (SAFE 2004), November 25–26, 2004, Veldhoven, The Netherlands, Publ. STW, 2004. ISBN 90-73461-43-X, 692–696.
[105] E.L. Memarzadeh, W.P. Kalisvaart, A. Kohandehghan, B. Zahiri, C.M.B. Holt and D. Mitlin, Silicon nanowire core aluminum shell coaxial nanocomposites for lithium ion battery anodes grown with and without a TiN interlayer, *Journal of Materials Chemistry*, 2012, 22(14), 6655.
[106] M. Gowtham, L. Eude, B. Marquardt, A.Q.L. Quang, C.S. Cojocaru, P. Legagneux and D. Pribat, Confined and controlled growth of silicon nanowire for planar type devices, Oral presentation, JNTE 08, Toulouse, France, November 19–21, 2008.
[107] J. Liu, S.-H. Huang, L.-P. Chen and L. He, Tin catalyzed silicon nanowires prepared by magnetron sputtering, *Materials Letters*, 2015, 151, 122–125.
[108] Y.L. Bunimovich, G. Ge, K.C. Beverly, R.S. Ries, L. Hood and J.R. Heath, Electrochemically programmed, spatially selective biofunctionalization of silicon wires, *Langmuir*, 2004, 20 (24), 10630.
[109] Y.L. Bunimovich, Chapter 4: *Electrochemically Programmed, Spatially Selective Functionalization of Silicon Nanowires*, pp. 90–127, in *Silicon nanowires as biological sensors and highly efficient thermoelectric materials*, Dissertation (Ph.D.), 2007, California Institute of Technology. http://resolver.caltech.edu/CaltechETD:etd-05242007-151027
[110] G. Zheng, F. Patolsky, Y. Cui, W.U. Wang and C.M. Lieber, Multiplexed electrical detection of cancer markers with nanowire sensor arrays, *Natural Biotechnology*, 2005, 23(10), 1294.
[111] E. Stern, R. Wagner, F.J. Sigworth, R. Breaker, T.M. Fahmy and M.A. Reed, Importance of the debye screening length on nanowire field effect transistor sensors, *Nano Letters*, 2007, 7(11), 3405.
[112] A. Kim, C.S. Ah, H.Y. Yu, J.H. Yang, I.B. Baek, C.G. Ahn, C.W. Park, M.S. Jun and S. Lee, Ultrasensitive, label-free, and real-time immunodetection using silicon field-effect transistors, *Applied Physics Letters*, 2007, 91(10), 3901.

[113] W.U. Wang, C. Chen, K.H. Lin, Y. Fang and C.M. Lieber, Labelfree detection of small-molecule–protein interactions by using nanowire nanosensors, *Proceedings of National Academy of Science USA*, 2005, 102(9), 3208.
[114] E. Stern, J.F. Klemic, D.A. Routenberg, P.N. Wyrembak, D.B. Turner-Evans, A.D. Hamilton, D.A. LaVan, T.M. Fahmy and M.A. Reed, Label-free immunodetection with CMOS-compatible semiconducting nanowires, *Nature*, 2007, 445(7127), 519.
[115] Y.L. Bunimovich, Y.S. Shin, W.S. Yeo, M. Amori, G. Kwong and J.R. Heath, Quantitative real-time measurements of DNA hybridization with alkylated nonoxidized silicon nanowires in electrolyte solution, *Journal of American Chemical Society*, 2006, 128(50), 16323.
[116] G.J. Zhang, G. Zhang, J.H. Chua, R.E. Chee, E.H. Wong, A. Agarwal, K.D. Buddharaju, N. Singh, Z. Gao and N. Balasubramanian, DNA sensing by silicon nanowire: Charge layer distance dependence, *Nano Letters*, 2008, 8(4), 1066.
[117] A. Bansal, X. Li, I. Lauermann, N.S. Lewis, S.I. Yi and W.H. Weinberg, Alkylation of Si surfaces using a two-step halogenation/grignard route, *Journal of American Chemical Society*, 1996, 118(30), 7225.
[118] P.T. Hurley, E.J. Nemanick, B.S. Brunschwig and N.S. Lewis, Covalent attachment of acetylene and methylacetylene functionality to Si(111) Surfaces: Scaffolds for organic surface functionalization while retaining Si–C passivation of Si(111) surface sites, *Journal of American Chemical Society*, 2006, 128(31), 9990.
[119] R.D. Rohde, H.D. Agnew, W.S. Yeo, R.C. Bailey, J.R. Heath and A. Non-Oxidative, Approach toward chemically and electrochemically functionalizing Si(111), *Journal of American Chemical Society*, 2006, 128(29), 9518.
[120] M. Molaie, Field effect transistor nanobiosensors: State-of-the-art and key challenges as point of care testing devices, *Nanomedicine Journal*, 2016, 3(2), 69.
[121] H. Wang, X. Zhang, X. Meng, S. Zhou, S. Wu, W. Shi and S. Lee, Bulk preparation of Si–SiO_x hierarchical structures: High-density radially oriented amorphous silica nanowires on a single-crystal silicon nanocore, *Angewandte Chemie International Edition*, 2005, 44(42), 6934.
[122] S. Choi, H. Lee and D.-W. Park, Synthesis of silicon nanoparticles and nanowires by a nontransferred arc plasma system, *Journal of Nanomaterials*, 2016, Article ID 5849041.
[123] O. Salihoglu and Y.E. Kahlout, Doped silicon nanowires for lithium ion battery anodes, *Materials Research*, 2019, 22(2), e20180303.
[124] M.N. Mankin, R.W. Day, R. Gao, Y.-S. No, S.-K. Kim, A.A. McClelland, D.C. Bell, H.-G. Park and C.M. Lieber, Facet-selective epitaxy of compound semiconductors on faceted silicon nanowires, Nano Letters, 2015, 15, 4776–4782.
[125] T. Ozel, B.A. Zhang, R. Gao, R.W. Day, C.M. Lieber and D.G. Nocera, Electrochemical deposition of conformal and functional layers on high aspect ratio silicon micro/ nanowires, Nano Letters, 2017, 17, 4502–4507.
[126] C.K. Chan, R.N. Patel, M.J. O'Connell, B.A. Korgel and Y. Cui, Solution-grown silicon nanowires for lithium-ion battery anodes, *ACS Nano*, 2010, 4(3), 1443.
[127] G. Zheng, X.P.A. Gao and C.M. Lieber, Frequency domain detection of biomolecules using silicon nanowire biosensors, *Nano Letters*, 2010, 10, 3179–3183.
[128] Y. Cui, Q.Q. Wei, H.K. Park and C.M. Lieber, Nanowire nanosensors for highly sensitive and selective detection of biological and chemical species, *Science*, 2001, 293(5533), 1289.
[129] J. Wan, S.R. Deng, R. Yang, Z. Shu, B.R. Lu, S.Q. Xie, Y.F. Chen, E. Huq, R. Liu and X.P. Qu, Silicon nanowire sensor for gas detection fabricated by nanoimprint on SU8/SiO2/PMMA trilayer, *Microelectronic Engineering*, 2009, 86(4–6), 1238.
[130] J.A. Howarter and J.P. Youngblood, Optimization of silica silanization by 3-aminopropyltriethoxysilane, *Langmuir*, 2006, 22(26), 11142.

[131] S. Fiorilli, P. Rivolo, E. Descrovi, C. Ricciardi, L. Pasquardini, L. Lunelli, L. Vanzetti, C. Pederzolli, B. Onida and E. Garrone, Vapor-phase self-assembled monolayers of aminosilane on plasma-activated silicon substrates, *Journal of Colloid and Interface Science*, 2008, 321(1), 235. DOI: 10.1016/j.jcis.2007.12.041.

[132] F.N. Crespilho, V. Zucolotto, O.N. Oliveira Jr and F.C. Nart, Electrochemistry of layer-by-layer films: A review, *International Journal of Electrochemical Science*, 2006, 1(5), 194.

[133] R. He, T.D. Day, M. Krishnamurthi, J.R. Sparks, P.J.A. Sazio, V. Gopalan and J.V. Badding, Silicon *p-i-n* junction fibers, *Journal of Advanced Materials*, 2013, 25, 1461–1467.

[134] S. Heysel, H. Vogel, M. Sanger and H. Sigrist, Covalent attachment of functionalized lipid bilayers to planar wave guides for measuring protein binding to biomimetic membranes, *Protein Science*, 1995, 4(12), 2532.

[135] O. Seitz, T. Bocking, A. Salomon, J.J. Gooding and D. Cahen, Importance of monolayer quality for interpreting current transport through organic molecules: Alkyls on oxide-free Si, *Langmuir*, 2006, 22(16), 6915.

[136] A.B. Sieval, R. Opitz, H.P.A. Maas, M.G. Schoeman, G. Meijer, F.J. Vergeldt, H. Zuilhof and E.J.R. Sudhölter, Monolayers of 1-alkynes on the *H*-terminated Si(100) surface, *Langmuir*, 2000, 16(26), 10359.

[137] A.B. Sieval, B. Van Den Hout, H. Zuilhof and E.J.R. Sudhölter, Molecular modeling of covalently attached alkyl monolayers on the hydrogen-terminated Si (111) surface, *Langmuir*, 2001, 17(7), 2172.

[138] E.J. Faber, L.C.P.M. De Smet, W. Olthuis, H. Zuilhof, E.J.R. Sudhölter, P. Bergveld and A. Van Den Berg, Covalently attached 1-alkynes on silicon surfaces provide superious insulators, *SAFE 2005, 8th Annual Workshop on Circuits, Systems and Signal Processing*, Veldhoven, The Netherlands, November 17–18, 2005.

[139] L. Scheres, M. Giesbers and H. Zuilhof, Self-assembly of organic monolayers onto hydrogen-terminated silicon: 1-Alkynes are better than 1-Alkenes, *Langmuir*, 2010, 26(13), 10924.

[140] M.N. Masood, S. Chen, E.T. Carlen and A. Van Den Berg, All-(111) surface silicon nanowires: Selective functionalization for biosensing applications, *ACS Applied Materials and Interfaces*, 2010, 2(12), 3422.

[141] F.M. Ross, J. Tersoff and M.C. Reuter, Sawtooth faceting in silicon nanowires, *Physical Review Letters*, 2005, 95(14), 146104.

[142] M.D. Kelzenberg, D.B. Turner-Evans, B.M. Kayes, M.A. Filler, M.C. Putnam, N.S. Lewis and H.A. Atwater, Photovoltaic measurements in single-nanowire silicon solar cells, *Nano Letters*, 2008, 8(2), 710.

[143] T. Stelzner, M. Pietsch, G. Andrä, F. Falk, E. Ose and S.H. Christiansen, Silicon nanowire-based solar cells, *Nanotechnology*, 2008, 19(29), 295203.

[144] D. Ramos, E. Gil-Santos, O. Malvar, J.M. Llorens, V. Pini, A. San Paulo, M. Calleja and J. Tamayo, Silicon nanowires: Where mechanics and optics meet at the nanoscale, *Scientific Reports*, 2013, 3, 3445.

[145] E. Baek, S. Pregl, M. Shaygan, L. Römhildt, W.M. Weber, T. Mikolajick, D.A. Ryndyk, L. Baraban and G. Cuniberti, Optoelectronic switching of nanowire-based hybrid organic/oxide/semiconductor field-effect transistors, *Nano Research*, 2015, 8(4), 1229.

[146] A.I. Boukai, Y. Bunimovich, J. Tahir-Kheli, J.-K. Yu, W.A. Goddard and J.R. Heath, Silicon nanowires as efficient thermoelectric materials, *Nature*, 2008, 451, 168.

[147] Y. Liu, G. Ji, J. Wang, X. Liang, Z. Zuo and Y. Shi, Fabrication and photocatalytic properties of silicon nanowires by metal-assisted chemical etching: Effect of H_2O_2 concentration, *Nanoscale Research Letters*, 2012, 7(1), 663.

10 Summary

A comprehensive review on preparation methods, properties, and practical applications of polysiloxanes (silicones) with various chemical structures and compositions, a wide family of polysiloxane copolymers, other silicon-containing polymers (polysilanes, polycarbosilanes, polysilazanes, etc.), ceramics derived from silicon-based polymers, composites containing silicones, fillers, and nanofillers modified with reactive silanes and siloxanes, as well as silicon nanoparticles and nanowires is presented in this book.

Silicon chips are most important semiconductors in an electronic and computer industry, and silicon-based polymers and polymeric materials as well as reactive silane coupling agents have been used in many fields – from industry, through everyday life commodity goods and cosmetics, to medicine. Their universal properties decide that they are very useful and attractive materials and components in a very wide range of products. Their production in the world is continuously growing, and perspectives of their future applications are still very optimistic. Silicon nanoparticles and nanocrystals are useful materials for solar cells, lithium ion batteries, optoelectronic devices, and biological markers. Silicon nanowires (SiNWs) are important components for electronic and sensor nanodevices, useful in many different fields of the materials science and chemical technology. SiNWs are promising ingredients of nanoelectronic, photonic, and photovoltaic and optoelectronic devices, field effect transistors, and chemical or gas and biological nanosensors.

https://doi.org/10.1515/9783110643671-010

Acknowledgments

The author is very grateful to the management staff of ŁUKASIEWICZ Research Network - Textile Research Institute in Łódź (Poland), for possibility to prepare the manuscript of this book, for financial support, and for encouraging, enthusiastic, and friendly atmosphere at workplace.

I would like to thank very much to Mrs. Helene Chavaroche (De Gruyter, formerly Smithers Rapra), for her invitation and encouragement to write this book. My special thanks are dedicated to De Gruyter's Team (and especially to Dr. Matthew A. Hood, Dr. Vivien Schubert, Mrs. Stella Müller, and Ms. Suruthi Manogaran) for a fruitful cooperation, understanding, patience and very helpful advices. The author is also especially grateful to his family for their encouragement, help, patience, cordial relations, and love.

https://doi.org/10.1515/9783110643671-011

Index

https://doi.org/10.1515/9783110643671-012